ELECTROMAGNETIC ANALYSIS USING TRANSMISSION LINE VARIABLES

ELECTROMAGNETIC ANALYSIS USING TRANSMISSION LINE VARIABLES

Maurice Weiner

United Silicon Carbide, New Jersey

World Scientific

Singapore • New Jersey • London • Hong Kong

Published by

World Scientific Publishing Co. Pte. Ltd.

P O Box 128, Farrer Road, Singapore 912805

USA office: Suite 1B, 1060 Main Street, River Edge, NJ 07661

UK office: 57 Shelton Street, Covent Garden, London WC2H 9HE

British Library Cataloguing-in-Publication Data
A catalogue record for this book is available from the British Library.

ISBN 981-02-4438-X

Printed in Singapore by World Scientific Printers

To Carole, Tammy, Josh, and Steve

Preface

Over the past one hundred years or so, we have witnessed remarkable progress in expanding the frontiers of scientific knowledge. During this scientific journey, researchers have discerned many trends and themes. One of the most important of these has been the relentless preference of nature to discretize. Whether we look to the biological sciences and the double helix, or the quantization of electromagnetic energy, or the existence of quarks in particle physics, it appears nature loves to count, to compartmentalize, and to express all phenomena in terms of some sort of unit. With its virtually limitless capacity to control vast arrays of such individual elements, nature has endowed us with an amazing range of materials, life forms, and variegated phenomena.

Nature's affinity for using fundamental building blocks is not the whole story, however. Its capability in the realm of the infinitesimal is superseded by its incredible ability to synthesize, to derive function, and to obtain meaning from among these immense arrays of discretized elements, whether they be material units or bits of information. Nature indeed is the supreme "system engineer". An oft cited analogy is that drawn from certain impressionistic paintings. If looked at very closely, a limited portion of the painting appears as nothing more than a random collection of colored dots. Looked at from a distance, however, the painting takes on meaning and substance.

We therefore can think of nature as providing us with two viewpoints. One reflects a fundamental, building block perspective, and the other a cooperative phenomenon, which ties together the discrete elements, and in so doing gives rise to form, function, and meaning. These viewpoints are of course simplifications but nevertheless useful ones. The examples in nature are boundless: a nugget of coal consists of 10^{23} atoms or so, an organism consists of many cells, and so forth. By themselves these basic units are usually unrecognizable; together these elements create new meaning. The designation of the building block is to some degree a matter of choice of course. The coal atoms consist of atomic as well as sub-actomic particles; the latter are not as useful in understanding the chemistry of coal and we therefore choose to ignore these units. On the other hand, we perceive atomic particles as having a decisive influence on the chemical properties.

Throughout the past century we have developed countless theories , many (but not all) of them containing fundamental units , or "building blocks", to help us understand natural phenomena. We have demanded of these theories self consistency, and of course the capability to explain known phenomena as well as to *predict* new phenomena. We have also insisted that they be consistent with other predictive theories. If the theory contained building blocks we have insisted that these units explain larger scale phenomena. In addition, although not absolutely necessary, we have attached extra value to theories which are mathematically "elegant".

The topic of classical electromagnetism, and the very important subset of electromagnetic wave propagation,   certainly qualifies as an eminently " superb theory", to use the labeling of C. Penrose [1]. Despite the lack of any apparent building block unit(within the classical domain), the theory has exhibited unsurpassed predictive capabilities. In addition, the classical electromagnetic theory (essentially, Maxwell's Equations) has forged very successful links with other theories, such as quantum electrodynamics and relativity. We then raise the issue as to whether physical building blocks exist in the realm of wave propagation, including the classical regime?   The use of the word "classical" would appear to contradict such an idea from the very outset. In this regard, we cannot point to any fundamental unit, or building block, except for its connections to quantum theory; in particular, the electromagnetic energy is quantized, with the elementary energy unit existing as a photon. In our discussion, however, we will forego the use of any units derived from quantum theory, assuming instead that the wavelength is relatively long and that the wave propagation remains in the classical regime.

Despite the lack of any apparent physical building block, we nevertheless proceed with the discretization of wave propagation phenomena. *Indeed, the discretization process, applied to the electromagnetic propagation medium, forms the basis of this book.* In view of the comments of the preceding paragraph, however, there are valid questions as to where and how the discretization process is to be introduced in the area of electromagnetics. In what follows we will imagine the propagation medium to be completely divided up into identical cells(for a given propagation velocity), with the electromagnetic energy confined to *transmission lines,* or "tracks", which separate the cells. Rather than relying on a fundamental building block derived from physics, the basic unit selected will be a mathematical one. The selection of the cell size is arbitrary, but we insist of course that as we reduce the size of the cell that the solution reduces to that of solving Maxwell's wave equation (including the effects of conductivity). We reiterate that the cell (and the cell size)is not an, irreducible, fundamental unit, but rather selected on the basis of mathematical convenience. The method is closely linked to numerical solutions of Maxwell's wave equation.

In view of the previous comments, a question which immediately comes to mind is the following: is it worthwhile to re-introduce an old theory, dressed in new clothing, with an accompanying computer code for solving Maxwell's Equations? The answer is affirmative, and for two reasons. First, the transmission line method offers an extremely intuitive means for dealing with a wide assortment of electromagnetic propagation problems. Equilibrium, transient conductivity, and antenna problems are readily amenable to this method. From the outset a particular problem is viewed as a transmission line grid, rather than a purely numerical grid. Secondly the method provides an opportunity to discretize the propagation region, which can offer many new insights and may also be useful as a bridge to examine small scale effects when the cell size is allowed to shrink further in size, leaving behind the classical regime.

Hopefully this book will fill a niche not presently satisfied by two related types of books. The first type, stressing numerical methods for solving electromagnetic problems, lacks any appeal to physical intuition ; very often the physics of electromagnetic propagation is lost because of the morass of mathematical detail, or the result of a physically unmotivated computer code. The second type of book, stressing transmission line techniques, has the desired physical appeal but is inadequate for solving two and three dimensional electromagnetic problems. This book bridges the gap between the two subject areas, illuminating the features common to both methods. Following the extensions and modifications to the transmission line theory, the model is applied to several illustrative electromagnetic problems. The book should have special appeal among electrical engineers and scientists involved with electromagnetic propagation and wideband transmitters/antennas. Although some background in transmission line theory is desirable, it is not essential since the needed background is provided.

What are some of the practical applications of the transmission line method? As alluded to in the previous paragraph, the method is particularly well suited for designing ultra -wideband or short pulse transmitters/ antennas. *The future will witness an inexorable drive to expand the bandwidth of transmitters/antennas, driven by the relentless needs of the information age.* Engineers and scientists will need a more sophisticated understanding of picosecond devices, and its interaction with the environment, and will require computer codes capable of accurately describing such situations. The transmission line method should provide the engineer with the analytic and software tools necessary for dealing with this new technology. Besides laying the foundation for the transmission line method, the book also provides examples of computer codes which illustrate the transmission line technique. The code for the two dimensional solution of a photonic switch is provided is provided as an example.

The Chapters, outlined in the following, contain a fair amount of "new" material. We enclose new in quotes since the basic assumptions, embodied in Maxwell's Equations and elementary symmetry arguments, always represent the starting point. In most cases, however, the results are straightforward extensions of transmission line theory.

In this book we will deal almost exclusively with transmission lines situated on the borders of cubical cells(or squares, in the case of 2D), mainly as a matter of convenience. Neat geometrical cells, such as squares or cubes, are more amenable to mathematically analysis, and this is the primary reason for adopting such cells. Other cell geometries are possible, and in fact random irregularly shaped cells may be more appropriate, as we shall see. In any event the transmission line approach leads to an iterative solution of Maxwell's equations, which in turn may be translated into computer code.

Chapter I begins with a discussion of the many types of electromagnetic problems one can solve using transmission line matrix (TLM) techniques, whether they involve antennas, ultra-wideband sources, or even static potential problems. This is followed by a review of basic TLM theory, starting with the well documented , exact correspondence of the one dimensional wave equation and the transmission line solution. Once one ventures into two and three dimensions, however, the correspondence between the wave equation and the TLM matrix is not as apparent. Chapter I addresses these issues, pointing out the close relationship between standard numerical methods and the TLM matrix. Mathematically, we will see that the TLM method may be regarded as a particular type of finite difference method. Besides applications to electromagnetic propagation and other branches of applied physics, more flexible versions of the TLM matrix may be used to describe other diverse phenomena, for example, the modelling of neurological activity. In Chapter II the geometries and notation adapted for the TLM matrix are discussed. In many ways they are similar to the symmetry elements used to describe solid state crystals, in which the unit cells occupy the entire space. A systematic procedure for mapping the electromagnetic properties onto the TLM matrix, i.e., the transmission lines and the nodes, is discussed. Once the basic framework is established , Chapter III examines the electromagnetic scattering equations for one, two, and three dimensions. The scattering equations are very important, of course, since they are the hub of any computer iteration used to describe electromagnetic propagation. In Chapter IV the TLM matrix is corrected for any plane wave properties which are usually present. The plane wave properties are studied by means of "correlations" between neighboring waves in adjacent transmission lines. This Chapter also deals with the inherent anisotropy present in the unit cells and the averaging procedures needed to remove this effect. Plane wave and anisotropy effects are often ignored in the standard numerical methods. Chapter V discusses the boundary conditions and dispersion. Dissimilar dielectrics, for example, will require different cell sizes, so the scattering at the interface becomes more complicated. The Chapter provides a systematic means for handling such boundary conditions. Embracing the effects of dispersion, as in Chapter V, does not involve any fundamental obstacle, but its incorporation does make great demands on computer requirements. The TLM method is typically aimed at solving purely electromagnetic problems, namely, fast transient phenomena. However the same TLM framework also may be used to incorporate other phenomena, such as carrier drift, recombination, and charge separation. Chapter VI addresses these issues. The main obstacle here is not so much the difficulty in incorporating such effects, but once again the substantial demands placed on computer memory and speed, brought about by combining fast and slow phenomena. In Chapter VII, an illustrative example is selected. We outline a computer program, based on the TLM method, for finding the transient solution of a 2D semiconductor switch, whose conductivity is induced by a light source. In addition to the static solution (Laplace's solution), the transient electric field profiles and the charge distribution are obtained. The actual program is also provided, allowing readers to gain a better understanding of the transmission line method. Finally, in Chapter VIII, we utilize existing software , such as SPICE. Such software may be applied to the TLM method when only a limited number of cells is required. Aside from the cell number limitations, other limitations inherent in SPICE involve the neglect of plane wave behavior, grid anisotropy, non-uniform propagation regions, boundary conditions, etc... Nevertheless, elementary but useful solutions may be obtained with SPICE, particularly one dimensional problems, and to a lesser extent, 2D problems. Several SPICE examples are discussed, which include RF transformers and pulse sources, as well as a simple description of a semiconductor switch. For comparative purposes, we also derive several 1D results, applicable to RF transformers and the like, using the TLM formulation.

The author's interest in transmission line methods, for solving electromagnetic problems, stems from his long and fruitful association with the Army Research Laboratory (ARL), Fort Monmouth, NJ. The Author has benefited from many discussions with scientists and engineers at ARL, and in particular, wishes to thank his many colleagues at the Pulse Power Laboratory for their interest and suggestions. The topics discussed in this book were initiated as a way of analyzing very fast conduction processes in semiconductor, for use in fast pulse generation and ultra- wideband sources. The author is presently with United Silicon Carbide Corp. in New Jersey.

REFERENCES

1. Penrose , R (1989) , *The Emperor's New Mind*, Oxford University Press.

CONTENTS

I. Introduction to Transmission Lines and Their Application to Electromagnetic Phenomena

Maurice Weiner

United Silicon Carbide, Inc.

In recent years, an exciting new branch of research activity has emerged, dealing with extremely fast phenomena in semiconductors and gases. The introduction of high speed instrumentation and devices, with time scales often in the 1 to 1000 picosecond range, has prompted the investigation of a variety of fast phenomena, including the generation of electromagnetic pulses and light, photoconductivity, avalanching, scattering, fast recombination, and many other physical processes. The research has been driven by several applications [1], [2]. These include ultra-wideband imaging and radar , as well as ultra-wideband communications(thus avoiding the use of wires or optical fibers). In addition, the availability of new,   high speed instrumentation has provided researchers with a valuable tool for learning the fundamental properties of materials. In all the aforementioned applications, a central feature is the generation of electromagnetic pulses with either a narrow pulse duration or a fast risetime(or both). The short time interval involved (in either the risetime or the pulse duration), insures that a wide frequency spectrum is produced, a property which is essential for the cited applications.

The understanding of fast phenomena and ultra-wideband electromagnetic sources is made more complicated by the very fast risetimes and by the fact that the wavelength of the signals being produced are often smaller or comparable to the characteristic length of the device or experimental configuration under study. As a result the use of lumped circuit variables is inappropriate and we must use either transmission line variables or Maxwell's equations directly.

Electromagnetic signals with very short wavelength may be generated by a sudden transition in the conductivity of the medium. Suppose, for example,   an electric field bias first is applied to the medium and that subsequently the conductivity of a portion of the medium is suddenly increased(for example, by photoconductivity or avalanching). The sudden change in conductivity will generate electromagnetic pulses with very steep risetimes, thus producing short wavelength signals. In cases where light is produced (for example, when carriers recombine in gallium arsenide), the wavelength naturally will be smaller or at least comparable to the device size. In any event , the analyses often used to describe devices and experimental configurations do not adequately address the short wavelength signals which are generated, and subsequently dispersed throughout the device and the surrounding space. One should not underestimate the importance of the electromagnetic energy dispersal(which includes   light signals). Often the physics of underlying processes are misunderstood because the electromagnetic energy dispersal, which delivers the physics to the detector, is not taken into account properly, particularly for fast phenomena. It is hoped the ensuing discussion will help to correct this deficiency and lead to a better understanding of the dispersal of ultra-wideband electromagnetic signals and associated phenomena.

In this volume we endeavor to describe fast electromagnetic phenomena, relying on iterative rate equations which use transmission line matrix(TLM) variables. As with comparable numerical techniques, such as the finite difference method, the transmission line element must be made very small in order to attain accuracy, and solutions at a given time step depend on a knowledge of solutions at a previous time step. In terms of physical interpretation and intuition , however, the TLM method is far superior to that of finite differences or other similar numerical techniques. The physical appeal of the TLM method may be viewed, in a conceptual way, from the two basic components which comprise the TLM matrix: the transmission lines and the nodes which form the intersection of the lines, as noted in Fig.1.1. With this model, we can conceptually separate the physics and energy dispersal of a given problem in electromagnetics. Accordingly, the nodes represent the physics, and physical processes(such as conductivity changes) are mapped onto the nodes, which then control the flow of energy in the lines. The other component, the transmission lines, are responsible for the energy distribution and storage.The

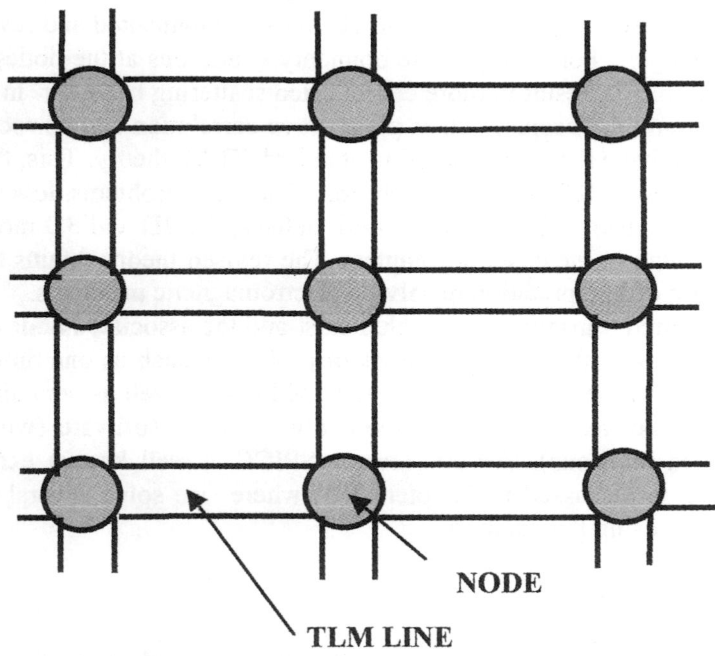

FIG.1. CONCEPTUAL VIEW OF TRANSMISSION LINE(TLM) MATRIX CONSISTING OF NODES AND TLM LINES. THE TWO COMPONENTS HAVE SEPARATE FUNCTIONS. THE TLM LINES DISTRIBUTE THE ELECTROMAGNETIC ENERGY AND THE NODES CONTROL THE PHYSICS.

TLM lines may be regarded as having spatial extent, whereas the nodes are regarded as infinitely small.

Within the electrical engineering community the use of transmission line variables to treat one dimensional electromagnetic problems has gained in popularity over the years. As a result, a certain comfort level has been attained by engineers in the use of transmission line terminology. The carryover of the TLM description to two and three dimensional electromagnetic problems, however, has been scanty. This may not be surprising, since the 2D and 3D treatments are more complicated and the 1D must be augmented and revised to a considerable degree. For example, the boundary conditions at the nodes, for the 2D and 3D problems, result in more complicated scattering behavior. In addition the incorporation of concepts such as plane wave correlation and corrections for grid anisotropy, have not been applied to standard TLM theory. This, therefore, would tend to make the TLM approach, for 2D and 3D problems, less valuable. The necessary revisions for removing these defects, for 2D and 3D models, are described in detail in the ensuing Chapters. The revised theory retains the usual benefits of ease of interpretation in solving electromagnetic problems.

If the required transmission line elements(and the associated nodes) are not too large in number, then certain classes of problems, such as one dimensional microwave transformers and non-uniform TLM lines, as well as very simple 2D problems, may be treated using commercially available software (without the aforementioned revisions). For this reason, SPICE, a well known example of such software, is discussed in Chapter VIII where we solve several types of TLM problems using this method.

1.1 Simple Experimental Example

The simple arrangement shown in Fig.1.2 will help to illustrate the concepts more easily. The Figure shows a side view of two electrodes separated by semiconductor material, with an electric field bias between the two electrodes. Suppose a limited region of the semiconductor, shown by the darkened region, is suddenly created (either with a light pulse or by a localized, fast avalanche breakdown). This will give rise to a electromagnetic disturbance and possibly a light pulse(depending on the medium), emanating from the conduction region. The dashed curve may be conceptually regarded as an equal amplitude contour of the electromagnetic disturbance at a given instant in time. As noted the disturbance is assumed to be asymmetric, since the amplitude will be more pronounced in the direction perpendicular to the bias field. One may regard the electromagnetic disturbance as a traveling wave created by that part of the initial field which is parallel to the surface of the activated conduction region. The wave, i.e., the disturbance, is reflected normally from the surface of the activation region while undergoing an electric field inversion, so that the total field, in

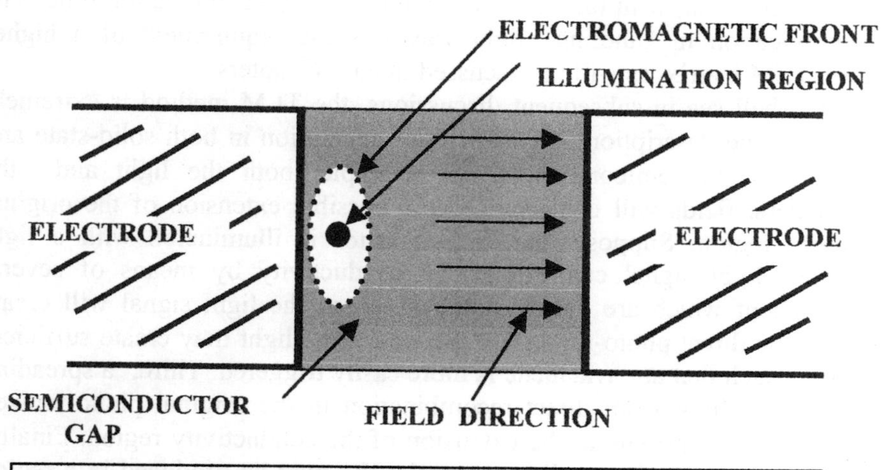

FIG.1.2 CONCEPTUAL VIEW OF DISPLACEMENT   CURRENT ARISING FROM ILLUMINATION OF A   LIMITED PORTION OF A SEMICONDUCTOR GAP.

the immediate vicinity of the initial conduction region, is partially or completely canceled (depending on the degree of conductivity). The situation changes , of course, if the initial wave disturbance undergoes an additional field reversal at the boundary of a nearby conductor, which can be one of the two electrodes or some auxiliary conductor(such as the grounded member of a transmission line). Changes in the wave disturbance can also take place, of course, due to the existence of a dielectric interface.

The final result is a complex array of waves throughout the region, in which the field configuration depends on the temporal and spatial properties of the conduction region, as well as the device geometry. Given such conditions, the field will change dramatically(compared to the initial uniform, static field) and regions of field diminution or field enhancement are likely to occur.

In case light is produced in the semiconductor, the situation is in some ways easier to describe, and in other ways it is more complicated. The light disturbance profile will be more symmetric, compared to that of the electromagnetic front, because of the shorter wavelength and random nature of the recombination process. However, the front of the light signal will lag the electromagnetic front, with the delay depending on the recombination time. The main complication in handling light waves is the requirement of a higher resolution TLM matrix, as will be discussed in later Chapters.

As we shall see in subsequent discussions, the TLM method is extremely well suited to the description of conductivity generation in both solid-state and gaseous media. In semiconductors, for example, both the light and the electromagnetic fields. will contribute to the possible extension of the original conductivity region. Suppose the semiconductor is illuminated with a light signal. The light signal can extend the conductivity by means of several processes, all of which are linked together. First, the light signal will create carriers via the direct photo-ionization. Second, the light may create sufficient seed carriers such that an avalanche is more easily triggered. Third, a spreading light signal, resulting from direct recombination in the original photo-ionized region, may also contribute to the extension of the conductivity region. Finally, even without direct carrier seeding, an avalanche may be produced in a region which is remote from the original site of the light impingement, caused by augmentation of the electromagnetic field (following the upset of the initial static field) in various regions. The TLM method is well suited for obtaining such a conductivity extension.

1.2 Examples of Impulse Sources

The types of electromagnetic problems which the TLM method can address are virtually without bound. It is worthwhile mentioning several generic impulse sources which have stimulated the development of the TLM method. It should be mentioned, however, that in virtually all the examples using the TLM method, cited here and throughout the book, the conductivity generation occurs by means of a light pulse impinging on a semiconductor, with the carriers produced by direct photo-ionization. This is merely a convenience, and it should be borne in mind that any other source of conductivity(such as avalanching or carrier injection)may be incorporated into the TLM formulation.

As a first example, Fig.1.3, consider a parallel plate transmission line in which a portion of the upper conductor has been removed and substituted with a

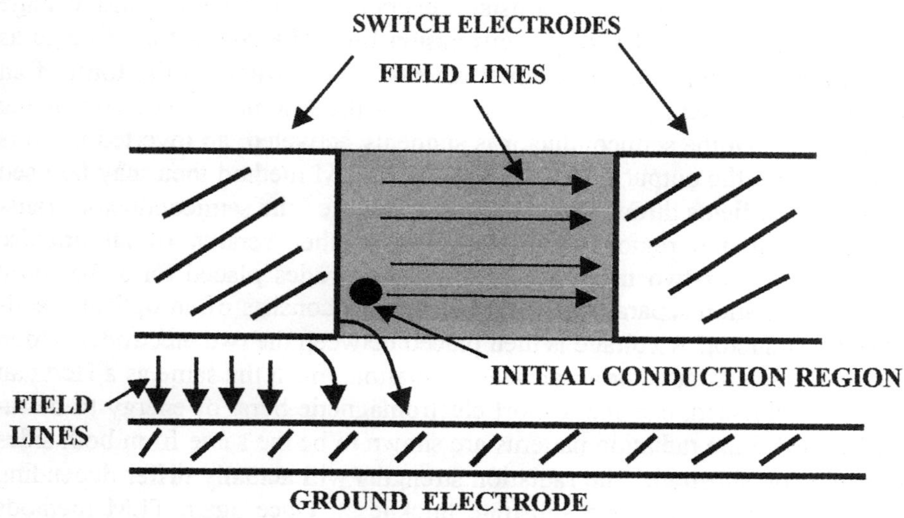

FIG. 1.3 EXTENSION OF CONDUCTIVITY REGION FOLLOWING ILLUMINATION NEAR ELECTRODE. SEMICONDUCTOR GAP INTERRUPTS UPPER CONDUCTOR OF TRANSMISSION LINE.

semiconductor. capable of holding off voltage. The line is then charged up to voltage, producing a field, as indicated in the semiconductor, as well as a fringing field between the two conductors. The sudden creation of carriers with a light pulse, throughout the entire semiconductor, or even in a limited portion of the semiconductor, then produces a fast risetime pulse which travels down the line toward the output , typically an antenna. The TLM method may the be used to determine the instantaneous field profiles throughout the entire device region, which includes the dielectric as well as the semiconductor regions. If so desired we also may apply the TLM analysis to the dielectric region above the semiconductor as well. Indeed, during the "commutation" time,   some electromagnetic energy will radiate out from the semiconductor. The TLM analysis and computation may be extended in   straightforward fashion to determine the radiated signal during commutation. Another impulse source, which combines the functions of the energy storage, antenna, and the switch is shown in Fig.1.4. The initial electrostatic energy is stored with a bias voltage between the conductors of a strip transmission line. The conductors diverge as shown to form a composite transformer/antenna. The switch, in the form of an optically activated semiconductor, is situated at the low impedance end of the transformer. When the semiconductor is suddenly activated, an inverted pulse is launched toward the output of the antenna. The TLM method then may be used to determine the fields throughout the entire space, i.e., the semiconductor, transformer and radiation regions. Fig.1.5 shows another version of an impulse source consisting of two tapered conducting electrodes placed on a dielectric substrate. The region separating the two electrodes consists of an optically activated semiconductor. A voltage is then placed between the two electrodes. Upon activating the semiconductor a transient oscillation, much the same as a Hertzian dipole , is established, causing a short electromagnetic burst of energy to be radiated(In Fig.1.5 the radiation patterns are shown to be the same from both sides of the dielectric substrate; the radiation strengths will actually differ depending on the dielectric constant and substrate thickness). Once again TLM methods may be employed for the various regions of interest.

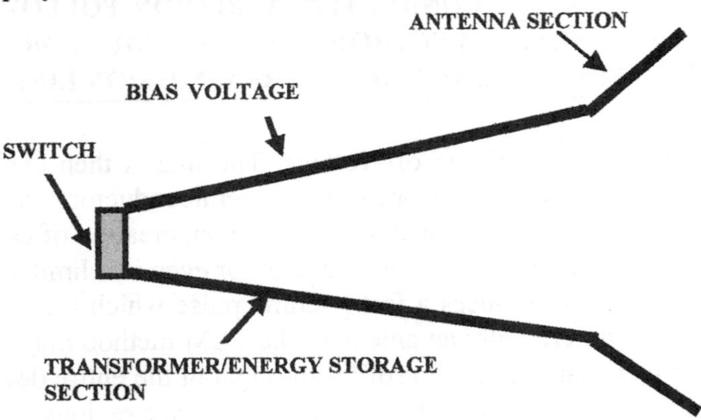

FIG. 1.4 PULSE SOURCE CONSISTING OF A FAST SWITCH AT THE LOW IMPEDANCE SIDE OF A TRANSFORMER , WHICH ALSO STORES THE ELECTROSTATIC ENERGY AND COUPLES TO AN ANTENNA.

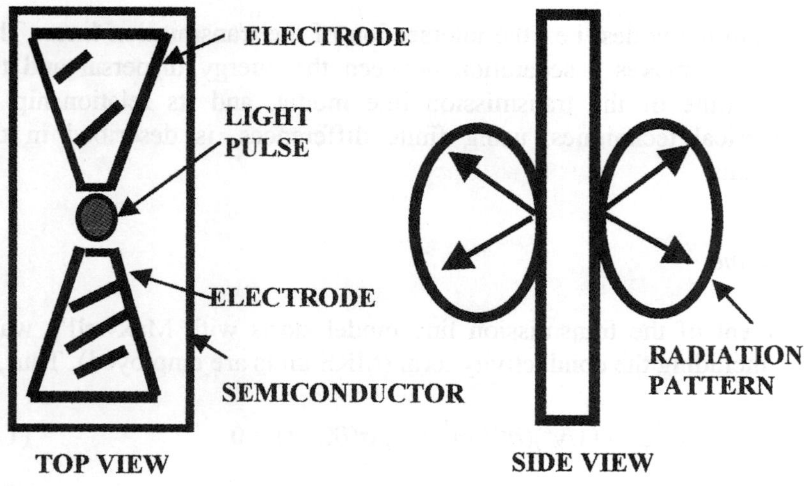

FIG.1.5 PULSE SOURCE CONSISTING OF PHOTOCONDUCTIVE SWITCH AND BOWTIE ANTENNA.

One can surmise that obtaining practical, quantitative, and accurate electromagnetic solutions for the above configurations would appear to be a formidable task, especially if one must rely solely on numerical analysis , based on Maxwell's equations, the boundary conditions, as well as the physics underlying various phenomena such as phtoconductivity, avalanching, recombination, etc... The purpose of this book is to invoke an alternative method, the TLM approach, which from a mathematical point of view has close links to standard numerical techniques but which is far superior in terms of its ease of physical interpretation and flexibility. The new method relies on transmission line variables, concepts which are readily familiar among many research workers, to describe the behavior of the medium. As mentioned previously, the medium is represented by an imaginary matrix of transmission lines, wherein the energy storage and dispersal is taken into account by the transmission lines and the all physical processes(except for the energy dispersal) are mapped onto the nodes, i.e., the intersection of the transmission lines. This method, therefore, makes a separation between the energy dispersal and the physics. An outline of the transmission line model, and its relationship to standard numerical techniques, using finite differences, is described in the following sections.

1.3 Model Outline

The development of the transmission line model starts with Maxwell's wave equation [3], including the conductivity term (MKS units are employed). Thus,

$$\nabla^2 E - (1/v^2)(\partial^2 E/\partial t^2) - \mu\sigma(\partial E/\partial t) = 0 \tag{1.1}$$

where:

E = Electric Field
t = time
μ = permeability
v = propagation velocity
σ = conductivity

Eq.(1.1) assumes there is no true charge in the medium. A similar equation also holds for the magnetic field. As mentioned previously, numerical methods may be used to obtain solutions to Eq.(1.1). Instead of following this path, however, a transmission line approach is investigated. Toward this goal, it is convenient to first consider the one dimensional case. Accordingly, the wave equation then becomes

$$\partial^2 E/\partial x^2 - (1/v^2)(\partial^2 E/\partial t^2) - \mu\sigma(\partial E/\partial t) = 0 \tag{1.2}$$

where x is the distance along the propagation direction and E is transverse to the propagation. An important time constant associated with the wave equation is the relaxation time, ε/σ, where ε is the permittivity. The relaxation time will determine the choice of our cell size, denoted by length Δl. In order to minimize losses over the cell length, and thus provide the necessary resolution and numerical convergence, the cell propagation time, $\Delta l/v$, should be much smaller than the relaxation time, ε/σ. Δl therefore should satisfy $\Delta l \ll \varepsilon v/\sigma$. In addition, Δl should be much smaller than any characteristic length associated with , for example, the geometry or other experimental condition. Obviously, the smaller the size of the cell size, the greater the resolution, although this places a greater burden on computer speed and memory.

Eq.(1.2) is identical to the well known transmission line equation, which governs the voltage V for a one dimensional transmission line using circuit variables, as shown Fig.1.6,

$$\partial V^2/\partial x^2 - (L'C')(\partial^2 V/\partial t^2) - L'G'(\partial V/\partial t) = 0 \qquad (1.3)$$

Eq.(1.3) is derived from the usual relationships between voltage and current, taking into account the circuit parameters of capacitance, inductance, and conductance(see for example reference [4]). Besides the inductance per unit length, L' , and capacitance per unit length, C' , there is included a shunt conductance per unit length, G'. G' accounts for losses between conductors, e.g., the presence of carriers in a semiconductor medium. Although somewhat redundant, Eq.(1.3) is important because of the familiarity of the circuit variables, several of which we will use during the subsequent discussion.

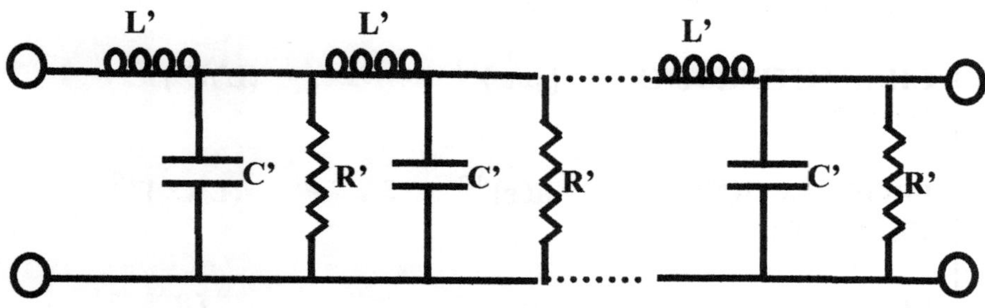

FIG.1.6 ONE DIMENSIONAL CIRCUIT CELL CONSISTING OF LUMPED CIRCUIT PARAMETERS.

Table 1.1 exhibits the relationship between the wave equation using field variables, Eq.(1.2), and that using circuit variables, Eq.(1.3). An important simplification occurs when we select a small transmission line element (or cell) of length Δl. It is useful to state the total capacitance , inductance, and conductance associated with the line element, which we identify as

$$C = C'\Delta l = \varepsilon\Delta l \tag{1.4a}$$
$$L = L'\Delta l = \mu\Delta l \tag{1.4b}$$
$$G = G'\Delta l = \sigma\Delta l \tag{1.4c}$$

<table>
<tr><td colspan="3" align="center">TABLE 1.1 CORRESPONDENCE BETWEEN WAVE
AND CIRCUIT VARIABLES</td></tr>
<tr><td></td><td align="center">ELECTROMAGNETIC
VARIABLE</td><td align="center">CIRCUIT
VARIABLE</td></tr>
<tr><td>FIELD</td><td align="center">E</td><td align="center">V</td></tr>
<tr><td>PERMMITIVITY</td><td align="center">ε</td><td align="center">C'</td></tr>
<tr><td>PERMEABILITY</td><td align="center">μ</td><td align="center">L'</td></tr>
<tr><td>CHARACTERISTIC
IMPEDANCE</td><td align="center">$\{\mu/\varepsilon\}^{1/2}$</td><td align="center">$\{L'/C'\}^{1/2}$</td></tr>
<tr><td>PROPAGATION
VELOCITY</td><td align="center">$\{\mu\varepsilon\}^{-1/2}$</td><td align="center">$(L'C'\}^{-1/2}$</td></tr>
<tr><td>LOSS</td><td align="center">σ</td><td align="center">G'</td></tr>
<tr><td>RELAXATION
TIME</td><td align="center">$\{\varepsilon/\sigma\}$</td><td align="center">C'/G'</td></tr>
</table>

The total resistance R of the element is thus $R=1/G$. The relaxation time, ε/σ, is equal to the "RC" discharge time for the cell, as noted from Eq.(1.4). We also identify the impedance of the line as

$$Z_o = (\mu/\varepsilon)^{1/2} = (L'/C')^{1/2} \tag{1.5}$$

At this point we can quantify the selection of Δl. In order to obtain accuracy, we select Δl such that

$$\Delta l << v(\varepsilon/\sigma) = v(C'/G') \tag{1.6}$$

where

$$v = (\mu\varepsilon)^{-1/2} = (L'C')^{-1/2} \tag{1.7}$$

Eqs(1.6)-(1.7) state that the transit time delay in cell Δl, equal to $\Delta l/v$, is much smaller than the RC time of the cell. An equivalent statement is that the lumped resistance, R, of the element Δl is much larger than the characteristic impedance, Z_o, or

$$R >> Z_o \tag{1.8}$$

By virtue of previous equations, L', C' may be combined into a lossless transmission element, Z_o, and the conductance may be combined into *two* resistors, R, located at the ends of the transmission line, as shown in Fig.1.7, where R is given by

$$R = 2/\sigma\Delta l = 2/G'\Delta l \tag{1.9}$$

A two factor appears in Eq.(1.9) since each of the two resistors, R , may be considered in parallel. Another way to view the introduction of the two factor is the following. By focusing an a single TLM line element, we ignore the adjoining TLM elements, each with similar end resistors; since such adjoining resistors are in parallel, a two factor should be introduced when "extracting " a single element from the chain.

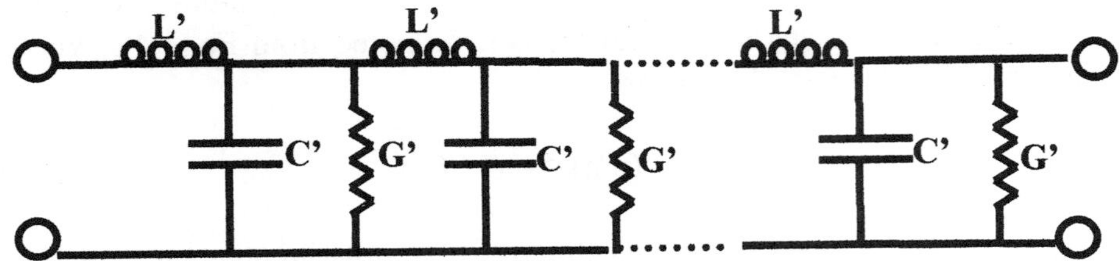

ONE DIMENSIONAL CELL CONSISTING OF LUMPED CIRCUIT PARAMETERS

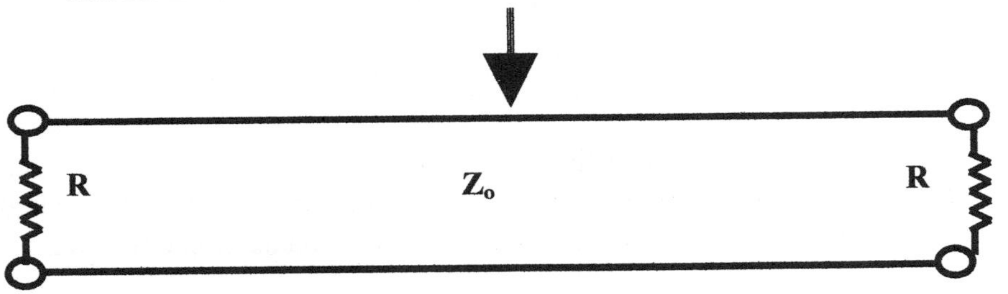

ONE DIMENSIONAL CELL CONSISTING OF A TRANSMISSION ELEMENT Z_0 OF LENGTH Δl WITH SHUNT RESISTORS $R = 2/G'\Delta l \gg Z_0$

FIG. 1.7 TRANSITION FROM LUMPED CIRCUIT CELL TO TRANSMISSION LINE CELL.

The same transmission line element may be viewed conceptually in 3D terms, Fig.1.8, as a TLM line with electrodes on two opposite surfaces and relatively large, terminal sheet resistors appearing at both ends of the TLM line, normal to the direction of propagation. The TLM height and width are identical As such the TLM element resembles a parallel plate transmission line. The fringing fields(perpendicular to page) are ignored. In a sense we regard the TLM lines as surrounded by a high permeability, high impedance medium. The sole function of the TLM line is to simulate the field in the medium. In any event, even without this assumption, the fringing fields in neighboring TLM lines will

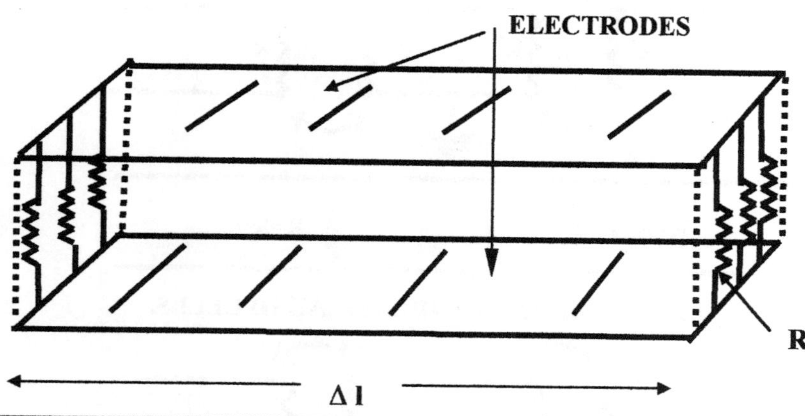

FIG 1.8 ELEMENTARY TLM LINE. THE TLM LINE IS SITUATED (CONCEPTUALLY) AT THE BOUNDARY BETWEEN CELLS. IN THE 3D CASE, A PAIR OF ORTHOGONAL TLM LINES IS PRESENT. $R=2/\sigma\Delta l$, $C=\varepsilon\,\Delta l$, $L=\mu\Delta l$, $R>>Z_0 =[\,\mu\,/\,\varepsilon\,]^{1/2}$.

tend to cancel one another as the cell size is diminished. The existence of fields perpendicular to the page, of course, is taken into account by a separate TLM line, similar to that in Fig.1.8, but rotated 90° (i.e., with the conductors parallel to the page). This then becomes a 3D problem; the TLM formulation for 3D configurations will be given in Chapters II and III.

The previous discussion has focused on a single isolated cell. We must now consider a linear chain of such cells as in Fig.1.9, consisting of coupled transmission line elements with lumped end resistors, $R'=2/\sigma\Delta l$, where we append a prime to indicate R' belongs to that of an isolated line element. For uniform line segments, Fig,1.9(a), the coupled cells will have identical end resistors and at each junction, as alluded to previously, the parallel resistors are combined into R $=R'/2=1/\sigma\Delta l$ where R' is given by Eq.(1.9). Now suppose the cell chain is non-uniform , noted in Fig1.9(b),so that adjacent cells have unequal resistivities(but the same dielectric constant). For this line, the conductances at each node will add and thus the corresponding resistors will add in parallel fashion, so that the

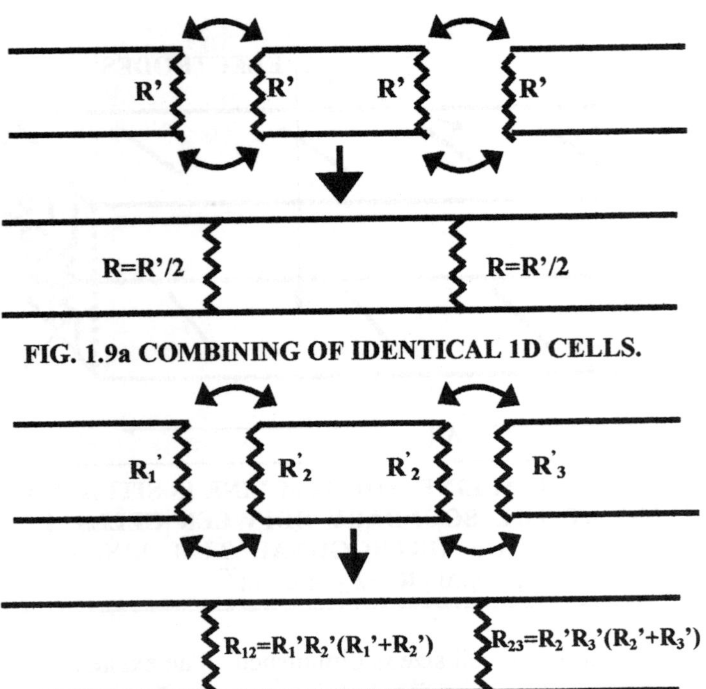

FIG. 1.9a COMBINING OF IDENTICAL 1D CELLS.

FIG. 1.9b   COMBINING OF CELLS WITH UNEQUAL RESISTIVITY.

combined resistance, R_{12} , satisfies $R_{12} = R_1'R_2'/(R_1'+R_2')$. As with the uniform line, R_1' and R_2' satisfy the isolated cell expressions, $R_1' = 2/\sigma_1\Delta l$ and $R_2' = 2/\sigma_2\Delta l$. At this point the characterization of the chain is complete, and one may invoke the transient theory for transmission lines. The resistance, R, initially very large, is activated when, for example, the effects of photoconductivity or avalanching occur. Once the cells are activated, propagation losses become important and we may then utilize transmission line theory to determine the response. These types of problems will be discussed quantitatively in the ensuing Chapters.

Although the circuits in Fig.1.9 are useful for understanding the basic concepts, two important flaws exist. For one thing the circuit is one dimensional and it does not take into account conductivity and electromagnetic spreading in the transverse directions. The second flaw, again related to the one dimensional nature of the circuit, has to do with the lack of a stable solution during equilibrium, i.e., prior to the activation of the cells, when R or R_{12}, R_{23}, etc... are very large. Suppose each of the cells in Fig.1.9, which are connected in series, is initially charged to a different voltage. The cells will discharge into one another, unless artificial means are taken to prevent such a discharge, such as the artificial insertion of a series switch. A self consistent way to preserve equilibrium, prior to activation , is to insert an orthogonal transmission line in series; this then converts the one dimensional circuit into a 2D one. A similar extension to 3D also preserves the equilibrium. Before proving these assertions, we briefly describe the 2D and 3D arrays. This will be followed by background discussion in transmission line theory, which will allow us to place our previous claims on a firmer footing.

Fig.1.10 shows the circuit matrix used to describe electromagnetic and conductivity spreading in two dimensions. One way to view the circuit cell is to note that there are four square cells, with each region constant in voltage(V_1,V_2, V_3,and V_4) but generally differing in value from the neighboring cell . The square regions ,therefore, may be considered as conductors. Separating the constant voltage regions are the transmission lines. In this case the line impedances, Zo, are the same. Initially the lines will charged up to a voltage value equal to the voltage difference between adjacent cells. Note that the node resistance R is located at the hub of the matrix and actually consists of four identical resistors, R, each terminating one of the four TLM lines. It is worthwhile to realize that any signal arriving at the node will be equally scattered among the four transmission lines(in the absence of any significant conductivity). This property is similar to that in electromagnetics, in which each region of the wave front may be regarded as a point source. A similar extension of the circuit may be made to three dimensions. In 3D, however, the iso-potential regions are cubical, and there are eight cubes centered about each node point(Fig.1.11). As mentioned previously, the nodal resistors contain the bulk of the physics, since these time varying elements represent photoconductivity, avalanching, recombination, charge transfer, and myriad other phenomena. Also, for 3D, there are two independent , orthogonal fields associated with each cube edge.

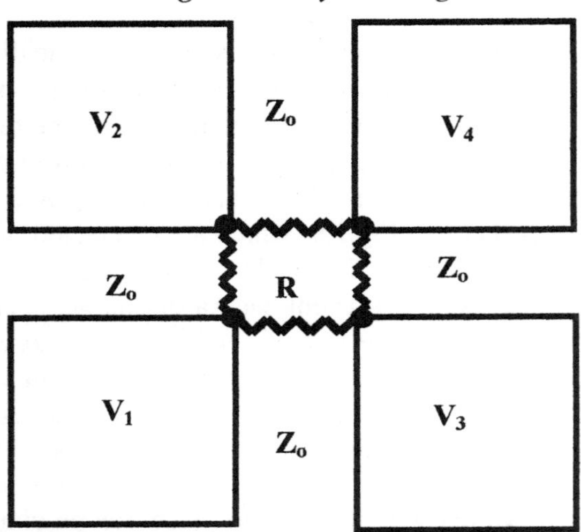

FIG. 1.10 TWO DIMENSIONAL MATRIX CONSISTING OF NODE AT CENTER OF FOUR ISO-POTENTIAL CELLS, SEPARATED BY TLM LINES Z_O.

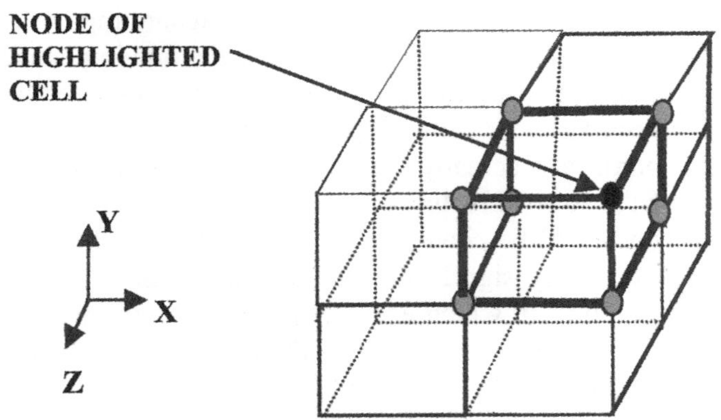

FIG 1.11 THREE DIMENSIONAL MATRIX SHOWING EIGHT CELLS AND THEIR ASSOCIATED NODES . THE DARKENED NODE IS ASSOCIATED WITH THE HIGHLIGHTED CELL.

Similarly there are two transmission lines transmission line associated with each cube edge, which renders the analysis more complicated. A simplified description of the 2D matrix is given later in this Chapter. A detailed description of both the 2D and 3D matrices, in *cellular* notation, is given in Chapters II-IV.

1.4 Application of Model for Small Node Resistance

We have adopted a model in which the cell node resistance is much larger than the cell line impedance. Stated another way, we impose the condition that the transit time of the cell Δt is much smaller than the RC time constant. Bear in mind that we can always satisfy this requirement by choosing a small enough cell size, Δl, for a given conductivity in the medium. Once conductivity is introduced among voltage biased cells, the cell voltages decay in a uniform fashion. In a simplified way, the situation may be depicted in terms of decaying waves (in both directions), which eventually vanish once a sufficient number of cells are traversed.

Whenever an intense conductivity is introduced into the medium, the model prompts us to examine the cell matrix size, and to employ appropriately small cells. The small cell size naturally implies larger array sizes, with subsequent complexities in the computer simulations. What are the implications of retaining larger cells, which do not satisfy $R >> Z_o$? The use of larger cells would of course simplify the computer process.

When R is small compared to Z_o in the cell matrix, the wave energy will slosh back and forth in the cell line, changing polarity and decaying with each successive time step. Within the high conductivity region, the solution using a large cell matrix cannot be accurately determined, although it is reassuring that the fields do dissipate after several time steps. So far as the region outside the high conductivity region is concerned, however, the waves may still be obtained with accuracy from the TLM model. The outside waves are repelled from the high conductivity region at the boundary, where the low resistance nodes exist.

The previous comments, concerning the case when $R < Z_o$, will be made more quantitative after we discuss additional aspects of the TLM model. The matter is then taken up in Chapter VII where we perform simulations in which

R$\ll$ Z_o regions exist, and also in App.7A.4 where we discuss the field dissipation in terms of the elementary TLM waves.

1.5 Transmission Line Theory Background

The discussion in the previous sections will be quantified and reconciled with transmission line theory. Before proceeding to this goal, however, we present a brief overview. The literature on transmission line theory is quite extensive(see, e.g., Reference [4]). In this discussion we limit ourselves to only those relationships which are deemed necessary for describing the technique. Before continuing, it is useful to describe the normal electromagnetic modes in a single section of transmission line, not coupled to any other line elements, in which the terminating resistors are extremely large, i.e., an open circuit. The transmission line is biased to the voltage *difference* V_o and is in an equilibrium state, as shown in Fig.1.12(a) The analysis proceeds by first choosing the correct set of normal modes which describe the standing waves during the off-state, when the line is biased to voltage V_o. This is not difficult to obtain, since we know that the general solution to the wave equation (Eq.(1.2)), or the equivalent Eq.(1.3) . Discarding the conductivity term, the solutions are a pair of waves traveling in opposite directions with velocity, v. The simplest set of modes which satisfy the boundary conditions, during the off-state, are two waves each with constant amplitude, each equal to half the bias voltage, $V_o/2$. The two waves travel in opposite directions, and are designated ^+V and ^-V in Fig.1.12. ^+V designates the wave traveling in the plus x direction while ^-V is the backward wave traveling in the negative x direction. We adopt the convention that the voltage waves , as indicated by the vertical arrows, point in the direction of increasing voltage(potential). The direction of the electric field is of course opposite to that of the voltage wave.

The voltage waves fill the entire transmission element and are constrained by the open circuit at both ends, where the waves are reflected so that ^+V converts to ^-V at one end, and vice versa at the other end. The waves obey the symmetry requirement and of course the waves superimpose to give the correct voltage at all times and at all points in the line during equilibrium, i.e., $V_o = {^+V} + {^-V} = {^+(V_o/2)} + {^-(V_o/2)}$. Thus, the general solution for the voltage consists of

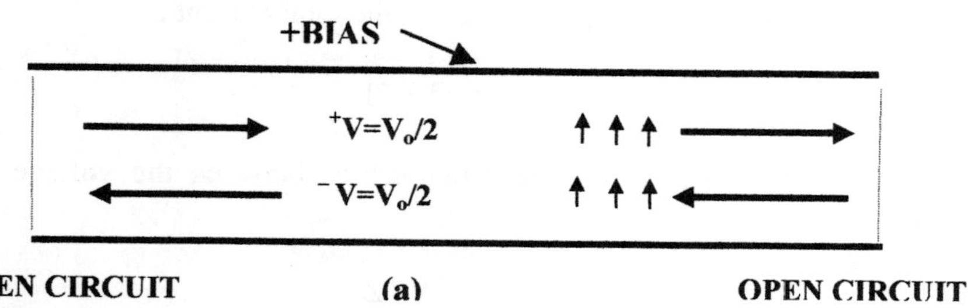

Fig 1.12(a) STATIC FIELD IN A SINGLE TLM ELEMENT, EXPRESSED IN WAVE VARIABLES. ^+V AND ^-V ARE REFLECTED AT ENDS WITH NO CHANGE IN POLARITY OR AMPLITUDE.

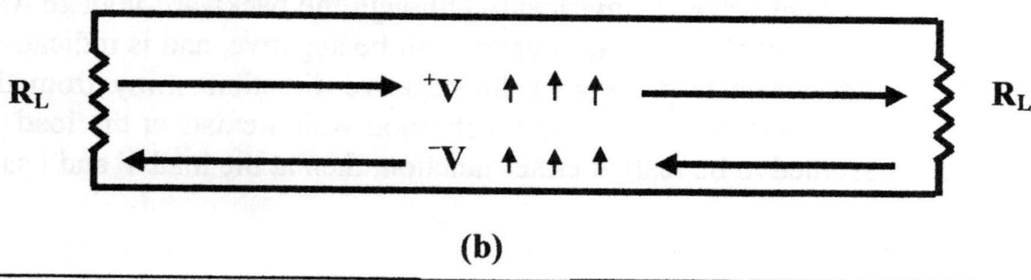

FIG. 1.12(b) TLM CELL IS ACTIVATED AND OPEN CIRCUIT IS REPLACED WITH R_L. LOSS OF AMPLITUDE OCCURS UPON REFLECTION.

a forward wave(positive direction), and a backward wave(negative direction), and the total voltage, which is the equilibrium bias voltage, equal to the sum of the two waves. What happens when we depart from equilibrium, i.e., we allow the open circuit resistance to suddenly decrease in value, as in Fig.1.12(b)? Just as before , there will exist forward and backward waves, but their amplitude will no longer be constant in time. As before the total voltage (which is no longer constant in time), will be the sum of the two waves,

$$V = {}^+V + {}^-V \qquad (1.10)$$

In a similar manner, the associated forward and backward current amplitudes are denoted by ^+I and ^-I, respectively, and the total current is

$$I = {}^+I + {}^-I \qquad (1.11)$$

Next we write down the relationships between the voltage and current waves:

$$^+I = {}^+V / Z_o \qquad (1.12)$$

$$^-I = -{}^-V / Z_o \qquad (1.13)$$

where Z_o is the characteristic impedance of the line. The minus sign for the backward current wave is significant. Although the backward voltage wave may be positive, the backward wave current will be negative, and is indicative of the fact that the current is moving in the negative direction, away from the load. Next, we take note of the boundary condition which exists at the load. If R_L is the load(assumed to be real) at either junction, then at the load V and I satisfy

$$V = R_L I \qquad (1.14)$$

At this point we define two types of voltage coefficients. The reflection coefficient, B, relates the reflected wave ^-V to the incident wave ^+V, while the transfer coefficient , T, relates the load voltage V to the incident wave. Thus,

$$B = {}^-V / {}^+V \qquad (1.15)$$

$$T = R_L I / {}^+V \qquad (1.16)$$

If Eqs.(1.10)-(1.14) are utilized then these coefficients become

$$B = (R_L - Z_o)/(R_L + Z_o) \qquad (1.17)$$

and

$$T= 2R_L/(R_L+Z_o) \tag{1.18}$$

These coefficients play a pivotal role in the dispersal of the electromagnetic signal due to the conductivity. One should point out that R_L may include not only the terminating resistors, but also the characteristic impedances of any adjoining transmission lines. If an adjoining cell has an impedance, Z_1, and the junction resistance is R, then the load impedance seen by a wave in Z_o, is the parallel combination given by $R_L = Z_1R/[R+Z_1]$. Note that when $Z_1 =Z_o$ and R $\sim \infty$ the cell impedances are matched , T=1,B=0, and the wave flows to the adjoining cell unimpeded.

It is natural to ask whether the transfer and reflection coefficients are related, based on energy flow considerations. This conjecture is indeed confirmed by considering the following relationship:

$$(T^2/R_L) +(B^2/Z_o) =1/Z_o \tag{1.19}$$

Substitution of Eqs.(1.17)-(1.18) into Eq.(1.19) verifies the relationship. Eq.(1.19) has a very simple interpretation. If a wave with unit amplitude, propagating in line Z_o, encounters a node with load impedance R_L, then Eq.(1.19) merely states that the incident energy flow, $1/Z_o$, is equal to the energy flow delivered to R_L (i.e., T^2/R_L) plus the energy flow reflected from the load(i.e., B^2/Z_o). Knowing one of the coefficients, however, does not automatically provide us with the other. For example, calculating B from Eq.(1.19) (based on a knowledge of T) leaves us with an ambiguity as to the sign of B; the original definition of B, Eq.(1.17), must then be used. From Eq.(1.17) we see , therefore, that B is positive when R_L exceeds Z_O and negative when R_L is less than Z_o.

In the previous paragraphs we depicted a situation in which a single wave in a transmission line was incident on the load impedance. Now suppose the load is nothing more than another transmission line with a differing impedance, and further suppose the second line also has an signal incident on the node separating the two transmission lines. The situation is illustrated in Fig.1.13 where the two TLM lines are Z_A and Z_B and the corresponding incident wave voltages are ^+V_A and ^-V_B. When we apply the TLM theory to this situation, does our interpretation of results change in any significant manner ? We proceed

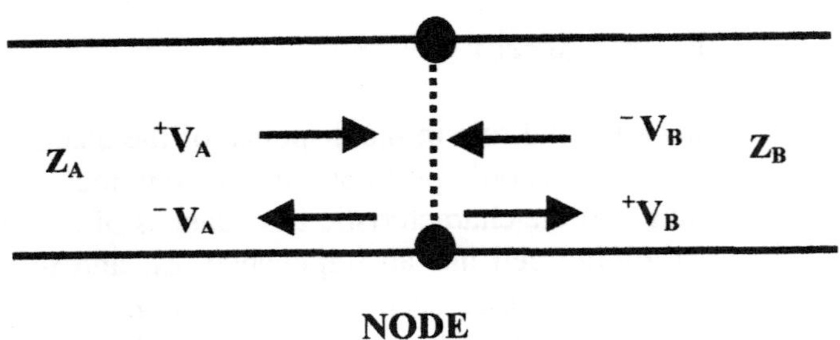

NODE

FIG. 1.13 WAVES ^+V_A AND ^-V_B INCIDENT ON NODE SEPARATING Z_A AND Z_B . CONTRIBUTIONS TO ^-V_A AND ^+V_B (THE REFLECTED AND TRANSMITTED TERMS) ADD LINEARLY.

by applying the TLM formulation individually to each of the two incident waves ^+V_A and ^-V_B. In the case of ^+V_A , $R_L = Z_B$, and this produces a transmitted wave of $[2Z_B/(Z_A+Z_B)] * {}^+V_A$ and a reflected wave $[(Z_B-Z_A)/Z_A+Z_B)] * {}^+V_A$. Likewise , in the case of ^-V_B, the load seen by ^-V_B is Z_A, and the transmitted and reflected waves are $[2Z_A/(Z_A+Z_B)] * {}^-V_B$ and $[(Z_A-Z_B)/Z_A+Z_B)] * {}^-V_B$ respectively. Adding the results(linearly) in each line for both waves one can check out that the continuity of voltage at the node is preserved in the presence of two incident waves. We first obtain the two waves moving away from the node. The total backward wave in Z_A will be

$$^-V_A = [(Z_B-Z_A)/Z_A+Z_B)] * {}^+V_A + [2Z_A/(Z_A+Z_B)] * {}^-V_B \qquad (1.20)$$

while the forward wave in Z_B will be

$$^+V_B = [(Z_A-Z_B)/Z_A+Z_B)] * {}^-V_B + [2Z_B/(Z_A+Z_B)] * {}^+V_A \qquad (1.21)$$

We can then verify that the total voltage at the node is $V = {}^+V_A + {}^-V_A = {}^+V_B + {}^-V_B$. It is worthwhile to make sure that the wave energies carried away from the node , associated with ^-V_A and ^+V_B , preserve the energy flow. We know that the total energy flow incident on the node , E_T , is given by

$$E_T = (1/Z_A) \, (^+V_A)^2 + (1/Z_B) \, (^-V_B)^2 \tag{1.22}$$

It is then easy to verify that the energy carried away from the node satisfies

$$E_T = (1/Z_A) \, (^+V_A)^2 + (1/Z_B) \, (^-V_B)^2 = (1/Z_A) \, (^-V_A)^2 + (1/Z_B) \, (^+V_B)^2 \tag{1.23}$$

using Eqs.(1.20) and (1.21) . We should take care to note the changes in wave direction in Eqs.(1.22) and (1.23). We have not included losses, represented by R at the node. If we do so, then in addition to the wave energies carried away by ^-V_A and ^+V_B , there will be resistive losses given by V^2/R, but just as before $V = {}^+V_A + {}^-V_A = {}^+V_B + {}^-V_B$.

1.6 Initial Conditions of Special Interest

Before we apply the transmission line concepts and relationships to the iterative methods, we first check the consistency of the model under initial conditions of special interest. First we consider whether a 1D   transmission line chain , as shown in Fig.1.14, is stable, i.e., remains in equilibrium even when the node resistors, R, are not activated and the cell lines are each biased to the *same* voltage. Toward this goal we consider the end resistors to be much larger than the characteristic impedance of the lines. As noted in Fig.1.14 the three adjoining lines have the same characteristic impedance. For t > 0, the total load impedance seen by the forward wave in A will be $R_L=Z_o$, since the parallel combination of R and Z_o is simply Z_o, since R is very large. Thus the forward wave going from line A to line B will be unimpeded(i.e., "matched" to B). Similarly, the backward wave in B will be matched to A. Similar comments apply to lines B and C. The net effect is that the equilibrium conditions in line B(or any other cell in the chain) remain the same, with the same forward and backward waves (and with equal amplitudes)as in the previous time step. The equilibrium also is preserved when the adjoining cell impedances are unequal. Thus there is no net transfer of

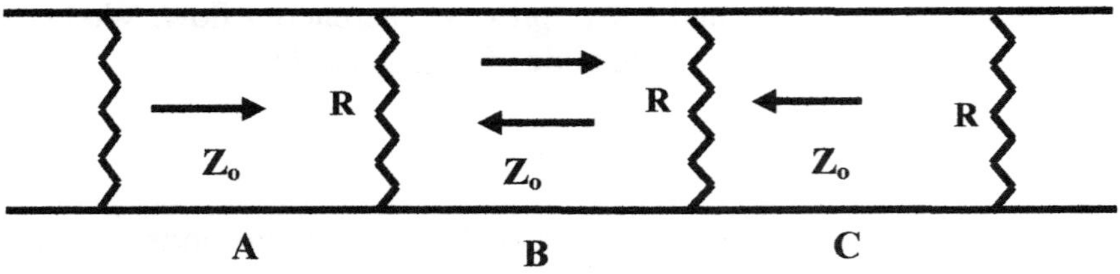

FORWARD WAVE IN B FOR (K+1)TH TIME STEP:
$$^+V_B{}^{k+1} = T^+V_A{}^K + B^-V_B{}^k$$

BACKWARD WAVE IN B FOR (K+1)TH TIME STEP:
$$^-V_B{}^{k+1} = T^-V_C{}^K + B^+V_B{}^k$$

$$T = 2R_L/(Z_o+R_L), \quad B=(R_L-Z_o)/(Z_o+R_L), \quad R_L=Z_oR/(Z_o+R)$$

FIG 1.14   WAVES   IN A AND C CELLS CONTRIBUTE TO THE FIELD IN B DURING THE ENSUING K+1 TIME STEP ACCORDING TO THE ABOVE RELATIONS. LOSS AND IMPEDANCE IN EACH CELL ARE ASSUMED IDENTICAL.

---

energy from one cell to the other.

It is also of interest to determine the decay of the various cells for a uniform 1D chain in which each cell is initially biased to V_o and the node resistance R is finite and uniform throughout the chain. Under these circumstances one may use Eqs.(1.17)-(1.18) to determine the decay within each cell. For large R (relative to the line impedance) the forward and backward waves decay by an amount $(1-Z_o/R)$ with each time step. Because the node resistance and line impedance are uniform, however, we may adopt the view that there is no net transfer of energy from one cell to another. The cell voltage in each cell declines, to be sure, but the decline may be regarded as internal to the cell. The situation changes of course when the adjacent node resistors differ, even when the initial bias voltage is uniform for all the cells. Under these circumstances the fields will redistribute themselves among the cells and a net transfer of energy from one cell to another will occur.

What happens when the adjoining lines are initially biased to *differing* voltages, even when the node resistors are infinitely large and the cell impedances are uniform? The forward wave going from A to B will not be compensated by the backward wave going from B to A. Equilibrium therefore is not maintained for the one dimensional array. As mentioned before, artificial changes may be introduced, for example , by placing series switches between line elements and activating these switches at the same time that the node resistors is activated. Such changes are not based on any physical considerations, and cannot be seriously considered. The only self consistent solution is to resort to either 2D or 3D arrays, which we do in a later Section of this Chapter. Despite the lack of an initial equilibrium solution(for spatial variations in the voltage), however, the 1D circuit is nevertheless very useful for a host of 1D problems as well as for gaining insight into the more complicated 2D and 3D arrays. In the following we compare the TLM and finite difference methods for obtaining 1D solutions.

One Dimensional TLM Analysis. Comparison with Finite Difference Method

1.7 TLM Iteration Method

We begin the comparison of the 1D TLM and finite difference methods by first considering the TLM iterations. For simplicity we assume the same resistance, R, separating the adjoining lines, which have the same characteristic impedance Z_o. During a given time interval the sum of the forward and backward waves, ^+V and ^-V, comprise the total field V. As noted previously, the field waves may be written in terms of the fields belonging to the prior interval. Using Fig.1.14 as a guide, the iterative equations for the forward and backward waves in cell B, during the (k+1)th step, are

$$^+V_B{}^{k+1} = T\ ^+V_A{}^k + B\ ^-V_B{}^k \qquad (1.24)$$

$$^-V_B{}^{k+1} = T\ ^-V_C{}^k + B\ ^+V_B{}^k \qquad (1.25)$$

where

$$T = 2R_L/\ (Z_o + R_L) \qquad (1.26a)$$
$$B = (R_L - Z_o)/(R_L + Z_o) \qquad (1.26b)$$

and

$$R_L = Z_o R/(Z_0 + R) \qquad (1.27)$$

The integer superscripts k, k+1, attached to the cell fields, denote the time step. The reflection coefficient B should not be confused, of course, with the label of the middle cell. In order to facilitate the comparison with the finite difference method, we introduce the loss parameter,

$$\alpha = Z_o/2R \qquad (1.28)$$

The scattering coefficients then become

$$T = 1/(1+\alpha) \qquad (1.29)$$

$$B = -\alpha/(1+\alpha) \qquad (1.30)$$

Adding $^+V_B{}^{k+1}$ and $^-V_B{}^{k+1}$ we obtain the forward iteration for the B cell, or

$$V_B{}^{k+1} = [(1/(1+\alpha)]\ [^+V_A{}^k + {}^-V_C{}^k] - [\ \alpha/(1+\alpha)]\ V_B{}^k \qquad (1.31)$$

In order to proceed further with the comparison with numerical methods, we shall need not only the forward iteration, but the "reverse " one as well, for which we shall have to digress.

1.8 Reverse TLM Iteration

Numerical techniques involving partial differential equations often require time elements of $t-\Delta t$ as well as $t+\Delta t$, in order to obtain correct solutions. It should come as no surprise, therefore, that in order to compare the numerical methods with the TLM method, we need to take into account reverse as well as forward iterations. The generic form of the two types of TLM iterations are shown in Eqs.(1.32)- (1.33), applicable not only to 1D but 2D and 3D as well.

$$\text{FORWARD}$$
$$^{+}V^{k+1} \;=\; \Sigma[\; S.C.\;]\, ^{+}V^{\,k} \;+\Sigma[\; S.C.\;]\, ^{-}V^{k} \tag{1.32}$$

$$^{-}V^{k+1} \;=\Sigma[\; S.C.\;]\, ^{+}V^{\,k} \;+\Sigma[\; S.C.\;]\, ^{-}V^{k} \tag{1.32b}$$

$$\text{REVERSE}$$
$$^{+}V^{k-1} \;=\; \Sigma[\; S.C.\;]\, ^{+}V^{\,k} \;+\Sigma[\; S.C.\;]\, ^{-}V^{k}] \tag{1.33a}$$

$$^{-}V^{k-1} \;=\; \Sigma[\; S.C.\;]\, ^{+}V^{\,k} \;+\Sigma[\; S.C.\;]\, ^{-}V^{k}] \tag{1.33b}$$

where [S.C.] are the scattering coefficients (evaluated during the kth step) and the summation is over wave contributions from adjacent lines. Note the important fact that in the reverse iteration, the (k-1)th wave is determined from waves existing during the ensuing kth step, whereas the forward iteration relates V^{k+1} to waves existing during the prior kth step. Although Eqs.(1.32) and (1.33) appear to be superficially the same, they are different. First of all, the scattering coefficients will differ , especially if losses are present. In addition, the two iterations will have different node locations for the forward and reverse iterations, which we discuss later.

It seems somewhat strange to consider reverse TLM iterations. One naturally asks the question whether it is truly necessary examine such a topic. There are at least two reasons, however, why it is important to take into account the reverse iteration. The first is that , using such an iteration, it may be possible to determine an earlier physical state based on the present state. The second reason is that the reverse iteration provides us with additional information, and when we combine both forward and reverse iterations we are able to make a more accurate comparison with other numerical techniques, such as the finite difference technique. In fact we shall see that the finite difference equation may be "decomposed" into forward and reverse TLM iterations, both of which have immediate physical significance.

Although the forward iteration is fairly straightforward, , it is not at all clear how one may obtain the reverse type. Table 1.2 provides a prescription , based on a knowledge of the waves during the kth step. Thus, we assume the

TABLE 1.2

HOW TO GO BACK IN TIME VIA THE REVERSED TLM ITERATION

1) **ASSUME WAVE $^+V^k$ AND $^-V^k$ ARE KNOWN(FOR THE FORWARD ITERATION)**

2) **REVERSE DIRECTION OF ALL WAVES**
$$(^+V^k)^* \rightarrow {}^-V_R{}^k , (^-V^k)^* \rightarrow {}^+V_R{}^k$$

3) **AS IN FORWARD ITERATION, CALCULATE THE SCATTERING AMONG THE WAVES AND OBTAIN $^+V_R{}^{k+1}$ AND $^-V_R{}^{k+1}$. TIME DEPENDENT NODES ARE EVALUATED AT THE k TH STEP**

4) **DURING SCATTERING , ANY NODE RESISTORS ARE CONSIDERE NEGATIVE, I.E., SIGNALS ARE AMPLIFIED**

5) **UPON COMPLETION OF SCATTERING , REVERT TO ORIGINAL DIRECTION,**
$$(^+V_R{}^{k+1})^* \rightarrow {}^-V^{k-1} , (^-V_R{}^{k+1})^* \rightarrow {}^+V^{k-1}$$

forward and backward waves, $^+V^k$ and $^-V^k$, are known in all cells, where the superscript denotes the time step. We then reverse the direction of all the waves; we denote this operation by *. Thus $(^+V^k)^*$ and $(^-V^k)^*$ denotes the wave reversal of $^+V^k$ and $^-V^k$ during the kth step. We therefore may write the waves which have been reversed as $(^+V^k)^* = {}^-V_R{}^k$ and $(^-V^k)^* = {}^+V_R{}^k$, and we should also take note of the obvious relationships $((^+V^k)^*)^* = (^-V_R{}^k)^* = {}^+V^k$, $((^-V^k)^*)^* = (^+V_R{}^k)^* = {}^-V^k$. The subscript R is added for clarity in this Section and in Table 1.2, to denote the reversed wave. In the ensuing Sections, however, we will drop the subscript since the reversed waves are always identified as such.

The next phase is to calculate the scattering among the reversed TLM waves, treating these waves in the same manner as the usual forward iteration. We then proceed to calculate the waves for the next step, but with an important difference. During the scattering any node resistors are considered negative, i.e., the signals are amplified. This is to be expected, since we are going back in time and any attenuated signals must regain their former strength by being amplified. We should also note that the nodes used in the scattering are evaluated at the same k th time step (with negative node resistance). The final step , after the completion of the scattering, is to revert to the original directions of the waves, so that $(^+V_R{}^{k+1})^* \rightarrow {}^-V^{k-1}$ and $^-V_R{}^{k+1} \rightarrow {}^+V^{k-1}$.

Suppose we have knowledge of the present and future TLM fields and nodes, but are ignorant of any prior states. Is it possible to obtain a similar knowledge of the previous states? The reversion to an earlier state , by means of the reversed iteration, is certainly achievable provided the nodes are time independent. With time independent nodes we presumably have a knowledge of the nodes throughout the medium at all times and therefore we can go back in time as far as we wish.

With time *dependent* nodes, however, there is in general no way to determine the prior states, even when the node resistance is much larger than the line impedance. In order to attempt such an effort we must first resort to an examination of the fields and nodes , as well as the time and spatial gradients of the fields and nodes , during the present and future states. With such information it is conceivable we can make educated "guesses" as to the prior states. Certainly the first few states just prior to the present state may be estimated by analytically continuing the node values back in time. Also, if we somehow know(or guess) the static field configurations of the initial state existed in a given, limited region, then this information may possibly provide clues as to such prior states(in this case the unknown prior states occur after activation). Further scrutiny of this topic, as well as questions concerning the uniqueness of the solution , will take us far beyond the scope of the present discussion.

1.9 Example of Reverse Iteration for Non-Uniform Line

Although we have considered adjacent cells to have the same impedance, Z_o, the procedure for finding V^{k-1} is exactly the same when the cell impedances differ. A simple example is given in Fig.1.15 where the impedance of cells A,B,and C are Z_0 , $2Z_o$, and $3Z_o$ respectively, and the existing waves during the kth step are as shown: $^-(1/3)_A{}^k$ and $^+(4/3)_B{}^k$ and zero wave in cell C. The dotted arrows indicate the reversed waves . If we proceed to the (k+1)th step with the reversed waves , we obtain $^-(1)_A{}^{k+1}$. Reversing this wave then results in $^+(1)_A{}^{k-1}$, as shown. We may confirm the result by proceeding with the forward iteration, which then produces the correct fields for the kth step.

1.10   Derivation of Scattering Coefficients For Reverse Iteration

We now return to the case where loss is included at the nodes. In our prescription for finding V^{k-1} , we said that we replace R with -R in the scattering coefficients, or, alternatively, α with $-\alpha$. We now verify this assumption. Fig.1.16 shows the status of the waves during the (k-1)th step where $V^{k-1}{}_1$ and $V^{k-1}{}_2$ are the waves in cells 1 and 2 respectively and the coefficients T and B are given by Eqs.(1.29) and (1.30). During the kth time step the backward wave in cell 1 is

$$^-V_1{}^k = (-\alpha\,^+V_1{}^{k-1} + \,^-V_2{}^{k-1})/(1+\alpha) \tag{1.34}$$

and the reversed wave is

$$(^-V_1{}^k)^* = \,^+V_1{}^k = (-\alpha\,^+V_1{}^{k-1} + \,^-V_2{}^{k-1})^*/(1+\alpha) \tag{1.35}$$

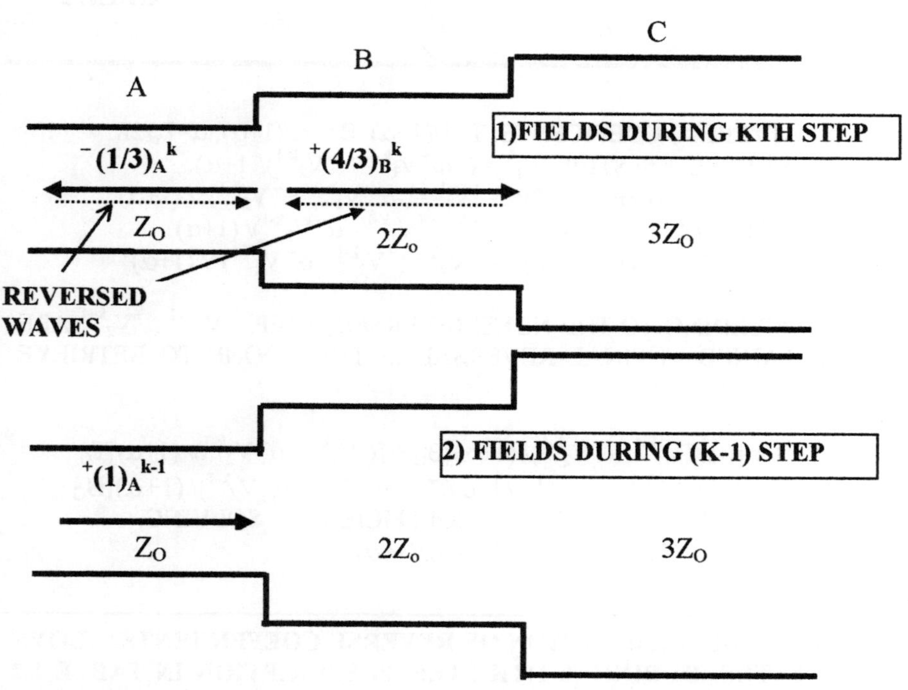

FIG 1.15 EXAMPLE OF REVERSE ITERATION FOR NON-UNIFORM LINE. TO FIND WAVES FOR (K-1)TH STEP, REVERSE ALL FIELDS, $\{^-(1/3)_A{}^K\}* = {}^+(1/3)_A{}^K$, $\{^+(4/3)_B{}^K\}* = {}^-(4/3)_B{}^K$, THENCE PROCEED TO THE (K+1)TH STEP. FINALLY REVERSE ALL FIELDS AGAIN TO OBTAIN THE FIELD S FOR THE (K-1)TH STEP.

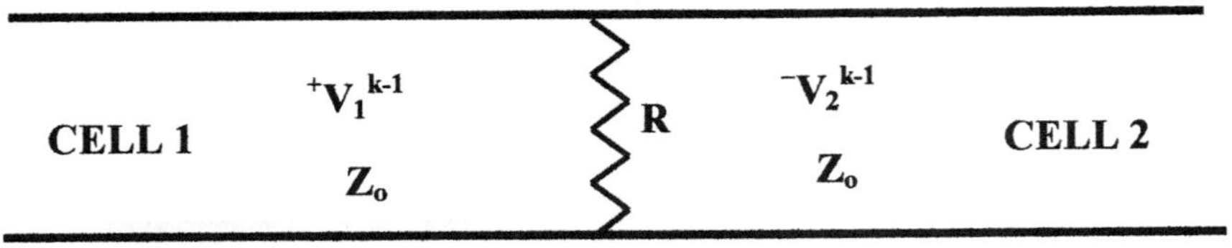

$$^+V_1^{k-1} \qquad R \qquad ^-V_2^{k-1}$$

CELL 1 $\qquad Z_o \qquad \qquad Z_o \qquad$ CELL 2

SCATTERING COEFICENTS: $T=1/(1+\alpha)$, $B= -\alpha/(1+\alpha)$, $\alpha=Z_o/2R$
FIRST CELL, Kth STEP: $^-V_1^k = (-\alpha\ ^+V_1^{k-1} + {}^-V_2^{k-1})/(1+\alpha)$
REVERSED WAVE $=(^-V_1^k)^* = {}^+V_1^k =(-\alpha\ ^+V_1^{k-1} + {}^-V_2^{k-1})^*/(1+\alpha)$
SECOND CELL, kTH STEP: $^+V_2^k=(^+V_1^{k-1} - \alpha^-V_2^{k-1})/(1+\alpha)$
REVERSED WAVE$= (^+V_2^k)^* = {}^-V_2^k =\{^+V_1^{k-1} - \alpha^-V_2^{k-1})^*/(1+\alpha)\}$

PROCEEDING TO THE NEXT(REVERSED) STEP, $^-V_1^{k+1}$, $^+V_2^{k+1}$ ARE OBTAINED. A FINAL REVERSAL IS THEN DONE TO RETRIEVE $^+V_1^{k-1}$ AND $^-V_2^{k-1}$:

$$^+V_1^{k-1} = [(-\alpha\ ^+V_1^{k-1} + {}^-V_2^{k-1})/(1+\alpha)]B_G + [(^+V_1^{k-1} - \alpha^-V_2^{k-1})/(1+\alpha)]T_G$$
$$^-V_1^{k-1} = [(-\alpha\ ^+V_1^{k-1} + {}^-V_2^{k-1})/(1+\alpha)]T_G +[(^+V_1^{k-1} - \alpha^-V_2^{k-1})/(1+\alpha)]B_G$$

T_G , B_G ARE UNKNOWN COEFFICIENTS. SOLVING,
$$T_G = 1/(1+\alpha)] , \quad B_G = \alpha/(1+\alpha)$$

FIG. 1.16 DETERMINATION OF REVERSE COEFFICIENTS. ABOVE IS STATUS DURING (k-1)TH STEP. PRESCRIPTION IN TABLE 1.2 IS FOLLOWED TO OBTAIN THE REVERSE COEFFICIENTS.

Similarly the the forward field in the second cell is

$$^+V_2^k = (^+V_1^{k-1} -\alpha^-V_2^{k-1})/(1+\alpha) \tag{1.36}$$

After reversal this wave becomes

$$(^+V_2^k)^* = {}^-V_2^k =(^+V_1^{k-1} -\alpha^-V_2^{k-1})^*/(1+\alpha) \tag{1.37}$$

We then use the reversed fields, $^+V_1^k$ and $^-V_2^k$, and proceed to the next(reverse) step, using the usual iterative Eqs.(1.24) and (1.25)(with as yet unknown scattering coefficients) to obtain $^-V_1^{k+1}$ and $^+V_2^{k+1}$. We then perform a final reversal , so that $^-V_1^{k+1}$ and $^+V_2^{k+1}$ become $^+V_1^{k-1}$ and $^-V_2^{k-1}$. The final equations are

$$^+V_1^{k-1} = [(-\alpha\ ^+V_1^{k-1} + ^-V_2^{k-1})/(1+\alpha)]\ B_G + [(^+V_1^{k-1} - \alpha^-V_2^{k-1})/(1+\alpha)]T_G \quad (1.38)$$

$$^-V_2^{k-1} = [(-\alpha^+V_1^{k-1} + ^-V_2^{k-1})/(1+\alpha)]\ T_G + [(^+V_1^{k-1} - \alpha^-V_2^{k-1})/(1+\alpha)]B_G \quad (1.39)$$

In Eqs(1.38)-(1.39), T_G, B_G are the as yet unknown coefficients which allow us to regain $^+V_1^{k-1}$ and $^-V_2^{k-1}$. Solving for these coefficients gives us

$$T_G = 1/(1-\alpha) \quad\quad\quad (1.40a)$$

$$B_G = \alpha/(1-\alpha) \quad\quad\quad (1.40b)$$

Both Eqs.(1.38) and (1.39) yield the same expressions for T_G and B_G , as required for consistency. Note that T_G and B_G are identical to Eqs.(1.29)-(1.30) when we substitute $-\alpha$ for α, i.e., -R for R. As promised the reverse iterations require that we substitute negative values for the node resistance in the scattering coefficients. This will enable us to perform the reverse iteration we need to complement the forward iteration in Eq.(1.31).

It is a curious fact that to a certain degree we possess the mathematical tools(certainly for time independent nodes) to go back in time, as well as forward. The reverse iteration may be used to view earlier events, based on present events. Does a preference exist for one or the other process, based on experience.? Based on observation, the answer is yes. Nature appears to favor the forward direction; in the present context this corresponds to the tendency of electromagnetic energy to spread out in the available spatial directions.

1.11 Complete TLM Iteration (Combining Forward and Reverse Iterations)

Having verified the scattering coefficients for the reverse wave, Eqs (1.40a)-(1.40b), we can obtain the reverse iteration , which will enable us to determine the complete TLM equation. Referring to Fig.1.14 , we now must now focus on $^-V_A^{\ k}$ and $^+V_C^{\ k}$ (instead of $^+V_A^{\ k}$ and $^-V_C^{\ k}$) in cells A and C. We then resort to the our oft-stated method, and reverse the direction of waves $^-V_A^{\ k}$ and $^+V_C^{\ k}$, allow the iteration to proceed for one step, using Eqs.(1.40a)-(1.40b), and finally revert to the original direction. The reverse iteration for cell B is then

$$V_B^{\ k-1} = [(1/(1-\alpha)]\ [^-V_A^{\ k} + {}^+V_C^{\ k}] + [\ \alpha/(1-\alpha)]\ V_B^{\ k} \tag{1.41}$$

If we compare the above with $V_B^{\ k+1}$, Eq.(1.31), we see that a shortcut method for obtaining $V_B^{\ k-1}$ exists: In Eq.(1.31) we simply replace α with $-\alpha$ and also replace any forward wave with a backward wave, and similarly any backward wave with a forward one.

    For purposes of comparison with numerical methods, we add $V_B^{\ k-1}$ and $V_B^{\ k+1}$ obtaining

$$V_B^{\ k+1} + V_B^{\ k-1} = [(V_A^{\ k} + V_C^{\ k}) + \alpha(-\ ^+V_A^{\ k} - {}^-V_C^{\ k} + {}^-V_A^{\ k} + {}^+V_C^{\ k}) + 2\alpha^2 V_B^{\ k}]/(1-\alpha^2) \tag{1.42}$$

We will see shortly that the above is compatible with the iteration obtained by finite difference methods. We thus may regard the finite difference iteration as a combination of forward and backward TLM iterations.

1.12 Finite Difference Method. Comparison with TLM Method

We now turn to the conventional numerical technique for solving the wave equation, which relies on the use of finite differences(see, e.g., Reference [5]) . A rectangular grid for the distance x and time t coordinates is first established. We then perform a Taylor expansion of the second order spatial derivative of $E(x,t)$ about x, and evaluated at two locations: $x-\Delta x$ and $x+\Delta x$, where Δx represents a small excursion from x. The results for the two locations are subtracted,

$$\partial^2 E(x,t)/\partial x^2 = [\ E(x+\Delta x,t) - 2E(x,t) + E(x-\Delta x,t)\ ]/\Delta x^2 + \text{higher order terms} \qquad (1.43)$$

where Δx is the difference in the x coordinate. Similar expansions of $\partial^2 E(x,t)/\partial t^2$ and $\partial E(x,t)/\partial t$ yield

$$\partial^2 E(x,t)/\partial t^2 = [\ E(x,t+\Delta t) - 2E(x,t) + E(x,t-\Delta t)\ ]/\Delta t^2 + \text{higher order terms} \qquad (1.44)$$

and

$$\partial E(x,t)/\partial t = [E(x,\ t+\Delta t)\ -E(x,\ t-\ \Delta t)\ ]/2\ \Delta t\ +\ \text{higher order terms} \qquad (1.45)$$

where Δt is the difference in the time. Substituting Eqs.(1.43)-(1.45) into the wave equation and solving for $E(x,y,t+\Delta t)$ yields the iterative equation,

$$E(x,t+\Delta t) = \ E(x,t)\ -\ E(x,\ t-\Delta t)\ +\ \lambda^2[E(x+\Delta x,t)\ +\ E(x-\Delta x,\ t)\ -\ 2E(x,t)]$$
$$-\alpha[E(x,t+\Delta t)\ -\ E(x,t-\Delta t)] \qquad (1.46)$$

where
$$\lambda^2 = v^2 \Delta t^2 /\ \Delta x^2 \qquad (1.47a)$$
$$\alpha\ =\ \Delta t \sigma/2\varepsilon\ =\ Z_o/2R \qquad (1.47b)$$

Eq.(1.46) simplifies by setting $\lambda=1$, or $\Delta x=v\Delta t$. The value for λ is within the allowable range needed to insure stability for the finite difference solution. Stability is assured when the "numerical" velocity, $\Delta x/\Delta t$, is greater than or *equal* to the wave velocity v (Reference [5]). Since $\Delta x/\Delta t$ is set equal to v, the finite difference solution is automatically stable and Eq(1.46) becomes

$$E(x,t+\Delta t) = -\ E(x,\ t-\Delta t)\ +[E(x+\Delta x,t)\ +E(x-\Delta x,\ t)]\ -\alpha[E(x,t+\Delta t)-E(x,t-\Delta t)] \qquad (1.48)$$

In order to compare the above with the TLM iteration we convert field variables to TLM voltage variables in the above, using the following correspondence:

$$E(x,t) \rightarrow V_B^{\,k}, \quad E(x,t+\Delta t) \rightarrow V_B^{\,k+1}, \quad E(x-\Delta x,\, t) \rightarrow V_A^{\,k}, \text{ etc..} \qquad (1.49)$$

The finite difference equation then becomes

$$V_B^{\,k+1} + V_B^{\,k-1} = V_A^{\,k} + V_C^{\,k} - \alpha[\, V_B^{\,k+1} - V_B^{\,k-1}\,] \qquad (1.50)$$

Substitution of the TLM iterations for $V_B^{\,k+1}$, $V_B^{\,k-1}$, Eqs.(1.31), (1.41) , into the above then *yields an identity*. The TLM and finite difference iterations are therefore completely compatible. We should add, that from a mathematical point of view, there are innumerable combinations of $V_B^{\,k+1}$ and $V_B^{\,k-1}$ which satisfy Eq(1.50). Eqs.(1.31) and (1.41) , however , are the only ones which satisfy the physical requirements imposed by the motion of the forward and backward TLM waves. We assert, therefore , that we can break up the 3 tier finite difference iteration into a pair of two tier forward and reverse iterations, corresponding to the motions of forward and reverse TLM waves, and given by Eqs.(1.31) and (1.41). As a practical matter, of course, the forward iteration is most often used, although the reverse iteration may on occasion be invoked.

Two Dimensional TLM Analysis. Comparison With Finite Difference Method

The transition from 1D to 2D exposes the potential advantages and flaws of the transmission line matrix method. One advantage, alluded to before, is that the 2D matrix allows for static solutions without the artificial insertion of any components (such as a switch). There is therefore a smooth passage from the static solution to the transient one, once the equilibrium conditions are upset. Another advantage has to with the fact that the transmission lines border a symmetry element, in this case a square(or in 3D, a cube). The symmetry elements occupy the entire space. which of course is essential, and in addition the elements can be labeled in a natural way. The notation used for these elements is discussed in Chapter II.

One shortcoming of the 2D (and 3D)matrices, which is repairable, stems from the lack of isotropy when using symmetry elements. Thus when using a 2D TLM matrix the electromagnetic energy is constrained to move along the lines surrounding the square element. As a result , if a signal source emanates from the source region O, the *first arrival time* of a signal reaching point B will exceed that for point B, as shown in Fig.1.17. This is because the path OA is a straight line while that of OB must proceed along a zigzag path. In an isotropic medium there is no reason why these two directions should differ. This effect is not surprising since the square element(or any other symmetry element, for that matter) is not isotropic. This and other related topics are discussed in Chapters II-IV , where we also modify the technique to compensate for the inherent

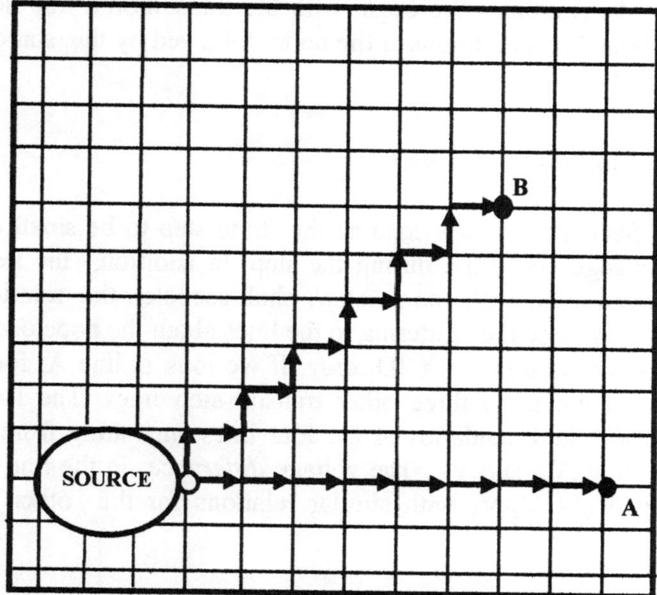

FIG.1.17 DIRECTIONAL ANISOTROPY OF SQUARE MATRIX.

anisotropy of the square symmetry element. One should also note that as the cell size is made smaller, the width of the earliest arriving signal is likewise smaller, A second shortcoming has to do with the fact that the 2D and 3D matrices do not possess plane wave properties. We correct the cell matrix , so as to account for plane wave effects, in Chapter IV.

Despite the aforementioned limitations, we will show that the two dimensional matrix of transmission lines is closely linked with the finite difference solutions of the two dimensional wave equation. For the treatment of the two dimensional problem, we compare a two dimensional matrix of transmission lines with the two component wave equations in the x and y directions. In the following discussion, the 2D comparison of the finite difference and TLM methods does not include losses, but this does not alter the generality of the conclusions in any way. The ensuing Chapters of course include loss terms in the TLM iterations. We begin with the transmission line description, first discussing boundary conditions at the node, followed by the static and non-static behavior.

1.13 Boundary Conditions at 2D Node

In the ensuing Sections we will assume the time step to be small enough so that the fields change very little during the step. In addition, the fields at the node are considered to be irrotational. As we shall see later, the rotational properties come into play once the scattering to the lines about the node occurs.

Fig.1.18 shows the node of a 2D array. If we look at line A, for example, we see that it is coupled to three other transmission lines. The iso-potential regions which form the boundaries of the four lines emanating from the node, are denoted by $V_1, V_2, V_3,$ and V_4 . The voltage *difference* in the line separating cells 1 and 2 is $V_A = V_2 - V_1,$ with similar relations for the other lines. The complete set is:

$$V_A = V_2 - V_1 \tag{1.51a}$$
$$V_B = V_4 - V_3, \tag{1.51b}$$
$$V_C = V_3 - V_1, \tag{1.51c}$$
$$V_D = V_4 - V_2. \tag{1.51d}$$

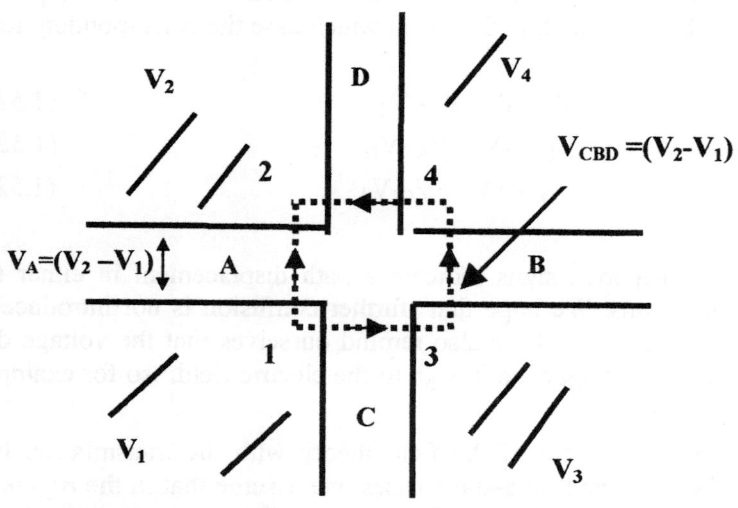

FIG. 1.18 BOUNDARY CONDITION AT THE 2D NODE
IS $V_A=V_C+V_B-V_D$, WHERE $V_A=V_{CBD}=V_2-V_1$, $V_C=V_3-V_1$,
$V_B=V_4-V_3$, $V_D=V_4-V_2$.

Since we assume the node region is small compared to the wavelengths being generated, we consider the field to be conservative about the node. Thus, the voltage path from 1 to 2 is equivalent to 1 to 3 followed by 3 to 4 and then 4 to 2. We therefore take as our boundary condition

$$V_A = V_B + V_C - V_D \qquad (1.52a)$$

It is important to note that the line voltages in the above represent the total voltage, i.e., the sum of the backward and forward voltage waves. Also note the negative sign for V_D, which stems from the fact that the path displacement is in the negative direction. and therefore the voltage wave is negative. The negative sign for V_D is completely dependent on the fact that we have selected V_A as our initial wave, dictating that the other three waves match V_A We could just as well have selected 1 to 3, 3 to 4, or 2 to 4 in which case the corresponding relations would be

$$V_C = V_A + V_D - V_B \qquad (1.52b)$$

$$V_B = -V_C + V_A + V_D \qquad (1.52c)$$

$$V_D = +V_B + V_C - V_A \qquad (1.52d)$$

Again note that the negative signs indicate a path displacement in either the negative x or y directions. We hope that further confusion is not introduced , concerning sign conventions, if we also remind ourselves that the voltage difference variable is always opposite in sign to the electric field, so for example, $V_A = V_2 - V_1 = \Delta V = -E\Delta l$.

To justify the replacement of the field theory with the transmission line matrix, for both the static and non-static cases, we assume that in the *region of the node* , both electric and magnetic fields , as well as the time, are slowly varying (compared to the TLM element Δl and the time step Δt) , so that *on the average* , the fields are irrotational. This implies that the two line integral paths from point 1 to point 2 , as shown in the Figure, are identical. In the case of the electric field, therefore, the fields are assumed to be concentrated in the TLM lines, and we replace the field variables with the line voltages. One should stress that the fields within the transmission lines themselves are rotational, as they must be in order to propagate as a wave. The other boundary condition at the node has to do with the magnetic vector associated with the electric field. The magnetic fields of the lines are all perpendicular to the electric fields , and thus the magnetic fields are perpendicular to the paper. As a boundary condition we insist that the magnetic field be continuous at the junction. Thus , for example, in line A the magnetic field will be equal to magnetic field in each of the lines B, C, and D.   In terms of transmission line variables , of course, the magnetic field translates into a line current, and the boundary condition requires continuity of current at the node. The total current in each line , therefore , is identical.

The above node boundary conditions are applicable to either static or non-static problems, and the boundary conditions remain the same whether the node resistors are activated or not. The curl properties are automatically satisfied , as will be described in a later Section. In addition, there is no constraint on the values of the characteristic impedances of the lines surrounding the node.

1.14 Static Behavior About 2D Node

The static behavior of the 2D array may be examined, once again using Fig.1.18 We assume the end resistors are much larger than the characteristic impedances, and therefore do not interfere with the assumed static conditions. The approach adapted is suggested by the 1D TLM line segment, where we know that a single isolated cell at voltage V contains forward and backward waves each with amplitude V/2. We determine whether such an arrangement in each of the 2D line segments is self consistent, and remains the same once the waves in each line are allowed to couple to other lines. To explore this, we look at line A. Initially, the forward wave amplitude is $^+V_A= (V_2-V_2)/2$ and it sees a load impedance of $(Z_B+Z_C +Z_D)$. The backward in A , ^-V_A , consists of two parts. First there is the contribution caused by the reflection at the node, denoted by $^-V_{A,R}$, and is given by

$$^-V_{A,R} =[(V_2 -V_1)/2][Z_B+Z_C+Z_D-Z_A] /[Z_A+Z_B+Z_C+Z_D] \qquad (1.53)$$

Next we consider the waves in B, C, and D headed in the direction of the node. and ask what part these waves are transferred to line A. Although one can consider each line individually, it is easiest to consider B,C and D as forming a composite line with impedance $(Z_B+Z_C+Z_D)$. The voltage transferred from this composite line into A, denoted by $^-V_{BCD}$, is

$$^-V_{BCD} =(V_2-V_1) Z_A /[Z_B+Z_C+Z_D+Z_A] \qquad (1.54)$$

If we add Eqs.(1.53) and(1.54) then we obtain the total backward wave in A,

$$^-V_A =^-V_{A,R} +^-V_{BCD} = (V_2-V_1)/2 \qquad (1.55)$$

But this is simply the reflected wave one would expect from an open circuit. The same result is obtained if we consider the voltage transfer contributions from each of the individual lines, B, C, and D. Thus the initial fields remain constant, i.e., the solution remain stable so long as the resistors are not activated. Unlike the one dimensional matrix, therefore, adjoining cells with differing voltage will remain at the same voltage without discharging into one another.

1.15 Non-Static Example: Wave Incident on 2D Node

Before deriving the iterative equations for the transmission line matrix, we consider a simple non-static case of a solitary forward wave incident on a 2 D node , as in Fig.(1.19). As we have mentioned before, the same boundary conditions and scattering equations apply to the transient situation.

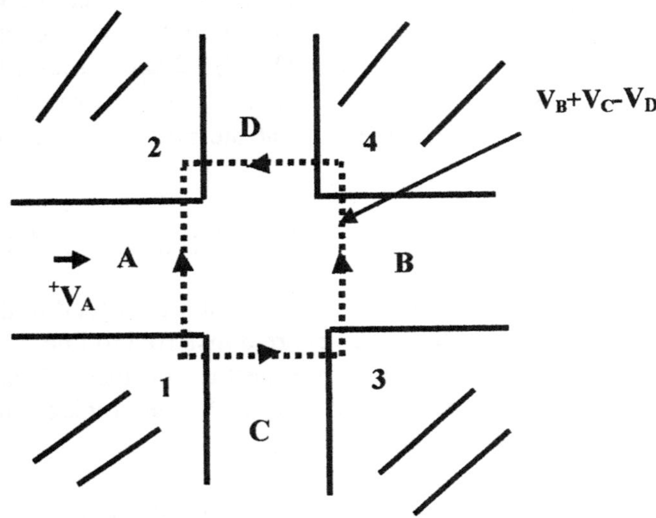

FIG. 1.19   EXAMPLE OF NON-STATIC BEHAVIOR. FORWARD WAVE ^+V_A IS LAUNCHED IN A, DIRECTED TOWARD NODE. $V_A = {}^+V_A - {}^-V_A = V_B + V_C - V_D$ WHEN WAVE REACHES NODE.

The boundary conditions are further illustrated by supposing that a wave ^+V_A is launched in line A , directed toward the node. Upon reaching the node , the field description dictates the conservation of voltage at the node. If V_A is the voltage associated with the A line, and likewise $V_B, V_C, V_D,$ are the node voltages in the other lines, then at the node we expect the following to be satisfied during the time step *following* the arrival of ^+V_A at the node.

$$V_A = {}^+V_A + {}^-V_A \tag{1.56a}$$

$$V_A = V_B + V_C - V_D \tag{1.56b}$$

Next, we determine how the wave scatters among the various lines. Although transmission line theory automatically satisfies the requirement of voltage conservation at the nodes, it is instructive, to verify this property, i.e., Eq.(1.56), taking into account both reflections and energy transfers into adjoining lines. To simplify matters we assume the lines all have the same impedance, $Z_o.$ Starting with the reflected wave , since $R_L = 3Z_o$,

$$B = (Z_L - Z_o)/(Z_L + Z_o) = 1/2 \tag{1.57}$$

and thus

$$^-V_A = B^+V_A = (1/2)\,{}^+V_A \tag{1.58}$$

Eqs.(1.57)-(1.58) allow us to calculate the total load voltage V . Thus the sum of the reflected and incident waves is

$$V_A = {}^+V_A + {}^-V_A = (3/2)\,{}^+V_A \tag{1.59}$$

The wave transmitted from A to B ,for example, is calculated from

$$^+V_B = \quad T^+V_A \tag{1.60}$$

T may be obtained from

$$T = [2R_L/(R_L+Z_o)] \, (Z_o/R_L) = 1/2 \tag{1.61}$$

where the (Z_o/R_L) is appended to T since the voltage transfer to B represents only a portion of the voltage delivered to R_L. Combining ,

$$^+V_B = (1/2) \, ^+V_A \tag{1.62a}$$

We should note that $^+V_B = V_B$ (the total field) since there is as yet no backward wave in B. Similarly, the voltage transfer to lines C and D are

$$^-V_C = (1/2) \, ^+V_A \tag{1.62b}$$
$$^+V_D = -(1/2) \, ^+V_A \tag{1.62c}$$

Again we should note that ^-V_C and ^+V_D represent the total fields V_C , V_D in lines C and D. We should also note the minus sign in Eq.(1.62c) , since ^+V_D is directed in the negative x direction. The previous equations are in agreement with the boundary condition, Eq.(1.56b), as expected, i.e., the total field in lines B,C,D is then (3/2)V. Although we have considered a solitary wave in one of the lines, the situation does not fundamentally change when waves from the other lines (C,B, or D) are simultaneously incident on the node. Under these circumstances the waves moving away from the node , in each line, will not only consist of the reflected wave , but will receive contributions from the   incident waves in the other lines, which add in linear fashion to the reflected wave, just as in the 1D case.

In the previous discussion, we focused on a single wave launched in one of the lines. The behavior of the 2D nodes when multiple coherent waves exist in *parallel* lines, however, is an important issue. Referring to Fig.1.20, we inquire whether identical waves, launched simultaneously in lines R, S, T , etc..., will ultimately simulate a plane wave. In other words, we ask whether the waves in R, S , T, transfer completely intact to lines D, E, F, without transverse scattering or reflection. In Chapters III and IV we examine in detail the question

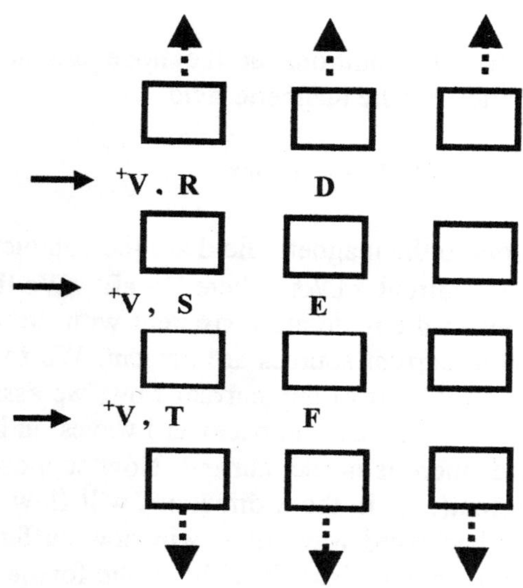

FIG.1.20 PLANE WAVE SIMULATION WITH IDENTICAL FORWARD WAVES, $^+$V. UNDER PLANE WAVE CONDITIONS , $^+$V IS TRANSFERRED INTACT FROM LINE R TO LINE D, S TO E, ETC...

of plane waves in 2D and 3D TLM matrices and describe the modifications which must be made to the TLM theory.

1.16 Integral Rotational Properties of Field about the Node

Although we rely on the rotational properties of the field within the TLM lines , the fields at the node are considered irrotational. We should therefore be concerned as to whether the fields about the node reflect the necessary rotational properties as expressed in Maxwell's Equations. In order to examine such compatibility we utilize the integral representations of the curl equations. To account for the curl properties , we will have to jump ahead slightly and allow, in a simplified way, for 3D scattering. Chapters III and IV describe the 3D scattering in detail.

The rotational boundary conditions at the node are addressed by first considering the curl equation for the magnetic field

$$\nabla \times H = j + \partial D / \partial t \tag{1.63}$$

where the two contributors to the magnetic field are the conduction current density j and the displacement current $\partial D / \partial t$ where D= εE. We first show that the TLM boundary conditions at the node are consistent with the above curl equation when only conduction current sources are present. We focus our attention on Fig.1.21 where, before the start of any current flow, we assume electrostatic conditions prevail. Thus the forward and backward waves in lines A and B are equal to one another and there is no net current. Now suppose the node resistance is activated. A current(e.g., in the x direction) will flow via the node resistance as shown. The backward wave in A will now suffer a diminution in amplitude, resulting from the node loss. In addition the forward wave in B ,i.e., the waves moving away from the node will also be lowered in amplitude , again because of the node activation. As a result of the differing forward and backward waves in A and B, a net current wave will flow in these lines , proportional to the *difference* between the initial wave moving toward the node , and the reduced field moving away from the node. In line A this net field is $(^+V_A - ^-V_A)$ and in line B it is $(^-V_B - ^+V_B)$. In fact , if we convert the voltage waves to electric field waves we can immediately determine from the Poynting vector that the net magnetic field is pointed into the paper in line A and emerges from the paper in line B. This is in accordance with the usual rule for determining the circular field due to a current. A more quantitative grasp of the situation may be obtained if we use the integral formulation of the curl equation, which for the conduction current is

$$\int H \cdot dl = \int j \cdot ds \tag{1.64}$$

where dl and ds are the usual line and area differentials . Consider for example all the elements of j in the x direction, with the line integral in the yz plane.

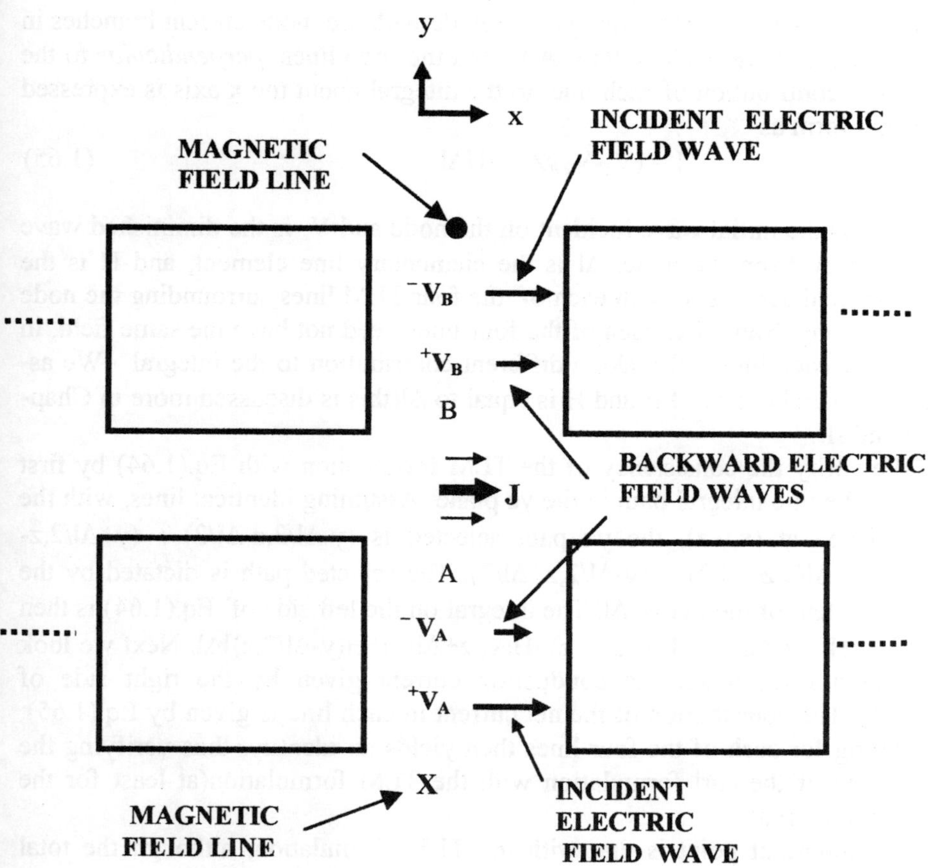

FIG .1.21 GENERATION OF CURLED MAGNETIC FIELD(YZ PLANE) ABOUT NODE IN THE TLM MATRIX, IN RESPONSE TO CURRENT DENSITY J IN THE X DIRECTION. THE MATRIX RESPONDS IN SIMILAR FASHION TO THE DISPLACEMENT CURRENT.

We can think of the curl equation as representing four separate contributions, one for each TLM line converging at the node with the node current branches in the x direction; these include lines A,B, and the two lines *perpendicular* to the paper. The contribution of each line to the integral about the x axis is expressed in general form as

$$I = (V_1 - V_2)/Z_o = H\Delta l \qquad (1.65)$$

where V_1 is the initial wave incident on the node and V_2 is the diminished wave moving away from the node, Δl is the elementary line element, and H is the magnetic field associated with each of the four TLM lines surrounding the node in the yz plane. Naturally, each of the four lines need not have the same field, in which case each line will make a different contribution to the integral. We assume the spatial extent of E and H is equal to Δl(this is discussed more in Chapters II and III.

We verify the consistency of the TLM formulation with Eq.(1.64) by first selecting the line integral path in the yz plane. Assuming identical lines, with the node located at (x,y,z), the yz path selected is (y-Δl/2,z-Δl/2)$\rightarrow$ (y+Δl/2,z-Δl/2)$\rightarrow$ (y+Δl/2,z+Δl/2)$\rightarrow$ (y-Δl/2,z+Δl/2). The selected path is dictated by the assumed extent of the fields, Δl. The integral on the left side of Eq.(1.64) is then given by [H(y,z-Δl/2) +H(y+Δl/2,z)+H(y, z+Δl/2)+H(y-Δl/2,z)]Δl. Next we look at the area integral for the conduction current given by the right side of Eq.(1.64). The contribution of the net current in each line is given by Eq.(1.65); substituting for each of the four lines then yields an identity, thus verifying the consistency of the curl formulation with the TLM formulation(at least for the conduction current).

An important point is that with the TLM formulation, although the total field is irrotational at the node, the formulation does in fact yield the correct rotational properties for a current source, in integral format, about the node. In Chapters II - IV we will describe the 3D notation needed to exactly calculate all the contributions to the line integral for the conduction current, as well the displacement current which follows.

The same type of analysis can also be applied to the displacement current. The integral representation of the curl equation with displacement current sources is

$$\int H \cdot dl = \int \partial (\varepsilon E)/\partial t \cdot ds \qquad (1.66)$$

As before suppose we consider the displacement current sources in the x direction with the line integral in the yz plane. Initially we assume the lines surrounding the node are not occupied with any fields. Now suppose a wave with amplitude ^-E_B (with $^-E_B = V_1/\Delta l$) moves into line B. During the next step the wave will move into the other three lines in the yz plane and in addition there will be a reflected wave in line B. The four waves generated in these lines will give rise to the magnetic fields which circle the displacement currents. As before we can break up the above integral into four contributions. For example in line B the displacement current is obtained from the *reflected* wave ^+E_B, with the displacement current given by $\varepsilon_B{}^+E_B /\Delta t$ (the reflected wave represents the *change* in the field in line B , and therefore the displacement current) The contribution to the line integral for this line is then $\{\varepsilon_B{}^+E_B /\Delta t\}\Delta l^2$. It is easily verified

$$^+H_B\Delta l = \{\varepsilon_B{}^+E_B /\Delta t\}\Delta l^2 \qquad (1.67)$$

using the standard TLM relationships for Δt, ^+E_B , etc... The contributions of the other three lines (line A and the two lines perpendicular to the paper) are similar but with ^+E_B replaced by the transmitted field to the particular line. The curl property for the displacement current is then verified using the same path applied to the conduction current as before. The addition of the fields, in the four lines, is equivalent to the curl property for the magnetic field. Note that the fields in these four lines are all moving away(or radiated) from the node. In the preceding we have only considered compatibility with the TLM formulation for only one of the two curl equations. The other, $\nabla \times E = - \mu \partial H/\partial t$, has a similar , straightforward interpretation using the TLM description.

An intriguing feature of Maxwell's Equations, and in particular the curl equations, has to do with its flexibility. In a certain sense we may regard the equations as a "loose" fitting garment which may be filled out to its proper exterior form by any one of several embodiments. Equivalent modes of description include the usual vector analysis, and the iterative equation approach, based on numerical methods or , as in our case, the TLM matrix.

1.17 2D TLM Iteration Method for Nine Cell Core Matrix

As with the 1D approach, we obtain an important relationship, derived from the nature of the 2D node, which relates the field in each line, at a particular time step, to the fields existing in the in the prior time step, both in the same cell and in the surrounding cells. The relationship is obtained with the help of Fig.1.22 which shows a matrix of lines, using as a backdrop a finite difference grid, which we will use later for comparative purposes.

We shall often refer to the nine cell matrix in Fig.1.22 as the *core matrix* . The four TLM lines immediately surrounding the center cell will be referred to as inner core TLM lines. The remaining eight TLM lines which are labeled we shall refer to as outer core TLM lines. It will be important to observe that the outer core lines form common boundaries with neighboring core matrices. Later, when we compare the TLM results with the finite difference results, we will make use of the fact that the outer core TLM lines will carry less weight than those in the inner core.

In examining the TLM formulation of the core matrix, we start with the voltage, V_1, in line at time $t+\Delta t$. This voltage is the result of a forward wave, $^{+}V_1$, and a backward wave, $^{-}V_1$, with $V_1 = {}^{+}V_1 + {}^{-}V_1$, evaluated at the $t+ \Delta t$ time step. How do these voltages relate to the voltages in the prior step at time t? The forward wave is a result of the of the backward, wave reflected from the left node, and the waves directly transmitted from lines 1L, 3, and 3U. Thus, since the reflection and transfer coefficients are both equal to 1/2 for a lossless 2D node,

$$^{+}V_1 (t+\Delta t) = (1/2)\,{}^{+}V_1 (t)+(1/2)[\,{}^{+}V_{1L} (t)- {}^{+}V_3 (t)+ {}^{-}V_{3U} (t)] \qquad (1.68)$$

A minus sign in front of $^{+}V_3 (t)$ is inserted since , in coupling to line 3, the field direction becomes is negative. A similar relation applies to the backward wave in line 1,

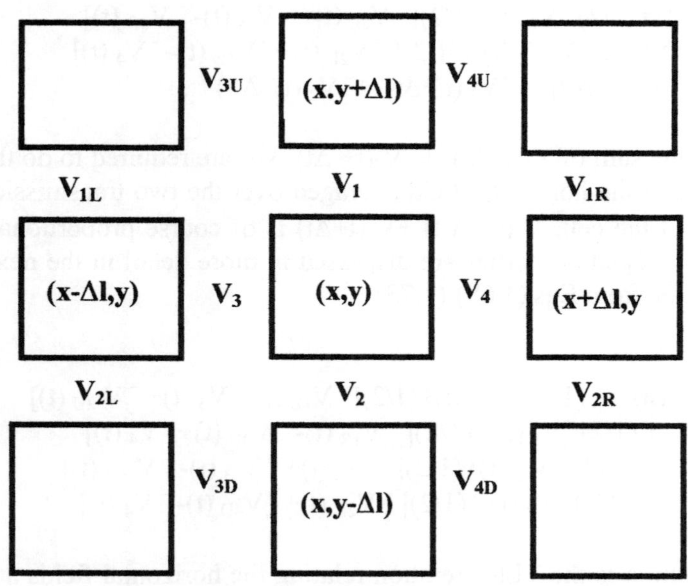

FIG. 1.22 SUPERPOSITION OF TLM MATRIX AND FINITE DIFFERENCE GRIDS USING CORE MATRIX SET.

$$^{-}V_1\,(t+\Delta t) = (1/2)\,^{+}V_1\,(t) + [(1/2)[\,^{-}V_{1R}\,(t) - \,^{-}V_{4U}\,(t) + \,^{+}V_4\,(t)] \qquad (1.69)$$

The total field at t+Δt is

$$V_1(t+\Delta t) = \,^{+}V_1\,(t+\Delta t) + \,^{-}V_1\,(t+\Delta t) \qquad (1.70)$$

In like manner, we calculate $V_2(t+\Delta t)$ in line 2, which is the neighboring line. The two fields, representing lines 1 and 2, are needed in order to make a comparison with the results from the finite difference method. The forward , backward, and total fields in line 2 are:

$$^+V_2\,(t+\Delta t) = (1/2)\,^-V_2\,(t)+(1/2)[\,^+V_{2L}\,(t)+\,^-V_3\,(t)-\,^-V_{3D}\,(t)] \qquad (1.71)$$

$$^-V_2\,(t+\Delta t) = (1/2)\,^+V_2\,(t)+(1/2)[\,^-V_{2R}\,(t)+\,^+V_{4D}\,(t)-\,^-V_4\,(t)] \qquad (1.72)$$

$$V_2(t+\Delta t) = \,^+V_2\,(t+\Delta t) +\,^-V_2\,(t+\Delta t) \qquad (1.73)$$

We now form the sum of $V_1(t+\Delta t) + V_2(t+\Delta t)$. We are required to do this since our interest lies in the horizontal field averaged over the two transmission lines, on either side of the cell. $V_1(t+\Delta t) + V_2(t+\Delta t)$ is of course proportional to this average (cell averaged properties are discussed in more detail in the next Chapter). The result is, from Eqs.(1.68)-(1.73),

$$\begin{aligned}
V_1(t+\Delta t) +V_2(t+\Delta t) = {}& (1/2)\,^-V_1\,(t)+(1/2)[\,^+V_{1L}\,(t)-\,^+V_3\,(t)+\,^-V_{3U}\,(t)] \\
&+ (1/2)\,^+V_1\,(t)+[(1/2)[\,^-V_{1R}\,(t)-\,^-V_{4U}\,(t)+\,^+V_4\,(t)] \\
&+ (1/2)\,^-V_2\,(t)+(1/2)[\,^+V_{2L}\,(t)+\,^-V_3\,(t)-\,^-V_{3D}\,(t)] \\
&+ (1/2)\,^+V_2\,(t)+(1/2)[\,^-V_{2R}\,(t)+\,^+V_{4D}\,(t)-\,^-V_4\,(t)] \qquad (1.74)
\end{aligned}$$

The above equation is the TLM iteration relating the horizontal fields at $t+\Delta t$ to those at t. The expression simplifies if we assume the system is slightly perturbed, in which case the forward and backward waves in each line are close in value. We can achieve the same results, however, without making such an approximation, by incorporating the reverse iteration. As we have discussed previously, we obtain the reverse iteration by first reversing the propagation directions everywhere, then allowing the iteration to proceed forward one step, and finally reverting to the original propagation direction. The result is that $V_1(t-\Delta t) + V_2(t-\Delta t)$ is identical to Eq.(1.74) except that wherever we see a forward wave it is replaced by a backward one, and wherever we see a backward wave it is replaced by a forward one. The result is

$$\begin{aligned}
V_1(t-\Delta t) +V_2(t-\Delta t) = {}& (1/2)\,^+V_1\,(t)+(1/2)[\,^-V_{1L}\,(t)-\,^-V_3\,(t)+\,^+V_{3U}\,(t)] \\
&+ (1/2)\,^-V_1\,(t)+[(1/2)[\,^+V_{1R}\,(t)-\,^+V_{4U}\,(t)+\,^-V_4\,(t)] \\
&+ (1/2)\,^+V_2\,(t)+(1/2)[\,^-V_{2L}\,(t)+\,^+V_3\,(t)-\,^+V_{3D}\,(t)] \\
&+ (1/2)\,^-V_2\,(t)+(1/2)[\,^+V_{2R}\,(t)+\,^-V_{4D}\,(t)-\,^+V_4\,(t)] \qquad (1.75)
\end{aligned}$$

In order to facilitate the comparison of the TLM result with the finite difference method, we add $V_1(t+\Delta t)+V_2(t+\Delta t)$ to $V_1(t-\Delta t)+V_2(t-\Delta t)$, giving us

$$[V_1(t+\Delta t)+V_2(t+\Delta t)]+[V_1(t-\Delta t)+V_2(t-\Delta t)] = [V_1(t)+V_2(t)]$$
$$+(1/2)[V_{1L}(t)+V_{3U}(t)+V_{1R}(t)-V_{4U}(t)]$$
$$+[1/2][V_{2L}(t)-V_{3D}(t)+V_{2R}(t)+V_{4D}(t)] \qquad (1.76)$$

Eq.(1.76) is the three tier TLM iteration. Note that the equation does not depend on any forward or backward waves, but only the sums of the two waves in each line. We also remind ourselves that the previous equations apply only to the horizontal lines. In the following we enumerate the similar relations for the transverse lines. Corresponding to Eqs.(1.68)-(1.73),

$$^{+}V_3(t+\Delta t) = (1/2)\,^{-}V_3(t)+(1/2)[\,^{+}V_{3D}(t)-\,^{+}V_{2L}\,^{+}V_3(t)+\,^{-}V_2(t)] \qquad (1.77)$$

$$^{-}V_3(t+\Delta t) = (1/2)\,^{+}V_3(t)+(1/2)[\,^{-}V_{3U}(t)+\,^{+}V_{1L}(t)-\,^{-}V_1(t)] \qquad (1.78)$$

$$V_3(t+\Delta t) = \,^{+}V_3(t+\Delta t)+\,^{-}V_3(t+\Delta t) \qquad (1.79)$$

$$^{+}V_4(t+\Delta t) = (1/2)\,^{-}V_4(t)+(1/2)[\,^{+}V_{4D}(t)-\,^{+}V_2(t)+\,^{-}V_{2R}(t)] \qquad (1.80)$$

$$^{-}V_4(t+\Delta t) = (1/2)\,^{+}V_4(t)+(1/2)[\,^{-}V_{4U}(t)+\,^{+}V_1(t)-\,^{-}V_{1R}(t)] \qquad (1.81)$$

$$V_4(t+\Delta t) = \,^{+}V_4(t+\Delta t)+\,^{-}V_4(t+\Delta t) \qquad (1.82)$$

As with $V_1(t+\Delta t)$ and $V_2(t+\Delta t)$, we form the sum of $V_3(t+\Delta t)$ and $V_4(t+\Delta t)$:

$$V_3(t+\Delta t)+V_4(t+\Delta t) = (1/2)\,^{-}V_3(t)+(1/2)[\,^{+}V_{3D}(t)-\,^{+}V_{2L}(t)+\,^{-}V_2(t)]$$
$$+(1/2)\,^{+}V_3(t)+(1/2)[\,^{-}V_{3U}(t)+\,^{+}V_{1L}(t)-\,^{-}V_1(t)]$$
$$+(1/2)\,^{-}V_4(t)+(1/2)[\,^{+}V_{4D}(t)-\,^{+}V_2(t)+\,^{-}V_{2R}(t)]$$
$$+(1/2)\,^{+}V_4(t)+(1/2)[\,^{-}V_{4U}(t)+\,^{+}V_1(t)-\,^{-}V_{1R}(t)] \qquad (1.83)$$

The corresponding reverse iteration is

$$V_3(t-\Delta t) + V_4(t-\Delta t) = (1/2)\,{}^+V_3(t) + (1/2)[\,{}^-V_{3D}(t) - {}^-V_{2L}(t) + {}^+V_2(t)]$$
$$+ (1/2)\,{}^-V_3(t) + (1/2)[\,{}^+V_{3U}(t) + {}^-V_{1L}(t) - {}^+V_1(t)]$$
$$+ (1/2)\,{}^+V_4(t) + (1/2)\,[\,{}^-V_{4D}(t) - {}^-V_2(t) + {}^+V_{2R}(t)]$$
$$+ (1/2)\,{}^-V_4(t) + (1/2)[\,{}^+V_{4U}(t) + {}^-V_1(t) - {}^+V_{1R}(t)] \qquad (1.84)$$

We then form the three tier iteration by adding $V_3(t+\Delta t) + V_4(t+\Delta t)$ to $V_3(t-\Delta t) + V_4(t-\Delta t)$ to give

$$[V_3(t+\Delta t) + V_4(t+\Delta t)] + [V_3(t-\Delta t) + V_4(t-\Delta t)] = [V_3(t) + V_4(t)]$$
$$+ (1/2[V_{3D}(t) + V_{3U}(t) - V_{2L}(t) + V_{1L}(t)]$$
$$+ (1/2)\,[V_{4D}(t) + V_{4U}(t) + V_{2R}(t) - V_{1R}(t)] \qquad (1.85)$$

This completes the TLM iterations, consisting of Eqs.(1.76) and (1.85) and their component two tier counterparts, Eqs.(1.74),(1.75) ,and (1.83) and (1.84).

1.18 2D Finite Difference Method . Comparison with TLM Method

The treatment of the two dimensional problem starts out with the two component wave equations in the x and y directions(without loss). We again use the same core matrix in Fig.1.22,   but now we consider the TLM lines as a backdrop while the finite difference grid is in the foreground. This will help facilitate the comparison of the TLM and finite difference methods. Once the finite difference analysis is completed , we will convert the results to TLM variables, using the core matrix in Fig.1.22.

The two components of the wave equation are:

$$\partial^2 E_Y/\partial x^2 + \partial^2 E_Y/\partial y^2 - (1/v^2)\partial^2 E_Y/\partial t^2 = 0 \qquad (1.86)$$

$$\partial^2 E_X/\partial x^2 + \partial^2 E_X/\partial y^2 - (1/v^2)\partial^2 E_X/\partial t^2 = 0 \qquad (1.87)$$

Note that the two equations are not decoupled, i.e., because of the presence of the $\partial^2 E_Y/\partial y^2$ and $\partial^2 E_X/\partial x^2$ terms we are no longer dealing with two independent plane wave equations in the x and y directions. A finite difference analysis, similar to that for the 1D case, is carried with the help of the core matrix. Thus,

$$[E_Y(x+\Delta l, y, t) +E_Y(x-\Delta l,y,t)+E_Y(x,y+\Delta l,t)+E_Y(x,y-\Delta l,t)-4E_Y(x,y,t)]/\Delta l^2$$
$$-(1/v^2\Delta t^2)[\ E_Y(x,y,t+\Delta t) + E_Y(x,y,t-\Delta t)]\ -2E_Y(x,y,t)] =0 \tag{1.88}$$

$$[E_X(x+\Delta l, y, t) +E_X(x-\Delta l,y,t)+E_X(x,y+\Delta l,t)+E_X(x,y-\Delta l,t)-4E_X(x,y,t)]/\Delta l^2$$
$$-(1/v^2\Delta t^2)[\ E_X(x,y,t+\Delta t) + E_X(x,y,t-\Delta t)]\ -2E_X(x,y,t)] =0 \tag{1.89}$$

In the above equations the difference element is assumed to be the same in both the x and y directions and we set $\Delta x =\Delta y =\Delta l$ with $v=\Delta l/\Delta t$. In order to compare the iterations using the finite difference and transmission line techniques, we solve for the elements $E_x(x,y, t+\Delta t)$ and $E_y(x,y,t+\Delta t)$, giving

$$E_Y(x,y,t+\Delta t) =E_Y(x+\Delta l,y,t) +E_Y(x-\Delta l,\ y,t)+ E_Y(x,y+\Delta l,t) +E_Y(x,y-\Delta l,t)$$
$$-2E_Y(x,y,t) - E_Y(x,y,t-\Delta t) \tag{1.90}$$

$$E_X(x,y,t+\Delta t) =E_X(x+\Delta l,y,t) +E_X(x-\Delta l,\ y,t)+ E_x(x,y+\Delta l,t) +E_x(x,y-\Delta l,t)$$
$$-2E_X(x,y,t) - E_X(x,y,t-\Delta t) \tag{1.91}$$

Eqs.(1.90)-(1.91) give the field elements at time $t+\Delta t$ in terms of prior time elements occurring at $t= t$ and at $t-\Delta t$, and we have chosen $v\Delta t= \Delta l$ in order to simplify the iterations. The selection of Δl is within the allowable range for a stable solution(Reference [5]). The numerical velocity, going from (x,y) to $(x+\Delta l.y+\Delta l)$, is $2^{1/2}\Delta l/\Delta t$. In order to assure a stable solution the wave velocity v must not exceed the numerical velocity. In this case the numerical velocity exceeds the wave velocity by a factor $2^{1/2}$, thus assuring a stable solution.

The following approximations will further simplify the iterative equations,

$$2E_Y(x,y,t) = E_Y(x, y+\Delta l, t) + E_Y(x,y-\Delta l,t) \qquad (1.92a)$$
$$2E_Y(x,y,t) = E_Y(x+\Delta l, y, t) + E_Y(x-\Delta l,y,t) \qquad (1.92b)$$
$$2E_X(x,y,t) = E_X(x+\Delta l,y,t) + E_X(x-\Delta l,y\ t) \qquad (1.92c)$$
$$2E_X(x,y,t) = E_X(x,y+\Delta l,t) + E_X(x.,y-\Delta l,t) \qquad (1.92d)$$

Eqs.(1.92a)-(1.92b) state that $E_Y(x,y,t)$ is the average of the fields at $x+\Delta l$, $x-\Delta l$, $y+\Delta l$, and, and $y-\Delta l$, as suggested by the core matrix. The same type averaging applies to $E_X(x,y,t)$ as well. Substitution of Eqs.(1.92a)-(1.892d) into Eqs.(1.90)-(1.91) then gives

$$E_Y(x,y,t+\Delta t) = 2E_Y(x,y,t) - E_Y(x,y,t-\Delta t) \qquad (1.93a)$$

$$E_X(x,y,t+\Delta t) = 2E_X(x,y,t) - E_X(x,y,t-\Delta t) \qquad (1.93b)$$

The interpretation of Eqs.(1.93a)-(1.93b) is quite simple. It states that the field at x,y,t is the time average of the fields at times $t+\Delta t$ and $t-\Delta t$. In fact, one may regard the iterative equation as the sum of two independent averaging processes. The first is the spatial averaging of the four cells surrounding (x,y,t), and the second is the temporal averaging of $t+\Delta t$ and $t-\Delta t$.

At this point, it is convenient to switch over from field variables to TLM variables, in order to facilitate the comparison. The field is assumed to be concentrated in the center of each cell. In order to switch variables, we imagine the cells to be separated from one another by the transmission lines, and for the fields to be concentrated now in the lines, while insuring that the averaging of the fields in the lines reproduces the original field at the center of the cell. Referring to the core matrix in Fig.1.22, we begin with the fields at the center, $E_x(x,y,t)$ and $E_y(x,y,t)$, which are averaged over the lines, V_3, V_4 and V_1, V_2, respectively, or

$$E_X(x,y,t) = -[V_3(x,y,t) + V_4(x,y,t)]/2\Delta l \qquad (1.94a)$$
$$E_Y(x,y,t) = -[V_1(x,y,t) + V_2(x,y,t]/2\Delta l \qquad (1.94b)$$

Note that in each case, a distance of $2\Delta l$ is used since the field is averaged over two transmission lines(or two cells). The same definitions also apply to times t+Δt and t-Δt, in which case, we merely substitute the appropriate time in Eqs.(1.94a)-(1.94b). Using the TLM variables, the simplified iterative equations become

$$V_1(t+ \Delta t) + V_2(t+\Delta t) = 2[V_1(t)+ V_2(t)] - [V_1(t-\Delta t)+ V_2(t-\Delta t)] \qquad (1.95a)$$

$$V_3(t+ \Delta t) + V_4(t+\Delta t) = 2[V_3(t)+ V_4(t)] - [V_3(t-\Delta t)+ V_4(t-\Delta t] \qquad (1.95b)$$

Eqs.(1.95a)-(1.95b) , however are not very useful since they do not contain any of the "outer" lines such as V_{1R}, V_{3U} , etc... , which contribute to the TLM iteration. The outer terms were "lost" when imposing the approximation given in Eq.(1.92). The omission of these terms is an oversimplification. Thus, if we do nothing further we will have succeeded in "approximating away" the problem!

In order to gather in the outer terms, we make a key assumption regarding the fields in the vicinity of the cell. If we select our length parameter, Δl, sufficiently small then we may regard the fields as quasi- conservative over the cell region(rather than the node region), and the TLM variables, which we have substituted for the fields, will indeed behave as voltages. This will allow us to state several important boundary conditions for the core matrix. In order to proceed we utilize Fig.1.23, which applies to the fields $V_1(t)+V_2(t)$. This Figure is similar to Fig.1.22, but now shows three alternate paths going from the (x,y-Δl) to (x,y+Δl) cell, each of which presumably are equivalent to $[V_1(t) +V_2(t)]$. Each path will have an associated weight factor, with the weights adding up to one. The three paths are listed below.

$$\text{PATH A : } W_1 \{V_1(t) +V_2(t)) = W_1\{A\}$$
$$\text{PATH B: } W_2\{ V_{4D}(t) +V_{2R}(t) +V_{1R}(t) - V_{4U}(t)\} = W_2\{B\}$$
$$\text{PATH C: } W_3\{-V_{3D}(t) + V_{2L}(t) +V_{1L}(t) +V_{3U}(t)\} = W_3\{C\}$$

W_1, W_2, W_3 are the weights associated with the three paths. We use $\{A\}$, etc... as a shorthand to denote the particular path. We should note that path A is in the inner core of the matrix while B and C are in the outer core. We now make the important assumption that the paths B and C , since they are in the outer core, share equally with its neighbors and therefore W_2 and W_3 have weights half that of W_1 .This is best seen by viewing Fig.1.23 and considering the adjoining 9 cell matrix whose center cell is $(x+2\Delta l, y)$, instead of (x,y) , i.e, consider the new matrix formed by shifting the old matrix two cell length to the right. In this case, the path B TLM lines of the old matrix are shared with the corresponding path C lines belonging to the new matrix. Similar sharing occurs for the original path C. B and C , therefore, should have half the weight of A and thus

$$W_2 = (1/2)\ W_1, \quad W_3 = (1/2)W_1 \qquad (1.96a),(1.96b)$$

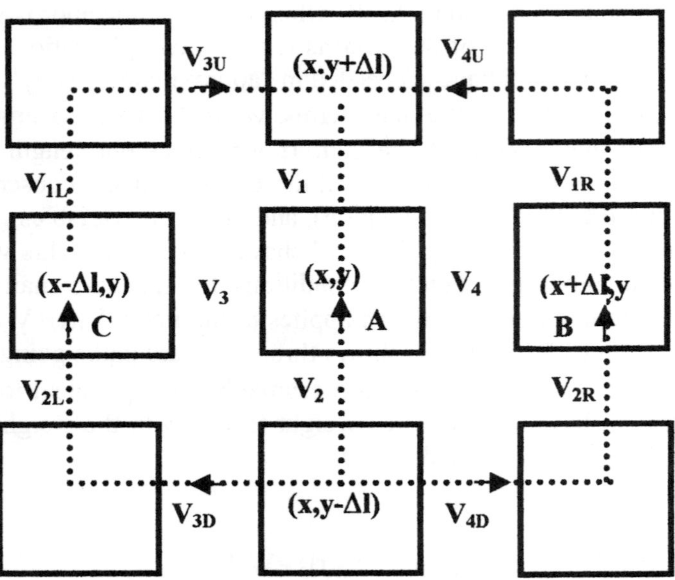

FIG.1.23   SUPERPOSITION   OF   TLM   AND   FINITE DIFFERENCE GRIDS SHOWING THREE POSSIBLE PATHS FOR THE Y DIRECTED FIELD, $V_1(t)+V_2(t)$. PATH A HAS TWICE THE WEIGHT OF B AND C IN THE CORE MATRIX.

where we have assumed that paths B and C are symmetric and thus $W_2 = W_3$. What we still lack , however, are the actual values of the weights. These we can attain from the normalization requirement, by which we demand that the sum of weights equal unity Thus

$$1= W_1 + W_2 + W_3 \tag{1.97}$$

In view of Eqs.(1.96a) and (1.96b)

$$W_1 = 1/2 \ , \ W_2 = W_3 = (1/4) \tag{1.98a), (1.98b}$$

The weight selection has a loose connection to the path lengths of A, B, and C. Since B and C have a path length twice that of A, and the corresponding average fields are therefore half that of A, and this invites the argument that the corresponding weights of B and C should also be half that of A. Although this happens to be the case here, the key factor in obtaining the path weight is related to the neighboring core matrices and the sharing of the path with its neighbors, discussed further in App(1A.1).

Before proceeding to the decomposition of $[V_1(t) + V_2(t)]$, based on the weights in Eq.(1.98), we should point out that the fact that A,B, and C are not the only possible paths in going from the $(x,y-\Delta y)$ to the $(x,y+\Delta y)$ cells. For example, one possible path not mentioned involves going from $(x, y-\Delta y)$ to $(x+\Delta x, y-\Delta y)$ thence to $(x+\Delta x, y)$ thence to (x,y) and finally to $(x, y+\Delta y)$. Indeed if one allows the path length to grow without limit the there are innumerable number of possible paths. App.1A.1 takes into account the alternative paths, and describes the weighing process involved; the decisive factor in so far as the path weight is concerned, is whether a particular line segment is in the inner core, the outer core, or completely outside the core. Taking into account the higher order paths, however, does not change any of the results. Thus the decomposition of $V_1(t)+V_2(t)$, based on the weights of alternative paths(which go beyond A,B, and C) , is identical. We therefore utilize the results of the three paths to decompose $V_1(t)+V_2(t)$. In view of the weights given by Eq.(1.98), the distribution is as follows

$$[V_1(t) + V_2(t)] \rightarrow (1/2)[\ [V_1(t) + V_2(t)] + (1/4)[\ V_{4D}(t) + V_{2R}(t) + V_{1R}(t) - V_{4U}(t)]$$
$$+ (1/4)[-V_{3D}(t) + V_{2L}(t) + V_{1L}(t) + V_{3U}(t)] \tag{1.99}$$

Substituting the above into Eq.(1.95a) then yields

$$[V_1(t+\Delta t) + V_2(t+\Delta t)] + [V_1(t-\Delta t) + V_2(t-\Delta t)] = [V_1(t) + V_2(t)]$$
$$+ (1/2)[V_{1L}(t) + V_{3U}(t) + V_{1R}(t) - V_{4U}(t)]$$
$$+ (1/2)\ [V_{2L}(t) - V_{3D}(t) + V_{2R}(t) + V_{4D}(t)] \tag{1.100}$$

But Eq.(1.100) is exactly the TLM iteration given in Eq.(1.76). Also, the substitution of the TLM relations, $V_1(t+\Delta t) + V_2(t+\Delta t)$ and $V_1(t-\Delta t) + V_2(t-\Delta t)$, produces an identity , just as in the 1D case. Again these choices are not mathematically unique. However they are the only functions which physically justify the proper wave motions of the forward and backward waves.   A similar averaging leads to the same conclusions for the fields in the x direction, $V_3(t) + V_4(t)$. For this we use Fig.1.24, which is the horizontal field counterpart of Fig.1.23.

1.18(a). Inclusion of 2D Losses and Final Comments

We have not included losses in the 2D treatment, which is very similar to the 1D case. As before, the TLM iterations for the forward and backward motions will not only involve reversing the propagation direction, but will also involve, in the case of the reverse iteration, the conversion of the resistive loss into a gain factor. As with the 1D case, the TLM and Finite Difference methods for 2D, including losses, lead to identical results. We have also not discussed what happens when the region is non-uniform and thus TLM lines in the same vicinity will have different line lengths(due to differing dielectric constants). Although the basic approach is the same, this issue cannot be addressed unless concepts such as "cluster cells" and "nearest nodes", discussed in Chapter V, are introduced .

The identity of the TLM and finite difference results should not be surprising, since we   insisted that the cell size(or, alternatively, the time step) is sufficiently small , so that, at least locally, the fields are conservative about the node region. To be sure the curl equations are satisfied, as indicated in Section

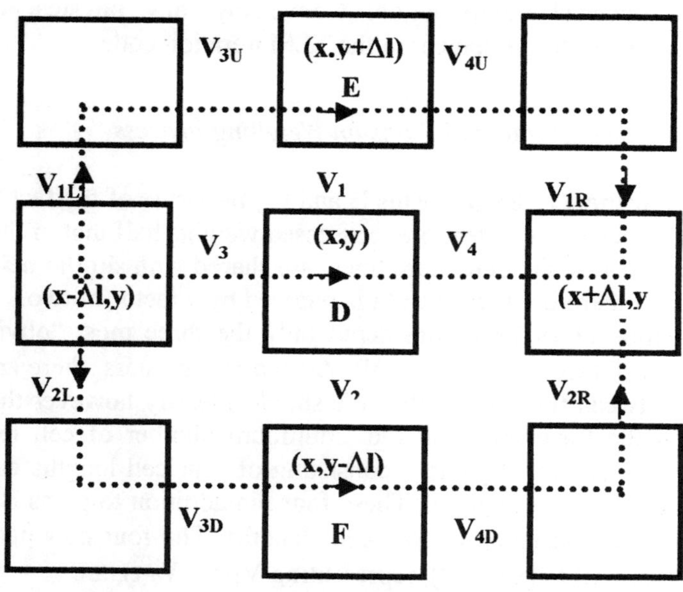

FIG.1.24 SUPERPOSITION OF TLM AND FINITE DIFFERENCE GRIDS SHOWING THREE POSSIBLE PATHS FOR THE X DIRECTED FIELD, $V_3(t)+V_4(t)$. PATH D HAS TWICE THE WEIGHT OF E AND F IN THE CORE MATRIX.

1.16. The numerical technique of finite differences is therefore consistent with the iteration based on transmission line concepts. In a mathematical sense, the transmission line approach is closely related to a particular finite difference method, of which there are many, differing mainly in the speed of convergence or in program complexity. An important advantage of the TLM iteration is the fact alluded to previously, namely, it involves only two time steps and is therefore easier to apply. Other advantages of the TLM method, as mentioned previously, are the powerful physical and intuitive understanding   which may be brought to bear on a wide spectrum of electromagnetic problems. With the

TLM method, any mathematical changes made in a computer code have an immediate physical interpretation. Conversely, any physical changes in a problem are easier to implement in the TLM iteration code.

App.1A.1 Effect of Additional Paths on Weighing Process

We saw in Section(1.18) that paths B and C , by virtue of the fact that they traversed TLM lines in the outer core, possessed weights half that of the core path A. This arises because the outer core lines are shared with similar neighboring core matrices and therefore their weight is reduced by a factor of two.

In Section (1.18) we considered only the three most "obvious" paths in going from cell $(x, y - \Delta l)$ to $(x, y + \Delta l)$. As one might guess, there are any number of paths between these two cells. We should identify however the paths which accomplish the traversal using the minimum number of cell lengths. Indeed there are four additional paths,   consisting of four cell lengths each, which we have thus far not yet specified. These four , in addition to paths B and C , make up all the paths consisting of four cell lengths. The four new paths , which go from cell $(x, y - \Delta l)$ to $(x, y + \Delta l)$, representing $V_1(t) + V_2(t)$, are

Path G: $(x, y-\Delta l) \rightarrow (x+\Delta l, y-\Delta l) \rightarrow (x+\Delta l, y) \rightarrow (x, y) \rightarrow (x, y+\Delta l)$ (1A.1a)

Path H: $(x, y-\Delta l) \rightarrow (x.y) \rightarrow (x+\Delta l, y) \rightarrow (x+\Delta l, y+\Delta l) \rightarrow (x, y+\Delta l)$ (1A.1b)

Path I: $(x, y-\Delta l) \rightarrow (x-\Delta l, y-\Delta l) \rightarrow (x-\Delta l, y) \rightarrow (x, y) \rightarrow (x, y+\Delta l)$ (1A.1c)

Path J: $(x, y-\Delta l) \rightarrow (x.y) \rightarrow (x-\Delta l, y) \rightarrow (x-\Delta l, y+\Delta l) \rightarrow (x, y+\Delta l)$ (1A.1d)

These four paths will actually have more weight than B or C since in each case one of the segments involves crossing an inner core TLM line.

We now consider all seven paths (A,B,C,G,H,I,J) in the weighing process. We symbolically make the following decomposition of $V_1(t) + V_2(t)$,

$$V_1(t)+V_2(t)=W_1\{A\}+W_2(B)+W_2\{C\}+W_3\{G\}+W_3\{H\}+W_3\{I\} +W_3\{J\} \quad (1A.2)$$

We have implicitly made the assumption that the weights for G.H,I,J are all equal, given by W_3. As we have noted previously the paths B and C traverse only outer core lines and therefore $W_2 = W_1/2$. The remaining four paths are slightly more complicated. As noted in Eq.(1A.1)), the paths traverse *both* inner and outer core lines. In particular one of the four line segments comprising each path belongs to an inner core while the other three belong to the outer core. We therefore define the weight of each one of these paths as

$$W_3 = AV[W_3] = (1/4)[(W_1/2)+(W_1/2)+(W_1/2)+(W_1)] = (5/8)W_1 \qquad (1A.3)$$

Note that the weight of these paths exceeds that of B or C , which arises from the fact that these paths traverse an inner core line rather than an outer one. Also note that the differences in weight exist despite the identical   length(four cell lengths)for both types of paths.

Having assigned weights, relative to W_1 , we must now determine W_1, and therefore all the other weights, by requiring the weights to be normalized. Thus

$$1 = W_1+[(W_1/2)+(W_1/2)] + [(5/8)W_1+(5/8)W_1+(5/8)W_1+(5/8)W_1] \qquad (1A.4)$$

where W_1 is the weight assigned to A, $W_1/2$ is weight for B and C, and $(5/8)W_1$ is the weight for G,H,I, and J. Solving

$$W_1 = 2/9 \qquad (1A.5)$$

Since $W_2 = (1/2)W_1$ and $W_3 = (5/8)W_1$

$$W_2 = 1/9 \; ; \quad W_3 = 5/36 \qquad (1A.6a),(1A.6b)$$

Before determining the distribution of $V_1(t)+V_2(t)$, resulting from the path weights, it will be convenient to note that the combination of certain paths simplifies the results. In particular,

$$\{G\}+\{H\} \to \{A\}+\{B\} \qquad (1A.7a)$$

$$\{I\}+\{J\} \to \{A\}+\{C\} \qquad (1A.7b)$$

Eq.(1A.7a) comes about because of the cancellation of $V_4(t)$ when adding paths G and H, leaving {A} and {B} terms. Similarly, $V_3(t)$ cancels in Eq.(1A.7b) when adding paths I and J, leaving {A} and {C}. At this point we can ascertain the distribution of $V_1(t) + V_2(t)$. Combining Eqs.(1A.2)-(1A.7), we have

$$V_1(t)+V_2(t) \rightarrow [W_1+2W_3]\{A\}+[W_2+W_3]\{B\}+[W_2+W_3]\{C\} \qquad (1A.8)$$

Combining above and the weight values, $W1=2/9$ $W_2=1/9$ $W_3=5/36$, we have

$$V_1(t)+V_2(t) \rightarrow (1/2)\{A\}+(1/4)\{B\}+(1/4)\{C\} \qquad (1A.9)$$

The distribution for all seven paths is therefore the same as that obtained solely from the paths {A} {B},and {C}

Use of Paths G.H,I,J Alone as an Independent Set

In simply adding the four paths G, H, I , J to A, B, and C , and then going through the normalization process to obtain the weight distribution, we have actually done twice the amount of work that is really necessary. This is because G,H,I, and J form an independent set so far as describing the various paths $(x,y-\Delta l) \rightarrow (x,y+\Delta l)$ is concerned. This is because of the relationships , Eqs.(1A.7a)-(1A.7b)), which show that A,B, and C m ay be expressed by combinations of G,H,I and J. As before we first find the weight W of each of the four paths; in this case the job is simple since they are all identical. Thus

$$W =1/4 \quad [\text{ Paths G.H,I,J only}] \qquad (1A.10)$$

If we now simply express $(x,y-\Delta y) \rightarrow (x,y+\Delta y)$ in terms of these variables we have

$$[V_1(t) +V_2(t)] \rightarrow (1/4)[\{G\}+\{H\}+\{I\}+\{J\}] \qquad (1A.11)$$

or, referring to Eq.(1A.7),

$$[V_1(t)+V_2(t)] \rightarrow (1/4)[2\{A\}+\{B\}+\{C\}] \rightarrow (1/2)\{A\}+(1/4)\{B\}+(1/4)\{C\} \qquad (1A.12)$$

But this is exactly the same result used when considering only A,B and C (or all seven paths at the same time). The result leads once more directly to Eqs(1.76) and analogously to Eq(1.85).

We have thus exhausted all the possibilities for paths up to four cell lengths. We can then perform the same type of procedure for path lengths of six cell lengths, but no new results or information are obtained. We should also add that as we go to paths of six cell segments or longer, then some of the paths will go outside the core matrix; for these paths any line segments outside the core matrix are considered to have zero weight.

App. 1A.2 Novel Applications of TLM Method: Description of Neurological Activity Using the TLM Method

Electromagnetics, of course, is not the only field where the TLM method may be employed. Acoustic wave motion (or any phenomena governed by a wave equation)and heat diffusion are examples of other technologies where the TLM matrix method may be utilized. In fact, the electromagnetics application is probably more difficult to incorporate into the TLM formulation because of the vector nature of the field and because of the dual polarization associated with the wave. Aside from the area of physics, however, there is a branch in the biological sciences to which the TLM method appears to have a natural affinity. The possible application is the functioning of the brain, which relies on a vast array of nerve fibers and synapses, analogous to the transmission lines and nodes of the TLM matrix.

Analogy of TLM Method and Neurological Activity

We *speculate* here how the TLM method may be employed to describe neurological activity, in particular that of the brain.. As we have mentioned, in the area of nerve cells the nerve fibers and synapses appear to play a role similar to transmission lines and nodal switches in the TLM model. Nerve impulses, which are believed to be electromagnetic in origin , are conveyed along the fibers . The synapses exist at the juncture of two or more fibers and they serve to control the flow of the impulses from one fiber to another.

In the case of the brain, the degree of activation of a particular node, and its location, are probably central to the thought process. The activation of a particular region of synapses , or else the simultaneous activation of several regions, may be interpreted by the brain as a sensation or thought. Further the activation of one or more regions may contribute to the activation of an entirely new region. During non-waking hours, particular(or even most)synapses may be activated or "tuned up" periodically to maintain functionality. The "recharging" of a node(and its associated fibers)also is of considerable importance. Ultimately, how fast these nodes can be recharged will affect the speed with which thought processes can be handled Other factors affecting speed, in analogy with ordinary physical processes, include the speed with which signals are conducted along the neurons, and the "switching speed" of the synapses. One can only speculate about the creative process within the brain. It is quite possible that random and frequent activations of various regions in the brain occur all the time. A healthy and robust brain has the ability to recognize a particular type of node activation(i.e., a thought or idea)as a "solution to a problem" or as the starting point of an entirely new concept. Most of the time these activations come and go without being recognized for their potential value. Memory storage may be regarded, within the TLM framework, as the charging up of a particular cell(node and line), or even a region of the TLM matrix, beyond some threshold value. Stored information is retrieved by activation of the node or region of nodes. The activation of a particular region may trigger other near-by regions, thus giving rise to a flood of related memories. The brain also appears to rely on redundancy, so that multiple regions throughout the brain produce the same memory when activated. It is quite possible that the strength of a particular memory may depend as much on the number of redundant sites, rather than the strength of a particular site. The simultaneous activation of these regions , acting in "parallel", then contributes to the strength of the memory. With time, the particular memory fades unless periodically activated. The memory dimming may be brought about perhaps by the decline of the number of sites as well as the strength of the individual sites , which produce the memory.

There is also the conjectural possibility of correlation among signals traveling along neighboring nerve fibers. Chapter IV deals with the topic of correlation between waves in neighboring TLM lines; such correlations drastically affect the wave scattering, wherein the correlated signals move as a group and with less transverse scattering . By analogy, correlations within the brain(assuming they exist) may allow neighboring signals to travel as a group along paths which represent a kind of "shortcut". We may regard the wave correlations as giving rise to "hunches" or intuitive ideas. Again, we stress that wave correlation among nerve fibers is an untested idea, although Penrose(see reference cited in Preface) has speculated on the role of a revised quantum theory in explaining brain activity.

These and other speculations concerning impulses in the brain are very adaptable to the TLM matrix model. The speculations mentioned in the previous discussion may be quantified using the TLM formulation. The application of the TLM matrix may be regarded either as a model or alternatively as a handy and adaptable mathematical formulation for the transmission of information contained in nerve impulses. Regarded as a model, one must then obtain predictions of TLM model and compare these with experimental observations. .

Of course the nerve fibers do not form neat geometrical shapes , such as cubes or hexagons, as we assume in TLM analysis. The fibers appear as a tangled array with irregular shapes and with varying fiber lengths(Fig.1A.1). Despite these differences, the same type of analysis may be applied to nerve impulses, taking into account the random nature of the fiber shape and length. In some ways the irregularity of the fibers is an advantage since it removes the anisotropy associated with the symmetry elements of a cube or hexagon, where the energy is constrained to flow in only certain directions. With an irregular cell matrix, we are not bound to a preferred direction. We will see in later Chapters how this anisotropy is removed from the symmetry elements.

FIG 1A.1   RANDOMLY GENERATED CELLS. RANDOMNESS PROVIDES ISOTROPIC BEHAVIOR.

REFERENCES

1. H Bertoni, L. Carin, and L. Felsen, *Ultra-Wideband , Short-Pulse Electromagnetics,* Plenum Press, New York, 1993.

2. L. Carin and L. Felsen, *Ultra-Wideband , Short-Pulse Electromagnetics 2* Plenum Press, New York, 1995.

3. W. Panofsky, and M. Philips , *Classical Electricity and Magnetism,* Addison-Wesley,Reading, Mass., 1962

4. W. Johnson, *Transmission Lines and Networks,* McGraw Hill , New York, 1950.

5. Haberman, *Elementary Applied Differential Equations* , Prentice Hall , Englewood Cliffs, N.J. , 1987 .

II.   Notation and Mapping of Physical Properties

In this Chapter we proceed to develop a notation, using space symmetry and numbering methods, to describe the cell and transmission line elements which occupy the medium, as well as to describe the mapping of the fields and dynamic physical properties onto the cells. The adopted notation is fully capable of describing the iterative rate equations of the interlocking transmission lines which represent the medium. The notation will make it possible to develop computer iterations to describe the electromagnetic behavior of the TLM matrix, or equivalently, the medium.

We show that the combination of transmission lines and node resistors can simulate the complete behavior of the medium. As mentioned before, the transmission lines both convey and store electromagnetic (including light) energy while the nodes simulate any changes in the electrical properties(particularly the conductivity). The transmission lines and nodes are attached to the cell; the cells completely occupy the space, i.e., they belong to spatial symmetry elements. Regardless of the type of cell used, the transmission lines surround the cell, and form the boundaries with the neighboring cells. The nodes are formed at the intersection of the transmission lines. The selection of the actual cell geometry is, to some degree, arbitrary. If the medium is isotropic, then at the very least we impose the requirement that the distance between adjacent nodes remain the same. This insures that the delay time between nodes is unchanged, regardless of the particular transmission line. This limits the cells to equilateral shapes. Examples of equilateral cells are shown in Fig.2.1 for the case of two dimensions. The examples shown, a square and hexagon, obviously satisfy the equilateral requirement. The nodes, shown as the small darkened circles, are located at the corners (or intersections) of the cells. The square cell is easiest to handle, however, both conceptually and from a bookkeeping point of view, so that the models to be described here will be focused almost entirely on this element(or cube, in the case of three dimensions). Other cell geometries should not be ignored,

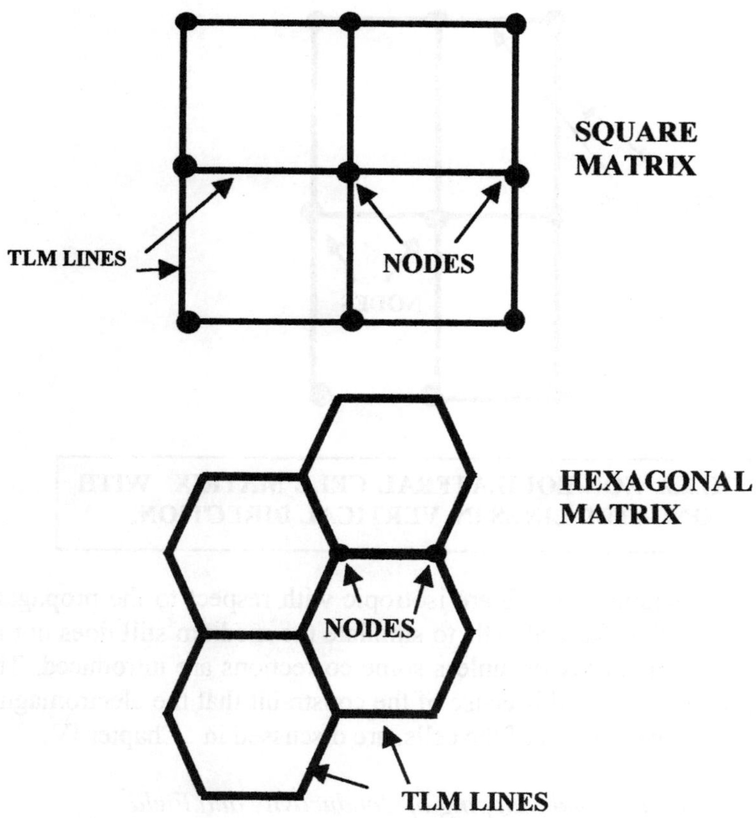

FIG.2.1 POSSIBLE CELL STRUCTURES FOR USE IN TLM MATRIX.

however, since they may possibly offer possible advantages. Certainly, in the case of anisotropic crystals, one is prompted to use non- equilateral cells. In this regard, we note in Fig.2.2 the same square cells as in Fig.(2.1), but elongated in the vertical direction. The elongated lengths of the cell are along directions in which the propagation velocity is proportionately faster. The elongated sections insure that the same delay time is maintained between nodes.

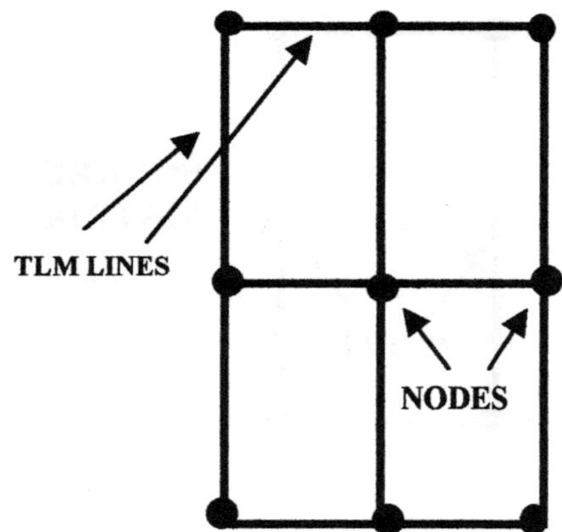

FIG.2.2. NON-EQUILATERAL CELL MATRIX WITH ELONGATED LINES IN VERTICAL DIRECTION.

Even with mediums which are isotropic with respect to the propagation velocity, the use of equilateral cells to simulate the medium still does not produce complete isotropic behavior, unless some corrections are introduced. The corrections, which are needed because of the constraint that the electromagnetic energy flow along the borders of the cells, are discussed in   Chapter IV.

2.1   1D Cell Notation and Mapping of Conductivity and Field

Before proceeding to the 2D and 3D matrices it will be instructive to look at the 1D circuit briefly and to apply the appropriate notation to such a system. As noted in Fig.2.3 the cells are numbered consecutively, with n being a positive integer. The forward and backward waves in each cell are given by $^{+}V^{k}(n)$ and $^{-}V^{k}(n)$ respectively. The pre-superscripts , + and - , indicate the wave direction, + for the forward wave (increasing n) and − the backward wave. The time step associated with a particular quantity, such as the field, is designated with a k index and added as a superscript. Thus $^{+}V^{k}(n)$ is the horizontal forward wave in the nth cell during the kth time step. The superscripts normally will be omitted

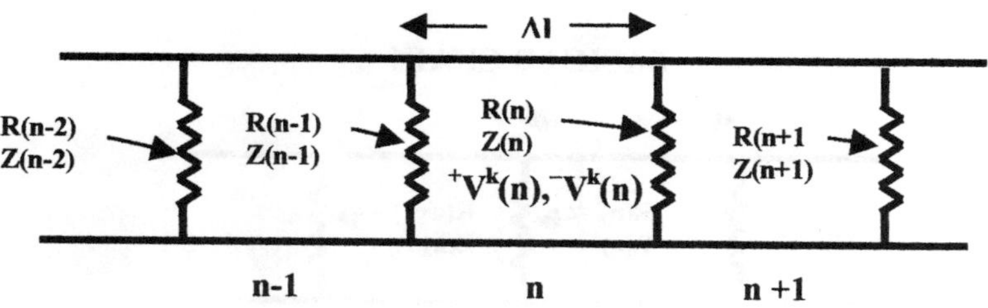

FIG.2.3 NOTATION FOR 1D COUPLED CELLS. BY CONVENTION THE nth NODE, ATTACHED TO THE NTH CELL, IS IN THE DIRECTION OF INCREASING n (TO THE "RIGHT" OF THE Z(n) CELL).

unless they are germane to the discussion. The characteristic impedance of each cell is labeled by Z(n). The n label for the cell Z(n) is used to indicate not only the cell impedance, but is also used to locate(at least for identical cells) the location of the cell within the chain. In Fig.2.3 the cell impedances are assumed to be the same and thus cell lengths are identical. Later, as well as in subsequent Chapters, the labeling also will allow us to consider differing values of cell impedance(and different line lengths), which will be necessary when treating nonuniform dielectrics, dispersion, and boundaries between differing dielectrics. Note that each cell shares two node resistors, R(n-1) and R(n). By convention the node resistor R(n), corresponding to the nth cell, is located in the direction of increasing n (i.e., located on the "right hand side "of the cell) while R(n-1) is the node resistor located in the direction of decreasing n.

Having outlined the 1D notation we can proceed to the calculation of the node resistance. There are alternate means for obtaining the effective node resistance all of which are more or less equivalent. One approach, already alluded to in Chapter 1, relies on first calculating the end resistors for each TLM element. As shown in Fig.2.4, we first focus on the nth and (n+1)th isolated cells, for which the end resistors are (see Section 1.3, Eq.(1.9)),

$$R'(n) = 2\rho(n)/\Delta l \tag{2.1a}$$
$$R'(n+1) = 2\rho(n+1)/\Delta l \tag{2.1b}$$

BASIC 1D CHAIN

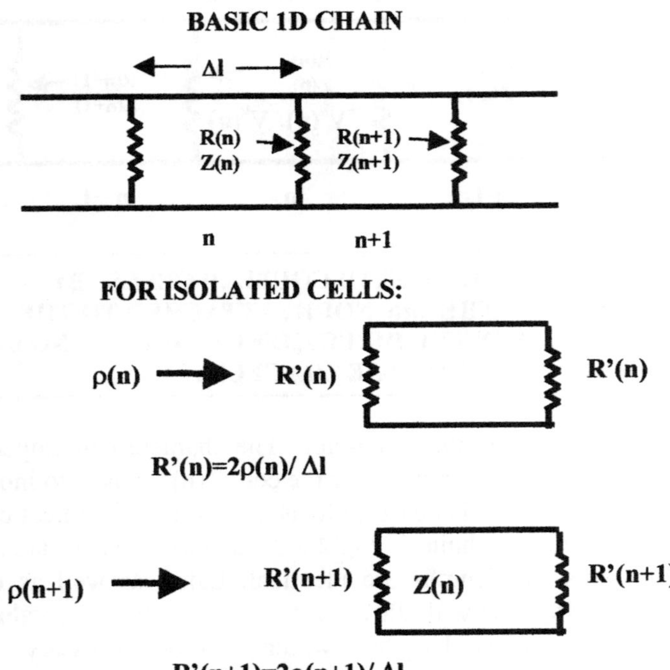

FOR ISOLATED CELLS:

$$R'(n)=2\rho(n)/\Delta l$$

$$R'(n+1)=2\rho(n+1)/\Delta l$$

FOR A CHAIN OF CELLS, R'(n), R'(n+1) ARE PARALLEL:

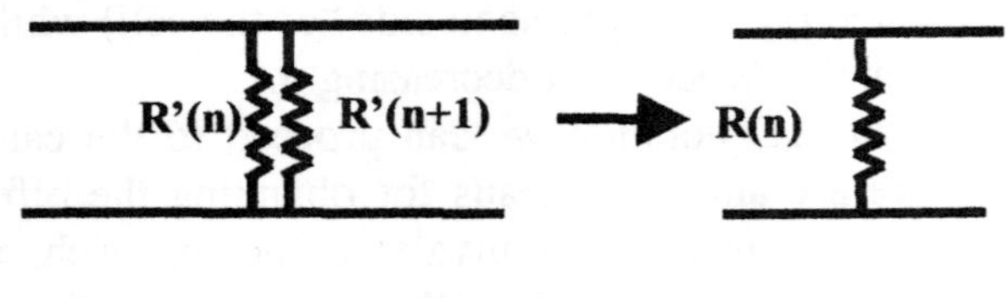

$$R(n)=(2/\Delta l)[\ \rho(n)\ \rho(n+1)]/\ [\rho(n)+\ \rho(n+1)]$$

FIG. 2.4. CALCULATION OF R(n).

where prime has been added to indicate this is an end resistor of an isolated cell, in contrast to that of a chain of cells in which the adjoining end resistors are combined in parallel fashion. Indeed, when considering the cell chain, the combined node resistance between the nth and (n+1)th cells is then

$$R(n) = R'(n)R'(n+1)/[R'(n)+R'(n+1)] \tag{2.2}$$

Using Eqs.(2.1a)-(2.1b), R(n) becomes

$$R(n) = (2/\Delta l)[\rho(n)\rho(n+1)/[\rho(n)+\rho(n+1)] \tag{2.3}$$

Note that for cells with the same resistivity,

$$R(n) = \rho(n)/\Delta l \tag{2.4}$$

and therefore the factor of two no longer appears.

A second technique for calculating R(n) involves taking the average of the conductivities of the two adjoining cells. Thus

$$\sigma_{AV}(n) = [\sigma(n)+\sigma(n+1)]/2 \tag{2.5}$$

There exists an equivalent *auxiliary* cell for this average conductivity, situated midway between the two cells with its center at the actual node location, as noted in Fig.2.5. Note that the auxiliary cell does not coincide with any of the original cells. The total node resistance for the auxiliary cell (without the two factor) is therefore

$$R(n) = 1/\sigma_{AV}(n)\Delta l \tag{2.6}$$

and the substitution of Eq.(2.5) gives

$$R(n) = (2/\Delta l)[1/(\sigma(n)+\sigma(n+1))] \tag{2.7}$$

or, $$R(n) = (2/\Delta l)[\rho(n)\rho(n+1)/[\rho(n)+\rho(n+1)] \tag{2.8}$$

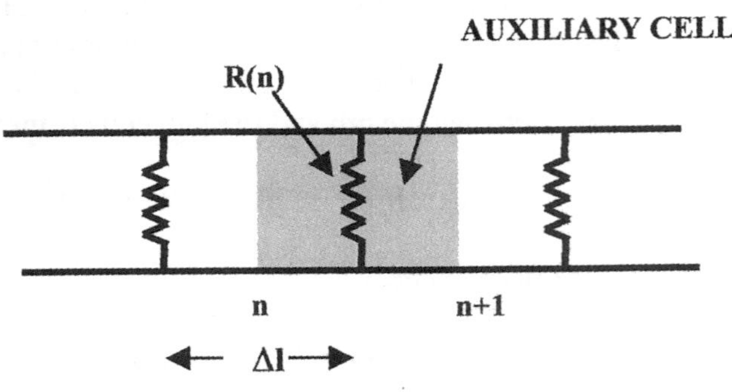

$$\sigma_{AV}(n) = [\sigma(n) + \sigma(n+1)]/2)$$
$$R(n) = 1/\sigma_{AV}(n)\Delta l = (2/\Delta l)[\rho(n)\rho(n+1)]/[\rho(n) + \rho(n+1)]$$

FIG.2.5 USE OF AUXILIARY CELL TO OBTAIN R(n).

Note that R(n) in Eq.(2.8) is identical to that of Eq.(2.3). Thus finding the end resistors of each cell and then combining them in parallel, is equivalent to the node resistance of the single auxiliary cell formed from the neighboring cells.

2.2 Neighboring 1D Cells With Unequal Impedance

In the previous discussion we allowed the conductivities of neighboring cells to differ, but constrained the line lengths to be the same. Suppose two neighbors have both different line length , as well as different conductivity. This can occur at a boundary between two different dielectrics, A and B where the line lengths are Δl_A and Δl_B., as shown in Fig.2.6. We note now that region B has the longer cell length. The notation adapted here is to employ two indices, n_A and n_B, to describe the cell location in each region and to specify the boundary between regions. Actually the indices n_A and n_B are related to one another, as we shall make note in the following paragraph and in Chapter V. Also, as noted in Fig.2.6 we assume a simple boundary in which cell A fits perfectly with cell B, in the sense that no truncated cells are required to form the boundary.

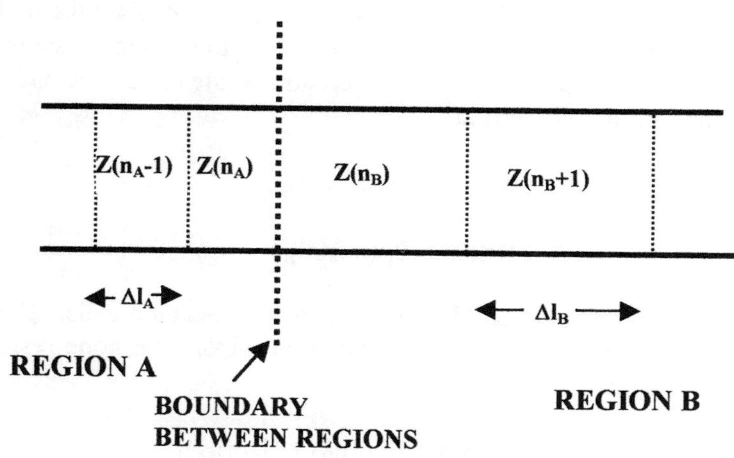

FIG.2.6 NOTATION FOR NEIGHBORING CELLS WITH UNEQUAL IMPEDANCE VALUES.

With the two dissimilar dielectric regions, the use of two separate indices, one for each region, adds complexity. If possible, the use of a single index for both regions would simplify matters. As one possibility, for example, suppose we continued to use the same index in B as in A. At the boundary, the B cell would satisfy $n_B = n_A + 1$ and in fact we would simply continue to use the   same integral index for the cells in region B. With this label we would know the number of transit time intervals required for a signal to reach the cell in question, but otherwise this labeling would not be very useful. As will be pointed out in Section 5.4 , it is more useful to employ a different single index as a label for the cells in both regions, one which would directly pinpoint the *location* for each type of cell, regardless of the region, in much the same fashion as a position coordinate.   In describing the B region, however, such a cell index will in general involve non- integral numbers and , in addition, the index will increase by a non-integral amount for consecutive cells . We do not dwell here on this type of single index cell notation, however, which is taken up further in Chapter V. For the present we use the two index notation, designated by n_A and n_B , to describe the adjoining dielectrics as depicted in Fig.2.6.

If we focus on cells n_A and n_B , we can proceed to calculate the node resistance at the boundary As before there are two ways to calculate the node resistance, the first of which combines the end resistors at the node. The end resistors for the adjoining n_A and n_B cells are

$$R'(n_A) = 2\rho(n_A)/\Delta l_A \qquad (2.9a)$$

$$R'(n_B) = 2\rho(n_B)/\Delta l_B \qquad (2.9b)$$

where the prime again indicates the resistors are the isolated cells. Combining these resistors in parallel fashion (for a chain)we obtain the node resistance at the boundary, designated by R_{AB} , or

$$R_{AB} = 2\rho(n_A)\rho(n_B)/ [\Delta l_A\rho(n_B) +\Delta l_B\rho(n_A)] \qquad (2.10)$$

Note that when $\Delta l_A = \Delta l_B$, the node resistance reduces to Eq.(2.8) as expected.

The second method involves taking the average of the two conductivity cells at the boundary. Denoting this average by $\sigma_{AV,AB}$, we have

$$\sigma_{AV,AB} = [\Delta l_A\sigma(n_A) + \Delta l_B\sigma(n_B)]/(\Delta l_A +\Delta l_B) \qquad (2.11)$$

The conductivity differs from Eq.(2.5) in as much as it is weighted by the length of the cell as well. The node resistance is then defined as

$$R_{AB} = 1/(\sigma_{AV,AB} [(\Delta l_A+\Delta l_B)/2]) \qquad (2.12)$$

where we have replaced the Δl in the identical cell situation, Eq.(2.6), with the average of the two cell lengths at the boundary. Substituting Eq.(2.11) into the above and replacing $\sigma(n_A)$, $\sigma(n_B)$ with $1/\rho(n_A)$ and $1/\rho(n_B)$ we obtain

$$R_{AB} = 2\rho(n_A)\rho(n_B)/ [\Delta l_A\rho(n_B) +\Delta l_B\rho(n_A)] \qquad (2.13)$$

which is identical to Eq(2.10).

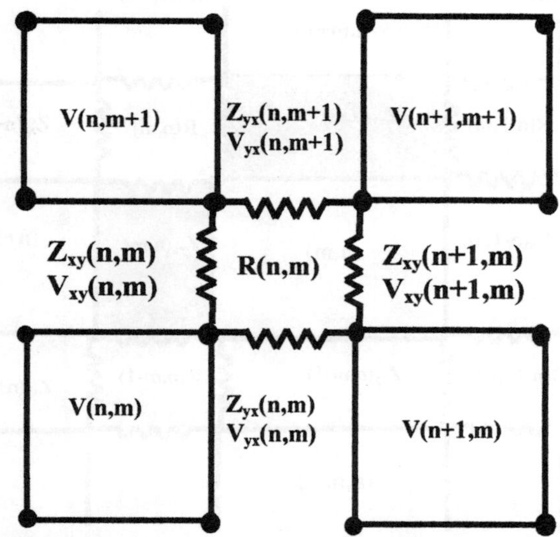

FIG.2.7 TWO DIMENSIONAL MATRIX CONSISTING OF CIRCUIT NODE AT CENTER OF FOUR ISO-POTENTIAL CELLS. CELL NOTATION IS EMPLOYED.

2.3 2D Cell Notation . Mapping of Conductivity and Field

In this section we outline the sequential numbering system to describe the cells, transmission lines ,and nodes which simulate the 2D medium. Fig.2.7 shows the system used for a 2D, identical square cell matrix. Fig.2.8 is identical to Fig.2.7, but with a larger view, taking into account additional cells. In Figs.2.7 and 2.8 , and in the many similar figures which follow, the matrix cells will be drawn as though the TLM lines are identical(equal length), with the same dielectric constant and propagation velocity. In fact, we always allow for non-uniformity among the TLM lines, but for visual simplicity the lines will be drawn with equal length. Non-uniformity among the TLM lines also will require us to use node coupling approximations, which are discussed in Chapter V.

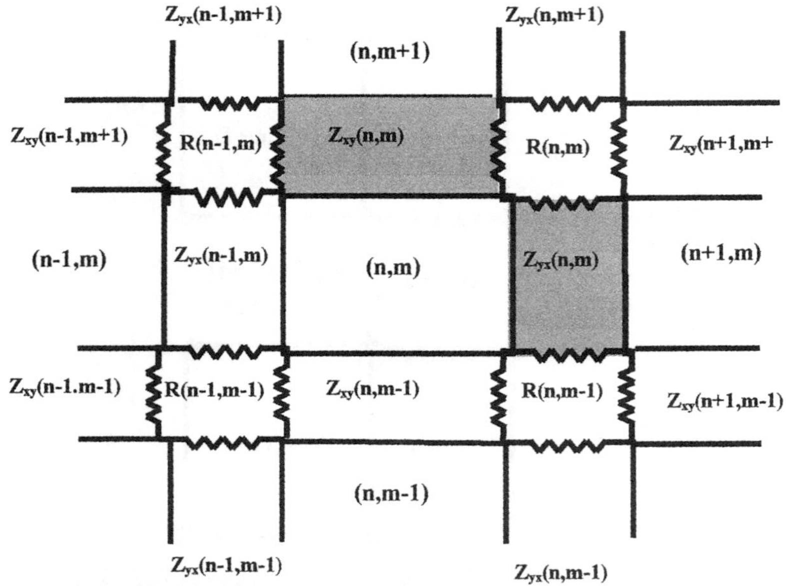

FIG, 2.8   2D TLM NOTATION CONVENTION:$Z_{xy}(n,m)$ AND $Z_{yx}(n,m)$ ARE THE TWO LINES(SHADED) ASSOCIATED WITH THE (n,m) CELL. R(n,m) IS AT THE INTERSECTION OF $Z_{xy}(n,m)$ AND $Z_{yx}(n,m)$.

Each cell is labeled with a pair of number indices, (n,m), corresponding to the x and y directions respectively. We note that the potential of each iso-potential cell region is labeled by V(n,m), V(n+1,m), and so forth. If Δl is the length of the cell, and n_o, m_o are the upper limits of n, m, then $n_o\Delta l$ and $m_o\Delta l$ represent the dimensions of the medium. Next, we associate the transmission lines and node resistors with a given cell. Which transmission lines are to be associated with a particular cell, to some degree, is a matter of choice. Certainly one cannot associate all four lines surrounding the cell, since this will leave the neighboring cells deficient. Associating one or three lines violates the basic symmetry. Associating two transmission lines with each cell, one horizontal and the other vertical, satisfies the spatial symmetry. However, we must choose the same two lines for each cell. As shown in Figure (2.8), the two lines selected , the horizontal $Z_{xy}(n,m)$ and the vertical $Z_{yx}(n,m)$, belong to the (n,m) cell, and represent the lines located in the direction of increasing x and y (or, equivalently, in the increasing n and m directions). A similar notation is adopted for the voltage waves. The voltages in the horizontal and vertical lines are designated by $V_{xy}(n,m)$ and $V_{yx}(n,m)$ respectively. The subscripts, for both the lines and voltages, designate the propagation and polarization directions respectively. To cite an example, $V_{xy}(n,m)$, located in the $Z_{xy}(n,m)$ line,   propagates in the x direction with the field in the y direction. The subscripts are of course *absent* when specifying the cell potential. It is important to realize that $V_{xy}(n,m)$ and $V_{yx}(n,m)$ represent the *difference* in cell voltages, $V(n, m+1)-V(n,m)$ and $V(n+1,m)-V(n,m)$, respectively.   Another important point is that there are forward and backward waves associated with each line and these are designated by $^{+}V_{xy}(n,m)$ and $^{-}V_{xy}(n,m)$, respectively, with $V_{xy}(n,m) = {}^{+}V_{xy}(n,m) + {}^{-}V_{xy}(n,m)$ and similar notation for $V_{yx}(n,m)$,   i.e.,   the two waves in $Z_{yx}(n,m)$ are $^{+}V_{yx}(n,m)$, $^{-}V_{yx}(n,m)$ with $V_{yx}(n,m) = {}^{+}V_{yx}(n,m) + {}^{-}V_{yx}(n,m)$.

In the case of the node resistors, we need only associate a single node with each cell. For consistency we must choose the same node for each cell, and adopt the convention that the node is located in the direction of increasing n and m in each cell(i.e., in the upper right corner of the cell) as shown in Fig.2.8. Each of the four nodal resistors, which are assumed to be equal, is designated by $R(n,m)$. The question naturally arises whether it is desirable to provide an additional label for each of the four resistors, perhaps to allow for the possibility that the resistors are unequal. Such a differentiation may be useful, perhaps in the case of boundary conditions. For the present discussion, however, we forego the opportunity to differentiate the four nodal resistors. Besides the issue of added complexity, the node, by its very nature, has no spatial extent and, therefore, the four resistors should be equal. The transmission lines account for any spatial changes while any time varying changes in the medium(unrelated to propagation effects)are presumed to occur in the node resistors. Boundary conditions are handled by embedding the entire node in the medium which forms the boundary, to be discussed later.

Next we relate the resistivity to the values of the nodal resistors. This is done by averaging the cell conductivities surrounding the lines and nodes and then using the results to obtain the average of the inverse node resistors. The calculation is done for one particular direction(representing a particular set of node resistors), but the results are the same when the averaging is done over all resistors and directions. We illustrate the calculation, using Fig.2.9 , in which we have selected the y direction. This implies that we focus on the lines $Z_{xy}(n,m)$ and $Z_{xy}(n+1,m)$, since the fields are y directed in these lines. We must then calculate and then average two resistors, which we designate as $R_A(n,m)$ and $R_B(n+1,m)$, each of which is associated with *auxiliary* cells centered about lines $Z_{xy}(n,m)$ and $Z_{xy}(n+1,m)$, respectively(we temporarily assign differing labels to the resistors in order to perform the averaging process). In order to obtain $R_A(n,m)$ we first obtain the average conductivity in which the $Z_{xy}(n,m)$ line is contained. Designating this conductivity by $\sigma_A(n,m)$ we have

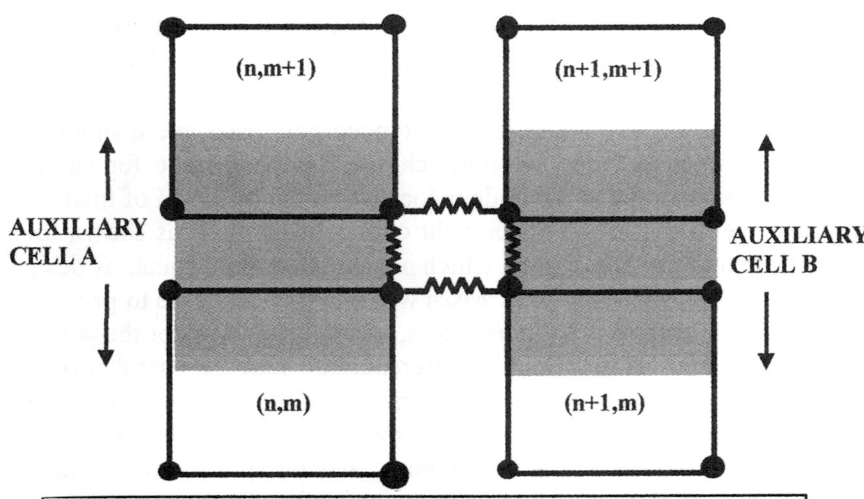

FIG.2.9 AVERAGING OF CONDUCTIVITIES IN AUXILIARY CELLS A AND B CENTERED ABOUT $Z_{XY}(n,m)$ AND $Z_{XY}(n+1,m)$. AVERAGING LEADS TO $2/\Delta\sigma_{AV}$

$$\sigma_A = [\sigma(n,m) + \sigma(n,m+1)]/2 \qquad (2.14)$$

Note that $\sigma(n,m)$ and $\sigma(n,m+1)$ are simply the conductivities of the two cells surrounding the $Z_{xy}(n,m)$ line. The resistor associated with this conductivity, from the previous discussion, is

$$R_A(n,m) = 2 / \Delta l\, \sigma_A \qquad (2.15)$$

where the two factor appears, as mentioned before, because the loss is represented by a pair of node resistors in (one at each end of the line). We next obtain the corresponding quantities for the $Z_{xy}(n+1,m)$ line, which is the second line with the y directed field. Designating the average conductivity about this line as σ_B, the corresponding equations are

$$\sigma_B = [\sigma(n+1,m)+\sigma(n+1,m+1)]/2 \qquad (2.16)$$

$$R_B(n+1,m) = 2 / \Delta l\, \sigma_B \qquad (2.17)$$

We next form the average of the inverse $R_A(n,m)$ and $R_B(n+1,m)$, which is equivalent to taking the average of the corresponding conductivities. This average defines the node resistance $R(n,m)$, or actually $1/R(n,m)$, so that

$$(1/R(n,m)) \equiv (1/2)\, [\, (1/R_A(n,m) +1/R_B(n,m)] = (1/4)\Delta l\, [\sigma_A +\sigma_B] \qquad (2.18)$$

We also note , however, that the average conductivity about the node resistor, designated by $\sigma_{AV}(n,m)$, is

$$\sigma_{AV}(n,m) = [\sigma(n,m) +\sigma(n,m+1)+ \sigma(n+1,m)+\sigma(n+1,m+1)]/4 \qquad (2.19)$$

Combining the previous equations,

$$(1/R(n,m)) = (1/2)\Delta l \sigma_{AV}(n,m) \qquad (2.20)$$

or
$$R(n,m) = 2/\Delta l \sigma_{AV}(n,m) \qquad (2.21)$$

The node resistance R(n,m) is the sought after result which relates R(n,m) to the average conductivity in much the same manner as the 1D relation, except for the factor of two, which is now present since adjoining end resistors are no longer combined. The result in Eq.(2.21) may seem premature since we have done the averaging in only in the y direction , while ignoring the x direction. We must therefore consider the $Z_{yx}(n,m+1)$ and $Z_{yx}(n,m)$ lines as well and perform a similar averaging. The reader will quickly verify, however, that the same average, given by Eq.(2.21) is obtained. An "intuitive" method for obtaining the same expression for R(n,m), which works for 2D, is as follows. The average conductivity about the node, $\sigma_{AV}(n,m)$, is first obtained, as seen in Fig.2.10 where we use an auxiliary cell centered about the node. We then assume an equivalent circuit where a pair of parallel resistors , each equal to R(n,m), is directed in , say, the y direction. The combined resistance, R(n,m)/2, is then set equal to the equivalent resistance of the center cell, or, $R(n,m)/2 = 1/\Delta l \sigma_{AV}(n,m)$, which is identical to Eq.(2.21).

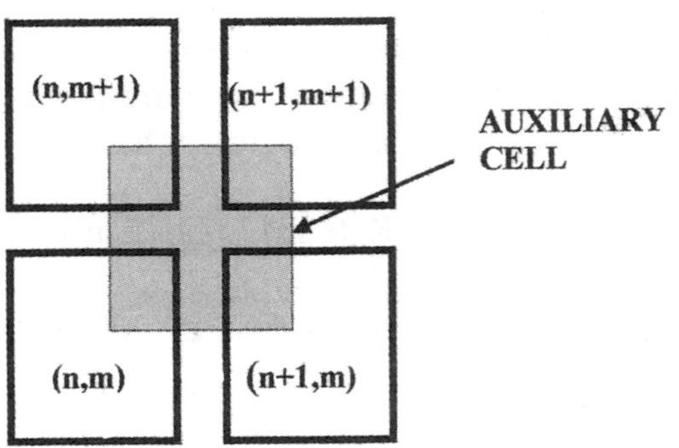

$$\sigma_{AV}(n,m)=[\ \sigma(n,m)+\sigma(n+1,m)+\sigma(n,m+1)+\sigma(n+1,m+1)]/4$$
$$R(n,m)=2/\Delta l \sigma_{AV}(n,m)$$

FIG.2.10 EQUIVALENT R(n,m) WITH CONDUCTIVITY CENTERED AT NODE.

One thing to note is the special case in which the node is situated adjacent a conducting electrode. Under these circumstances $\sigma_{AV}(n,m)$ will be dominated by the electrode, even if the node itself is not embedded in the electrode, but rather in the dielectric. Given Eq.(2.21), the node resistor will be shorted out, which is counter to our intuition if the node is located outside the electrode. In the applications which follow, we will often disregard the effect of the conducting electrode on the node resistance; in the limit of large cell densities there will no difference in the two choices.

Next we interpret the fields in the medium, based on the transmission line description. Note that there are four iso-potential square regions surrounding the node, designated by $V(n,m)$, $V(n,m+1)$, $V(n+1,m)$, and $V(n+1,m+1)$. The TLM voltage in each of the four lines , which intersect the node , are given by the voltage difference between appropriate cells. Thus, in line $Z_{xy}(n,m)$, the TLM voltage $V_{xy}(n,m)$ is given by $V(n,m+1)-V(n,m)$, and similar relationships apply to the other lines.

An important point to emphasize is that the width of the TLM line is purely conceptual, and is ideally interpreted as an extremely narrow line, concentrating the energy that is actually distributed in the adjoining cells. Suppose we wish to obtain the average field distributed throughout the medium (rather than the concentrated field in the TLM line).   First we obtain the fields corresponding to the TLM lines, as mentioned previously. How do we interpret the voltage difference in the line, so far as the distributed field is concerned?   Referring to Fig.2.11(a), as an example, the voltage difference in line $Z_{xy}(n,m)$ is assumed to be the same as that of the auxiliary cell shown by the dashed line. The size of the auxiliary cell is identical to the elementary cell, but is centered about the $Z_{xy}(n,m)$. One may regard the voltage difference in the line, $V_{xy}(n,m)$, as uniformly distributed throughout the auxiliary cell centered about $V_{xy}(n,m)$ line. A similar auxiliary cell is shown for $V_{yx}(n,m)$ in Fig.2.11(b).

Often we require the average field in the (n,m) cell itself (rather than the auxiliary cell centered about the line) Following the lead in obtaining the resistivity , this field is simply averaged over its neighbors. Thus , for example, the y directed field in the (n,m) cell, Fig.2.12, is the average of the fields in the $Z_{xy}(n,m)$ and $Z_{xy}(n,m-1)$ lines, with fields $V_{xy}(n,m)$ and $V_{xy}(n,m-1)$, respectively. We define the average field, $V_{AV,xy}(n,m)$ for the (n,m) cell as

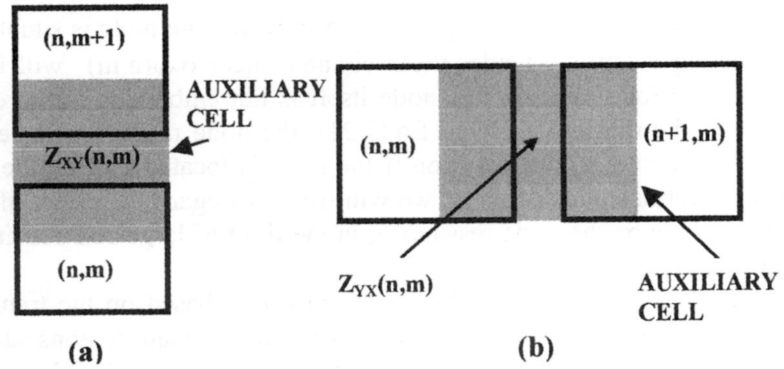

FIG.2. 11(a) EQUIVALENT VERTICAL FIELD BASED ON AUXILIARY CELL CENTERED ABOUT $Z_{XY}(n,m)$ LINE, EQUAL TO V(n,m+1)-V(n,m). SIMILARLY, IN (b) THE HORIZONTAL FIELD IS V(n+1,m)-V(n,m).

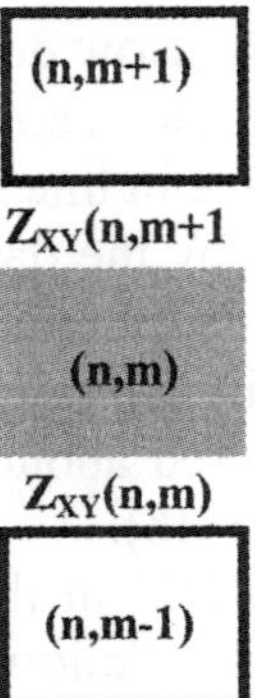

FIG. 2.12 FIELD AVERAGED THROUGHOUT (n,m) CELL IS $V_{AV,\,XY}(n,m)= (1/2)[V_{XY}(n,m)+V_{XY}(n,m-1)]$. SIMILARLY, $V_{AV,\,YX}(n,m)= (1/2)[V_{YX}(n,m)+V_{YX}(n-1,m)]$.

$$V_{AV,xy}(n,m) \;=\; (1/2)[V_{xy}(n,m) + V_{xy}(n,m-1) \tag{2.22}$$

where the first subscript AV denotes this is an average field for the (n,m) cell. It should be noted of course that each term on the right side is the sum of forward and backward waves, e.g., $V_{xy}(n,m) = {}^{+}V_{xy}(n,m) + {}^{-}V_{xy}(n,m)$, etc.. A similar averaging of the x directed field for the (n,m) cell gives

$$V_{AV,yx}(n,m) \;=\; (1/2)[V_{yx}(n,m) + V_{yx}(n-1,n)] \tag{2.23}$$

2.4 3D Cell Notation . Mapping of Conductivity and Field

We next consider a 3D matrix , in which case the elementary cell is cubic. Each cube, as shown in Fig.2.13(a), is labeled by the set of positive integers (n,m,q), corresponding to the discretization of the x,y,and z axes, respectively. The node associated with the (n,m,q) cell is located at a particular corner of the cube, corresponding to the maximum values of x,y, z within the cube. The three line segments of the cube, emanating from the node, indicate the locations of the transmission lines associated with the cell. Actually there are a total of six such transmission lines, two for each of the three line segments. This is so since there are two transverse fields(two possible polarization directions) for each line segment, i.e., for each direction of propagation. The six transmission lines and fields associated with the (n,m,q) cell are shown in the following:

Propagation Direction	Transmission Lines	Fields
x	$Z_{xy}(n,m,q)$, $Z_{xz}(n,m,q)$	$V_{xy}(n,m,q)$, $V_{xz}(n,m,q)$
y	$Z_{yx}(n,m,q)$, $Z_{yz}(n,m,q)$	$V_{yx}(n,m,q)$, $V_{yz}(n,m,q)$
z	$Z_{zx}(n,m,q)$, $Z_{zy}(n,m,q)$	$V_{zx}(n,m,q)$, $V_{zy}(n,m,q)$

An alternate way of classifying the impedance lines is according to the plane of propagation(defined by the field direction and the propagation direction), as shown below:

Propagation Plane	Transmission Lines	Fields
xy	$Z_{xy}(n,m,q)$, $Z_{yx}(n,m,q)$	$V_{xy}(n,m,q), V_{yx}(n,m,q)$
yz	$Z_{yz}(n,m,q), Z_{zy}(n,m,q)$	$V_{yz}(n,m,q),\ V_{zy}(n,m,q)$
zx	$Z_{zx}(n,m,q),\ Z_{xz}(n,m,q)$	$V_{zx}(n,m,q),\ V_{xz}(n,m,q)$

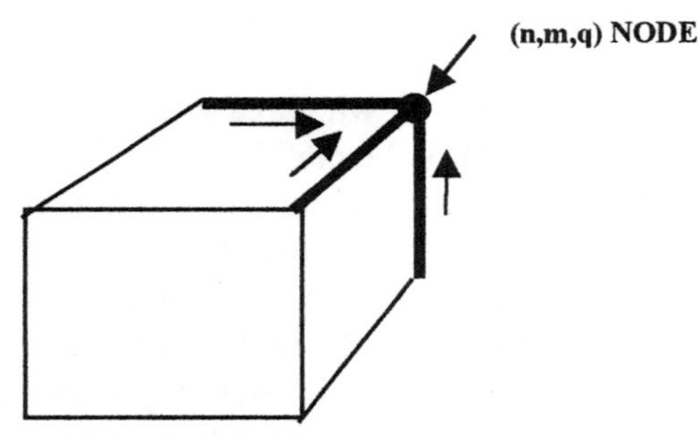

FIG. 2.13(a) ELEMENTARY CELL OF TL MATRIX.

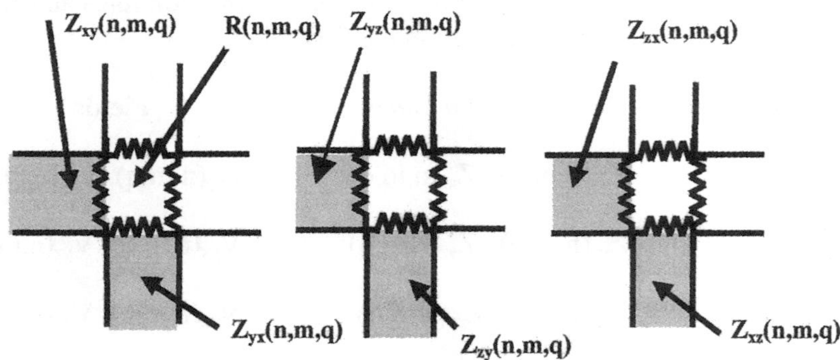

FIG.2.13(b) PROJECTION OF 3D TLM MATRIX ONTO 2D GRIDS.

Fig.2.13(b) shows the 2D projections of the TLM lines for the (n,m,q) cell. The shaded lines are those belonging to the (n,m,q) cell. Each of the lines in Fig.2.13(b) contains the associated voltage wave, e.g., $Z_{xy}(n,m)$ contains $V_{xy}(n,m)$, etc...

The notation for the designation of the TLM lines and voltages is similar to that of 2D. We cite $Z_{xy}(n,m,q)$ and $Z_{xz}(n,m,q)$ as examples. In the case of $Z_{xy}(n,m,q)$, (n,m,q,) of course refers to the particular cell. As with the 2D, the first subscript, x, indicates the direction of propagation and the second subscript, y, indicates the field direction. xy also defines the propagation plane Similarly, $Z_{xz}(n,m,q)$, has the same direction of propagation but whose field is in the z direction. The other four lines, $Z_{yx}(n,m,q)$, $Z_{yz}(n,m,q)$, $Z_{zx}(n,m,q)$, and $Z_{zy}(n,m,q)$, follow the same designation. Note that the remaining nine line segments of the cubical cell(Fig.2.13(a)), corresponding to twelve transmission lines, "belong" to other cells. Thus, for example, the line segment giving rise to $Z_{xy}(n,m,q-1)$ line, which runs parallel to the Z_{xy} (n,m,q) line, belongs to the (n,m,q-1) cell. By constructing the neighboring cells, one may observe that the space symmetry is observed, i.e., all space is fully utilized. The notation for the voltage wave difference in each transmission line follows the same pattern, so that in the $Z_{xy}(n,m,q)$ line, for example, the voltage wave is $V_{xy}(n,m,q)$, and the voltage waves in the other lines follow in the same way. Fig.2.14 shows a 3D depiction of four of the eight cells surrounding the (n,m,q) node. The six TLM lines ($Z_{xy}(n,m,q)$, $Z_{yx}(n,m,q)$, etc..) associated with the (n,m,q) cell are highlighted. Although only the edges of the (n,m,q) cell are highlighted, each 3D TLM line should be regarded as an infinitesimal line running along the common border of the four surrounding cells. Thus, for example, $Z_{xy}(n,m,q)$ and $Z_{xz}(n,m,q)$ run along the common border of the (n,m,q), (n,m,q+1), (n,m+1,q), and (n,m+1,q+1) cells.

Having identified the location of the nodes and the transmission lines for the 3D case, we now turn our attention to the mapping of the physical properties, such as the resistivity and the fields. As alluded to previously, the 3D problem may be separated into three 2D constructions ,as noted in Fig.2.13(b). On occasion , we may wish to label the node resistors so as to identify to which line the node resistor the belongs. Hence, we may employ the designations of $R_{xy}(n,m,q)$, $R_{zx}(n,m,q)$, and $R_{yz}(n,m,q)$. We stress , however , that the three node resistors of the cell are entirely equal, and unless there is the possibility of confusion, we

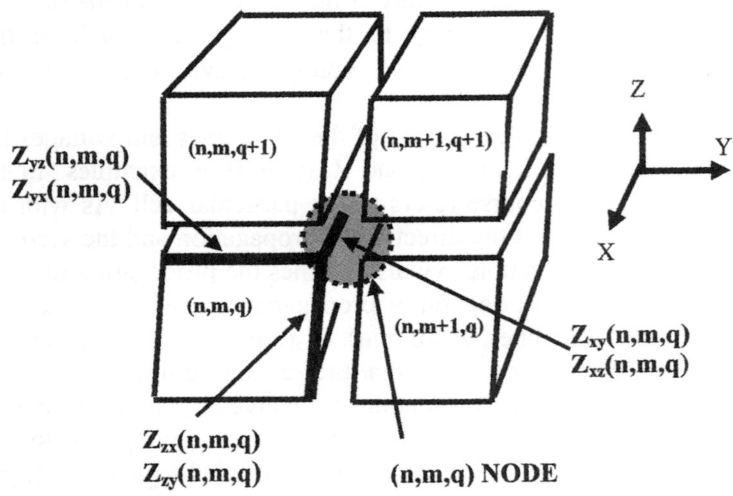

FIG.2.14 DEPICTION OF FOUR OF THE EIGHT CELLS SURROUNDING THE (n,m,q) NODE. HIGHLIGHTED EDGES CONTAIN THE SIX TLM LINES BELONGING TO THE (n,m,q) CELL.

omit the subscripts.

The first property to be mapped is the node resistance. The node resistor may be calculated by methods analogous to the 2D matrix. As before, we select a field direction and then calculate the average conductivities surrounding the lines, from which the average inverse node resistance may be calculated. If we select the z direction then there will be four transmission lines, converging on the (n,m,q) node, which will have a field in the z direction. These are: $Z_{xz}(n,m,q)$, $Z_{xz}(n+1,m,q)$, $Z_{yz}((n,m,q)$, and $Z_{yz}(n,m+1,q)$. We temporarily label the equivalent end resistors for each of the four lines, as $R_A(n,m,q)$, $R_B(n,m,q)$, $R_C(n,m,q)$, and $R_D(n,m,q)$. In order to obtain these end resistors we need to obtain the average conductivity in the line, which may be estimated from the average of the conductivity of the *four* cells surrounding each line. For line $Z_{xz}(n,m,q)$, for example, the four conductivity cells are $\sigma(n,m,q)$ $\sigma(n,m+1,q)$

$\sigma(n,m,q+1)$ and $\sigma(n,m+1,q+1)$. The average conductivity for this line, designated by σ_A, is

$$\sigma_A = [\sigma(n,m,q)+\sigma(n,m+1,q) +\sigma(n,m,q+1) + \sigma(n,m+1,q+1)]/4 \qquad (2.24)$$

From our previous discussion, the resistor $R_A(n,m,q)$, which is at each end of the $Z_{zx}(n,m,q)$ line, is

$$R_A(n,m,q) = 2/\Delta l \sigma_A \qquad (2.25)$$

Similar relationships , corresponding to Eqs.(2.24) and (2.25) , may be obtained for the other three lines. For $Z_{xz}(n+1,m,q)$ the average conductivity and corresponding node resistance, designated by σ_B and $R_B(n,m,q)$ are

$$\sigma_B =[\sigma(n+1,m,q)+\sigma(n+1,m+1,q) +\sigma(n+1,m,q+1) + \sigma(n+1,m+1,q+1)]/4 \qquad (2.26)$$

$$R_B(n,m,q) = 2/\Delta l \sigma_B \qquad (2.27)$$

Similar relationships apply to $Z_{yz}(n,m,q)$ and $Z_{yz}(n,m+1,q)$, labeled with the subscripts C and D respectively, are

$$\sigma_C =[\sigma(n,m,q)+\sigma(n,m,q+1) +\sigma(n+1,m,q) +\sigma(n+1,m,q+1)]/4 \qquad (2.28)$$

$$R_C(n,m,q) = 2/\Delta l \sigma_C \qquad (2.29)$$

$$\sigma_D=[\sigma(n,m+1,q)+\sigma(n,m+1,q+1) +\sigma(n+1,m+1,q)+\sigma(n+1,m+1,q+1)]/4 \qquad (2.30)$$

$$R_D(n,m,q) = 2/\Delta l \sigma_D \qquad (2.31)$$

As with the 2D we now form the average of the inverse end resistors, which then defines $R(n,m,q)$ (or more precisely, $1/R(n,m,q)$). Thus

$$1/R(n,m,q) \equiv [(1/R_A +1/R_B+ 1/R_C+ 1/R_D)]/4 = (1/8)\Delta l\, [\sigma_A+\sigma_B+\sigma_C+\sigma_D] \qquad (2.32)$$

We now recall that the average conductivity centered about the node , $\sigma_{AV}(n,m,q)$, is

$$\sigma_{AV}(n,m,q)=[\sigma(n,m,q)+\sigma(n,m+1,q)+\sigma(n,m,q+1)+\sigma(n,m+1,q+1)+$$

$$\sigma(n+1,m,q)+\sigma(n+1,m+1,q) +\sigma(n+1,m,q+1)+\sigma(n+1,m+1,q+1)]/8 \qquad (2.33)$$

Combining Eqs(2.24)-(2.32) then gives for the node resistance

$$R(n,m,q) \quad = 2/\Delta l \sigma_{AV}(n,m,q) =2\rho_{AV}(n,mq)/\Delta l \qquad (2.34)$$

which is identical to the 2D result. The above result is the same if we average over other directions(using other sets of end resistors) or, equivalently, if we average over all the end resistors.

Next we determine the voltages in the transmission lines, and from this the fields centered about the TLM lines, based on the difference in voltage between neighboring cells. As an example, we consider the $Z_{xz}(n,m,q)$ lines, which is a line, associated with the (n,m,q) cell, propagating in the x direction with the field in the z direction. The voltage difference for this line is $V_{xz}(n,m,q)$.This difference is calculated by first taking the average of the voltages of the (n,m,q) and $(n,m+1,q)$ cells followed by the average of those for the $(n,m,q+1)$ and $(n,m+1,q+1)$ cells and then taking the difference between these two averages. This is more complicated than the 2D case where the voltage difference in the line is simply the difference between adjacent cell voltages. Table 2.1 gives the averaging for the various transmission lines belonging to the (n,m,q) cell.

Just as we defined average fields throughout the 2D cell, we also do the same for 3D cell as well. For example, the y directed field in the (n,m,q) cell is the average of the fields of the eight lines : $Z_{xy}(n,m,q)$, $Z_{xy}(n,m-1,q)$, $Z_{xy}(n,m-1,q)$, $Z_{xy}(n,m-1,q-1)$, $Z_{zy}(n,m,q)$, $Z_{zy}(n,m-1,q)$, $Z_{zy}(n-1,n,q)$, and $Z_{zy}(n-1,m-1,q)$. The average y directed field for the (n,m,q) cell defined as $V_{AV,y}(n,m,q)$ is thus

$$V_{AV,y}(n,m,q)=(1/8)[V_{xy}(n,m,q)+V_{xy}(n,m-1,q)+V_{xy}(n,m,q-1)+V_{xy}(n,m-1,q-1)+$$

$$V_{zy}(n,m,q)+V_{zy}(n,m-1,q)+V_{zy}(n-1,m,q)+V_{zy}(n-1,m-1,q)] \qquad (2.35)$$

TABLE 2.1

LINE	AVERAGE FIELD IN LINE BASED ON CELL VOLTAGES
$V_{xz}(n,m,q)$	$(1/2)[V(n,m,q+1)+Vn,m+1,q+1)-V(n,m,q)-V(n,m+1,q)]$
$V_{yz}(n,m,q)$	$(1/2)[V(n,m,q+1)+V(n+1,m,q+1)-V(n,m,q)-V(n+1,m,q)]$
$V_{xy}(n,m,q)$	$(1/2)[V(n,m+1,q)+V(n,m+1,q+1)-V(n,m,q)-V(n,m,q+1)]$
$V_{zy}(n,m,q)$	$(1/2)[V(n,m+1,q)+V(n+1,m+1,q)-V(n,m,q)-V(n+1,m,q)]$
$V_{yx}(n,m,q)$	$(1/2)[V(n+1,m,q)+V(n+1,m,q+1)-V(n,m,q)-V(n,m,q+1)]$
$V_{zx}(n,m,q)$	$(1/2)[V(n+1,m+1,q)+V(n+1,m,q)-V(n,m,q)-V(n,m+1,q)]$

As mentioned before , each term on the right side of Eq.(2.35) consists of forward and backward waves. In addition , as we shall see in Chapter 4, each term may be split into plane wave and symmetric components, but this need not concern us for the moment.

For completeness we write down the average x and z fields for the cell:

$$V_{AV,x}(n,m,q) = (1/8)[V_{zx}(n,m,q)+V_{zx}(n,m-1,q)+V_{zx}(n-1,m,q)+V_{zx}(n-1,m-1,q)$$

$$+V_{yx}(n,m,q)+V_{yx}(n-1,m,q)+V_{yx}(n,m,q-1)+V_{yx}(n-1,m,q-1)] \qquad (2.36)$$

$$V_{AV,z}(n,m,q) = (1/8)[V_{xz}(n,m,q)+V_{xz}(n,m,q-1)+V_{xz}(n,m-1,q)+V_{xz}(n,m-1,q-1)$$

$$+V_{yz}(n,m,q)+V_{yz}(n-1,m,q)+V_{yz}(n,m,q-1)+V_{yz}(n-1,m,q-1)] \qquad (2.37)$$

Other Node Controlled Properties

Besides the conductivity, what other properties may we ascribe to the nodes? This is a wide ranging question which can only be answered by the physics underlying the wave propagation The nodes embody the physics of the medium. As such they control the scattering of the electromagnetic waves in the medium. Fig.2.15 indicates the generic node function, denoted by $S(n,m)$. In most cases this will amount to the conductivity control, in which case $S(n,m) = R(n,m)$. However many other properties, such as signal gain, signal generation, mode conversion, and plane wave correlation(discussed in Chapter IV)may be controlled by the nodes. The following gives examples of several node controlled properties, assuming 2D to simplify the discussion.

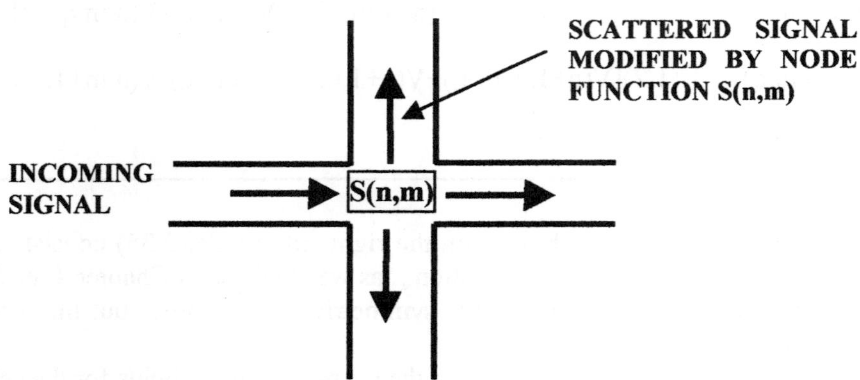

FIG. 2.15 ANY PHYSICAL PROPERTY MAY BE ASSIGNED TO THE NODE FUNCTION S(n,m) PROVIDED THE SCATTERING COEFFICIENTS ARE PROPERLY DEFINED. EXAMPLES ARE, BESIDES LOSS, SIGNAL GAIN AND GENERATION, MODE CONVERSION, AND PLANE WAVE CORRELATION.

2.5 Node Control of 2D Scattering Coefficients Due to Finite Node Resistance

In Chapter I we briefly described the 2D scattering process but with the stipulation that the node resistance was infinitely large. Before considering other types of node functions, it is useful to generalize the node scattering when R is finite. We suppose a forward wave is incident on the node in the $Z_{xy}(n,m)$ line and then proceed to calculate the transfer of energy to any of the other three lines as well as the reflected energy. For this calculation Fig.2.7 is useful and to simplify matters we assume the TLM lines surrounding the node are all the same, denoted by Z_O. To obtain the energy dispersal we will need the scattering coefficients , representing the wave transfer and reflection respectively. To obtain the coefficients we require the load impedance viewed by ${}^+V_{xy}(n,m)$, and designated by $RL1_{xy}(n,m)$. From inspection of the circuit, $RL1_{xy}(n,m)$ is given by

$$RL1_{xy}(n,m,) = \ [R(n,m) \ R_P(n,m)]/[R(n,m) +R_P(n,m,)] \tag{2.38}$$

where

$$R_P = \ 3R(n,m)Z_o/[R(n,m)+Z_o] \tag{2.39}$$

$RL1_{xy}(n,m)$ may be interpreted as the parallel combination of the two resistors terminating $Z_{xy}(n,m)=Zo$. One is the node resistance $R(n,m)$, and the other is $R_P(n,m)$, which is the series connected impedances of the three other lines converging at the node. Having found $RL1_{xy}(n,m)$ we can write down the transfer coefficient, to any of the other three lines, as well as the reflection coefficient:

$$T_{xy}(n,m) \ = \ (1/3)2RL1_{xy}(n,m)Z_o)/[RL1_{xy}(n,m)+Z_o] \tag{2.40}$$

$$B_{xy}(n,m) \ = [RL1_{xy}(n,m)-Z_o]/[\ RL1_{xy}(n,m)+Z_o] \tag{2.41}$$

The subscripts for the coefficients are the same as that for the incident wave. The generalized notation for the scattering coefficients and node parameters is given in the next Chapter , but this need not concern us in the present situation. Note also that a (1/3) factor appears since we are interested in the energy transfer to only one of the three lines comprising the output. To interpret the results more readily, it is useful to assume $R(n,m)$ is small compared to Z_o. The above then become(after expanding in powers of $Z_O/R(n,m)$)

$$T_{xy}(n,m) = (1/2) - (1/8)Z_o/R(n,m) \qquad (2.42)$$

$$B_{xy}(n,m) = (1/2) - (3/8)Z_o/R(n,m) \qquad (2.43)$$

We see from the above the obvious result that the wave amplitudes are diminished by the node resistance so that the both the transfer and reflection coefficients are reduced from their values of 1/2 when no loss is present.

2.6 Simultaneous Conductivity Contributions

We next consider simultaneous contributions to the conductivity. Strictly speaking this is not a new property but the formulation will prove useful when we consider semiconductors or any other medium with a background conductivity. For concreteness we assume the two contributions. One stems from a background conductivity[1], whose equivalent node resistance is denoted by $R_B(n,m)$(excluding the other conductivity) and a conductivity induced by either light or avalanching, whose equivalent node resistance (again, excluding the other conductivity) is denoted by $R_O(n,m)$. The total node resistance is then obtained by combining the two contributions in parallel, or

$$R(n,m) = [R_B(n,m,)R_O(n,m,)/[R_B(n,m,)+R_O(n,m,)]] \qquad (2.44)$$

Eq.(2.44) allows us to control each mechanism separately. The equivalent circuit at the node is shown in Fig.2.16, and as expected each $R(n,m)$ element is replaced by the parallel combination of $R_B(n,m)$ and $R_O(n,m)$. It is worthwhile pointing out the limiting behavior of Eq.(2.44) since some confusion might result if interpreted incorrectly. If the background losses are predominant, then $R_B(n,m)$ is much *lower* than $R_O(n,m)$ and $R(n,m)=R_B(n,m)$. On the other hand, if the loss due to light/avalanching is predominant, $R_O(n,m,)$ is much *lower* than $R_B(n,m)$ and $R(n,m) \sim R_O(n,m)$.

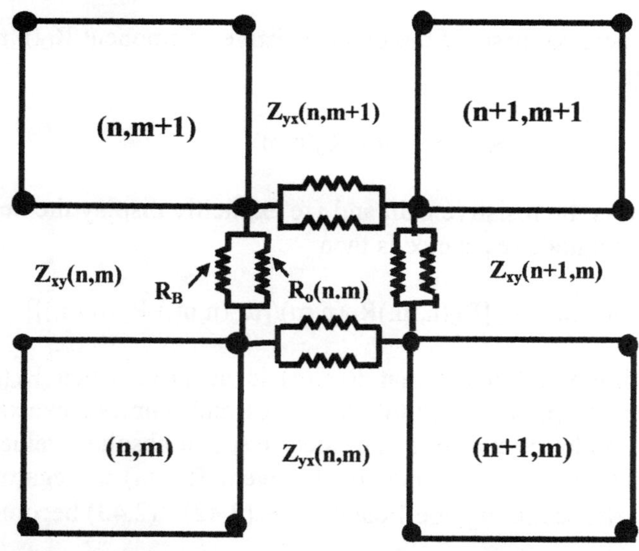

FIG. 2.16 DIVIDING OF NODE RESISTANCE, R(n,m), INTO A LIGHT ACTIVATED PART, $R_O(n,m)$, AND A BACKGROUND PART, R_B. $R(n,m)= R_B * R_O(n,m)/[R_B+ R_O(n,m)]$.

2.7 Signal Gain

We have seen that the nodal resistors serve to dissipate the electromagnetic signal propagating throughout the medium. These same elements, as we have seen in Chapter I, in connection with reverse iterations, can serve also to amplify electromagnetic signals, i.e., provided the gain mechanism at the nodes exceeds any losses. In this discussion, we do not inquire as to the origin of the gain but merely assume that it exists. The presence of gain will of course alter the transfer and gain coefficients. To illustrate the gain, we again consider the 2D case and once more assume the TLM lines about the (n,m,) node are identical, equal to Zo, with the wave $^+V_{xy}(n,m)$ incident on the (n,m) node. We then assume the gain can be represented by a negative node resistance. This notion has already been reinforced previously when we saw that gain occurs in the reverse iteration, replacing R(n,m,) with -R(n,m). In this case we again assume a background resistance, but assume the other resistive component $R_O(n,m)$ is responsible for the gain or

$$R_O(n,m) \rightarrow - R_G(n,m) \qquad (2.45)$$

where $R_G(n,m)$ is the resistive gain and we explicitly display the negative sign in $R(n,m)$. The total node resistance is then

$$R(n,m) = [R_B(n,m,)R_G(n,m)/[R_G(n,m,)-R_B(n,m,)]] \qquad (2.46)$$

Note the important result that $R(n,m)$ is negative when $R_B(n,m)$ exceeds $R_G(n,m)$. When $R_G(n,m) \sim R_B(n,m)$ the losses and gain are exactly counterbalanced and the node resistance is extremely large in absolute value and thus has very little effect on the scattering. In the event $R(n,m)$ is negative and $Z_O \ll |R(n,m)|$ then the scattering coefficients, Eqs.(2.42) - (2.43) become

$$T_{xy}(n,m) = (1/2) + (1/8) Z_o/R(n,m) \qquad (2.47)$$

$$B_{xy}(n,m) = (1/2) + (3/8)Z_o/R(n,m) \qquad (2.48)$$

and we see that there is gain in the scattered waves. We should caution that in many situations involving signal gain, the wave is a plane wave type and the scattering to the transverse lines is minimal. Under plane wave conditions the values of $T_{xy}(n,m)$ and $B_{xy}(n,m)$ become one dimensional in nature and the coefficients are equal to $T_{xy}=1+Zo/2R$ and $B_{xy}=Zo/2R$ (or equal to to $T_{xy}=1-Zo/2R$ and $B_{xy}= -Zo/2R$ when there is loss instead of gain). The plane wave modification of scattering coefficients is given in Chapter IV.

2.8 Signal Generation. Use of Node Coupling

Another property which may be mapped onto the nodes has to do with the generation of a signal rather than the amplification of one. One example is the generation of a light signal in the semiconductor. In the case of direct bandgap semiconductors, for example, the light will be generated by direct recombination. One must exercise great caution in transferring this phenomena to the same transmission line matrix as the one originally defined . This is because in general the propagation velocities of the low frequency electromagnetic signal and the light signal will differ and, in addition, the dependence of the node resistors on the higher frequency light signal also will differ. This is nothing more than dispersion, of course, which will be discussed in detail in Chapter V. In addition , if the generated signal has a short wavelength (relative to the cell length) then the field propagating in the TLM line may be nonuniform, resembling an "ac" signal, unless we select an extremely small time step. This of course will impose great demands on computer capacity. At this point, nevertheless, some idea of the approach needed to handle the generation of signals(the creation and distribution of light , in this case) with a different wavelength, node resistance, and propagation velocity, is useful.

One approach is to create a "parallel" line (or matrix in the case of 2D) which occupies the same space as the original line (or matrix), as shown conceptually in Fig.2.17, which shows a side view of the two 1D lines. In this case,

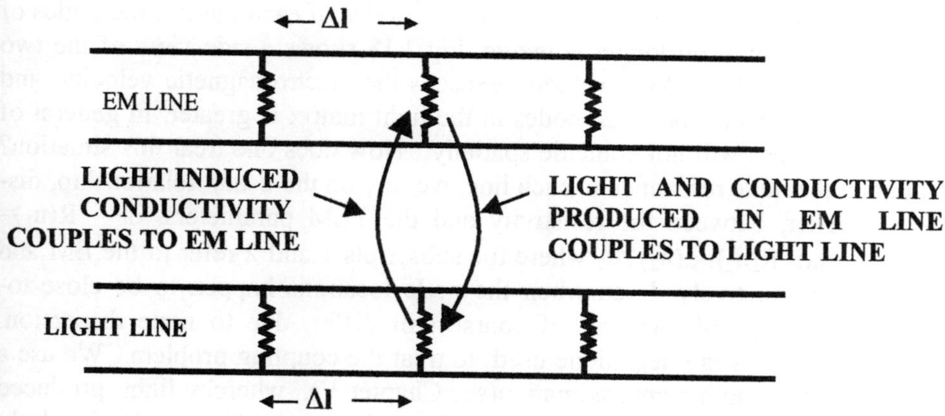

FIG. 2.17. INTERACTION OF LIGHT AND EM SIGNALS(1D) SHOWN AS PARALLEL LINES WITH NO VELOCITY DISPERSION, BUT WITH DISPERSION IN THE NODES.

since the cell length is the same, the light velocity and the electromagnetic velocity are equal. Although we assume the lines are non-dispersive, we can still consider the nodes as dispersive, with the light and EM signals viewing completely different values of node resistance; in fact the light signal may view under certain circumstances a negative resistance , due to additional light produced by avalanche effects(brought about by the field in the EM grid).

The interplay of the matrices at each node requires some explanation. Light created at each EM node is fed into the light matrix. In addition, conductivity may be produced in the EM line due to avalanching (caused by an intense EM field). This conductivity will likewise be transferred to the light line (with a differing value due to dispersion). On the other hand, light is attenuated in the light matrix, which in turn creates carriers and additional conductivity, which then changes the values of the resistor nodes in the electromagnetic matrix. We can also allow for the possibility that the light line produces a "low frequency" signal , which would then enter the EM matrix at the node.

A further complication arises in situations involving different propagation velocities, as well as differing node resistors. This is the more typical case. In order to maintain the same time step between nodes of each matrix, the nodes of the two signals will no longer coincide. Fig.2.18 shows a side view of the two 1D lines, in which the light velocity exceeds the electromagnetic velocity. and therefore the distance between nodes in the light matrix is greater. In general of course, the nodes will not coincide spatially. How does one treat this situation? To obtain the node resistance in each line, we rely on the usual relationship, discussed earlier, between the resistivity and the TLM parameters, or $R(n_1)= \rho(n_1)/\Delta l_1$ and $R(n_2)=\rho(n_2)/\Delta l_2$ where the subscripts 1 and 2 refer to the EM and light lines respectively. Even when the node locations happen to be close together, the node resistors will of course still differ, due to node dispersion. Fig.2.18 illustrates the technique used to treat the coupling problem. We use a multiple node coupling approximation(see Chapter V), whereby light produced in the EM line is routed from the EM node to the nearest light node. Similarly any conductivity produced in the EM line, due to avalanching, may be transferred (allowing for dispersion)to the light line, using the same coupling path. Conversely, the node conductance in the light line node carries over to its nearest neighbors in the EM line, adding to any preexisting conductance in the EM

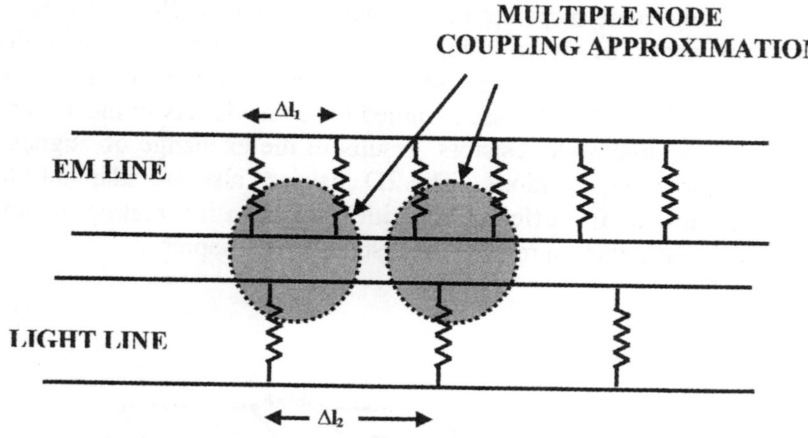

FIG. 2.18 INTERACTION OF LIGHT AND EM SIGNALS(1D) WITH DIFFERING WAVE VELOCITIES.

line or to conduction from other sources (such as avalanching). All the nodes , in both the EM and light lines , participate in the coupling process, with every node in the EM line having a nearest neighbor in the light line and vice-versa. This means that in the EM line (the slower line) two or more nodes may be attached to the same light node as noted in Fig.2.18. Combining multicoupled waves will require techniques discussed in Chapter IV.

For two interacting mediums occupying the same space, multiple node coupling is the most appropriate method; however, a simpler (but somewhat less accurate) method also is available, namely, the nearest node method. This method may be used to treat *spatial* boundary conditions, i.e., the interface between mediums with differing propagation velocities. Here it is possible to have unattached "partial nodes ", as well as one to one "nearest nodes", at the interface. This is in contrast to the multiple node coupling, where there are no partial nodes and all nodes participate in the coupling. Nearest nodes, appearing at the dielectric interface, are treated in detail in Chapter V.

Although we have alluded to only 1D situations, it is easy to visualize the 2D interaction between the light and electromagnetic("low frequency") signals. For treating node dispersion , e.g., we use two parallel matrices which occupy the same space, as shown in Fig.2.19, which assumes no velocity dispersion. Each of the   2D matrices represents a different frequency(the light and EM frequencies) , but the nodes are assumed to coincide. As in the 1D case of Fig. 2.17, the dispersive node resistors results in the exchange of signals and conductivity at the node locations. The 2D matrices also are adaptable to the node coupling approach for differing velocities, and also for treating boundaries between differing propagation regions, as discussed in Chapter V.

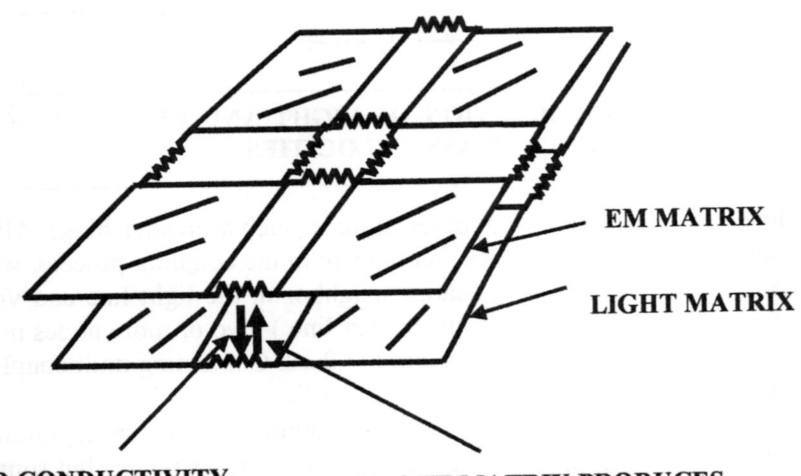

FIG. 2.19 2D INTERACTION OF LIGHT AND EM SIGNALS SHOWN AS PARALLEL MATRIX ARRAYS(NO VELOCITY DISPERSION).

2.9 Mode Conversion

Another property which we may assign to the node is that of mode conversion. This is unique to the 3D case, and has to do with the fact that for a given direction of propagation there are two orthogonal fields associated with the wave. With mode conversion we allow for the possibility that a net portion of the wave energy in one mode is transferred to the orthogonal mode. Thus for example a portion of the wave $V_{xy}(n,m,q)$ will be transferred to $V_{xz}(n,m,q)$. and vice versa. During this process, for a lossless node, the total energy among the nodes is conserved. In order to proceed further we need to resort to wave partitioning , which is discussed in Chapter IV.   We should also mention that the mode conversion is often accompanied by dispersion, i.e., the two modes may have differing velocities in which case we must resort to techniques discussed in Chapter V.

Example of Mapping: Node Resistance in Photoconductive Semiconductor

2.10 Semiconductor Switch Geometry(2D)

At this point we provide a very simple example in which the electromagnetic and conductivity properties of a semiconductor sample are mapped onto the transmission line matrix. We first consider the field in a semiconductor slab filling the space between a pair of electrodes separated by length l_o. A constant voltage is superimposed on the electrodes. A portion of the top view matrix is shown in Figure.2.21(a) and the side view in Fig.2.21(b). To simplify, we assume the semiconductor is extremely wide(in the z direction) , so that there is no variation in that direction. We also assume there is little field fringing, and the field is initially uniform across the sample. We arbitrarily select 10 cells to span the semiconductor gap, as noted in Fig.2.20. Since the field is uniform, the field likewise will be distributed uniformly among the lines $Z_{yx}(n,m,q)$ and $Z_{zx}(n,m,q)$. The fields in the horizontal lines, $Z_{xy}(n,m,q)$ and $Z_{xz}(n,m,q)$ will be zero of course since we assume there is no vertical field to begin with.

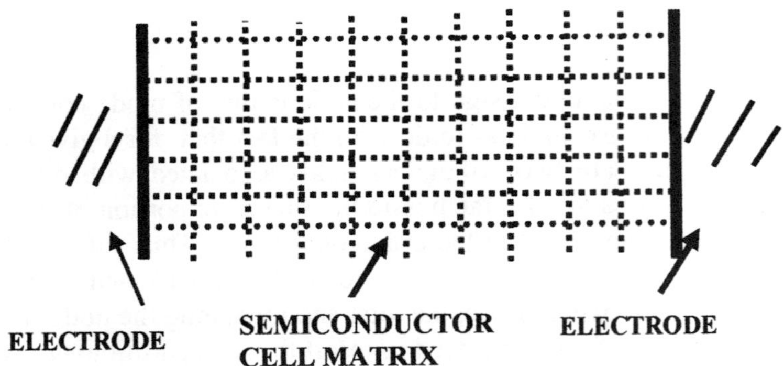

FIG, 2.20(a) TOP VIEW OF SEMICONDUCTOR SWITCH.

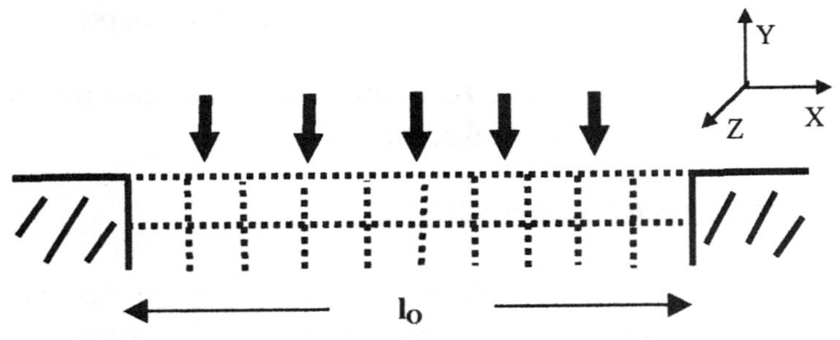

FIG.2.20(b) SIDE VIEW OF SEMICONDUCTOR SWITCH.

Next we map the conductivity properties. In the case of semiconductors, phenomena such as photoconductivity, avalanche breakdown, recombination, etc..., are directly linked to the conductivity[1]. As an illustration of the mapping we consider a single mechanism, light activated conductivity, which we simulate with the cell matrix in Fig.2.20. The amplitude of the light pulse, directed perpendicular to the face of the semiconductor, is assumed to be turned at time t=0, and to be constant in time thereafter. We assume a spatial dependence in the x direction, however, for the light impinging on top of the semiconductor. The spatial dependence for the incident light power $P(x)$ falling on each of the top cells , each with area Δl^2 , is taken to be gaussion , and given by

$$P(x) = P_oEXP[-a\{ n\Delta l/l_o -1/2\}^2] \qquad (2.49)$$

where a is the spread factor of the light pulse and $n\Delta l = x$ is the distance measured from the elecrode. The total semiconductor length l_o is equal to $N_o\Delta l$ where N_o is the total number of cells between electrodes. Po is the constant(in time) light power falling on the center cell, where the light is a maximum, and the power level tapers away from the center. In addition, we assume the light signal attenuates as it enters the semiconductor. If we assume an exponential attenuation then the signal follows $EXP\{-m\Delta l/h_o\}$ where h_o is the absorption depth of the light signal. Thus, the actual power deposited in the mth cell will follow an exponential decay in the particular cell. We can now calculate the energy deposited in the (n,m,q) cell during the kth time step. We simplify matters somewhat by assuming the semiconductor is dispersionless, so that the light signal velocity is identical to that of the electromagnetic signals in the transmission lines. If we denote the energy deposited in the (n,m,q) cell by γ(n,m,q), we have

$$\gamma (n,m,q) = P_oEXP[-a\{ n\Delta l/l_0 -1/2 \}^2] D(m)\Delta t \qquad (2.50)$$

where

$$D(m) = 0 \ \ if \ \ k<m \qquad (2.51)$$

and

$$D(m) = [EXP-(m\Delta l/h_0)][EXP (\Delta l/h_0)-1] \ \ for \ k\geq m \qquad (2.52)$$

The delay nature of D(m) expressed by Eq.(2.51) is understandable, of course, since no conductivity will be produced until the light signal reaches the cell in question. Once $k\geq$ m then the wave energy is deposited in the cells. Eq.(2.52) is the *difference* in the decay factor , $EXP\{-m\Delta l/h_o\}$, at depths $m\Delta l$ and $(m-1)\Delta l$. The difference, D(m) , is thus proportional to the energy deposited in the mth cell. If U is the photon energy then the number of photons deposited in (n,m,q) is γ(n,m,q)/U. To simplify the effect of the light signal we assume there is no lateral scattering of the light as it is absorbed in the semiconductor We now relate the number of carrier pairs produced , during Δt, to the deposited photons. If the conversion of photons to carrier pairs is is done with efficiency, ξ, then during Δt the incremental change in carrier pairs is

$$\text{No. carrier pairs added in } (n,m,q) \text{ cell} = \Delta N(n,m,q) = \xi\, \gamma\,(n,m,q)/U \qquad (2.53)$$

From simple semiconductor transport theory[1], the added conductivity $\Delta\sigma(n,m,q)$ in the cell is

$$\Delta\sigma(n,m,q) = e\Delta N(n,m,q)(1/\Delta l^3)\{\,\mu_h(n,m,q) + \mu_n(n,m,q)\} \qquad (2.54a)$$

where e is the electron charge, $\Delta N(n,m,q)/\Delta l^3$ is the added number density of holes and electrons pairs, and $\mu_h(n,m,q)$, $\mu_n(n,m,q\,)$ are the hole and electron mobilities. We are assuming that the photons create equal numbers of holes and electrons, and that they remain equal throughout the time scale of interest. During this time scale, we reiterate that the carriers do not venture(either by drift or diffusion) outside the cell in which they were created, nor do they recombine. In Chapter 6 we indicate how to incorporate transport phenomena into the iteration. Combining the previous equations and definitions, the added conductivity is written as(MKS units)

$$\Delta\sigma(n,m,q)=[e\{\mu_h(n,m,q)+\mu_n(n,m,q)\}\xi P_o EXP[-a\{n\Delta l/l_o-1/2\}^2]D(m)(1-r)\Delta t]/\Delta l^3 U$$
$$(2.54b)$$

We have included a reflection coefficient, r, indicating that a portion of the light will reflect off the semiconductor surface. If k is larger than m the *accumulated* conductivity is simply

$$\sigma(n,m,q) = \Delta\sigma(n,m,q)(k+1-m), \qquad (2.55)$$

Again, Eq.(2.55) is justified because of the assumption that the carriers being produced are "frozen" in place. We make the approximation that $\sigma(n,m,q) \approx \sigma_{AV}(n,m,q)$, which should apply for *large* cell densities(recall that $\sigma_{AV}(n,m,q)$ applies to the auxiliary cell centered about the node). We now can convert

Eq.(2.55) into a node resistance using the relationship $R(n,m) = 2/\sigma_{AV}(n,m,q)\Delta l$. The result is

$$R(n,m) = 2/\sigma_{AV}(n,m,q)\Delta l \approx 2/\sigma(n,m,q)\Delta l =$$
$$2\Delta l^2 U[e\{\mu_h(n,m,q)+\mu_n(n,m,q)\}\xi P_o EXP[-a\{n\Delta l/l_o-1/2\}^2] D(m)(1-r)(k+1-m)\Delta t]^{-1}$$

$$(2.56)$$

The above is the sought after result for the node resistance for each cell (subject to the approximations used), which may then be used in any computer iteration, such as described in Chapter VII.

2.11 Node Resistance Profile in Semiconductor

To reinforce the above we select a very simple example in which the experimental conditions resemble those of a silicon semiconductor activated with a short pulse from a Nd:Yag laser(either Q switched or mode locked), and proceed to calculate the profile of the node resistance across the length of the semiconductor gap. In order to simplify, we set (U/e), the photon energy (expressed in volts), equal to unity and further assume a conversion efficiency $\xi=1$. We assume a one centimeter gap between electrodes, and we arbitrarily divide up the semiconductor into 0.1 cm cells , so that the total length is spanned by 10 cells. A gaussian shaped (as described previously) light pulse, with a spread factor of $a=4$, is assumed. For illustrative purposes, 10^{-7} Joules of pulse energy is assumed to fall on the auxiliary cell surrounding the center node. For simplicity we assume the energy pulse has a constant power amplitude, and thus the energy is equal to $P_o k \Delta t$. Regarding the absorption of light in the semiconductor, we assume $h_o \sim 0.1$ cm and therefore the bulk of the light energy will be deposited in the top cells, i.e., m=1, since the cell length itself is 0.1 cm. In fact the light energy deposition in the top cell is proportional to D(1) which in view of Eq.(2.52) is

$$D(1) = 1-EXP-(\Delta l/h_o)$$

$$(2.57)$$

The above is ~ 0.63 when $\Delta l/h_o$ =1.0. For the second row, we need D(2), which we obtain by multiplying the above, D(1), by EXP-($\Delta l/h_o$) with the result that D(2)=0.23.

For our example suppose we wish to calculate the node resistance for the top row, R(n,1,m). To do this we cannot assume that $\sigma_{AV}(n,m,q)$ is approximated by $\sigma(n,m,q)$, as indicated in Eq.(2.56), since the cell resolution is not sufficiently large. In other words, we employ $\sigma_{AV}(n,m,q)$ and apply the right side of Eq.(2.56) to each of the four cells which contribute to $\sigma_{AV}(n,m,q)$. Thus we must first calculate the conductivity in each cell for rows 1 and 2, using the cell conductivity formula of the previous Section, Eq.(2.55). We then find $\sigma_{AV}(n,m,q)$ using the average conductivity formula given by Eq.(2.19), which takes the average of the four surrounding cell conductivities. From this the node resistance may be calculated, using Eq.(2.21). In order to make use of the formulae, however, we must first specify additional semiconductor and light properties.

Additional numerical values are as follows. The value of the TLM delay time Δt(for the given cell size of 0.1cm) is specified if we know the propagation velocity in the semiconductor. Selecting silicon as the semiconductor (which is consistent with the small absorption depth), then Δt for the 0.1cm cell becomes 11.5ps. The value of k at the end of the pulse is unspecified as yet since we have not yet specified P_o. If we assume a light signal pulsewidth of 46ps, then k=4, and $P_o = 10^{-7}/(46 \times 10^{-12}) = 2174$ W on the center cell. Assuming $\mu_h(n,m,q) + \mu_n(n,m,q)$ =1900 cm^2/V-s for silicon, and r=0.3, we then calculate the node resistance for n=0 to n=10 at t= 46ps. At this time most transport phenomena still play a very minor role. For example, if we assume a drift velocity of ~ 10^7 cm/sec , then at the end of the 46 ps pulse, the electron carriers will have moved a distance of .0046 cm, a relatively small fraction of the 0.1cm cell size. Fig.2.21 shows the profile of the node resistance across the gap. Note of course that the resistance is lowest in the center where the light intensity is greatest and proportionately more carriers are being produced. As expected , therefore, the spatial profile essentially follows that of the light intensity profile of the incoming signal. The dotted arrows indicate the vanishing of the node resistances adjacent the electrodes, if we strictly follow the definition of $\sigma_{AV}(n,m,q)$ in cells bordering the electrodes.

FIG. 2.21 DEPENDENCE OF NODE RESISTANCE ON CELL INDEX NUMBER n. DASHED LINE PORTIONS TAKE INTO ACCOUNT ELECTRODES IN THE AVERAGING OF R(n,m,q).

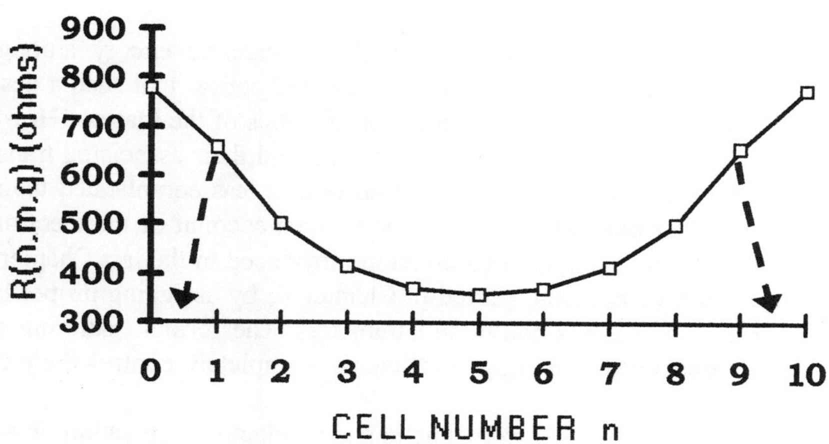

REFERENCES

1. S.M. Sze , *Semiconductor Devices* ,John Wiley and Sons, New York, 1985

III. Scattering Equations

In this Chapter we describe the flow of electromagnetic energy among the transmission lines, as controlled by   nodes. It would appear that such a description is a daunting task, given the potential complexities of the matrix . However, since we have provided unique labels to the cells and their associated transmission lines and nodes (Chapter II), the actual task is less complicated than one might suppose. Our main goal here will be to keep account of the electromagnetic energy dispersal, using the cell notation introduced in the last Chapter. We also finish the task of notation, started in Chapter II, by assigning proper labels to the scattering coefficients and node parameters. The formal scattering equations, and the associated scattering coefficients, completely control the electromagnetic dispersal.

In order to describe the electromagnetic and conductivity spreading it will be sufficient to consider a single cell, (n,m), and to relate the associated fields at time $t + \Delta t$ to fields in the previous time element, t. The fields considered at time t include not only the (n,m) cell but the surrounding cells as well. This is followed by iterations over all the cells which occupy the space. In this regard, the description here overlaps to some extent with that of Chapter 1; the discussion however differs in two important respects. First the approach described herein is generalized, especially with regard to the 3D treatment. Secondly the notation is rendered more useful by employing numerical indices. This is an important issue, since the notation must be adaptable to iterative methods for use in computer programs. The generalized notation will then make it possible to develop computer codes , which in turn will determine the detailed field profiles throughout the entire space, extending over a multitude of cells. The result is the time evolution of the entire system.

3.1 1D Scattering Equations

As in previous discussions it is easiest, by way of illustration, to first consider the scattering in the 1D case. Toward this goal we utilize Fig.3.1, borrowed from Chapter II, and seek the fields in the nth cell. We assume the delay time in each cell is identical, but do not require the cell lengths to be the same(i.e., we allow differing cell impedances). The forward and backward waves in the n cell, during the kth time interval, are denoted by $^{+}V^{k}(n)$ and $^{-}V^{k}(n)$, respectively. Similarly the same waves in the (n-1)th and (n+1) th cells are $^{+}V^{k}(n-1)$, $^{-}V^{k}(n-1)$ and $^{+}V^{k}(n+1)$, $^{-}V^{k}(n+1)$ respectively.

Our goal is to determine the fields scattered into the nth cell during the (k+1)th time step, based on the fields existing in the nth, (n-1)th, and (n+1)th cells during the kth time step. We first determine $^{+}V^{k+1}(n)$, i.e. , the forward wave in the nth cell during the (k+1)th interval. This wave will be the sum of two waves , consisting of a transmitted wave from the (n-1) cell as well as the reflected backward wave in the nth cell. Thus

$$^{+}V^{k+1}(n) = \ T^{k}(n\text{-}1,1) \ ^{+}V^{k}(n\text{-}1) + \ B^{k}(n\text{-}1,2) \ ^{-}V^{k}(n) \qquad (3.1)$$

where $T^{k}(n\text{-}1,1)$ and $B^{k}(n\text{-}1,2)$ are the transmission and reflection coefficients, respectively, of $^{+}V^{k}(n\text{-}1)$ and $^{-}V^{k}(n)$ at the (n-1)th node. The additional argument of one or two, in the scattering coefficient, is adapted to denote the fact that waves incident on the node are in the forward or backward directions, respectively. The coefficients, by definition, are:

$$T^{k}(n\text{-}1,1) = 2 \ RL1^{k}(n\text{-}1)/[RL1^{k}(n\text{-}1) +Z(n\text{-}1)] \qquad (3.2)$$

$$B^{k}(n\text{-}1,2) = [RL2^{k}(n\text{-}1)\text{-}Z(n)]/ \ [RL2^{k}(n\text{-}1)+Z(n)] \qquad (3.3)$$

where $RL1^{k}(n\text{-}1)$ and $RL2^{k}(n\text{-}1)$ are the load impedances seen by the forward and backward waves , incident on the (n-1) node. These impedances are easily calculated from

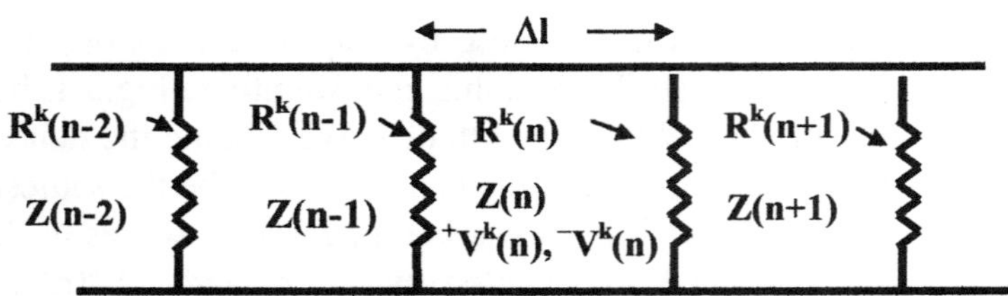

1D SCATTERING COEFFICIENTS *INTO* Z(n) LINE

$T^k(n-1,1)$: **TRANSFER OF WAVE FROM Z(n-1) LINETO Z(n) LINE.
n-1 INDICATES NODE. 1 DESIGNATES THE FORWARD
(POSITIVE) INCIDENT WAVE , AND k THE TIME STEP.**

$T^k(n,2)$: **TRANSFER OF WAVE FROM Z(n+1) LINE TO Z(n) LINE AT
nth NODE. 2 INDICATES THE BACKWARD(NEGATIVE)
WAVE.**

$B^k(n,1)$: **REFLECTION OF WAVE IN Z(n) LINE AT nth
NODE. 1 INDICATES FORWARD INCIDENT WAVE.**

$B^k(n-1,2)$: **REFLECTION OF WAVE IN Z(n) LINE AT (n-1)th
NODE . 2 INDICATES BACKWARD INCIDENT WAVE.**

**FIG 3.1 NOTATION FOR 1D SCATTERING COEFFICIENTS AND
TLM COUPLED CELLS. n AND k ARE INTEGERS. n DESIGNATES
THE CELL LOCATION AND k THE TIME STEP.**

$$RL1^k(n-1) = R^k(n-1)Z(n)/[R^k(n-1)+Z(n)] \qquad (3.4)$$

$$RL2^k(n-1) = R^k(n-1)Z(n-1)/[R^k(n-1)+Z(n-1)] \qquad (3.5)$$

where the load impedance represents the total impedance seen by the waves, and is made up of the parallel combination of $R^k(n-1)$ and the characteristic impedance. Note that we have added the superscript k to R, RL1 , and RL2 as well as the scattering coefficients to denote the time step. This completes the calculation of the forward wave. A comparable calculation is made to determine the backward wave in the nth cell during the (k+1)th interval. This wave is the result of a backward wave transmitted from the (n+1) cell as well as a reflection of the forward wave. in the nth cell. The result is

$$^-V^{k+1}(n+1) = \ T^k(n,2)\ ^-V^k(n+1) + \ B^k(n,1)\ ^+V^k(n) \qquad (3.6)$$

where $B^k(n,1)$ and $T^k(n,2)$ and the load impedances are

$$T^k(n,2) = \ 2RL2^k(n)/[RL2^k(n)+Z(n+1)] \qquad (3.7)$$

$$B^k(n,1) = RL1^k\ (n)-Z(n)]/\ [RL1^k(n)+Z(n)] \qquad (3.8)$$

$$RL1^k(n) = \ R^k(n)Z(n+1)/[R^k(n)+Z(n+1)] \qquad (3.9)$$

$$RL2^k(n) = \ R^k(n)Z(n)/[R^k(n)+Z(n)] \qquad (3.10)$$

We have completed our task, therefore, of finding the forward and backward waves (which comprise the total field) in the nth cell during the (k+1)th time step, in terms of fields belonging to the kth time step, using cell notation. The one dimensional approach is most useful in cases when the field gradient is dominant in a particular direction. A frequent application involves an impedance transformation in a non-uniform, one dimensional transmission line, which is amenable to SPICE software. One dimensional problems will be discussed later in Chapter VIII, which covers the topic of SPICE applications.

3.2 2D Scattering Equations

With the 2D matrix (and particularly the 3D) the scattering equations will naturally become more involved. However, by continuing to use the notation introduced in the previous Chapter, we will be able to describe the scattering in a very compact manner. We proceed by viewing a portion of the matrix in the vicinity of the (n,m) cell, repeated in Fig.3.2. The (n,m) cell and the surrounding cells and lines, pertinent to the scattering process, are shown. Our aim will be to determine the fields in the $Z_{xy}(n,m)$ and $Z_{yx}(n,m)$, for these are the two transmission lines associated with the (n,m) cell. As before, we will express the fields at time $t + \Delta t$ in terms of fields (in the surrounding lines) at time t. In the ensuing

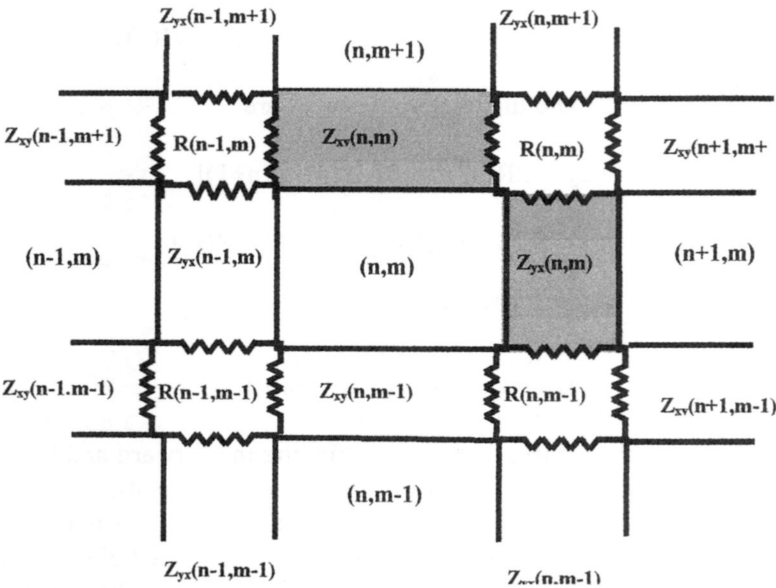

FIG. 3.2 2D TLM NOTATION CONVENTION:$Z_{xy}(n,m)$ AND $Z_{yx}(n,m)$ ARE THE TWO LINES(SHADED) ASSOCIATED WITH THE (n,m) CELL. R(n,m) IS AT THE INTERSECTION OF $Z_{xy}(n,m)$ AND $Z_{yx}(n,m)$.

discussion we suppress the superscript k, designating the time element, until such situations when it is significant, as in the iterative equations

Take for example the $Z_{xy}(n,m)$ line. The forward wave, $^+V_{xy}(n,m)$, during the (k+1)th interval, results from several waves converging on the R(n-1,m) node during the prior interval. First, there are the transfer of waves into $Z_{xy}(n,m)$ from three neighboring lines: $Z_{xy}(n-1,m)$, $Z_{yx}(n-1,m)$, and $Z_{yx}(n-1,m+1)$. Second, there is the reflection of the backward wave at R(n-1,m) in $Z_{xy}(n,m)$. Similarly, during the (k+1)th step, the forward wave in the $Z_{yx}(n,m)$ line, $^+V_{yx}(n,m)$, will be the result of waves scattering at the R(n,m-1) node. The backward waves, on the other hand, for both the $Z_{xy}(n,m)$ and $Z_{yx}(n,m)$ lines, will involve scattering at the R(n,m) node.

One can see that the scattering coefficients(i.e., both the transfer and the reflection type)will have many possible values, depending on the transmission line from which the wave emanates and line to which it is directed. It will be convenient to label these coefficients before proceeding further with the analysis. Since we are dealing with only the $Z_{xy}(n,m)$ and the $Z_{yx}(n,m)$ lines, we need only consider the nodes bounding these lines, which are, from Fig.3.2, (n,m), (n-1,m), and (n,m-1). Table 3.1 lists the 16 scattering coefficients associated with the (n,m) cell, 12 of which are transfer type and 4 reflection type. For example, the first transfer coefficient listed is $T_{xy}(n-1,m,1)$. The subscripts follow that of the incident wave (or TLM line), indicating the propagation and field directions respectively. The first two arguments n-1,m identify the node while the third argument, 1, simply indicates, by definition, that the wave is being coupled from the $Z_{xy}(n-1,m)$ line to the $Z_{xy}(n,m)$ one. As an another example, the third transfer coefficient listed , $T_{yx}(n-1,m,3)$, has the same node, but the third argument, 3, denotes the transfer of the wave from the $Z_{yx}(n-1,m+1)$ line to the $Z_{xy}(n,m)$ line. In the case of the reflection coefficients , the first one listed, $B_{xy}(n-1,m,1)$ refers to the n-1,m node and the argument 1, indicates the reflection from that node takes place for a backward wave in the $Z_{xy}(n,m)$ line.(the labeling of the 2D reflection coefficient differs from that used in the 1D notation). We emphasize that Table 3.1 represents the scattering *into* the lines $Z_{xy}(n,m)$ and $Z_{yx}(n,m)$, associated with the (n,m) cell. It will also be useful to consider the scattering *about* the node (n,m) , i.e., the scattering of all waves convergent on a particular (n,m)

TABLE 3.1 2D SCATTERING COEFFICIENTS <u>INTO</u> UNIT CELL LINES $Z_{xy}(n,m)$, $Z_{yx}(n,m)$ (XY PLANE)

TRANSFER TYPE

<u>COEFFICIENT</u>	<u>FROM</u>	<u>TO</u>
$T_{xy}(n-1,m,1)$	$Z_{xy}(n-1,m)$	$Z_{xy}(n,m)$
$(-)T_{yx}(n-1,m,2)$	$Z_{yx}(n-1,m)$	$Z_{xy}(n,m)$
$T_{yx}(n-1,m,3)$	$Z_{yx}(n-1,m+1)$	$Z_{xy}(n,m)$
$T_{xy}(n,m,4)$	$Z_{xy}(n+1,m)$	$Z_{xy}(n,m)$
$(-)T_{yx}(n,m,5)$	$Z_{yx}(n,m+1)$	$Z_{xy}(n,m)$
$T_{yx}(n,m,6)$	$Z_{yx}(n,m)$	$Z_{xy}(n,m)$
$(-)T_{xy}(n,m-1,7)$	$Z_{xy}(n,m-1)$	$Z_{yx}(n,m)$
$T_{yx}(n,m-1,8)$	$Z_{yx}(n,m-1)$	$Z_{yx}(n,m)$
$T_{xy}(n,m-1,9)$	$Z_{xy}(n+1,m-1)$	$Z_{yx}(n,m)$
$T_{xy}(n,m,10)$	$Z_{xy}(n,m)$	$Z_{yx}(n,m)$
$T_{yx}(n,m,11)$	$Z_{yx}(n,m+1)$	$Z_{yx}(n,m)$
$(-)T_{xy}(n,m,12)$	$Z_{xy}(n+1,m)$	$Z_{yx}(n,m)$

REFLECTION TYPE

$B_{xy}(n-1,m,1)$	$Z_{xy}(n,m)$	$Z_{xy}(n,m)$
$B_{xy}(n,m,2)$	$Z_{xy}(n,m)$	$Z_{xy}(n,m)$
$B_{yx}(n,m-1,3)$	$Z_{yx}(n,m)$	$Z_{yx}(n,m)$
$B_{yx}(n,m,4)$	$Z_{yx}(n,m)$	$Z_{yx}(n,m)$

node. Table 3.2 shows this scattering process. Note in Table 3.2 that the coefficient is always represented by the same (n,m), in contrast to Table 3.1 where the scattering is into the two lines associated with the (n,m) cell.

The issue of sign regarding the transfer coefficients requires some clarification. In Table 3.1, e.g., four of the transfer coefficients, $T_{yx}(n-1,m,2)$, $T_{yx}(n,m,5)$, $T_{xy}(n,m-1,7)$, and $T_{xy}(n,m,12)$, produce inverted waves with a negative polarity

<table>
<tr><td colspan="3" align="center">TABLE 3.2 2D SCATTERING COEFFICIENTS <u>ABOUT</u> (n,m) NODE (XY PLANE)</td></tr>
<tr><td colspan="3" align="center">TRANSFER TYPE</td></tr>
<tr><td><u>COEFFICIENT</u></td><td><u>FROM</u></td><td><u>TO</u></td></tr>
<tr><td>$T_{xy}(n,m,1)$</td><td>$Z_{xy}(n,m)$</td><td>$Z_{xy}(n+1,m)$</td></tr>
<tr><td>$(-)T_{yx}(n,m,2)$</td><td>$Z_{yx}(n,m)$</td><td>$Z_{xy}(n+1,m)$</td></tr>
<tr><td>$T_{yx}(n,m,3)$</td><td>$Z_{yx}(n,m+1)$</td><td>$Z_{xy}(n+1,m)$</td></tr>
<tr><td>$T_{xy}(n,m,4)$</td><td>$Z_{xy}(n+1,m)$</td><td>$Z_{xy}(n,m)$</td></tr>
<tr><td>$(-)T_{yx}(n,m,5)$</td><td>$Z_{yx}(n,m+1)$</td><td>$Z_{xy}(n,m)$</td></tr>
<tr><td>$T_{yx}(n,m,6)$</td><td>$Z_{yx}(n,m)$</td><td>$Z_{xy}(n,m)$</td></tr>
<tr><td>$(-)T_{xy}(n,m,7)$</td><td>$Z_{xy}(n,m)$</td><td>$Z_{yx}(n,m+1)$</td></tr>
<tr><td>$T_{yx}(n,m,8)$</td><td>$Z_{yx}(n,m)$</td><td>$Z_{yx}(n,m+1)$</td></tr>
<tr><td>$T_{xy}(n,m,9)$</td><td>$Z_{xy}(n+1,m)$</td><td>$Z_{yx}(n,m+1)$</td></tr>
<tr><td>$T_{xy}(n,m,10)$</td><td>$Z_{xy}(n,m)$</td><td>$Z_{yx}(n,m)$</td></tr>
<tr><td>$T_{yx}(n,m,11)$</td><td>$Z_{yx}(n,m+1)$</td><td>$Z_{yx}(n,m)$</td></tr>
<tr><td>$(-)T_{xy}(n,m,12)$</td><td>$Z_{xy}(n+1,m)$</td><td>$Z_{yx}(n,m)$</td></tr>
<tr><td colspan="3" align="center">REFLECTION TYPE</td></tr>
<tr><td>$B_{xy}(n,m,1)$</td><td>$Z_{xy}(n+1,m)$</td><td>$Z_{xy}(n+1,m)$</td></tr>
<tr><td>$B_{xy}(n,m,2)$</td><td>$Z_{xy}(n,m)$</td><td>$Z_{xy}(n,m)$</td></tr>
<tr><td>$B_{yx}(n,m,3)$</td><td>$Z_{yx}(n,m+1)$</td><td>$Z_{yx}(n,m+1)$</td></tr>
<tr><td>$B_{yx}(n,m,4)$</td><td>$Z_{yx}(n,m)$</td><td>$Z_{yx}(n,m)$</td></tr>
</table>

in the TLM lines. The issue then is whether to ascribe negative values to these coefficients or whether to simply display the negative signs in the scattering equations. We have selected the latter convention, so that the transfer coefficients are always positive, with the scattering equations expressly displaying the negative signs as appropriate. Under some circumstances, however, we may wish to explicitly identify coefficients which produce a negative wave. To do this we can use the symbol (-) immediately preceding the transfer coefficient, as shown in Tables 3.1 and 3.2. Thus , in Table 3.1, $(-)T_{yx}(n-1,m,2)$ indicates that this coefficient produces a negative polarity wave as it couples from the $Z_{yx}(n-1,m)$ line to the $Z_{xy}(n,m)$ line. To illustrate its use, Section 3.3 employs this notation in discussing symmetry properties among the scattering coefficients. In general, however, we choose not to explicitly display the (-) symbol, except for the following: Section 3.3, Tables 3.1 , 3.2, 3. 4, and the corresponding 3D Tables, 3.5, 3.7, and 3A.2.

The notation described allows us to identify the way in which wave energy is routed from one transmission line to another. The task of actually calculating the scattering coefficients, however, still remains. Toward this goal we define certain preliminary arrays, which relate to an arbitrary node $R(n,m)$. We start by defining four arrays $R1_{xy}(n,m)$, $R2_{yx}(n,m)$, $R3_{xy}(n,m)$, and $R4_{xy}(n,m)$. Each of these is a parallel combination of $R(n,m)$ and the characteristic impedance of one of the four lines(indicated in the expression) surrounding the (n,m) node. The four arrays thus are defined as:

$$R1_{xy}(n,m) = Z_{xy}(n,m)R(n,m)/[Z_{xy}(n,m)+R(n,m)] \tag{3.11}$$

$$R2_{yx}(n,m) = Z_{yx}(n,m)R(n,m)/[Z_{yx}(n,m)+R(n,m)] \tag{3.12}$$

$$R3_{xy}(n,m) = Z_{xy}(n+1,m)R(n,m)/[Z_{xy}(n+1,m)+R(n,m)] \tag{3.13}$$

$$R4_{yx}(n,m) = Z_{yx}(n,m)R(n,m+1)/[Z_{yx}(n,m+1)+R(n,m)] \tag{3.14}$$

The xy and yx subscripts are actually redundant for the 2D node parameters and will be deleted in the remainder of the 2D discussion(for the sake of clarity, however, the subscripts are retained for the scattering coefficients). When we take up the 3D analysis we will resume the display of these subscripts since they are needed to identify the particular scattering plane.

Next we define another set of four arrays, representing the total load impedance seen by each of the four waves (one for each line) incident on the (n,m) node. For example let us consider the load impedance seen by a wave incident on the (n,m) node in the $Z_{xy}(n,m)$ line. The value of such an impedance, denoted by RL1(n,m),will comprise the parallel combination of R(n,m) and the series combination of R4(n,m), R3(n,m), and R2(n,m), as noted in Fig.3.3. Thus,

$$RL1(,n,m)=R(n,m)[R2(n,m)+R3(n,m)+R4(n,m)]/[R2(n,m)+R3(n,m)$$
$$+R4(n,m)+R(n,m)] \qquad (3.15)$$

The load impedances for the other three lines may be calculated in like manner. RL2(n,m), RL3(n,m), and RL4(n,m) are the load impedances viewed by the incident wave in $Z_{yx}(n,m)$, $Z_{xy}(n+1,m)$, and $Z_{yx}(n,m+1)$, respectively. Thus,

$$RL2(,n,m)=R(n,m)[R1(n,m)+R3(n,m)+R4(n,m)]/[R1(n,m)+R3(n,m)$$
$$+R4(n,m)+R(n,m)] \qquad (3.16)$$

$$RL3(,n,m)=R(n,m)[R1(n,m)+R2(n,m)+R4(n,m)]/[R1(n,m)+R2(n,m)$$
$$+R4(n,m)+R(n,m)] \qquad (3.17)$$

$$RL4(,n,m)=R(n,m)[R1(n,m)+R2(n,m)+R3(n,m)]/[R1(n,m)+R2(n,m)$$
$$+R3(n,m)+R(n,m)] \qquad (3.18)$$

Having calculated all the preliminary quantities, we are now in a position to obtain the scattering coefficients listed in Table 3.1. We cite two examples, $T_{xy}(n,m,10)$ and $B_{yx}(n,m-1,3)$. $T_{xy}(n,m,10)$ represents , from the Table, the transfer of the wave from the $Z_{xy}(n,m)$ line to the $Z_{yx}(n,m)$ line. The arguments n,m indicate the transfer occurs at the (n,m) node while the 10 specifies that the transfer is from the $Z_{xy}(n,m,)$ line to the $Z_{yx}(n,m)$. The total load impedance seen by the wave in the $Z_{xy}(n,m)$ line is given by RL1(n,m), provided by Eq.(3.15). Using the standard expression for the transfer coefficient, $T_{xy}(n,m,10)$ may be written as

$$T_{xy}(n,m,10)=\{2RL1(n,m)/[RL1(n,m)+Z_{xy}(n,m)]\}$$
$$*\{R2(n,m)/[R2(n,m)+R3(n,m)+R4(n,m)]\} \qquad (3.19a)$$

where the expression in the second parenthesis represents the share of the voltage transfer to $Z_{yx}(n,m)$, i.e., the ratio of the voltage transfer to $Z_{yx}(n,m)$ to that of the sum of voltage transfers across the three lines $Z_{yx}(n,m)$, $Z_{yx}(n,m+1)$, and $Z_{xy}(n+1,m)$. Next we consider the reflection coefficient, $B_{yx}(n,m-1,3)$. The arguments n,m-1 denote the (n,m-1) node, while the 3 subscript indicates the reflection is for the backward wave in the $Z_{yx}(n,m)$ line. Utilizing the usual expression for the reflection coefficient then gives

$$B_{yx}(n,m-1,3) = [RL4(n,m-1)-Z_{yx}(n,m)]/ [RL4(n,m-1)+Z_{yx}(n,m)] \qquad (3.19b)$$

where RL4(n,m-1) is given by Eq.(3.18), but we replace m with m-1. The complete listing of scattering coefficients is given in Table 3.3, which represents the scattering *about* the node. The node parameters are given in Eqs.(3.11)-(3.14) and (3.15)-(3.18). When used in the scattering equations, the proper index is inserted in both the scattering coefficients and the node parameters.

Having enumerated the scattering coefficients, we are now in a position to write down the fields, at time t, in the lines $Z_{xy}(n,m)$, $Z_{yx}(n,m)$, in terms of the fields of the previous time step. Starting with the forward wave in $Z_{xy}(n,m)$, during the kth time step, the voltage is (we now use the k superscript to indicate the time step)

$$^{+}V^{k+1}{}_{xy}(n,m) = T^{k}{}_{xy}(n-1,m,1) \, {}^{+}V^{k}{}_{xy}(n-1,m) - T^{k}{}_{yx}(n-1,m,2) \, {}^{+}V^{k}{}_{yx}(n-1,m)$$
$$+ T^{k}{}_{yx}(n-1,m,3) \, {}^{-}V^{k}{}_{yx}(n-1,m+1) + B^{k}{}_{xy}(n-1,m,1) \, {}^{-}V^{k}{}_{xy}(n,m) \qquad (3.20)$$

We now explicitly describe each term in Eq.(3.20). The first, $T^{k}{}_{xy}(n-1,m,1)$ $^{+}V^{k}{}_{xy}(n-1,m)$ represents that portion of the forward wave, in the $Z_{xy}(n-1,m)$ line, that is transferred to the $Z_{xy}(n,m)$ line.The second term, $T^{k}{}_{yx}(n-1,m,2) {}^{+}V^{k}{}_{yx}(n-1,m)$ represents the energy coupled from the forward wave in the vertical $Z_{yx}(n-1,m)$ line to the $Z_{xy}(n,m)$ line, via the (n-1,m) node. A negative sign is present here since the $^{+}V^{k}{}_{yx}(n-1,m)$ wave couples to the $Z_{xy}(n,m)$ in the negative direction. The third , $T^{k}{}_{yx}(n-1,m,3) \, {}^{-}V^{k}{}_{yx}(n-1,m+1)$, corresponds to the energy transfer of a backward wave in the vertical $Z_{yx}(n-1,m+1)$ line to the $Z_{xy}(n,m)$ line. Finally the term $B^{k}{}_{xy}(n-1,m,1) \, {}^{-}V^{k}{}_{xy}(n,m)$ represents the reflection of the backward

TABLE 3.3 2D SCATTERING COEFFICIENT EXPRESSIONS ABOUT (n,m) NODE

$Txy(n, m, 1) = 2*RL1(n, m)*R3(n, m) / ((RL1(n, m)+Zxy(n, m))*(R2(n, m)+R3(n, m) + R4(n, m)))$

$Tyx(n, m, 2) = 2*RL2(n, m)*R3(n, m) /((RL2(n, m)+Zyx(n, m))*(R1(n, m) + R3(n, m) + R4(n, m)))$

$Tyx(n, m, 3) = 2*RL4(n,m)*R3(n-1,m)/((RL4(n, m)+Zyx(n,m+1)) * (R1(n,m) + R2(n, m) + R3(n, m)))$

$Txy(n, m, 4) = 2*RL3(n, m)*R1(n, m)/((RL3(n,m)+Zxy((n+1),m))*(R1(n, m)+R2(n, m)+R4(n,m)))$

$Tyx(n,m,5) = 2*RL4(n,m)*R1(n,m)/((RL4(n,m)+Zyx(n,(m+1)))*(R1(n,m)+R2(n, m) + R3(n, m)))$

$Tyx(n, m, 6) = 2*RL2(n, m)*R1(n, m) /((RL2(n, m)+Zxy(n, m))*(R1(n, m)+R3(n, m) + R4(n, m)))$

$Txy(n, m, 7) = 2*RL1(n, m)*R4(n, m)/((RL1(n, m)+Zxy(n, m)) * (R2(n,m) + R3(n, m) + R4(n, m)))$

$Tyx(n, m, 8) = 2* RL2(n, m)*R4(n, m)/((RL2(n, m)+Zyx(n, m))* (R1(n, m) + R3(n, m) + R4(n, m)))$

$Txy(n, m, 9) = 2*RL3(n, m)*R4(n, m)/((RL3(n, m)+Zxy((n +1),m))*(R1(n, m) + R2(n, m) + R4(n, m)))$

$Txy(n, m, 10) = 2*RL1(n, m)*R2(n, m)/((RL1(n, m)+Zxy(n, m))* (R2(n,m) +R3(n, m) + R4(n, m)))$

$Tyx(n, m, 11) = 2*RL4(n, m)*R2(n,m)/((RL4(n, m)+Zyx(n,(m+1)))*(R1(n, m) + R2(n, m)+R3(n, m)))$

$Txy(n, m, 12) = 2 *RL3(n,m)*R2(n,m)/((RL3(n,m)+Zxy((n+1),m))*(R1(n, m)+R2(n, m)+R4(n, m)))$

$Bxy(n, m, 1) = (RL3(n, m) - Zxy(n+1, m)) / (RL3(n, m) + Zxy(n+1, m))$

$Bxy(n, m, 2) = (RL1(n, m) - Zxy(n, m)) / (RL1(n, m) + Zxy(n, m))$

$Byx(n, m, 3) = (RL4(n, m) - Zyx(n,m+1)) / (RL4(n, m) + Zyx(n ,m+1))$

$Byx(n, m, 4) = (RL2(n, m) - Zyx(n,m)) / (RL2(n, m) + Zyx(n,m))$

backward wave in the $Z_{xy}(n,m)$ line, occurring at the $(n-1,m)$ node. Similar relationships may be expressed for the backward wave in the $Z_{xy}(n,m)$ line, as well as both the forward and backward waves in the $Z_{yx}(n,m)$ line. Thus,

$$^-V^{k+1}{}_{xy}(n,m) = T^k{}_{xy}(n,m,4)\ ^-V^k{}_{xy}(n+1,m) - T^k{}_{yx}(n,m,5)\ ^-V^k{}_{yx}(n,m+1)$$
$$+T^k{}_{yx}(n,m,6)\ ^+V^k{}_{yx}(n,m)+B^k{}_{xy}(n,m,2)\ ^+V^k{}_{xy}(n,m) \qquad (3.21)$$

$$^+V^{k+1}{}_{yx}(n,m) = -T^k{}_{xy}(n,m-1,7)\ ^+V^k{}_{xy}(n,m-1)+T^k{}_{yx}(n,m-1,8)\ ^+V^k{}_{yx}(n,m-1)$$
$$+ T^k{}_{xy}(n,m-1,9)\ ^-V^k{}_{xy}(n+1,m-1)+B^k{}_{yx}(n,m-1,3)\ ^-V^k{}_{yx}(n,m) \qquad (3.22)$$

$$^-V^{k+1}{}_{yx}(n,m) = T^k{}_{xy}(n,m,10)\ ^+V^k{}_{xy}(n,m) + T^k{}_{yx}(n,m+1,11)\ ^-V^k{}_{yx}(n,m+1)$$
$$-T^k{}_{xy}(n,m,12)\ ^-V^k{}_{xy}(n+1,m) + B^k{}_{yx}(n,m,4)\ ^+V^k{}_{yx}(n,m) \qquad (3.23)$$

Eqs.(3.20)-(3.23) are the key set of iterative relations which completely determine the behavior of the medium. If the fields throughout the medium are known at a particular time, then the evolution of the fields following that moment in time are determined by Eqs.(3.20)-(3.23). The most obvious starting time, of course, is during equilibrium , i.e., when the static solution is presumed known. As pointed out before, the node resistors during equilibrium are extremely large and the waves in each line segment are totally reflected at each node, which behaves as an open circuit. The amplitude of the forward and backward waves   in each line is then exactly half the equilibrium voltage difference between cells. The transient solution follows once the node resistance is activated.

In addition to solving transient problems, the iterative equations also may be used to seek out the static solution, i.e, the solution of Laplaces' equation. . As we will see in Chapter VII, the iterative equations produce the static solution when we charge up the electrodes, by means of long input transmission lines. The iteration therefore covers the full gamut of the various states experienced by the medium, from the transient "charge-up" process(during which time the nodes are unactivated), the establishment of the equilibrium state, and the activation of the conductivity, followed by recovery.

3.3 Effect of Symmetry on Scattering Coefficients[*]

As one might surmise, symmetry introduces tremendous simplifications in the scattering coefficients. Before considering the general case, in which node losses are present, we first consider 2D symmetries present when $R(n,m)=\infty$, i.e., when losses are absent. As an example, we examine $T_{xy}(n,m,1)$, which describes the wave transfer from $Z_{xy}(n,m)$ to $Z_{xy}(n+1,m)$. We obtain $T_{xy}(n,m,1)$ using Table 3.3 , with load impedance $RL1(n,m) = Z_{yx}(n,m) + Z_{yx}(n,m+1) + Z_{xy}(n+1,m)$, and with node parameters $R3(n,m)=Z_{xy}(n+1,m)$, $R2(n,m)= Z_{yx}(n,m)$,   and $R4(n,m)=Z_{yx}(n,m+1)$. This then gives

$$T_{xy}(n,m,1)=(2Z_{xy}(n+1,m))/[Z_{yx}(n,m)+Z_{xy}(n,m)+Z_{yx}(n,m+1)+Z_{xy}(n+1,m)] \quad (3.24)$$

An important point to note about Eq.(3.24) is that $Z_{xy}(n+1,m)$ is divided by the sum of all four line impedances surrounding the node, a symmetric function, which means that the same transfer coefficient is obtained when the scattering is from $Z_{yx}(n,m)$ or $Z_{yx}(n,m+1)$, represented by $T_{yx}(n,m,3)$ and $(-)T_{yx}(n,m,2)$. Thus

$$T_{xy}(n,m,1)=T_{yx}(n,m, 3)= (-)T_{yx}(n,m,2) \qquad (3.25a)$$

Here we illustrate the use of the (-) symbol , placing it in front of $T_{yx}(n,m,2)$ to indicate that the wave scattered from $Z_{yx}(n,m)$ into $Z_{xy}(n+1,m)$ is inverted. Eq.(3.25a) is a result of the symmetries ensuing from any contributions to the forward wave in the $Z_{xy}(n+1,m)$ line. Similarly, we can state the symmetries for the backward wave in $Z_{xy}(n,m)$, the forward wave in $Z_{yx}(n,m+1)$ and the backward wave in $Z_{yx}(n,m)$. This gives the result

$$T_{xy}(n,m,4)=T_{yx}(n,m, 6)= (-)T_{yx}(n,m,5) \qquad (3.25b)$$

$$T_{yx}(n,m,8)= T_{xy}(n,m,9)= (-)T_{xy}(n,m,7) \qquad (3.25c)$$

$$T_{yx}(n,m,11)= T_{xy}(n,m,10)= (-)T_{xy}(n,m,12) \qquad (3.25d)$$

---

[*] In this Section the symbol(-) is used to designate negative wave transfer

Without node loses, therefore, the number of transfer coefficients may be reduced from twelve to four. We stress however that these symmetries apply only to the 2D case, and they do not apply to the generalized 3D scattering(which includes scattering normal to the propagation plane), to be discussed later this Chapter

We now turn our attention to the symmetries present in 2D scattering when losses are present. In this situation the degree of symmetry will be determined by the degree of equality of the four TLM lines surrounding the (n,m) node. The results are tabulated in Table 3.4, which considers scattering about the node. The Table, it should be emphasized, illustrates the symmetrization for specific line equalities. The first grouping indicates the symmetry for $Z_{xy}(n,m) = Z_{xy}(n+1,m)$. We see that in this case the number of transfer coefficients is reduced from twelve to seven, and the number of reflection coefficients from four to three. To illustrate how the relationships are obtained, we look at the second entry, which states that $T_{yx}(n,m,3) = (-)T_{yx}(n,m,5)$. From Table 3.2 we see that $T_{xy}(n,m,3)$ represents the wave transfer from $Z_{yx}(n+1,m)$ while $T_{yx}(n,m,5)$ applies to the wave transfer from $Z_{yx}(n,m)$. Since $Z_{xy}(n,m) = Z_{xy}(n+1,m)$, however, the calculation of the load impedances , as well as the voltage divisions applicable to $Z_{xy}(n,m)$ and $Z_{xy}(n+1,m)$, will be identical. Thus $T_{yx}(n,m,3)$ and $T_{yx}(n,m,5)$ will be identical. The $(-)$ in front of $T_{xy}(n,m,5)$, as mentioned before, is used to indicate the inverted coupling of $Z_{yx}(n,m+1)$ to $Z_{xy}(n,m,)$. We again call attention to the arbitrary selection of the two equal TLM lines. Any other pair of lines, surrounding the node, could have been selected (such as $Z_{xy}(n,m)$ and $Z_{yx}(n,m+1)$), which will then generate a different set of symmetry relationships.   In   the   next grouping of symmetry relationships, we assume, arbitrarily, that the three lines $Z_{xy}(n,m)$, $Z_{yx}(n,m,+1)$, and $Z_{xy}(n+1,m)$ are identical As shown in the Table, this then reduces the number of independent trasfer coefficients from twelve to three (designated in Table 3.4 by $T_I(n,m)$, $T_{II}(n,m)$, $T_{III}(n,m)$) and the number of reflection coefficients from four to two(designated by $B_I(n,m)$ and $B_{yx}(n,m)$).

The simplifications due to symmetry are most apparent when all the lines are equal, i.e., when there is complete uniformity. In this case the twelve scat-

TABLE 3.4 SYMMETRY PROPERTIES OF 2D COEFFICIENTS (XY PLANE) FOR SPECIFIC TLM LINE EQUALITIES. SCATTERING IS ABOUT (n,m) NODE

A. TWO LINES EQUAL: $Z_{xy}(n,m) = Z_{xy}(n+1,m)$

$$T_{xy}(n,m,1) = T_{xy}(n,m,4)$$
$$T_{yx}(n,m,3) = (-)T_{yx}(n,m,5)$$
$$T_{xy}(n,m,9) = (-)T_{xy}(n,m,7)$$
$$T_{xy}(n,m,10) = (-)T_{xy}(n,m,12)$$
$$T_{yx}(n,m,6) = (-)T_{yx}(n,m,2)$$
$$B_{xy}(n,m,1) = B_{xy}(n,m,2), B_{yx}(n,m,3), B_{yx}(n,m,4)$$

B. THREE LINES EQUAL: $Z_{xy}(n,m) = Z_{xy}(n+1,m) = Z_{yx}(n,m+1)$

$$T_{xy}(n,m,1) = T_{yx}(n,m,3) = T_{xy}(n,m,4) = T_{xy}(n,m,9) \equiv T_I(n,m)$$
$$(-)T_{yx}(n,m,5) = (-)T_{xy}(n,m,7) = (-)T_I(n,m)$$
$$T_{yx}(n,m,6) = T_{yx}(n,m,8) \equiv T_{II}(n,m) , \ (-)T_{yx}(n,m,2) = T_{II}(n,m)$$
$$T_{xy}(n,m,10) = T_{yx}(n,m,11) \equiv T_{III}(n,m) , (-)T_{xy}(n,m,12) = (-)T_{III}(n,m)$$
$$B_{xy}(n,m,1) = B_{xy}(n,m,2) = B_{yx}(n,m,3) \equiv B_I(n,m) , B_{yx}(n,m,4)$$

C. ALL LINES EQUAL: $Z_{xy}(n,m) = Z_{xy}(n+1,m) = Z_{yx}(n,m) = Z_{yx}(n,m+1)$

$$T_{xy}(n,m,1) = T_{yx}(n,m,3) = T_{xy}(n,m,4) = T_{yx}(n,m,6) = T(n,m)$$
$$T_{yx}(n,m,8) = T_{xy}(n,m,9) = T_{xy}(n,m,10) = T_{yx}(n,m,11) = T(n,m)$$
$$(-)T_{yx}(n,m,2) = (-)T_{yx}(n,m,5) = (-)T_{xy}(n,m,7) = (-)T_{xy}(n,m,12) = (-)T(n,m)$$
$$B_{xy}(n,m,1) = B_{xy}(n,m,2) = B_{yx}(n,m,3) = B_{yx}(n,m,4) = B(n,m)$$

tering coefficients are all equal to one another, thus reducing the number of different transfer scattering coefficients to one, $T(n,m)$. Similarly, the four reflection coefficients are all equal, so that only a single reflection coefficient, $B(n,m)$, is needed.

3.4 3D Scattering Equations: Coplanar Scattering

We first start with the 3D case in which the scattered fields remain in the propagation plane, which of course is equivalent to the 2D matrix done previously. Since we have already obtained the scattering coefficients and equations which pertain to the 2D matrix , the results carry over to the coplanar 3D without the necessity of re-deriving equations. The main changes will be the addition of a third index to denote the extra dimension as well as in the interpretation of the results. Once these results are obtained for a particular scattering plane, the equations for the two other planes may be obtained by simple permutation. The 3D coplanar solution is by and large a 2D solution dressed in the clothing of 3D variables. The techniques are primarily useful for 2D and quasi 2D geometries, and where we may ignore scattering normal to the propagation plane.

The simplest approach is to divide the matrix into the three propagation planes, xy,yz, and zx. We first consider scattering in the xy plane As we noted previously in Chapter 2, the transmission line impedances and voltages for the xy plane are denoted by $Z_{xy}(n,m,q)$, $Z_{yx}(n,m,q)$ and $V_{xy}(n,m,q)$, $V_{yx}(n,m,q)$ respectively, while the node resistance is given by $R(n,m,q)$. As mentioned before, the first subscript refers to the direction of propagation while the second refers to the field direction. The q index, of course, represents the added z dimension. The procedure in the previous section is followed , using Fig.3.3 as a guide, and as before we first define a four element auxiliary array, consisting of the parallel combination of the node resistance and the characteristic impedance of each of the four lines. The four elements, denoted by $R1_{xy}(n,m,q)$, $R2_{yx}(n,m,q)$ $R3_{xy}(n,m,q)$, $R4_{yx}(n,m,q)$ are :

$$R1_{xy}(n,m,q) = Z_{xy}(n,m,q)R(n,m,q)/[Z_{xy}(n,m,q)+R(n,m,q)] \qquad (3.26a)$$

$$R2_{yx}(n,m,q) = Z_{yx}(n,m,q)R(n,m,q)/[Z_{yx}(n,m,q)+R(n,m,q)] \qquad (3.26b)$$

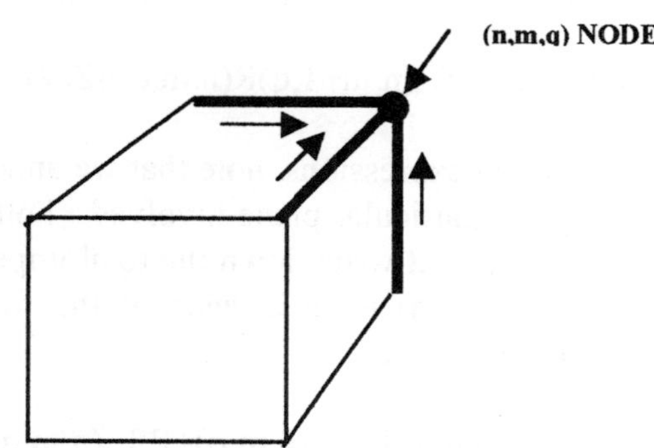

FIG. 3.3(a) ELEMENTARY CELL OF TLM MATRIX

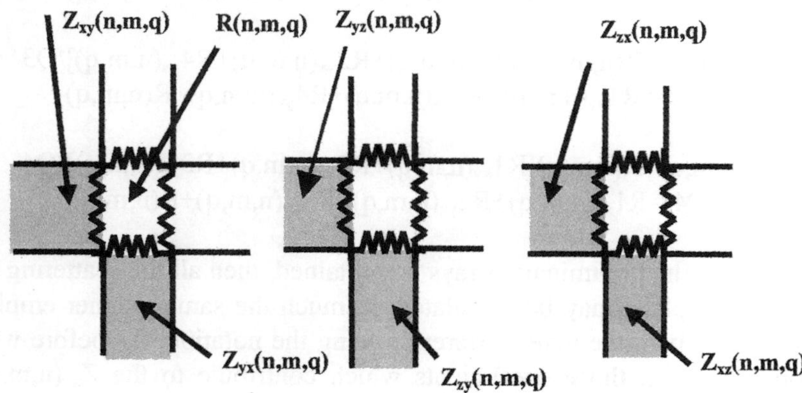

FIG. 3.3b PROJECTION OF 3D TLM MATRIX ONTO 2D GRIDS. THE ABOVE FACILITATES THE CALCULATION OF THE SCATTERING COEFFICIENTS.

$$R3_{xy}(n,m,q) = Z_{xy}(n+1,m,q)R(n,m,q)/[Z_{xy}(n+1,m,q)+R(n,m,q)] \qquad (3.26c)$$

$$R4_{yx}(n,m,q) = Z_{yx}(n,m+1,q)R(n,m,q)/[Z_{yx}(n,m+1,q)+R(n,m,q)] \qquad (3.26d)$$

Unlike the 2D expressions, note that we attach the subscripts xy ,etc.. in order to identify the particular plane involved. Following exactly the same procedure as before, we next write down the total impedance value seen by the wave incident on the (n,m,q) node in each of the four transmission lines. Using a similar notation as before,

$$RL1_{xy}(n,m,q) =R(n,m,q)[R2_{yx}(n,m,q)+R3_{xy}(n,m,q)+R4_{yx}(n,m,q)]/Q1$$
$$Q1=R2_{yx}(n,m,q)+R3_{xy}(n,m,q)+R4_{yx}(n,m,q)+R(n,m,q) \qquad (3.27)$$

$$RL2_{yx}(n,m,q)=R(n,m,q)[R1_{xy}(n,m,q)+R3_{xy}(n,m,q)+R4_{yx}(n,m,q)]/Q2$$
$$Q2=R1_{xy}(n,m,q)+R3_{xy}(n,m,q)+R4_{yx}(n,m,q)+R(n,m,q) \qquad (3.28)$$

$$RL3_{xy}(n,m,q)=R(n,m,q)[R1_{xy}(n,m,q)+R2_{yx}(n,m,q)+R4_{yx}(n,m,q)]/Q3$$
$$Q3= R1_{xy}(n,m,q)+R2_{yx}(n,m,q)+R4_{yx}(n,m,q)+R(n,m,q) \qquad (3.29)$$

$$RL4_{yx}(n,m,q)=R(n,m,q)[R1_{xy}(n,m,q)+R2_{yx}(n,m,q)+R3_{xy}(n,m,q)]/Q4$$
$$Q4= R1_{xy}(n,m,q)+R2_{yx}(n,m,q)+R3_{xy}(n,m,q)+R(n,m,q) \qquad (3.30)$$

Once the preliminary arrays are obtained, then all the scattering coefficients for the xy plane may be calculated, in much the same manner employed in the 2D matrix with the main difference being the notation. As before we need only concentrate on those coefficients which contribute to the $Z_{xy}(n,m,q)$ and $Z_{yx}(n,m,q)$ lines. Table 3.5 lists the scattering coefficients for the xy plane. To help with the understanding of the notation, we use the transfer coefficient $T_{xy}(n-1,m,q,1)$ as an example, given by

$$T_{xy}(n-1,m,q,1)=[2RL1_{xy}(n-1,m,q,1)Z_{xy}(n-1,m,q)/[RL1xy(n-1,m,q)$$
$$+Z_{xy}(n-1,m,q)]]*S \qquad (3.31a)$$
$$S = R3_{xy}(n-1,m,q)/[R2_{yx}(n-1,m,q)+R3_{xy}(n-1,m,q)+R4_{yx}(n-1,m,q)] \qquad (3.31b)$$

TABLE 3.5 3D COPLANAR SCATTERING COEFFICIENTS <u>INTO</u> UNIT CELL LINES $Z_{xy}(n,m,q)$ and $Z_{yx}(n,m,q)$

TRANSFER TYPE

<u>COEFFICIENT</u>	<u>FROM</u>	<u>TO</u>
$T_{xy}(n-1,m,q,1)$	$Z_{xy}(n-1,m,q)$	$Z_{xy}(n,m,q)$
$(-)T_{yx}(n-1,m,q,2)$	$Z_{yx}(n-1,m,q)$	$Z_{xy}(n,m,q)$
$T_{yx}(n-1,m,q,3)$	$Z_{yx}(n-1,m+1,q)$	$Z_{xy}(n,m,q)$
$T_{xy}(n,m,q,4)$	$Z_{xy}(n+1,m,q)$	$Z_{xy}(n,m,q)$
$(-)T_{yx}(n,m,q,5)$	$Z_{yx}(n,m+1,q)$	$Z_{xy}(n,m,q)$
$T_{yx}(n,m,q,6)$	$Z_{yx}(n,m,q)$	$Z_{xy}(n,m,q)$
$(-)T_{xy}(n,m-1,q,7)$	$Z_{xy}(n,m-1,q)$	$Z_{yx}(n,m,q)$
$T_{yx}(n,m-1,q,8)$	$Z_{yx}(n,m-1,q)$	$Z_{yx}(n,m,q)$
$T_{xy}(n,m-1,q,9)$	$Z_{xy}(n+1,m-1,q)$	$Z_{yx}(n,m,q)$
$T_{xy}(n,m,q,10)$	$Z_{xy}(n,m,q)$	$Z_{yx}(n,m,q)$
$T_{yx}(n,m,q,11)$	$Z_{yx}(n,m+1,q)$	$Z_{yx}(n,m,q)$
$(-)T_{xy}(n,m,q,12)$	$Z_{xy}(n+1,m,q)$	$Z_{yx}(n,m,q)$

REFLECTION TYPE

$B_{xy}(n-1,m,q,1)$	$Z_{xy}(n,m,q)$	$Z_{xy}(n,m,q)$
$B_{xy}(n,m,q,2)$	$Z_{xy}(n,m,q)$	$Z_{xy}(n,m,q)$
$B_{yx}(n,m-1,q,3)$	$Z_{yx}(n,m,q)$	$Z_{yx}(n,m,q)$
$B_{yx}(n,m,q,4)$	$Z_{yx}(n,m,q)$	$Z_{yx}(n,m,q)$

where $R1_{xy}(n,m,q)$, etc.., and $RL1_{xy}(n,m,q)$ are given by Eqs.(3.26)-(3.30) with n=n-1 The first three arguments of the coefficient, n-1,m,q , identify the node responsible for the scattering. xy indicates that the propagation is in the x direction, while the direction of the field is along y. Finally, the fourth argument, 1, indicates that the wave incident on the node emanates from the $Z_{xy}(n-1,m,q,)$ line and a portion of the wave is being transferred to the $Z_{xy}(n,m,q)$ line. We select the reflection coefficient $B_{xy}(n-1, m,q, 1)$ as another example. We see from the Table 3.5 that this represents the backward wave in the $Z_{xy}(n,m,q)$ line, which is then reflected at the (n-1,m,q) node and becomes a forward wave in the same line. The expression for $B_{xy}(n-1,m,q,1)$ is

$$B_{xy}(n-1,m,q)=[RL3_{xy}(n-1,m,q)-Z_{xy}(n,m,q)]/[RL3_{xy}(n-1,m,q)+Z_{xy}(n,m,q)] \quad (3.32)$$

where $RL3_{xy}(n,m,q)$ is given by Eq.(3.29) with n=n-1. The interpretation of the other coefficients in Table 3.5   is similar.

Having specified the scattering coefficients for the xy plane of the 3D matrix, we are now equipped to write down the iterative scattering equations which, except for the notation, will be very similar to Eqs.(3.20)-(3.23) of the 2D matrix. Fig.3.2 represents the xy portion of the cubic cell in Fig.3.4. As before, we focus on only two of the lines, $Z_{xy}(n,m,q)$, and $Z_{yx}(n,m,q)$, since these are the two lines associated with the (n,m,q) cell. The forward and backward waves in each of these lines, at time $t+\Delta t$, may be expressed in terms of waves existing in the surrounding lines at time t. At the risk of some redundancy we display the iterative equations, which are identical to Eqs.(3.20)-(3.23) except for the addition of the third index q.

$$^{+}V^{k+1}_{xy}(n,m)=T^{k}_{xy}(n-1,m,q,1)\,{}^{+}V^{k}_{xy}(n-1,m,q) - T^{k}_{yx}(n-1,m,q,2)\,{}^{+}V^{k}_{yx}(n-1,m,q)$$
$$+T^{k}_{yx}(n-1,m,q,3)\,{}^{-}V^{k}_{yx}(n-1,m+1,q)+B^{k}_{xy}(n-1,m,q,1)\,{}^{-}V^{k}_{xy}(n,m,q) \quad (3.33)$$

$$^{-}V^{k+1}_{xy}(n,m,q) = T^{k}_{xy}(n,m,q,4)\,{}^{-}V^{k}_{xy}(n+1,m,q) - T^{k}_{yx}(n,m,q,5)\,{}^{-}V^{k}_{yx}(n,m+1,q)$$
$$T^{k}_{yx}(n,,q,6)\,{}^{+}V^{k}_{yx}(n,m,q) + B^{k}_{xy}(n,m,q,2)\,{}^{+}V^{k}_{xy}(n,m,q) \quad (3.34)$$

$$^{+}V^{k+1}_{yx}(n,m,q)=-T^{k}_{xy}(n,m-1,q,7)\,{}^{+}V^{k}_{xy}(n,m-1.q)+T^{k}_{yx}(n,m-1,q,8)\,{}^{+}V^{k}_{yx}(n,m-1,q)$$
$$T^{k}_{xy}(n,m-1,q,9)\,{}^{-}V^{k}_{xy}(n+1,m-1,q)+B^{k}_{yx}(n,m-1,q,3)\,{}^{-}V^{k}_{yx}(n,m,q) \quad (3.35)$$

$$^-V^{k+1}{}_{yx}(n,m,q) = T^k{}_{xy}(n,m,q,10)\, {}^+V^k{}_{xy}(n,m,q) + T^k{}_{yx}(n,m+1,q,11)\, {}^-V^k{}_{yx}(n,m+1,q)$$
$$-T^k{}_{xy}(n,m,q,12)\, {}^-V^k{}_{xy}(n+1,m,q) + B^k{}_{yx}(n,m,q,4)\, {}^+V^k{}_{yx}(n,m,q) \qquad (3.36)$$

Eqs.(3.33)-(3.36), together with Table 3.5, complete about one third of the task needed to fully describe the 3D matrix. The listing of the scattering coefficients is identical to that in Table 3.3 , if we suppress the q index. Similar equations and Tables must be obtained for the yz and zx planes. Fortunately the remaining task is facilitated a great deal by symmetry conditions. For the yz plane, the transformation $x \to y$, $y \to z$, and $z \to x$ enables one to write down similar scattering coefficients and iterative equations and the Table for the scattering coefficients. In a similar manner we use transformation $x \to z$, $y \to x$, and $z \to y$ for the zx plane. The transformation properties are summarized in Table 3.6a. Extra care must be exercised when transforming wave voltages in the iterative equations, or scattering coefficients, containing the node arguments (n,m,q). Although the node location (n,m,q) does not change under transformation, any fields or coefficients which contain the indices ± 1 exhibit obvious changes during transformation. Thus , for example, in going from the xy plane to the yz plane, $V_{xy}(n-1,m,q) \to V_{yz}(n,m-1,q)$ and similarly $T_{xy}(n-1,m,q,1) \to T_{yz}(n,m-1,q,13)$. The transformation properties may be written in a more general form. Thus, for example, $V_{xy}(n+\Delta n, m+\Delta m, q+\Delta q) \to V_{yz}(n+\Delta q, m+\Delta n, q+\Delta m)$ as noted in Table 3.6, where Δm , Δn , Δq may be zero or plus or minus one. Also note the changes in the routing index s. When going from the xy plane to the yz plane the transfer coefficients change according to $s \to s+12$, while for the zx plane $s \to s+24$. In the case of the reflection coefficients $s \to s+4$ and $s \to s+8$ for the yz and zx planes respectively. Also note in Table 3.6b that the node parameters (e.g., $R1_{xy}(n-1,m,q)$, $RL1_{xy}(n-1,m,q)$) transform to the yz and zx planes in the same manner as the scattering coefficients. This completes the coplaner results, which is essentially a generalization of the 2D discussion. Next, we examine the 3D case, in which we allow scattering to any of the three spatial directions.

TABLE 3.6a. 3D COPLANAR SCATTERING COEFFICIENTS AND WAVE VOLTAGES FOR YZ AND ZX PLANES

SPECIAL CASE: Δn, Δm, Δq =0

__XY PLANE__		__YZ PLANE__	__ZX PLANE__
X	$\rightarrow$	Y	Z
Y	$\rightarrow$	Z	X
Z	$\rightarrow$	X	Y
$V_{xy}(n,m,q)$	$\rightarrow$	$V_{yz}(n,m,q)$	$V_{zx}(n,m,q)$
$V_{yx}(n,m,q)$	$\rightarrow$	$V_{zy}(n,m,q)$	$V_{xz}(n,m,q)$
$T_{xy}(n,m,q,s)$	$\rightarrow$	$T_{yz}(n,m,q,s+12)$	$T_{zx}(n,m,q,s+24)$
$T_{yx}(n,m,q,s$	$\rightarrow$	$T_{zy}(n,m,q,s+12)$	$T_{xz}(n,m,q,s+24)$
$B_{xy}(n,m,q,s)$	$\rightarrow$	$B_{yz}(n,m,q,s+4)$	$B_{zx}(n,m,q,s+8)$
$B_{yx}(n,m,q,s)$	$\rightarrow$	$B_{zy}(n,m,q,s+4)$	$B_{xz}(n,m,q,s+8)$

GENERAL TRANSFORMATION: Δn, Δm, Δq =0 or ± 1

$$V_{xy}(n+\Delta n, m+\Delta m, q+\Delta q) \rightarrow V_{yz}(n+\Delta q, m+\Delta n, q+\Delta m) \qquad V_{zx}(n+\Delta m, m+\Delta q, q+\Delta n)$$

$$V_{yx}(n+\Delta m, m+\Delta q, q+\Delta n) \rightarrow V_{zy}(n+\Delta q, m+\Delta n, q+\Delta m) \qquad V_{xz}(n+\Delta m, m+\Delta q, q+\Delta n)$$

$$T_{xy}(n+\Delta n, m+\Delta m, q+\Delta q, s) \rightarrow T_{yz}(n+\Delta q, m+\Delta n, q+\Delta m, s+12) \quad T_{zx}(n+\Delta m, m+\Delta q, q+\Delta n, s+24)$$

$$T_{yx}(n+\Delta n, m+\Delta m, q+\Delta q, s) \rightarrow T_{zy}(n+\Delta q, m+\Delta n, q+\Delta m, s+12) \quad T_{xz}(n+\Delta m, m+\Delta q, q+\Delta n, s+24)$$

$$B_{xy}(n+\Delta n, m+\Delta m, q+\Delta q, s) \rightarrow B_{yz}(n+\Delta q, m+\Delta n, q+\Delta m, s+4) \quad B_{zx}(n+\Delta m, m+\Delta q, q+\Delta n, s+8)$$

$$B_{yx}(n+\Delta n, m+\Delta m, q+\Delta q, s) \rightarrow B_{zy}(n+\Delta q, m+\Delta n, q+\Delta m, s+4) \quad B_{xz}(n+\Delta m, m+\Delta q, q+\Delta n, s+8)$$

TABLE 3.6b   3D COPLANAR TRANSFORMATION OF NODE PARAMETERS

GENERAL TRANSFORMATION : $\Delta n, \Delta m, \Delta q = 0$ or ± 1

__XY PLANE__	__YZ PLANE__	__ZX PLANE__
$RL1_{xy}(n+\Delta n, m+\Delta m, q+\Delta q) \rightarrow RL1_{yz}(n+\Delta q, m+\Delta n, q+\Delta m)$		$RL1_{zx}(n+\Delta m, m+\Delta q, q+\Delta n)$
$RL3_{xy}(n+\Delta n, m+\Delta m, q+\Delta q) \rightarrow RL3_{yz}(n+\Delta q, m+\Delta n, q+\Delta m)$		$RL3_{zx}(n+\Delta m, m+\Delta q, q+\Delta n)$
$RL2_{yx}(n+\Delta n, m+\Delta m, q+\Delta q) \rightarrow RL2_{zy}(n+\Delta q, m+\Delta n, q+\Delta m)$		$RL2_{xz}(n+\Delta m, m+\Delta q, q+\Delta n)$
$RL4_{yx}(n+\Delta n, m+\Delta m, q+\Delta q) \rightarrow RL4_{zy}(n+\Delta q, m+\Delta n, q+\Delta m)$		$RL4_{xz}(n+\Delta m, m+\Delta q, q+\Delta n)$
$R1_{xy}(n+\Delta n, m+\Delta m, q+\Delta q) \rightarrow R1_{yz}(n+\Delta q, m+\Delta n, q+\Delta m)$		$R1_{zx}(n+\Delta m, m+\Delta q, q+\Delta n)$
$R3_{xy}(n+\Delta n, m+\Delta m, q+\Delta q) \rightarrow R3_{yz}(n+\Delta q, m+\Delta n, q+\Delta m)$		$R3_{zx}(n+\Delta m, m+\Delta q, q+\Delta n)$
$R2_{yx}(n+\Delta n, m+\Delta m, q+\Delta q) \rightarrow R2_{zy}(n+\Delta q, m+\Delta n, q+\Delta m)$		$R2_{xz}(n+\Delta m, m+\Delta q, q+\Delta n)$
$R4_{yx}(n+\Delta n, m+\Delta m, q+\Delta q) \rightarrow R4_{zy}(n+\Delta q, m+\Delta n, q+\Delta m)$		$R4_{xz}(n+\Delta m, m+\Delta q, q+\Delta n)$

General Scattering , Including Scattering Normal to Propagation Plane

Until this point the discussion has concentrated exclusively on scattered waves which remain in the same propagation plane as the incident wave. We remind ourselves that the propagation plane is formed by the direction of propagation and by the field direction. We illustrate the directional behavior of the waves by looking at the by now familiar Fig.3.4(a). A wave $^{+}V_{xy}(n,m,q)$ is incident on the $R(n,m,q)$ node.   The scattered waves are then $^{+}V_{xy}(n+1,m,q)$, $^{-}V_{yx}(n,m,q)$, $^{+}V_{yx}(n,m+1,q)$, and $^{-}V_{xy}(n,m,q)$. As noted before, the scattered

waves are all in the xy plane, as is the incident wave, i.e., the scattering is coplanar. This type scattering is incomplete, however, and in fact is contradictory to
our experience with scattered electromagnetic waves; one also should expect
waves scattered normal to the propagation plane. This notion may be reinforced
using several different arguments. First and foremost, of course, the scattering
must be consistent with Maxwell's equations. As we observed in Section 1.16,
we required normal scattering in order to complete the curl property of Maxwell's equations. Conversely if we omit normal scattering, then we are con-
fronted with unphysical gaps in the magnetic field loops (and the same for the
displacement field) which are at variance with Maxwell's equations.

There are also related arguments for normal scattering, based on the self consistency of the TLM formulation. One such argument involves looking at the magnetic field vector associated with $^{+}V_{xy}(n,m,q)$. denoted by $^{+}H_{xz}(n,m,q)$, which of
course is perpendicular to $^{+}V_{xy}(n,m,q)$. and is in the xz plane. At the node we
make the same assumption regarding the magnetic field as that for the electric
field. For sufficiently small cell size, the magnetic field at the node, which includes any reflected wave, is equal to the sum of the waves scattered to the
three other lines(Fig.3.4(b)). This allows the magnetic field $^{+}H_{xz}(n,m,q)$ to scatter in coplanar fashion to $Z_{xz}(n+1,m,q)$, $Z_{zx}(n,m,q+1)$, and $Z_{zx}(n,m,q)$. The magnetic field scattered into the $Z_{xz}(n+1,m,q)$ line, $^{+}H_{xz}(n+1,m,q)$, is no surprise; this
is simply the field accompanying $^{+}V_{xy}(n+1,m,q)$. However, the magnetic fields
in lines $Z_{zx}(n,m,q+1)$ and $Z_{zx}(n,m,q)$, denoted by $^{+}H_{zx}(n,m,q+1)$ and $^{-}H_{zx}(n,m,q)$,
are precisely those which would accompany any electric field waves if it were
scattered perpendicular to the plane in Fig.3.4(a), and denoted by $^{+}V_{zy}(n,m,q+1)$
and $^{-}V_{zy}(n,m,q)$. The necessity for the latter waves is also compelling, based on
symmetry arguments. Both physically and mathematically coplanar scattering
is incomplete and there is no reason why $^{+}V_{xy}(n,m,q)$ should not scatter normally into the $Z_{zy}(n,m,q)$ and $Z_{zy}(n,m,q+1)$ lines as well. A similar argument
may also be made for the scattering of the incident magnetic field normal to the
propagation plane. The incident magnetic field, $^{+}H_{xz}(n,m,q)$ undergoes scattering into the $Z_{yz}(n,m,q)$ and $Z_{yz}(n,m+1,q)$ lines and these fields are precisely those
which accompany the electric field waves scattered into the. $Z_{yx}(n,m,q)$ and
$Z_{yx}(n,m+1,q)$ lines.

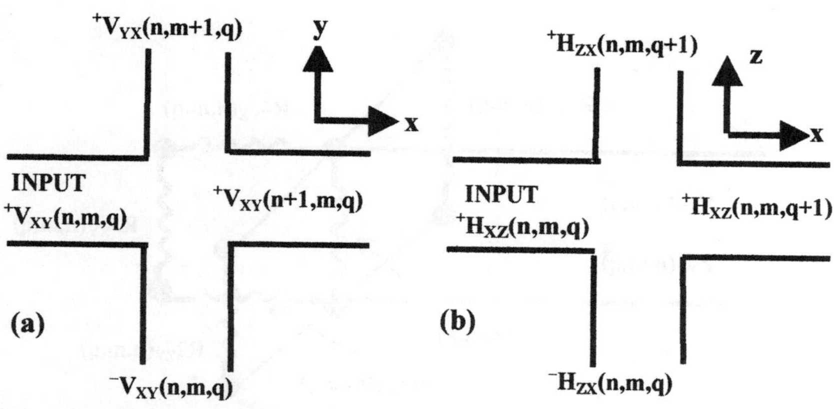

FIG.3.4 COPLANAR SCATTERING OF $^{+}V_{XY}(n,m,q)$ AND ASSOCIATED $^{+}H_{XZ}(n,mq)$. INCORPORATION OF MAGNETIC FIELD IMPLIES SCATTERING NORMAL TO PROPAGATION PLANE.

3.5 *Equivalent TLM Circuit . Quasi-Coupling Effect*

Based on the comments in the previous paragraph, it would appear that the scattering, including losses, is controlled by the equivalent node circuit in Fig.3.5 which shows a wave $^{+}V_{xy}(n,m,q)$ incident on the (n,m,q) node. The circuit applies if line impedances surrounding the node obey certain symmetry conditions; for example, if the TLM lines surrounding the node have the same impedance. In the event the such symmetry conditions are lacking, however, then the incident $^{+}V_{xy}(n,m,q)$ wave will exhibit aplanar scattering not only to the z direction but to the y direction as well. The equivalent circuit therefore must be revised. A first order revision of the circuit is shown in Fig.3.6 and is used to account for first order "quasi" coupling to the $Z_{yz}(n,m,q)$ and $Z_{yz}(n,m+1,q)$ lines. A quantitative description of quasi-coupling , and the conditions for its existence, are

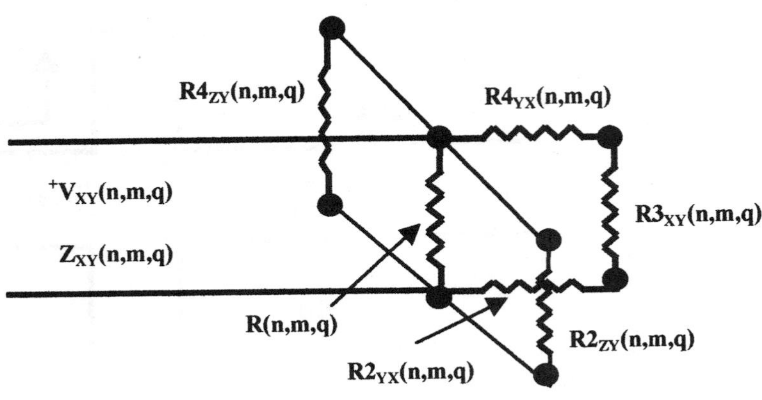

FIG. 3.5 EQUIVALENT CIRCUIT FOR SCATTERING OF BOTH COPLANAR AND APLANAR WAVES , WITH QUASI-COUPLING EXCLUDED (ZERO ORDER APPROXIMATION).

subsequently discussed in Sections 3.7 and 3.8. Regarding the circuit in Fig.3.6, we shall also see that the appropriate node parameters, $R1_{yz}(n,m,q)$ and $R3_{yz}(n,m,q)$ also must be taken into account to obtain the coupling to $Z_{yz}(n,m,q)$ and $Z_{yz}(n,m+1,q)$ and in order to properly calculate the load impedances and the scattering coefficients. These node parameters vanish in the absence of quasi-coupling to the $Z_{yz}(n,m,q)$ and $Z_{yz}(n,m+1,q)$ lines. The equivalent circuit of Fig.3.6, as well as a follow-on, modified version of the circuit, yield both the aplanar(including quasi-coupling) and coplanar scattering, and are discussed in detail The resultant scattering coefficients and iterative equations are then obtained. We should remark that the use of the quasi-coupling circuits does add complexity to the TLM formulation and , therefore, before discussing quasi-coupling in detail, we should describe the implications for neglecting this type scattering.

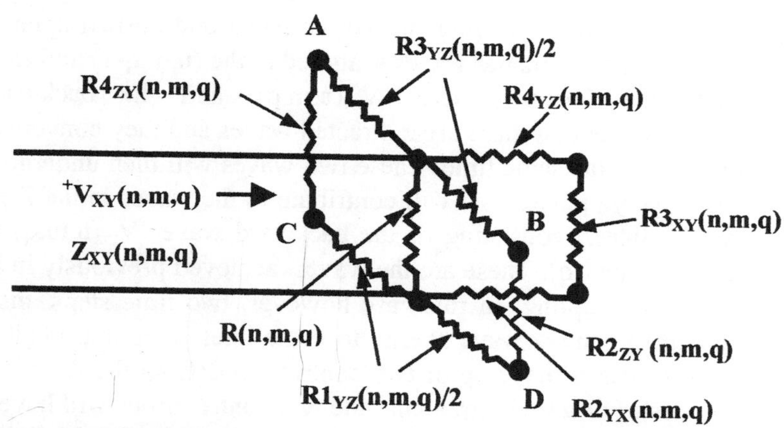

FIG. 3.6 EQUIVALENT CIRCUIT FOR SCATTERING OF BOTH COPLANAR AND APLANAR WAVES. NODAL PLANES INTERSECT AT MIDPOINTS. THE TERMINAL PAIRS, A,B AND C,D REPRESENT FIRST ORDER QUASI- COUPLING PORTS TO THE $Z_{YZ}(n,m,q)$ AND $Z_{YZ}(n,m+1,q)$ LINES.

3.6 *Neglect of Quasi-Coupling*

Under what conditions is it possible to neglect the quasi-coupling effect and to use the simple circuit in Fig.3.5, even when there is inequality in the surrounding lines, brought about by a gradient in the dielectric constant? We address this issue by first supposing that the cell size is much smaller than the characteristic length associated with the change in the dielectric constant(we can define this length, e.g., as the number of cell lengths needed for ε to change by an amount (1/e). We assume, for concreteness a small gradient of the dielectric constant in the z direction. We then suppose that waves traveling in the x direction reach the yz plane at about the same time. Consider these waves to be $^+V_{xy}(n,m,q)$, $^+V_{xy}(n,m,q+1)$ and $^+V_{xy}(n,m,q-1)$. They then impinge on the (n,m,q), $(n,m,q+1)$, and $(n,m,q-1)$ nodes respectively. If we assume that no quasi-coupling occurs, the only scattering is that belonging to the coplanar and normal aplanar type. We then focus on the two aplanar waves scattered at the $(n,m,q-1)$ and $(n,m,q+1)$ nodes, namely the forward wave $^+V_{zy}(n,m,q)$, and the backward one $^-V_{zy}(n,m,q+1)$. We note that these are z directed waves and they converge on the (n,m,q) node at about the same time. These two waves will then undergo coplanar scattering in the yz plane, and will contribute to the waves in the $Z_{yz}(n,m,q)$ and $Z_{yz}(n,m+1,q)$ lines, consisting of the backward wave $^-V_{yz}(n,m,q)$ and the forward one $^+V_{yz}(n,m+1,q)$. These are the waves achieved previously in a single time step via quasi-coupling. In this case however, two time steps, instead of one, are required to convey wave energy to these lines. It is important to note that each of the waves is made up of two contributions(from the $Z_{zy}(n,m,q)$ and $Z_{zy}(n,m,q+1)$ lines). For a small gradient the two contributions will have mostly opposite polarity, so that field cancellation occurs, and the total field delivered to $Z_{yz}(n,m,q)$ and $Z_{yz}(n,m+1,q)$ will be small.

What therefore, is the main impact in ignoring quasi-coupling? Without quasi-coupling, as mentioned previously, the outputs to $^-V_{yz}(n,m,q)$ and $^+V_{yz}(n,m+1,q)$ will lag behind, with the delay on the order of one transit time. Without doubt, however, the greatest difference occurs when one is presented with a well defined dielectric boundary, In this case , waves traveling in a TLM line, situated along the dielectric interface, will experience a large impedance gradient . In this case the quasi-coupling effects are magnified and the neglect of quasi-coupling will result in field distortion near the boundary. Usually, the effect of ignoring quasi-coupling can be made mitigated by further shrinking the size of the cell. As a practical matter, however, it is not always possible to select an arbitrarily small cell and in such cases it is better to incorporate quasi-coupling , especially when dielectric interfaces are present.

In the following we discuss the quasi-coupling circuit to improve the accuracy. Unfortunately the broken symmetry, caused by a non-uniformity in the dielectric constant, requires "messy" solutions involving successive approximations. If one utilizes the approach of using extremely small cells, thereby reducing the effect of ignoring quasi-coupling,   the reader may opt to bypass the ensuing discussion on quasi-coupling and go directly to Section 3.12, which describes the complete set of iterative equations. The quasi- coupling effects may be suppressed in the iterative equations by setting equal to zero any terms with the subscript Q.

3.7 Simple Quasi-Coupling Circuit: First Order Approximation

It is first useful to classify quasi-coupling according to the approximations used. With the zero order approximation, the circuit of Fig.3.5 applies and in this case both the node parameters and the voltages for quasi-coupling vanish. In the case of the first order quasi-coupling, we utilize the circuit in Fig.3.6 and we proceed with the help of Fig.3.7, which shows an on -axis view (x axis) of the $^+V_{xy}(n,m,q)$ wave. We concentrate for the moment on the coupling to waves perpendicular to the propagation plane of $^+V_{xy}(n,m,q)$. This includes the aplanar coupling to the $Z_{zy}(n,m,q+1)$ and $Z_{zy}(n,m,q)$, which has been discussed previously and is well understood. It is not at all clear, however, what degree of coupling exists, if any, to the $Z_{yz}(n,m,q)$ and $Z_{yz}(n,m+1,q)$ lines. In order to proceed, we make the reasonable assumption, based on symmetry arguments, that the incident field $^+V_{xy}(n,m,q)$ at the (n,m,q) node is located midway between the various lines. From Fig.3.7 there are two distinct paths which yield the same node voltage as $V_{xy}(n,m,q)$,   that given by A$\rightarrow$ B$\rightarrow$C$\rightarrow$D and that given by A$\rightarrow$F$\rightarrow$G$\rightarrow$D. In region A , the first path implies that the wave traverses half the width of the line $Z_{yz}(n,m,q)$, and thus the wave "sees" an impedance $Z_{yz}(n,m,q)/2$.   Similarly, the wave sees a node resistance of $R(n,m,q)/2$ in region A. Analogous reasoning in region D leads us to conclude that $^+V_{xy}(n,m,q)$ encounters a line impedance of $Z_{yz}(n,m+1,q)/2$ and again a load resistance of $R(n,m,q)/2$. The same statements apply to the second path, A$\rightarrow$F$\rightarrow$G$\rightarrow$D, except for the fact that the currents in regions A and D are opposite to that of the first path. This circuit must be carefully interpreted and the resultant equations examined, especially in regard to the coupling of the incident wave to the $Z_{yz}(n,m,q)$ and $Z_{yz}(n,m+1,q)$ lines, which we designate as quasi coupling. A

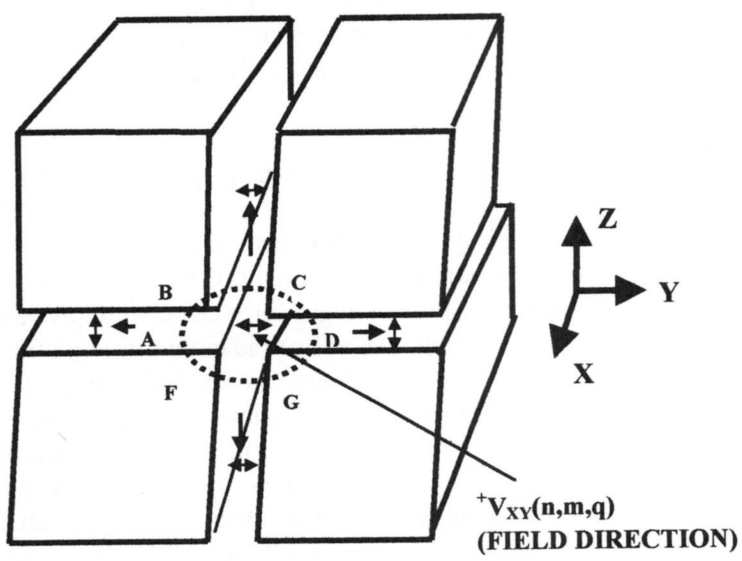

FIG.3.7 APLANAR SCATTERING AT A 3D NODE. BOTH THE DIRECTIONS OF PROPAGATION (SINGLE ARROW) AND FIELD (DOUBLE ARROW) ARE SHOWN. QUASI-COUPLING MAY OCCUR AT A AND D, i.e. , AT $Z_{YZ}(n,m,q)$ AND $Z_{YZ}(n,m+1,q)$.

separate calculation must be performed for this type coupling. If the line imped-ances surrounding a node are identical, one can surmise that the quasi coupling vanishes due to symmetry.

To determine first order coupling we examine the TLM lines in Fig.3.8, using simplified notation. Lines Z_1 and Z_2 in Fig.3.8 belong to the $R1_{yz}(n.,m,q)$ and $R3_{yz}(n,m,q)$ branches in Fig.3.6, while Z_2 and Z_4 belong to the $R2_{zy}(n,m,q)$ and $R4_{zy}(n,m,q)$ branches. We then allow for field ^+V, propagating perpendicu-lar to the plane(emerging from the page) and with the field direction as shown.

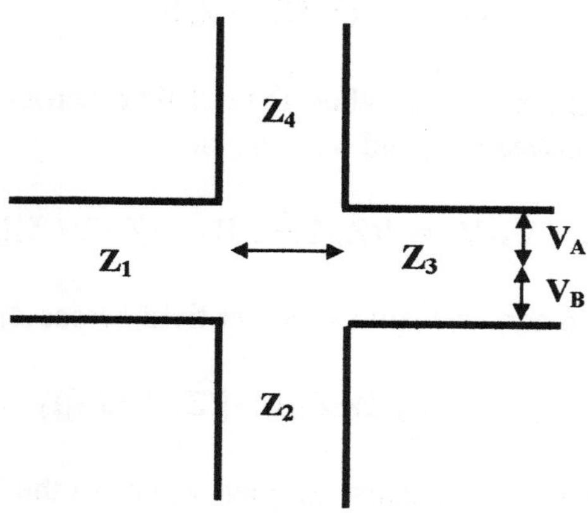

FIG. 3.8 APLANAR SCATTERING (INCLUDING QUASI-COUPLING) OF SIGNAL ^+V INTO LINES Z_1, Z_2, Z_3, AND Z_4 (^+V PROPAGATES PERPENDICULAR TO PAGE).

This wave will couple to the four lines(in addition to the co-planar coupling, which we ignore for the moment), assumed to be lossless for the moment. As implied in Fig.3.8, there are two possible current paths, I_1 and I_2, related to V (the total input node voltage) and the line impedances by

$$V = [(Z_1/2) + Z_4 + (Z_3/2)]I_1 \tag{3.37a}$$

$$I_1 = V/[(Z_1/2) + Z_4 + (Z_3/2)] \tag{3.37b}$$

$$V = [(Z_1/2) + Z_2 + (Z_3/2)]I_2 \tag{3.38a}$$

$$I_2 = V/[(Z_1/2) + Z_2 + (Z_3/2)] \tag{3.38b}$$

Next we obtain the partial voltages across Z_3, V_A and V_B, given by

$$V_A = V - (Z_1/2)I_1 - Z_4 I_1 \qquad (3.39)$$

$$V_B = V - (Z_1/2)I_2 - Z_2 I_2 \qquad (3.40)$$

Since V_A and V_B oppose one another, the net field across Z_3 is obtained by taking the difference between V_A and V_B, Δ_3, or

$$\Delta_3 = V_A - V_B = [(Z_1/2) + Z_2]I_2 - [(Z_1/2) + Z_4]I_1 \qquad (3.41)$$

A similar calculation across Z_1 gives the net field across Z_1

$$\Delta_1 = \ = \ [(Z_3/2) + Z_4]I_1 - [(Z_3/2) + Z_2]I_2 \qquad (3.42)$$

Δ_1 and Δ_3 are the first order quasi coupled waves to the Z_1 and Z_3 lines respectively. We can express these amplitude in terms of the assumed node voltage ,V, and the circuit parameters by combining with Eqs(3.37)-(3.40). Thus

$$\Delta_3 = V (Z_3/2)/ [(Z_1/2) + Z_4 + (Z_3/2)] - V(Z_3/2)/[(Z_1/2) + Z_2 + (Z_3/2)] \qquad (3.43)$$

$$\Delta_1 = V (Z_1/2)/ [(Z_1/2) + Z_4 + (Z_3/2)] - V(Z_1/2)/[(Z_1/2) + Z_2 + (Z_3/2)] \qquad (3.44)$$

Note that $\Delta_1 = \Delta_3 = 0$ when $Z_4 = Z_2$. Quasi- coupling is therefore suppressed under these conditions, which is a less restrictive condition than having all the TLM lines identical. Normally it is more useful to express the scattered waves in terms of the incident wave, V_{INC}, which is related to V by the usual expression

$$V = 2V_{INC}R_L/[R_L + Z_{INC}] \qquad (3.45)$$

where Z_{INC} is the impedance of the incident line, and R_L is the load impedance, which is a parallel combination of the coplanar circuit and the two "half" circuit branches just discussed.. The use of the equivalent resistors $Z_3/2$ and $Z_1/2$ is not straightforward, however, and we must modify the analysis, as indicated in the next Section,   so as to obtain the proper impedance values and to correctly calculate R_L.

3.8 *Correction to Quasi- Coupling Circuit :Second Order Approximation*

Although Eqs.(3.43)-(3.45) provide a reasonable initial estimate of the quasi-coupling to the Z_1 and Z_3 lines, the circuit in Fig.3.6, as it stands, it does not ac-curately portray the coupling to the lines Z_2, Z_4 and to the coplanar branch, as a simple example will demonstrate. Suppose there is no quasi coupling, a situation which occurs when , for example, the TLM lines are all identical or if $Z_2 = Z_4$. Despite this fact, Z_1 and Z_3 still appear in the outputs to the Z_2, Z_4 , and coplanar lines. Essentially the lines Z_1 and Z_3 siphon off energy even when the quasi coupling vanishes. This is because the load impedance for this circuit , R_L , in-correctly assumes that the wave energy is either dissipated or delivered to Z_1 or Z_3 even when the quasi-coupling is not present.

The problem is with the values assigned to the quasi-coupling parameters $(Z_1/2)$ and$(Z_3/2)$ in each branch. $(Z_1/2)$ and $Z_3/2)$ represent first order values for the coupling parameters. While the selection of $(Z_1/2)$ and $Z_3/2)$ may appear intuitive, these parameters are controlled by the energy coupled to the Z_1 , Z_3 lines. These quasi-coupling parameters, which control the degree of output to the Z_1 and Z_3 lines, must be modified. We allow $(Z_1/2)$ and $(Z_3/2)$ to be re-placed by a new pair of second order quasi-coupling resistance parameters, as yet undetermined, and designated by $\mathcal{R}3_Q$ and $\mathcal{R}1_Q$, with

$$(Z_1/2) \to \mathcal{R}1_Q \text{ and } (Z_3/2) \to \mathcal{R}3_Q \qquad (3.46a), (3.46b)$$

The lines Z_2 and Z_4 are themselves of course unchanged in the new circuit. With losses present , the replacements become

$$R1_{yz}(n,m,q)/2 \to \mathcal{R}1_Q \text{ and } R3_{yz}(n,m,q)/2 \to \mathcal{R}3_Q \qquad (3.46c),(3.46d)$$

We now proceed to determine these parameters based on symmetry and energy considerations.

Most important, we require that these quasi-coupling parameters vanish when the coupled voltages to the Z_1 and Z_3 lines, Δ_1 and Δ_3, vanish. The new circuit is shown in Fig.3.9, which is a modification of that in Fig (3.6). Fig.9 also uses obvious cell notation, which we discuss in a moment. Superficially the two circuits appear the same, but they differ with respect to the node parameters

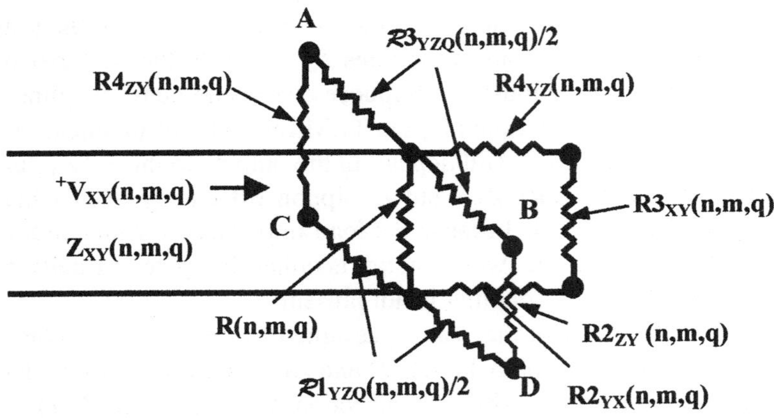

FIG. 3.9 EQUIVALENT CIRCUIT FOR SCATTERING OF BOTH COPLANAR AND APLANAR WAVES. THE TERMINAL PAIRS, A,B AND C,D REPRESENT SECOND ORDER QUASI-COUPLING TO THE Z_{YZ}(n,m,q) AND Z_{YZ}(n,m+1,q) LINES. QUASI –COUPLING NODE PARAMETERS $\mathcal{R}$1$_{yzQ}$(n,m,q) AND $\mathcal{R}$3$_{yzQ}$(n,m,q) ARE INDICATED.

for the lines Z_1 and Z_3. Specifically $\mathcal{R}1_Q$ will in general not be equal to half the node parameter $R1_{yz}$(n,m,q), shown in Fig.3.6. Similarly $\mathcal{R}3_Q$, will in general not be equal to half the node parameter $R3_{yz}$(n,m,q). Without losses, and using the simplified notation of the previous Section, this is equivalent to saying that in general $\mathcal{R}1_Q \neq Z_1/2$ and $\mathcal{R}3_Q \neq Z_3/2$.

As noted in Fig.3.9 It may appear at first sight that we have gone against our earlier admonition against selecting different node resistors at the same node; in fact the node resistors are still the same and we should regard $\mathcal{R}1_Q$, $\mathcal{R}3_Q$, as necessary coupling parameters which help facilitate the transfer of wave energy when quasi coupling is present. Once we obtain these new parameters we will be able to ascertain the coupling to all four lines. In order to obtain these parameters we first utilize the voltage relationships for the branches in Fig.3.8, in which the upper branch is associated with Z_4 and the lower one Z_2 Looking at the lower branch,

$$V = I_1{}^* [Z_4 + \mathcal{R}3_Q + \mathcal{R}1_Q] \tag{3.47}$$

from which we obtain

$$I_1 = V/[Z_4 + \mathcal{R}3_Q + \mathcal{R}1_Q] \tag{3.48}$$

A similar equation for the lower branch gives

$$I_2 = V/[Z_2 + \mathcal{R}3_Q + \mathcal{R}1_Q] \tag{3.49}$$

At this point we still lack sufficient information to determine the quasi-coupling parameters. However we have not yet made use of the total power radiated into the Z_1 and Z_3 lines, via the coupling parameters. We require that the power delivered to these lines satisfy the following

$$\Delta_3{}^2 /Z_3 = \mathcal{R}3_Q(I_1{}^2 + I_2{}^2) \tag{3.50}$$

$$\Delta_1{}^2 /Z_1 = \mathcal{R}1_Q(I_1{}^2 + I_2{}^2) \tag{3.51}$$

where in the above we use the first order voltages Δ_3 and Δ_1 , provided by Eqs.(3.43) and (3.44) respectively , while I_1 and I_2 are given by Eqs.(3.48) and (3.49). Note that Δ_3 and Δ_1 vanish when the coupling parameters go to zero. Substituting for Δ_3 , I_1, and I_2. Eq.(3.50) for example becomes

$$(Z_3/4)[1/\{(Z_1/2) + Z_4 + (Z_3/2)\} - 1/\{(Z_1/2) + Z_2 + (Z_3/2)\}]^2 [V(1)]^2 =$$

$$\mathcal{R}3_Q[\ 1/\{\ Z_4 + \mathcal{R}3_Q + \mathcal{R}1_Q\}^2\ + 1/\ \{\ Z_2 + \mathcal{R}3_Q + \mathcal{R}1_Q\}^2\ ]\ [V(2)]^2 \qquad (3.52)$$

Some explanation of Eq.(3.52) is required. On the left side of the equation, $V(1)$ is the total load voltage given by Eq.(3.45) with the load impedance given by Fig.3.6 , i.e., the coupling parameters are first order and are given by $Z_1/2$, $Z_3/2$. We have added the argument 1 to emphasize this fact. On the right side , $V(2)$ is again the load voltage given by Eq.(3.45) but we have now added the argument 2 to V, to indicate that the load impedance is now obtained from Fig.3.9 with second order coupling parameters $\mathcal{R}3_Q$ and $\mathcal{R}1_Q$.

Similarly, Eq.(3.51) becomes

$$(Z_1/4)[\ 1/\{(Z_1/2) + Z_4 + (Z_3/2)\}\ -\ 1/\{(Z_1/2) + Z_2 + (Z_3/2)\}\ ]^2\ [V(1)]^2 =$$
$$\mathcal{R}1_Q[\ 1/\{\ Z_4 + \mathcal{R}3_Q + \mathcal{R}1_Q\}^2\ + 1/\ \{\ Z_2 + \mathcal{R}3_Q + \mathcal{R}1_Q\}^2\ ]\ [V(2)]^2 \qquad (3.53)$$

We also note that the Eqs.(3.50), (3.51) lead to the simple relationship

$$(\mathcal{R}3_Q/\mathcal{R}1_Q) = Z_3/Z_1 \qquad (3.54)$$

Further evaluation of $\mathcal{R}3_Q$ and $\mathcal{R}1_Q$ necessitates the simultaneous solution of Eqs.(3.52) and (3.53). It is important to understand the implications of Eqs. (3.52) and (3.53). The second order coupling parameters $\mathcal{R}1_Q$ and $\mathcal{R}3_Q$, based on the circuit in Fig.3.9 , have been defined using the first order coupling voltages Δ_1 and Δ_3 , based on the circuit in Fig.3.6. With this bootstrap technique, one may obtain the modified circuit representation of quasi- coupling.

3.9 Calculation of Load Impedance with Quasi-Coupling

In order to proceed further we need to obtain the load impedances which appear implicitly in Eqs.(3.52)-93.53), due to the presence of $V(1)$ or $V(2)$. We thus calculate the load impedance seen by the incident wave using the circuits in Figs.3.6 and 3.9, so that we include the quasi-node parameters for the latter circuit. Assuming for the moment that no losses are present, the load impedance R_L is given by

$$R_L = Z_I Z_{II} Z_{III} / [Z_I Z_{II} + Z_I Z_{III} + Z_{II} Z_{III}] \qquad (3.55)$$

where Z_I, Z_{II} are the upper and lower impedance branches and Z_{III} is the coplanar branch. The impedance branches(and the resultant load impedance) will differ depending on whether the circuit is first order, second order, or zero order if quasi-coupling is ignored. Using Fig.3.6, for the first order coupling

$$Z_{III} = Z_4 + (Z_3/2) + (Z_1/2) \quad (Fig.3.6) \qquad (3.56)$$

while for the second order coupling of Fig.3.9

$$Z_{III} = Z_4 + \mathcal{R}3_Q + \mathcal{R}1_Q \quad (Fig.3.9) \qquad (3.57)$$

Similarly Z_{II} for the first and second order coupling is

$$Z_I = Z_2 + (Z_3/2) + (Z_1/2) \quad (Fig.3.6) \qquad (3.58)$$

$$Z_I = Z_2 + \mathcal{R}3_Q + \mathcal{R}1_Q \quad (Fig.3.9) \qquad (3.59)$$

The coplanar branch is the same for both first and second order circuits

$$Z_{II} = Z_{C1} + Z_{C2} + Z_{C3} \quad (Coplanar\ Lines) \qquad (3.60)$$

where Z_{C1}, Z_{C2}, Z_{C3} are the three coplanar lines seen by the incident wave.

For completeness we also provide the load impedance corresponding to that of Fig.3.5, the zero order approximation. In this case the three branches are

$$Z_{III} = Z_4 \qquad (3.61a)$$
$$Z_I = Z_2 \qquad (3.61b)$$
$$Z_{II} = Z_{C1} + Z_{C2} + Z_{C3} \quad (Coplanar\ Lines) \qquad (3.61c)$$

The previous discussion implies zero loss at the nodes. If loss is present the following transformation is substituted:

$$Z_1 \rightarrow R1 = RZ_1/(R+Z_1) \tag{3.62a}$$
$$Z_2 \rightarrow R2 = RZ_2/(R+Z_2) \tag{3.62b}$$
$$Z_3 \rightarrow R3 = RZ_3/(R+Z_3) \tag{3.62c}$$
$$Z_4 \rightarrow R4 = RZ_4/(R+Z_4) \tag{3.62d}$$

When we include losses we must be careful to include the node resistance in the calculation of the load resistance R_L. With loss Eq.(3.55) becomes

$$R_L = Z_I Z_{II} Z_{III} R/[Z_I Z_{II} R + Z_I Z_{III} R + Z_{II} Z_{III} R + Z_I Z_{II} Z_{III}] \tag{3.63}$$

where we note the inclusion of the node resistance R in the expression for R_L.

3.10 Small Coupling Approximation of Second Order Quasi- Coupling

When the difference in Z_2 and Z_4 is small compared to the line impedance values then an approximate solution for the coupling parameters, $\mathcal{R}3_Q$, $\mathcal{R}1_Q$ is easy to obtain. To distinguish between the loads in Figs.3.6 and 3.9, i.e., between first and second order, we will add the argument 1 or 2 respectively, $R_L(1)$ and $R_L(2)$, as was done with the voltage. We will also have occasion to use the circuit in Fig.3.5, in which the quasi coupling vanishes altogether; this of course is the zero order approximation. We designate this load voltage as $V(0)$ and the load impedance as $R_L(0)$.

We then expand the right side of Eqs.(3.52)-(3.53) in terms of $\mathcal{R}3_Q$, $\mathcal{R}1_Q$, If we retain only the linear term then

$$\mathcal{R}3_Q = (Z_3/4)[1/\{(Z_1/2)+Z_4+(Z_3/2)\} - 1/\{(Z_1/2)+Z_2+(Z_3/2)\}]^2/[(1/Z^2_2)+(1/Z_4^2)] \tag{3.64}$$

The above is obtained by neglecting the quadratic terms of $\mathcal{R}3_Q$ and $\mathcal{R}1_Q$, i.e.,

$$\mathcal{R}3_Q \mathcal{R}1_Q \rightarrow 0; \ \mathcal{R}1_Q \mathcal{R}1_Q \rightarrow 0; \ \mathcal{R}3_Q \mathcal{R}3_Q \rightarrow 0 \tag{3.65}$$

and also making use of

$$V(2)^2/V(1)^2 \rightarrow 1+0\{\text{quadratic terms of } \mathcal{R}3_Q, \mathcal{R}1_Q\} \qquad (3.66)$$

A similar calculation for $\mathcal{R}1_Q$ gives

$$\mathcal{R}1_Q=(Z_1/4)[1/\{(Z_1/2)+Z_4+(Z_3/2)\}-1/\{(Z_1/2)+Z_2+(Z_3/2)]^2 \, /[(1/Z_2^2)+(1/Z_4^2)] \qquad (3.67)$$

As noted before the same relationship between $\mathcal{R}3_Q$, $\mathcal{R}1_Q$ is maintained, given by

$$(\mathcal{R}3_Q / \mathcal{R}1_Q) = (Z_3/Z_1) \qquad (3.68)$$

With the second order quasi coupling parameters calculated we then perform a similar calculation for Δ_1 and Δ_3. The first order values of Δ_1 and Δ_3 have already been obtained from Eqs.(3.43)-(3.44), based on Fig.3.6. To obtain the second order approximation, we merely replace V with V(2), based on the circuit in Fig.3.9, and using Eq.(3.45) to relate V(2) to the incident wave and the circuit. We rewrite Eqs.(3.43) and (3.44) to explicitly show the appropriate argument for V, V(2), or

$$\Delta_3(2) = V(2) \, \mathcal{R}3_Q \, / \, [\mathcal{R}1_Q +Z_4+ \mathcal{R}3_Q] - V(2) \, \mathcal{R}3_Q \, /[\mathcal{R}1_Q +Z_2+ \mathcal{R}3_Q] \qquad (3.69)$$

$$\Delta_1(2) = V(2) \, \mathcal{R}1_Q \, / \, [\mathcal{R}1_Q +Z_4+ \mathcal{R}3_Q] - V(2) \, \mathcal{R}1_Q \, /[\mathcal{R}1_Q +Z_2+ \mathcal{R}3_Q] \qquad (3.70)$$

Eqs.(3.64),(3.67) and (3.69)-(3.70) provide us with the second order coupling parameters and voltages based on the circuit in Fig.3.9. We have appended the argument two to indicate the quasi voltages are second approximations. We again note that the quasi node parameters and voltages vanish when $Z_4 =Z_2$.

As one may surmise , we may continue the approximation process to obtain higher order solutions. Thus, to obtain a third order approximation, we first modify Eqs.(3.52)-(3.53). On the left side of the two equations, V(1) is replaced by V(2). Since V(2) involves second order node parameters, obtained previously, the left side of the equations represent known quantities. On the right side, however, $\mathcal{R}3_Q$ and $\mathcal{R}1_Q$ are replaced by $\mathcal{R}3_Q(3)$ and $\mathcal{R}1_Q(3)$, where we add the argument 3 to indicate these are the third order parameters. In addition V(2) , on the right side, is replaced by V(3), where the 3 argument again indicates that the circuit contains the third order parameters(as yet unknown). We then solve Eqs.(3.52) and (3.53) for R3$_Q$(3) and $\mathcal{R}1_Q$(3). As in Section 3.10 we may use the small coupling approximation to solve for the node parameters. We should point out that the circuit for the 3rd order approximation is identical to that of Fig.3.9 except for the fact that $\mathcal{R}3_Q$ (3) and $\mathcal{R}1_Q$(3) replace $\mathcal{R}3_Q$(2) and $\mathcal{R}1_Q$(2) . Once we have the third order parameters we can obtain the third order quasi voltages, $\Delta_3(3)$ and $\Delta_1(3)$ from Eqs.(3.69 and (3.70), remembering to replace V(2) by V(3) , which contains the previously solved third order node parameters. The procedure for higher order solutions is similar.

With the determination of the quasi-coupling parameters and voltages, we can now ascertain the scattering coefficients and the transfer of wave energy to the various lines(with and without quasi-coupling). Rather than treating this topic now, however, we avoid redundancy and postpone the topic until the next Section, in which the entire subject matter is cast into cell notation, an absolute necessity for handling the iterative equations.

3.11 General 3D Scattering Process Using Cell Notation

In much of the previous discussion on aplanar coupling, particularly quasi-coupling, cell notation was not utilized, and therefore the results therefore are not suitable for iterative processes. In the following we incorporate aplanar coupling into the formulation , using cell notation. Losses and quasi-coupling are included throughout. Following the circuit in either Figs.3.9, 3.6, or 3.5 (if we neglect quasi-coupling), we are now in a position to calculate the various scattering coefficients and iterative equations, using the generalized cell notation. The scattering process will now be complete since, in addition to the coplanar scattering described previously, we include scattering perpendicular to the propagation plane, which includes the usual aplanar scattering with or without quasi scattering. If we wish, quasi-coupling may ignored entirely by setting the quasi- node parameters (to be described explicitly) equal to zero. As we have done before, it is useful to define auxiliary node parameters consisting of the parallel combination of the line impedance and the node resistance. To illustrate the definitions, we focus on the (n,m,q) node and examine a wave $^+V_{xy}(n,m,q)$ incident on the node. For coplanar (xy plane) scattering the node parameter definitions remain the same as before and the relevant relationships are identical to Eqs.(3.26a)-(3.26d)   repeated here to give

$$R1_{xy}(n,m,q) = [R(n,m,q)Z_{xy}(n,m,q)]/[R(n,m,q)+Z_{xy}(n,m,q)] \qquad (3.71)$$

$$R2_{yx}(n,m,q) = [R(n,m,q)Z_{yx}(n,m,q)]/[R(n,m,q)+Z_{yx}(n,m,q)] \qquad (3.72)$$

$$R3_{xy}(n,m,q) = [R(n,m,q)Z_{xy}(n+1,m,q)]/[R(n,m,q)+Z_{xy}(n+1,m,q)] \qquad (3.73)$$

$$R4_{yx}(n,m,q) = [R(n,m,q)Z_{yx}(n,m+1,q)]/[R(n,m,q)+Z_{yx}(n,m+1,q)] \qquad (3.74)$$

Because of aplanar effects the wave $^+V_{xy}(n,m,q)$ will scatter unto the yz plane with coupling to the lines $Z_{zy}(n,m,q)$, $Z_{zy}(n,m,q+1)$, $Z_{yz}(n,m,q)$ and $Z_{yz}(n,m+1,q)$ lines. Because of the aplanar and quasi scattering the values of the node parameters will differ from the conventional type, following the circuit in either Fig..3.6 or Fig.3.9 and the discussions in the previous Sections. In the following it will be more useful and less confusing if we consider the cell notation of the node parameters for just the half impedance portions, i.e., for either $Z_1/2$, $Z_3/2$ (first order) or if we use the second or higher order parameters. In the following we always assume losses are present .

3.11(a): Quasi-Node Parameters in Cell Notation

We first describe the first order quasi-node parameters in cell notation. In the first order approximation we continue to use $Z_3/2$, $Z_1/2$ for the coupling parameters , using Fig.3.6 as the circuit. The cell notation for the node parameters involved in quasi-coupling are as follows.

$$R1_{yzQ}(1,n,m,q) = (1/2)(Z_{yz}(n,m,q))(R(n,m,q)/[Z_{yz}(n,m,q)+R(n,m,q)] \qquad (3.75)$$

where $R1_{yzQ}(1,n,m,q)$ is the quasi node coupling parameter and we have re-placed Z_1 with $Z_{yz}(n,m,q)$ to conform with the cell notation. Similarly the node parameter $R3_{yzQ}(n,m,q)$ is

$$R3_{yzQ}(1,n,m,q)=(1/2)(Z_{yz}(n,m+1,q))(R(n,m,q))/Z_{yz}(n,m+1,q)+R(n,m,q)] \qquad (3.76)$$

Note that we have now replaced Z_3 with $Z_{yz}(n,m+1,q)$. Note also that the m index is increased by one since the corresponding line belongs to the $(n,m+1,q)$ cell. We have also appended a one argument to $R1_{yzQ}(1,n,m,q)$ and $R3_{yzQ}(1,n,m,q)$ to indicate the first order approximation and to differentiate them from the second order parameters. To suppress quasi-coupling we set $R1_{yzQ}(1,n,m,q) = R3_{yzQ}(1,n,m,q)=0$.

Next we consider the second order node parameters, based on Fig.3.9 Without realizing it, however, we have already obtained these parameters and they are as follows

$$R1_{yzQ}(2,n,m,q) = \mathcal{R}1_Q \qquad (3.77)$$

$$R3_{yzQ}(2,n,m,q) = \mathcal{R}3_Q \qquad (3.78)$$

where of course $\mathcal{R}1_Q$ and $\mathcal{R}3_Q$ correspond , e.g., to the approximate expressions of Eqs.(3.64) and (3.67), but in cellular notation. Using these expressions , modified to include loss, we have

$$R1_{yzQ}(2,n,m,q)=(R1_{yzQ}(1,n,m,q)/2)[1/\{R1_{yzQ}(1,n,m,q)+R4_{zy}(n,m,q)+$$
$$R3_{yzQ}(1,n,m,q)\}- [1/\{ R1_{yzQ}(1,n,m,q)+ R2_{zy}(n,m,q) + R3_{yzQ}(1,n,m,q)\}]^2 *F$$
$$(3.79)$$

where

$$F = 1/[(1/ R2_{zy}(n,m,q))^2 +(1/ R4_{zy}(n,m,q))^2] \qquad (3.80)$$

and quadratic terms such as

$$(R1_{yz}(2,n,m,q))^2 , \text{ etc...} \rightarrow 0 \qquad (3.81)$$

As noted earlier, $R1_{yzQ}(2,n,m,q)$ is evaluated using first order quantities defined previously. The expression for $R3_{yzQ}(2,n,m,q)$ is obtained in exactly the same manner, or by analogy with Eq.(3.68),

$$R3_{yzQ}(2,n,m,q)= R1_{yzQ}(2, n,m,q) R3_{yzQ}(1,n,m,q)/ R1_{yzQ}(1,n,m,q)\} \qquad (3.82)$$

The final two node parameters are for the aplanar case but without quasi coupling. They assume the more familiar forms, and as with the coplanar parameters, there is no concern with half impedance coupling. Thus,

$$R2_{zy}(n,m,q) = [R(n,m,q)Z_{zy}(n,m,q)]/ [R(n,m,q)+Z_{zy}(n,m,q] \qquad (3.83)$$

$$R4_{zy}(n,m,q) = [R(n,m,q)Z_{zy}(n,m,q+1)]/[R(n,m,q)+Z_{zy}(n,m,q+1)] \qquad (3.84)$$

This completes the specification of the first set of node parameters Again note that in the above the subscript Q identifies quasi coupling.

3.11(b) : Calculation of Load Impedance in Cell Notation

The next step is to calculate the branch impedance seen by each of the incident waves, from which we can obtain the load impedance, using the circuit in either Fig.3.6 or Fig.3.9, or for that matter the zero order circuit in Fig.3.5. Note that the first and zero order load impedances are required to calculate the second order node parameters, as shown by Eq.(3.79). To illustrate , we again focus on

$^{+}V_{xy}(n,m,q)$. By way of preparation, it is first convenient to calculate the total impedance in each of the three parallel branches, denoted by $^{+}R_{Ixy}(n,m,q)$, $^{+}R_{IIxy}(n,m,q)$, and $^{+}R_{IIIxy}(n,m,q)$, where the subscript denotes the wave and the superscript removes the ambiguity as whether the wave is a forward or backward one. The I indicates the lowest branch, which is in the negative z direction, the II branch is the coplanar branch, and III is the elevated branch in the plus z direction. These additional node parameters are

$$^{+}R_{Ixy}(n,m,q) = R1_{yzQ}(n,m,q) + R2_{zy}(n,m,q) + R3_{yzQ}(n,m,q) \tag{3.85}$$

$$^{+}R_{IIxy}(n,m,q) = R2_{yx}(n,m,q) + R3_{xy}(n,m,q) + R4_{yx}(n,m,q) \tag{3.86}$$

$$^{+}R_{IIIxy}(n,m,q) = R1_{yzQ}(n,m,q) + R3_{yzQ}(n,m,q) + R4_{zy}(n,m,q) \tag{3.87}$$

The quasi-coupling parameters are obtained using either the first order expressions, Eqs.(3.75)-(3.76) or the second order ones from Eqs.(3.79) and (3.82). In the absence of either a 1 or 2 argument the above may assume either approximation , depending on which values for $R1_{yzQ}(n,m,q)$ and $R3_{yzQ}(n,m,q)$ are used. (The same applies to $RL1_{xy}(n,m,q)$, to be discussed). We reinforce the notation of the three branches by citing a wave $^{+}V_{yz}(n,m,q)$ propagating in the yz plane. The upper branch then will the be $^{+}R_{IIIyz}(n,m,1)$, which will be in the plus x direction with the $^{+}R_{IIIyz}(n,m,1)$ circuit existing in the xz plane.

The node parameters facilitate the calculation of the load impedance seen by the wave $^{+}V_{xy}(n,m,q)$, since we now know the parallel contributions, namely, $^{+}R_{Ixy}(n,m,q)$, $^{+}R_{IIxy}(n,m,q)$, $^{+}R_{IIIxy}(n,m,q)$, and $R(n,m,q)$, the latter contribution being the node resistor in the $Z_{xy}(n,m,q)$ line. Thus the load impedance seen by $^{+}V_{xy}(n,m,q)$, denoted $RL1_{xy}(n,m,q)$ is

$$RL1_{xy}(n,m,q) = [\{^{+}R_{Ixy}(n,m,q)\} * \{^{+}R_{IIxy}(n,m,q)\} * \{^{+}R_{IIIxy}(n,m,q)\} * R(n,m,q)]/D \tag{3.88}$$

where

$$D = \{^{+}R_{Ixy}(n,m,q)\} * \{^{+}R_{IIxy}(n,m,q)\} * \{^{+}R_{IIIxy}(n,m,q)\} + \{^{+}R_{Ixy}(n,m,q)\} *$$

$$\{^+R_{IIxy}(n,m,q)\}*R(n,m,q)+\{^+R_{IIxy}(n,m,q)\}*\{^+R_{IIIxy}(n,m,q)\}*R(n,m,q)$$
$$+\{^+R_{Ixy}(n,m,q)\}*\{^+R_{IIIxy}(n,m,q)*R(n,m,q) \tag{3.89}$$

As another example we cite the backward wave, $^-V_{xy}(n,m,q)$. The load impedance seen by this wave is $RL3_{xy}(n,m,q)$. The expression for $RL3_{xy}(n,m,q)$ has a form very similar to that of Eq.(3.88). The sole differences , as seen from the circuit, involves reversing the arrow directions for the branch impedances and replacing $R3_{xy}(n,m,q)$ with $R1_{xy}(n,m,q)$ in Eq.(3.86).

3.11(c):Transferred Quasi-Voltages in Cell Notation

Next we adopt the quasi- voltages themselves to cell notation, i.e., Δ_1 and Δ_3. The first and second order expressions for these voltages are identical except for the use of the proper load voltage, given by $^+V_{xy}(p,n,m,q)$ where p=1 (first order) or p=2(second order), and a similar notation for the node parameters. We utilize Eqs.(3.43)-(3.44) or (3.69) -(3.70), denoting the waves by $^-\Delta_{yz}(p,n,m,q)$ and $^+\Delta_{yz}(p,n,m+1,q)$. The first order voltage is then

$$^-\Delta_{yz}(1,n,m,q)=^+V_{xy}(1,n,m,q)R1_{yzQ}(1,n,m,q))/[(R1_{yzQ}(1,n,m,q)+R4_{zy}(n,m,q)+$$
$$R3_{yzQ}(1,n,m,q)]-^+V_{xy}(1,n,m,q)R1_{yzQ}(1,n,m,q))/$$
$$[(R1_{yzQ}(1,n,m,q)+R2_{zy}(n,m,q)+R3_{yzQ}(1,n,m,q)] \tag{3.90a}$$

$$^+\Delta_{yz}(1,n,m+1,q)=^+V_{xy}(1,n,m,q)R3_{yzQ}(1,n,m,q))/[(R1_{yzQ}(1,n,m,q)+R4_{zy}(n,m,q)+$$
$$R3_{yzQ}(1,n,m)]-^+V_{xy}(1,n,m,q)R3_{yzQ}(1,n,m,q))/$$
$$[(R1_{yzQ}(1,n,m,q)+R2_{zy}(n,m,q)+R3_{\ yzQ}(1,n,m,q)] \tag{3.90b}$$

where

$$^+V_{xy}(1,n,m,q)=2V_{xy}(n,m,q)RL1_{xy}(1,n,m,q)/[RL1_{xy}(1,n,m,q)+Z_{xy}(n,m,q)] \tag{3.91}$$

To obtain the second order expressions, $^-\Delta_{yz}(2,n,m,q)$, $^+\Delta_{yz}(2,n,m+1,q)$, we replace $^+V_{xy}(1,n,m,q)$ by $^+V_{xy}(2,n,m,q)$, and $RL1_{xy}(1,n,m,q)$ by $RL1_{xy}(2,n,m,q)$. The node parameters $R1_{yzQ}(1,n,m,q)$ and $R3_{yzQ}(1,n,m,q)$, in Eqs.(3.90a), (3.90b), are replaced by $R1_{yzQ}(2,n,m,q)$ and $R3_{yzQ}(2,n,m,q)$. We again emphasize that for non -quasi scattering the coefficients are calculated on the basis of the usual circuit representation.

With the adoption of the node parameters and quasi-voltages to cell notation, we can now do the same for the scattering coefficients, which now will include both aplanar and coplanar forms. A total of 96 coefficients exist, consisting of 84 transfer type and 12 reflection type. These numbers may be inferred from the fact that there are 12 possible waves incident on each node, which accounts for the 12 reflection coefficients. And since each line is connected to 7 other lines at the node, there are 84 transfer coefficients, thus accounting for the total of 96 coefficients. Thus for the transfer coefficient, $T_{**}(n,m,q, s)$, the routing index ,s, will range from 1 to 84, and for the reflection coefficient , $B_{**}(n,m,q,s)$, the s index will range from 1 to 12.

3.11(d): Scattering Coefficients in Cell Notation

We first illustrate the scattering, using the same node $(n-1,m,q)$, as in the pure coplanar case, discussed earlier, in which wave energy is scattered from an incident wave in the $Z_{xy}(n-1,m,q)$ line into the $Z_{xy}(n,m,q)$ line. The formal expression for this coefficient, $T_{xy}(n-1,m,q,1)$ is identical to that of Eq.(3.31), except that $RL1_{xy}(n-1,m,q)$ will differ since the equivalent circuit now is given by Fig. 3.9. Thus $RL1_{xy}(n-1,m,q)$ will be given by Eq.(3.88), but with n replaced by n-1. The fourth argument in $T_{xy}(n-1,m,q,1)$,   specifies the routing of the wave, which is the same as before, namely, s=1.

Next consider scattering at the (n,m,q) node in which energy is scattered from an incident wave in the $Z_{xy}(n,m,q)$ line to the $Z_{zy}(n,m,q)$ line, thus representing normal aplanar scattering, designated by $T_{xy}(n,m,q, s)$, and given by

$$T_{xy}(n,m,q,53)=\{2RL1_{xy}(n,m,q)/[RL1_{xy}(n,m,q)+Z_{xy}(n,m,q)]\}*$$
$$\{R2_{zy}(n,m,q)/R_{1xy}(n,m,q)\} \tag{3.92}$$

The first factor within the { } bracket is recognized as the total load voltage delivered by $Z_{xy}(n,m,q)$ to the load (for a unit amplitude input), while the second is the fraction allotted to $Z_{zy}(n,m,q)$ in the I branch. The identification of the routing index as s=53   (as well as for other examples to follow) makes use of Tables 3.7 and 3.8 , and will be discussed at the end of the Section. We again reiterate that in the above the load impedance, branch impedances, and node parameters may assume either first or second order values.

Another example is the scattering from the $Z_{xy}(n,m,q)$ line, via the (n,m,q) node, to the $Z_{yz}(n,m,q)$ line. This is an example of a quasi -scattering process. As such, we may use , for example, the expressions for the coupling voltage factor $^-\Delta_{yz}(p,n,m,q)$ to describe the wave transfer to the $Z_{yz}(n,m,q)$ line.   In cell notation, the incident wave $^+V_{xy}(n,m,q)$ is scattered into $^-V_{yz}(n,m,q)$ according to

$$^-V_{yz}(n,m,q) = \; ^+V_{xy}(n,m,q) \; T_{xyQ}(n,m,q,41) \qquad (3.93)$$

where $^-V_{yz}(n,m,q) = \; ^-\Delta_{yz}(p,n,m,q)$ represents the quasi-coupled voltage to the $Z_{yz}(n,m,q)$ line. If we set $^+V_{xy}(n,m,q)$ equal to one then $T_{xyQ}(n,m,q,41)$ is

$$T_{xyQ}(n,m,q,41)= \; ^-\Delta_{yz}(p,n,m,q) \qquad (3.94)$$

where $^-\Delta_{yz}(p,n,m,q)$ is given by Eq.(3.90a)(for second order 1 is replaced by 2). A similar scattering coefficient may be determined for the forward quasi-scattered wave

$$^+V_{yz}(n,m+1,q) = \; ^+V_{xy}(n,m,q) \; T_{xyQ}(n,m,q,34) \qquad (3.95)$$

where we note that scattered wave belongs to the $(n,m+1,q)$ cell. Again setting $^+V_{xy}(n,m,q)$ equal to one, we obtain

$$T_{xyQ}(n,m,q,34) = \; ^-\Delta_{yz}(p,n,m+1,q) \qquad (3.96)$$

where $^-\Delta_{yz}(p,n,m+1,q)$ is given by Eq.(3.90b) or its second order version. In the above equations, the scattering is to outside the (n,m,q) cell; to obtain the scattering into the (n,m,q) cell we use $m+1\rightarrow m$ and $m\rightarrow m-1$ in the above.

We then proceed to obtain the coefficients for scattering into the xy plane ; the coefficients as well as the fields for scattering into the other planes, yz and zx, follow by symmetry. Table 3.7 provides the listing of the 32 coefficients for the xy plane(i.e., the scattering of fields into the xy plane belonging to the (n,m,q) cell). Twenty eight of them are transfer type and four are reflection type.

TABLE 3.7 SCATTERING COEFICIENTS: BOTH COPLANAR AND APLANAR CONTRIBUTIONS <u>INTO</u> UNIT CELL LINES $Z_{xy}(n,m,q)$ AND $Z_{yx}(n,m,q)$ (XY PLANE).

TRANSFER TYPE

<u>COEFFICIENT</u>	<u>FROM</u>	<u>TO</u>
$T_{xy}(n-1,m,q,1)$	$Z_{xy}(n-1,m,q)$	$Z_{xy}(n,m,q)$
$(-)T_{yx}(n-1,m,q,2)$	$Z_{yx}(n-1,m,q)$	$Z_{xy}(n,m,q)$
$T_{yx}(n-1,m,q,3)$	$Z_{yx}(n-1,m+1,q)$	$Z_{xy}(n,m,q)$
$T_{zy}(n-1,m,q,4)$	$Z_{zy}(n-1,m,q)$	$Z_{xy}(n,m,q)$
$T_{zy}(n-1,m,q,5)$	$Z_{zy}(n-1,m,q+1)$	$Z_{xy}(n,m,q)$
$T_{zxQ}(n-1,m,q,6)$	$Z_{zx}(n-1,m,q)$	$Z_{xy}(n,m,q)$
$T_{zxQ}(n-1,m,q,7)$	$Z_{zx}(n-1,m,q+1)$	$Z_{xy}(n,m,q)$
$T_{xy}(n,m,q,8)$	$Z_{xy}(n+1,m,q)$	$Z_{xy}(n,m,q)$
$(-)T_{yx}(n,m,q,9)$	$Z_{yx}(n,m+1,q)$	$Z_{xy}(n,m,q)$
$T_{yx}(n,m,q,10)$	$Z_{yx}(n,m,q)$	$Z_{xy}(n,m,q)$
$T_{zy}(n,m,q,11)$	$Z_{zy}(n,m,q)$	$Z_{xy}(n,m,q)$
$T_{zy}(n,m,q+1,12)$	$Z_{zy}(n,m,q+1)$	$Z_{xy}(n,m,q)$
$T_{zxQ}(n,m,q,13)$	$Z_{zx}(n,m,q)$	$Z_{xy}(n,m,q)$
$T_{zxQ}(n,m,q,14)$	$Z_{zx}(n,m,q+1)$	$Z_{xy}(n,m,q)$
$(-)T_{xy}(n,m-1,q,15)$	$Z_{xy}(n,m-1,q)$	$Z_{yx}(n,m,q)$
$T_{yx}(n,m-1,q,16)$	$Z_{yx}(n,m-1,q)$	$Z_{yx}(n,m,q)$
$T_{(xy}(n,m-1,q,17)$	$Z_{xy}(n+1,m-1,q)$	$Z_{yx}(n,m,q)$
$T_{zx}(n,m-1,q,18)$	$Z_{zx}(n,m-1,q)$	$Z_{yx}(n,m,q)$
$T_{zx}(n,m-1,q,19$	$Z_{zx}(n,m-1,q+1)$	$Z_{yx}(n,m,q)$
$T_{zyQ}(n,m-1,q,20)$	$Z_{zy}(n,m-1,q)$	$Z_{yx}(n,m,q)$
$T_{zyQ}(n,m-1,q,21)$	$Z_{zy}(n,m-1,q+1)$	$Z_{yx}(n,m,q)$
$T_{xy}(n,m,q,22)$	$Z_{xy}(n,m,q)$	$Z_{yx}(n,m,q)$
$T_{yx}(n,m+1,q,23)$	$Z_{yx}(n,m+1,q)$	$Z_{yx}(n,m,q)$
$(-)T_{xy}(n,m,q,24)$	$Z_{xy}(n+1,m,q)$	$Z_{yx}(n,m,q)$
$T_{zx}(n,m,q,25)$	$Z_{zx}(n,m,q)$	$Z_{yx}(n,m,q)$
$T_{zx}(n,m,q,26)$	$Z_{zx}(n,m,q+1)$	$Z_{yx}(n,m,q)$
$T_{zyQ}(n,m,q,27)$	$Z_{zy}(n,m,q)$	$Z_{yx}(n,m,q)$
$T_{zyQ}(n,m,q,28)$	$Z_{zy}(n,m,q+1)$	$Z_{yx}(n,m,q)$

REFLECTION TYPE

$B_{xy}(n-1,m,q,1)$	$Z_{xy}(n,m,q)$	$Z_{xy}(n,m,q)$
$B_{xy}(n,m,q,2)$	$Z_{xy}(n,m,q)$	$Z_{xy}(n,m,q)$
$B_{yx}(n,m-1,q,3)$	$Z_{yx}(n,m,q)$	$Z_{yx}(n,m,q)$
$B_{yx}(n,m,q,4)$	$Z_{yx}(n,m,q)$	$Z_{yx}(n,m,q)$

As with the coplanar arrays, the elements may be regarded as four dimensional arrays, in which the indices (n,m,q) provide the node location , and the index s specifies the routing of the wave. In general the array values will change with each time step, since they depend on the time dependent conductivity. The previous discussion and Eqs.(3.92)-(3.96) illustrate the technique for obtaining the twenty eight scattering coefficients in terms of the explicit circuit parameters. A listing of the coefficients without quasi-coupling(about node in xy plane), expressed in their circuit parameters , is given in App.3A.3 . Once more, symmetry arguments may be invoked to obtain the expressions for the coefficients in the yz and zx planes, based on a knowledge of the coefficients for scattering in the xy plane. Table 3.8 indicates the transformations used to obtain the scattering coefficients, as well as the wave voltages and node parameters, for the yz and zx plane The transformations are similar to the coplanar type found in Table 3.6. We should note that under the column for the xy plane the fields V_{zx} and V_{xy} and the corresponding coefficients T_{zx} and T_{zy} appear as may be verified by inspecting the iterative equations, to be discussed in the next Section. This is due of course to the aplanar scattering which now occurs as part of the iterative process. The remaining coefficients and iterative equations for scattering into the yz and zx planes follow from symmetry using Tables 3.7 and 3.8, as will be discussed.

One should add that in the event that we wish to omit quasi coupling then we set all the scattering coefficients involving quasi-coupling (identified by the subscript Q) equal to zero. Alternatively we may also achieve the same effect by setting the appropriate node parameters(those containing a Q subscript, e.g., $R1_{xyQ}(n,m,q)$ or $R1_{xyQ}(n,m,q)$) equal to zero.

We now return to the issue of identifying the s index for scattering into the yz and zx planes, and cite the examples discussed earlier. The first was the aplanar example in which the forward wave $^{+}V_{xy}(n,m,q)$ is scattered into $Z_{zy}(n,m,q)$ in which the wave is a backward type, $^{-}V_{zy}(n,m,q)$. In order to proceed, based on the scattering results to the xy plane, we must first transform the scattered field $^{-}V_{zy}(n,m,q)$. back to $^{-}V_{yx}(n,m,q)$. and similarly the incident field $^{+}V_{xy}(n,m,q)$ is transformed back to $^{-}V_{zx}(n,m,q)$ (i.e., z→y, x→z, and y→x). We then "look up" Table 3.7 and determine that the only entry which corresponds to scattering of the field from the $Z_{zx}(n,m,q)$ to the $Z_{yx}(n,m,q)$ lines is given by the

TABLE 3.8 TRANSFORMATION OF 3D SCATTERING COEFFICIENTS AND WAVE VOLTAGES TO YZ AND ZX PLANES. BOTH COPLANAR AND APLANAR WAVES.

XY PLANE		YZ PLANE	ZX PLANE
X	$\rightarrow$	Y	Z
Y	$\rightarrow$	Z	X
Z	$\rightarrow$	X	Y
$V_{xy}(n,m,q)$	$\rightarrow$	$V_{yz}(n,m,q)$	$V_{zx}(n,m,q)$
$V_{yx}(n,m,q)$	$\rightarrow$	$V_{zy}(n,m,q)$	$V_{xz}(n,m,q)$
$V_{zx}(n,m,q)$	$\rightarrow$	$V_{xy}(n,m,q)$	$V_{yz}(n,m,q)$
$V_{zy}(n,m,q)$	$\rightarrow$	$V_{xz}(n,m,q)$	$V_{yx}(n,m,q)$
$T_{xy}(n,m,q,s)$	$\rightarrow$	$T_{yz}(n,m,q,s+28)$	$T_{zx}(n,m,q,s+56)$
$T_{yx}(n,m,q,s)$	$\rightarrow$	$T_{zy}(n,m,q,s+28)$	$T_{xz}(n,m,q,s+56)$
$T_{zx}(n,m,q,s)$	$\rightarrow$	$T_{xy}(n,m,q,s+28)$	$T_{yz}(n,m,q,s+56)$
$T_{zy}(n,m,q,s)$	$\rightarrow$	$T_{xz}(n,m,q,s+28)$	$T_{yx}(n,m,q,s+56)$
$B_{xy}(n,m,q,s)$	$\rightarrow$	$B_{yz}(n,m,q,s+4)$	$B_{zx}(n,m,q,s+8)$
$B_{yx}(n,m,q,s)$	$\rightarrow$	$B_{zy}(n,m,q,s+4)$	$B_{xz}(n,m,q,s+8)$

GENERAL TRANSFORMATION(Δn, Δm, Δq =0 or ±1)

$V_{xy}(n+\Delta n,m+\Delta m, q+\Delta q) \rightarrow$	$V_{yz}(n+\Delta q,m+\Delta n, q+\Delta m)$	$V_{zx}(n+\Delta m,m+\Delta q, q+\Delta n)$
$V_{yx}(n+\Delta n,m+\Delta m, q+\Delta q) \rightarrow$	$V_{zy}(n+\Delta q,m+\Delta n, q+\Delta m)$	$V_{xz}(n+\Delta m,m+\Delta q, q+\Delta n)$
$V_{zx}(n+\Delta n,m+\Delta m, q+\Delta q) \rightarrow$	$V_{xy}(n+\Delta q,m+\Delta n, q+\Delta m)$	$V_{yz}(n+\Delta m,m+\Delta q, q+\Delta n)$
$V_{zy}(n+\Delta n,m+\Delta m, q+\Delta q) \rightarrow$	$V_{xz}(n+\Delta q,m+\Delta n, q+\Delta m)$	$V_{yx}(n+\Delta m,m+\Delta q, q+\Delta n)$
$T_{xy}(n+\Delta n,m+\Delta m, q+\Delta q,s) \rightarrow$	$T_{yz}(n+\Delta q,m+\Delta n, q+\Delta m,s+28)$	$T_{zx}(n+\Delta m,m+\Delta q, q+\Delta n,s+56)$
$T_{yx}(n+\Delta n,m+\Delta m, q+\Delta q,s) \rightarrow$	$T_{zy}(n+\Delta q,m+\Delta n, q+\Delta m,s+28)$	$T_{xz}(n+\Delta m,m+\Delta q, q+\Delta n,s+56)$
$T_{zx}(n+\Delta n,m+\Delta m, q+\Delta q,s) \rightarrow$	$T_{xy}(n+\Delta q,m+\Delta n, q+\Delta m,s+28)$	$T_{yz}(n+\Delta m,m+\Delta q, q+\Delta n,s+56)$
$T_{zy}(n+\Delta n,m+\Delta m, q+\Delta q,s) \rightarrow$	$T_{xz}(n+\Delta q,m+\Delta n, q+\Delta m,s+28)$	$T_{yx}(n+\Delta m,m+\Delta q, q+\Delta n,s+56)$
$B_{xy}(n+\Delta n,m+\Delta m, q+\Delta q,s) \rightarrow$	$B_{yz}(n+\Delta q,m+\Delta n, q+\Delta m,s+4)$	$B_{zx}(n+\Delta m,m+\Delta q, q+\Delta n,s+8)$
$B_{yx}(n+\Delta n,m+\Delta m, q+\Delta q,s) \rightarrow$	$B_{zy}(n+\Delta q,m+\Delta n, q+\Delta m,s+4)$	$B_{xz}(n+\Delta m,m+\Delta q, q+\Delta n,s+8)$

TRANSFORMATION OF NODE PARAMETERS

XY PLANE	YZ PLANE	ZX PLANE
$RL1_{xy}(n+\Delta n,m+\Delta m, q+\Delta q) \rightarrow$	$RL1_{yz}(n+\Delta q,m+\Delta n, q+\Delta m)$	$RL1_{zx}(n+\Delta m,m+\Delta q, q+\Delta n)$
$RL3_{xy}(n+\Delta n,m+\Delta m, q+\Delta q) \rightarrow$	$RL3_{yz}(n+\Delta q,m+\Delta n, q+\Delta m)$	$RL3_{zx}(n+\Delta m,m+\Delta q, q+\Delta n)$
$RL2_{yx}(n+\Delta n,m+\Delta m, q+\Delta q) \rightarrow$	$RL2_{zy}(n+\Delta q,m+\Delta n, q+\Delta m)$	$RL2_{xz}(n+\Delta m,m+\Delta q, q+\Delta n)$
$RL4_{yx}(n+\Delta n,m+\Delta m, q+\Delta q) \rightarrow$	$RL4_{zy}(n+\Delta q,m+\Delta n, q+\Delta m)$	$RL4_{xz}(n+\Delta m,m+\Delta q, q+\Delta n)$
$R1_{xy}(n+\Delta n,m+\Delta m, q+\Delta q) \rightarrow$	$R1_{yz}(n+\Delta q,m+\Delta n, q+\Delta m)$	$R1_{zx}(n+\Delta m,m+\Delta q, q+\Delta n)$
$R3_{xy}(n+\Delta n,m+\Delta m, q+\Delta q) \rightarrow$	$R3_{yz}(n+\Delta q,m+\Delta n, q+\Delta m)$	$R3_{zx}(n+\Delta m,m+\Delta q, q+\Delta n)$
$R2_{yx}(n+\Delta n,m+\Delta m, q+\Delta q) \rightarrow$	$R2_{zy}(n+\Delta q,m+\Delta n, q+\Delta m)$	$R2_{xz}(n+\Delta m,m+\Delta q, q+\Delta n)$
$R4_{yx}(n+\Delta n,m+\Delta m, q+\Delta q) \rightarrow$	$R4_{zy}(n+\Delta q,m+\Delta n, q+\Delta m)$	$R4_{xz}(n+\Delta m,m+\Delta q, q+\Delta n)$

routing index s=25, i.e., the transfer coefficient is $T_{zx}(n,m,q,25)$. At this point we turn to Table 3.8 , which provides the the fields and coefficients for scattering into the yz and zx planes. Making use of the Table we regain the incident and scattered fields, $^{+}V_{xy}(n,m,q)$ and $^{-}V_{zy}(n,m,q)$, respectively, but in addition we also obtain the corresponding transfer coefficient which is $T_{xy}(n,m,q,s+25)$. Since s=25 the sought after coefficient is $T_{xy}(n,m,q,53)$, as stated previously.

Next we cite another example in which the forward wave $^{+}V_{xy}(n,m,q)$ is scattered , in quasi-coupling fashion , to the $Z_{yz}(n,m,q)$ line. The scattered wave in this case is a backward one, $^{-}V_{yz}(n,m,q)$. Just as before we transform the fields so that the scattered field is in the xy plane, i.e., $^{-}V_{xy}(n,m,q)$, while the incident field becomes as before $^{+}V_{zx}(n,m,q)$. Referring to Table 3.7 we see that the entry for s=13 satisfies the necessary scattering requirements, and therefore the transfer coefficient $T_{zxQ}(n,m,q)$ is the appropriate quantity. Next we resort to the transformation to the yz plane where we again retrieve the incident and scattered fields, $^{+}V_{xy}(n,m,q)$ and $^{-}V_{yz}(n,m,q)$ as well as the transfer coefficient $T_{xyQ}(n,m,q, 41)$, i.e., s=13+28 =41. Naturally quasi scattering in one plane, when transformed to another plane, remains a quasi scattering process.

Next we cite a two examples in which the indices change upon transformation. First, suppose $^{+}V_{xy}(n,m,q)$ is scattered, in normal aplanar fashion, as a forward wave, becoming $^{+}V_{zy}(n,m,q+1)$. Extra care must be exercised because of the fact that the scattered wave is in the (n,m,q+1) cell. When we transform back obtain the field in the xy plane we must reduce the q index by 1 since we require the scattered wave to be in the (n,m,q) cell. Thus we should consider the incident wave to be $^{+}V_{xy}(n,m,q-1)$ and the scattered wave to be $^{+}V_{zy}(n,m,q)$.When we transform the scattered wave to the xy plane, the scattered wave becomes $^{+}V_{yx}(n,m,q)$ while the incident wave becomes $^{+}V_{zx}(n,m-1,q)$. Note that the -1 change in the q index has moved over to the m index in accordance with the transformation. We then examine Table 3.7 and observe that the coefficient wth s=18 satisfies the scattering process and thus the coefficient is $T_{zx}(n,m-1,q,18)$. As before, to obtain the scattering in the yz plane we then go to transformation Table 3.8, from which we regain the incident and scattered fields $^{+}V_{xy}(n,m,q-1)$ and $^{+}V_{zy}(n,m,q)$ as well as the transfer coefficient $T_{xy}(n,m,q-1,s+28)$, and since s=18 for the xy plane, the new coefficient is $T_{xy}(n,m,q-1, 46)$ with s=46.

Finally we consider the quasi-coupling of the incident wave $^+V_{xy}(n,m,q)$, scattered to the forward wave $^+V_{yz}(n,m+1,q)$. As before, we want the scattering to be in the (n,m,q) cell, and therefore m should be reduced by one in both the incident and scattered waves, giving us $^+V_{xy}(n,m-1,q)$ and $^+V_{yz}(n,m,q)$ respectively. Transforming the scattering to the xy plane, the incident wave becomes $^+V_{zx}(n-1,m,q)$ and he scattered wave becomes $^+V_{xy}(n,m,q)$. Examining Table 3.7 we see that s=6 provides the correct routing and the transfer coefficient is $T_{zx}(n-1,m,q,6)$. We then use Table 3.8 to transfer to the scattering in the yz plane, which yields a coefficient index of s+28=34, and thus the coefficient is $T_{xy}(n,m-1,q,34)$.

From the previous examples we see how we may generate Tables comparable to 3.7 for the scattering coefficients in the yz and zx planes, allowing us to obtain the full set of 96 coefficients properly identified by the routing index. App. 3A.2 lists the additional coefficients for the yz and zx planes. We can then obtain the coefficients in terms of the circuit parameters as illustrated in Section 3.11d and stated in App.3A.3. The scattering to the various planes may then easily expressed in computer code.

3.12 Complete Iterative Equations

With the scattering coefficients now specified, we are now in a position to write down the iterative equations for the waves about a given cell, (n,m,q). As before, in the coplanar case, we need only concentrate on the six lines associated with the (n,m,q) node, namely, $Z_{xy}(n,m,q)$, $Z_{yx}(n,m,q)$, $Z_{yz}(n,m,q)$, $Z_{zy}(n,m,q)$, $Z_{zx}(n,m,q)$, and $Z_{xz}(n,m,q)$. It is convenient, once more , to examine the waves in a particular propagation plane, since many of the results may be carried over to the waves in the other two planes as well. We first look at the lines in the xy plane, i.e., the $Z_{xy}(n,m,q)$ and $Z_{yx}(n,m,q)$ lines. We wish to determine what waves from surrounding lines, together with reflected in the lines themselves, contribute to the fields in these lines to produce the follow-on fields for the next time step. Consider first the forward waves in the $Z_{xy}(n,m,q)$ line. The forward wave $^+V^{k+1}_{xy}(n,m,q)$, at time $t+\Delta t$, may be expressed in terms of the waves surrounding the line at time t. Thus

$$^+V^{k+1}{}_{xy}(n,m,q)=T^k{}_{xy}(n-1,m,q,1)\ ^+V^k{}_{xy}(n-1,m,q)-T^k{}_{yx}(n-1,m,q,2)\ ^+V^k{}_{yx}(n-1,m,q)$$

$$+T^k{}_{yx}(n-1,m,q,3)^-V^k{}_{yx}(n-1,m+1,q)+B^k{}_{xy}(n-1,m,q,1)\ ^-V^k{}_{xy}(n,m,q)$$

$$+\ T^k{}_{zy}(n-1,m,q,4)\ ^+V^k{}_{zy}(n-1,m,q)+\ T^k{}_{zy}(n-1,m,q,5)\ ^-V^k{}_{zy}(n-1,m,q+1)$$

$$+T^k{}_{zxQ}(n-1,m,q,6)^+V^k{}_{zx}(n-1,m,q)+T^k{}_{zxQ}(n-1,m,q,7)^-V^k{}_{zx}(n-1,m,q+1) \qquad (3.97)$$

The first four terms are nothing more than the coplanar terms encountered earlier in the Chapter. Formally the terms appear identical; the scattering coefficients will of course differ since the equivalent circuit has changed. The next four terms arise from scattering normal to the xy plane at the $(n-1,m,q)$ node. The fifth term, for example, represents the scattering of the $^+V^k{}_{zy}(n-1,m,q)$ wave into the $Z_{xy}(n,m,q)$ line, and similarly, the sixth term denotes the scattering of the $^-V^k{}_{zy}(n-1,m,q+1)$ wave. The last two terms represent the quasi-coupling of the $^+V^k{}_{zx}(n-1,m,q)$ and $^-V^k{}_{zx}(n-1,m,q+1)$ waves to the $Z_{xy}(n,m,q)$ line.

We complete the wave picture in the xy plane by stating the iterative equations for the backward wave $^-V^{k+1}{}_{xy}(n,m,q)$, and the waves in the $Z_{yx}(n,m,q)$ line. Thus

$$^-V^{k+1}{}_{xy}(n,m,q),=T^k{}_{xy}(n,m,q,8)\ ^-V^k{}_{xy}(n+1,m,q)-\ T^k{}_{yx}(n,m,q,9)\ ^-V^k{}_{yx}(n,m+1,q)$$

$$+T^k{}_{yx}(n,m,q,10)\ ^+V^k{}_{yx}(n,m,q)+\ B^k{}_{xy}(n,m,q,2)\ ^+V^k{}_{xy}(n,m,q)$$

$$+\ T^k{}_{zy}(n,m,q,11)\ ^+V^k{}_{zy}(n,m,q)+\ T^k{}_{zy}(n,m,q+1,12)\ ^-V^k{}_{zy}(n,m,q+1)$$

$$+\ T^k{}_{zxQ}(n,m,q,13)\ ^+V^k{}_{zx}(n,m,q)+\ T^k{}_{zxQ}(n,m,q+1,14)\ ^-V^k{}_{zx}(n,m,q+1) \qquad (3.98)$$

$$^+V^{k+1}{}_{yx}(n,m,q)=-T^k{}_{xy}(n,m-1,q,15)^+V^k{}_{xy}(n,m-1,q)+T^k{}_{yx}(n,m-1,q,16)^+V^k{}_{yx}(n,m-1,q)$$

$$+T^k{}_{xy}(n,m-1,q,17)\ ^-V^k{}_{xy}(n+1,m-1,q)+\ B^k{}_{yx}(n,m-1,q,3)\ ^-V^k{}_{yx}(n,m,q)$$

$$+\ T^k{}_{zx}(n,m-1,q,18)\ ^+V^k{}_{zx}(n,m-1,q)+\ T^k{}_{zx}(n,m-1,q,19)\ ^-V^k{}_{zx}(n,m-1,q+1)$$

$$+T^k_{zyQ}(n,m\text{-}1,q,20)\,^+V^k_{zy}(n,m\text{-}1,q)+T^k_{zyQ}(n,m\text{-}1,q,21)\,^-V^k_{zy}(n,m\text{-}1,q+1) \qquad (3.99)$$

$$^-V^{k+1}_{yx}(n,m,q)=T^k_{xy}(n,m,q,22)\,^+V^k_{xy}(n,m,q)+\,T^k_{yx}(n,m+1,q,23)\,^-V^k_{yx}(n,m+1,q)$$

$$-\,T^k_{xy}(n,m,q,24)\,^-V^k_{xy}(n+1,m,q)+B^k_{yx}(n,m,q,4)\,^+V^k_{yx}(n,m,q)$$

$$+T^k_{zx}(n,m,q,25)\,^+V^k_{zx}(n,m,q)+\,T^k_{zx}(n,m,q,26)\,^-V^k_{zx}(n,m,q+1)$$

$$+\,T^k_{zyQ}(n,m,q,27)\,^+V^k_{zy}(n,m,q)+\,T^k_{zyQ}(n,m,q,28)\,^-V^k_{zy}(n,m,q+1) \qquad (3.100)$$

Just as we did in the coplanar case, we may obtain the remaining iterative equations(as well as he scattering coefficients) from symmetry considerations. To do this, we transform each term in the iterative equation, whether it be a wave voltage or scattering coefficient, using the prescription in Table 3.8. App.3A.1 provides the remaining iterative equations for the yz and zx planes.

As mentioned previously, we can greatly simplify the iterative equations, at the expense of some resolution, by neglecting quasi-coupling. Thus any transfer coefficient with a Q subscript, displayed in the iterative equations, is set equal to zero. This reduces the number of transfer coefficients from 84 to 60. In addition we utilize the zero order expressions for the node parameters, or equivalently, set equal to zero any node parameters with a Q subscript(e.g., $R1_{yzQ}(1,n,m,q)$).

Eqs.(3.97)-(3.100) and the companion ones for the yz and zx planes in App.3A.1, represent the core of the iterative process. All the input information, via the scattering coefficients, is routed by means of these equations They perform the simple but critical task of relating the new state to the old one. Because of their importance, these equations may be regarded the "crown jewels" of any computer program using the TLM formulation.

3.13 Contribution of Electric and Magnetic Fields to Total Energy

Although the scattering equations provides us with the electric field amplitudes throughout the TLM space, it is often useful to compare the electric and magnetic field energies in the space at any one moment. This allows us to determine if there are regions dominated by either field, or whether a region contains significant contributions from both types of energy. In fact, we shall see in the following Sections, which describe plane waves incident on a cell matrix, that certain TLM lines will be dominated by magnetic energy while others will be dominated by the electric field. We know from previous comments that when the field is dominated by the electric field, usually the forward and backward waves are about equal,

$$^{+}V = {}^{-}V \tag{3.101}$$

(For this Section, we drop the cell designation). Such is the case, e.g., when static conditions prevail. On the hand, when the magnetic field dominates, the forward and backward waves have opposite amplitudes, i.e.,

$$^{+}V = -{}^{-}V \tag{3.102}$$

We employ a straightforward technique for finding the relative contributions to the total energy. We first calculate the energy levels residing in the current and voltage(i.e., in the magnetic and electric fields), designated respectively by U_h and U_e. These quantities may be estimated from

$$U_h = (1/2)({}^{+}I + {}^{-}I)^2 Z\, \Delta t = (1/2)(({}^{+}V/Z) - {}^{-}V/Z))^2 Z\, \Delta t \tag{3.103}$$

$$U_e = (1/2)({}^{+}V + {}^{-}V)^2 \Delta t / Z \tag{3.104}$$

The total energy U_t is given by

$$U_t = U_h + U_e = [{}^{+}V^2 /Z + {}^{-}V^2 /Z]\Delta t \tag{3.105}$$

using Eqs.(3.103) and (3.104). We may interpret Eq.(3.105) to say that the total energy is the sum of the energies associated with the forward and backward waves. It is often convenient to define energy partition parameters

$$r_h = (U_h / U_t) \tag{3.106a}$$
$$r_e = (U_e / U_t) \tag{3.106b}$$
$$r_h + r_e = 1 \tag{3.106c}$$

Certain special cases are of interest. If $^+V = \,^-V$, then $r_e = 1$ and $r_h = 0$. On the other hand if $^+V = -\,^-V$, then $r_e = 0$ and $r_h = 1$, as expected. A very important case occurs when $^+V = 0$ or $^-V = 0$. Here $r_e = r_h = 1/2$, i.e., in a plane wave the magnetic and electric energies share equally and the total energy in ^+V or ^-V (in which ^+V and ^-V do not co-exist) are

$$U_t = (^+V)^2 \, \Delta t/Z \quad \text{(forward wave)} \tag{3.107}$$

and similarly

$$U_t = (^-V)^2 \Delta t/Z \quad \text{(backward wave)} \tag{3.108}$$

Plane Wave Behavior

3.14 Response of 2D Cell Matrix to Input Plane Wave

The TLM scattering equations should enable one, in principle, to predict the dynamic electromagnetic behavior of a medium undergoing conductivity changes. Before applying this tool to a detailed problem, however, we must investigate whether the TLM method is compatible with a simple but obviously important phenomenon, that of a plane wave propagating throughout the space. We perform this investigation by finding out the response of a 2D cell matrix to an input plane wave.

To secure compatibility with the plane wave, certain kinds of behavior are required of the waves in a TLM matrix. Throughout a uniform region we insist that the reflected waves vanish, and conversely, that the amplitude front be maintained in the forward direction. Fig.3.11 indicates this more explicitly. We initially assume a 2D array in order to simplify the analysis. Assume a plane wave propagates in the x direction, and that during the kth time step the wave front occupies the n cells, i.e., (n, m-1), (n,m) (n,m+1), etc... In order for the plane wave to maintain itself, we require that that the reflected wave $^-V^k_{xy}(n,m)$

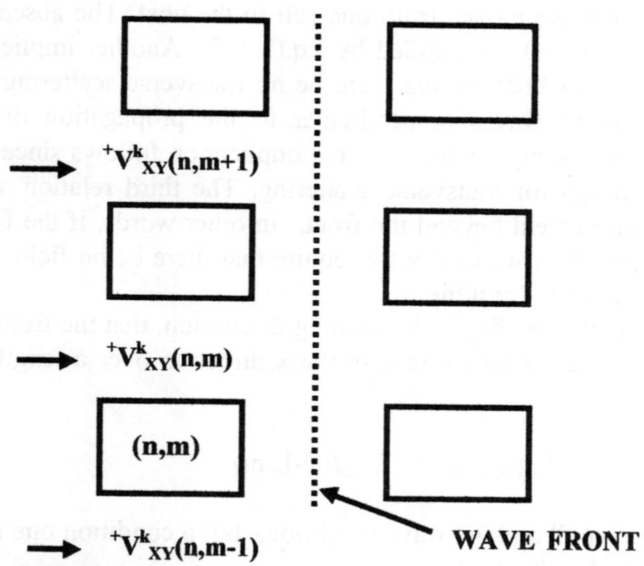

FIG.3.10 PLANE WAVE CONDITIONS IN UNIFORM REGION:
$^+V^k_{XY}(n,m-1) = {}^+V^k_{XY}(n,m) = {}^+V^k_{XY}(n,m+1)$, ETC...
$^+V^k_{XY}(n,m) = {}^+V^{k+1}_{XY}(n+1,m)$;
$^+V^k_{XY}(n,m) =0$ FOR n BEYOND FRONT

vanish and that the forward wave, expressed in the horizontal lines, be preserved as it moves from the nth cell to the (n+1)th cell for arbitrary values of m and n. The pertinent relationships in cell notation for the plane wave (2D)are, for a uniform region,

$$^+V^k_{xy}(n,m-1) = {}^+V^k_{xy}(n,m) = {}^+V^k_{xy}(n,m+1), \text{ etc....} \qquad (3.109)$$

$$^+V^k_{xy}(n,m) = {}^+V^{k+1}_{xy}(n+1,m) \qquad (3.110)$$

$$^+V^k_{xy}(n,m) =0 \text{ (n beyond front)} \qquad (3.111)$$

Eq.(3.109) simply states the uniformity of the field in the y direction. The second relation (3.110) requires that the forward wave in the horizontal lines remain intact as the wave moves from one cell to the next. The absence of wave reflection, $^-V^k_{xy}(n,m) = 0$, is implied by Eq.(3.110). Another implied condition stemming from Eq.(3.110) is that there be no transverse scattering of waves , i.e., no scattering of waves perpendicular to the propagation direction(e.g,, $^+V^k_{yx}(n,m) = 0$, $^-V^k_{yx}(n,m) = 0$, etc...). This implication follows since there is no leftover wave energy for transverse scattering. The third relation simply indicates the absence of field beyond the front. In other words, if the front is at n_F during the kth step then we obviously require that there be no fields beyond the front, or $^+V^k_{xy}(n,m) = 0$ for $n > n_F$.

Finally we may specify , in the ensuing discussion, that the front have some depth, i.e., that the field be uniform in the x direction over a length of several cells , or

$$^+V^k_{xy}(n_F, m) \cong {}^+V^k_{xy}(n_F-1, m) \tag{3.112}$$

The above is not at all a plane wave condition, but a condition one may impose to simplify particular situations.

In the previous discussion we focused on plane wave behavior in a uniform region. What happens when we maintain constancy in the yz plane, Eq.(109), but we allow for variations in the x direction? For example, the wave may encounter a dielectric or conducting semi-infinite yz plane, or else there may simply be variations of the dielectric constant in the x direction. In this case, the requirement set forth by Eq.(3.110) is obviously not valid, because of reflections due to impedance mismatch. Instead we must use the aforementioned, less stringent condition, which excludes transverse scattering ; thus for n at or behind the wave front:

$$^+V^k_{yx}(n,m) = 0 \; ; \; {}^-V^k_{yx}(n,m) = 0 \; ; \text{for all } m \tag{113}$$

while in the x direction, the scattering is controlled by the usual one dimensional relationships.

We now return to the original question posed at the start of this Section: does the present formulation of the TLM matrix allow for plane wave behavior?

To simplify the discussion, the remainder of the Chapter will consider only the case of the uniform region. We shall shortly observe, however, that even with this simplification, the plane wave relationships cannot be satisfied, using the TLM formulation, and significant modifications must be incorporated, as will be done in Chapter IV. It is extremely instructive, nevertheless, to determine the response of a plane wave launched in the present matrix. The results thus obtained will help guide us as to how to modify the TLM matrix in order to preserve the plane wave as it progresses in the matrix.

We first determine the behavior of a plane wave launched in a lossless, uniform 2D TLM matrix. Because of the zero loss, we simply apply the 2D scattering equations derived previously, setting the node resistors equal to very large values in the matrix. The plane wave properties introduce additional symmetries, which allows us to utilize semi-graphical techniques using only a limited portion of the matrix. Fig.3.11 shows a series of "snapshots" of the waves in the TLM matrix, up to k=3(for extra clarity in Fig.3.11, as well as in Fig.3.12, the wave direction is specified by placing the appropriate directional arrow immediately after the wave amplitude). Note that we need only consider the m,m-1, and m+1 lines which congregate about the m node From the aforementioned symmetry conditions, it is unnecessary to consider any other TLM lines in the y direction.

We impose certain initial boundary conditions on the input plane wave in Fig.3.11, as well as in the next three Sections. We assume the input horizontal lines(for which n=1), which contains the input forward plane wave, are much longer than any of the other lines. This enables the analysis to proceed without having to take into account reflections at the input, or back-scattering into the region behind n=1. Chapters V and VII further discuss this type of "long input" boundary condition. An alternative boundary condition makes use of a uniform, vertically directed, static electric field in the region x<0 , but with a high impedance boundary at the x=0 . Initially, therefore, the field is concentrated in the x<0 region , and there is no field in the x>0 region. The high impedance boundary is then suddenly removed and the TLM waves are allowed to enter the x>0 region. In any case, the response in the x>0 region, due to this boundary, is qualitatively the same as that due to long input lines.

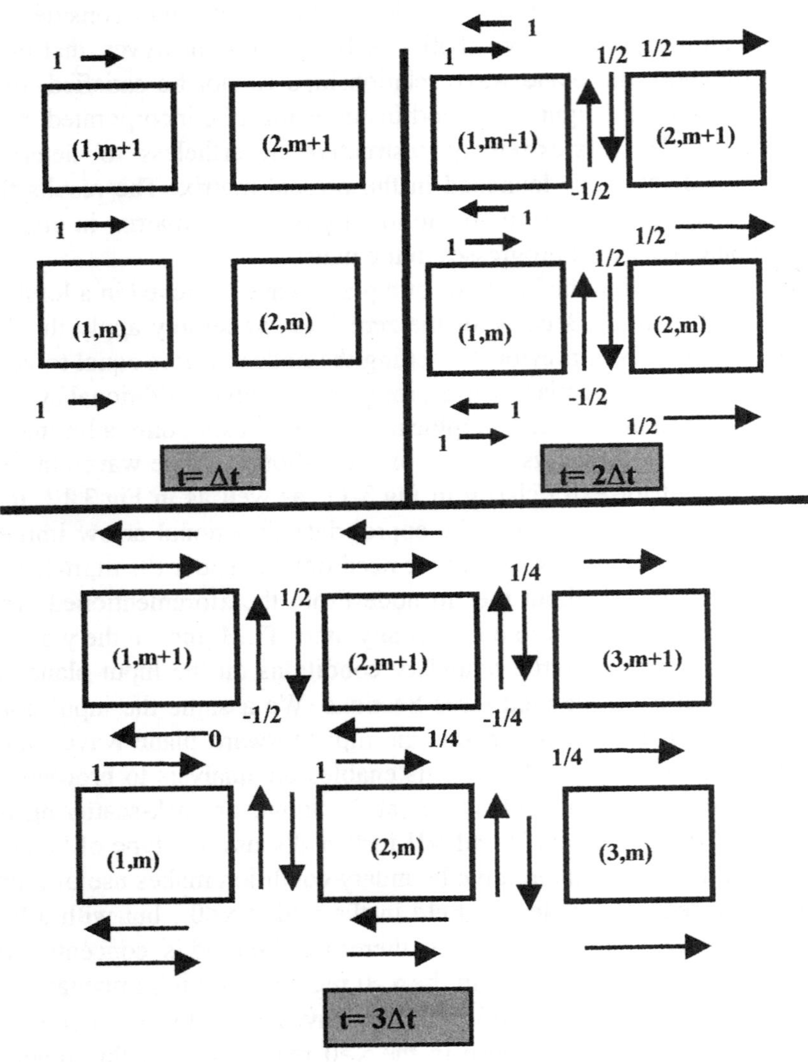

FIG. 3.11 FIELD EVOLUTION IN 2D MATRIX , DURING FIRST THREE TIME STEPS, DUE TO PLANE WAVE INPUT.   INPUT LINES(n=1) SHOULD BE CONSIDERED LONG COMPARED TO OTHER LINES.

We start the analysis using Fig.3.12 , which shows the wave $^+V^1_{xy}(1,m)$ occupying the $Z_{xy}(1,m)$ line, and representing a portion of the x directed plane wave front, with unity amplitude, at k=1(t = Δt).   We next consider the waves at t = 2 Δt in Fig.3.12. The incident wave is scattered into the $Z_{xy}(2,m)$, $Z_{yx}(1,m)$, and $Z_{yx}(1,m+1)$ lines, as well as reflected in the $Z_{xy}(n,m)$ line. We know from Chapter1 that for this type of 2D node, with identical line impedance values for each line,   each of the scattering coefficients(both the transfer and reflection type) is equal to 1/2. Thus the scattered waves from $Z_{xy}(1,m)$ are

$$^-V^2_{xy}(1,m) =1/2 \qquad (3.114a)$$
$$^+V^2_{xy}(2,m) = 1/2 \qquad (3.114b)$$
$$^-V^2_{yx}(1,m) =1/2 \qquad (3.114c)$$
$$^+V^2_{yx}(1,m+1) = -1/2 \qquad (3.114d)$$

We remind ourselves of the fact that $^+V^2_{yx}(1,m+1) = -1/2$ has a negative amplitude because of the boundary condition at the node. The wave picture is complete except for two additional waves incident on the (n,m) node in the transverse lines, $Z_{yx}(1,m,)$ and $Z_{yx}(1,m+1)$, as shown in Fig.3.11. These waves may be obtained from the aforementioned symmetry considerations, or simply by considering the input waves $^+V^1_{xy}(1,m-1)$ and $^+V^1_{xy}(1,m+1)$ in like manner. Thus,

$$^+V^2_{yx}(1,m) = {}^+V^2_{yx}(1,m+1) = -1/2 \qquad (3.115a)$$
$$^-V^2_{yx}(1,m+1) = {}^-V^2_{yx}(1,m) = 1/2 \qquad (3.115b)$$

Eqs.(3.114)-(3.115) complete the wave description at the end of t= 2 Δ t. We perform one additional iteration, t=3Δt, to reinforce the graphical technique. The waves at t = 3Δt will of course rely completely on the t= 2Δt results. We begin with the $Z_{xy}(1,m)$ line, which will consist of the unity incident wave and a backward wave, $^-V^1_{xy}(1,m)$. The backward wave will consist of a reflected wave and contributions from the $Z_{yx}(1,m)$ and $Z_{yx}(1,m+1)$ lines. Thus

$$^-V^3_{xy}(1,m) =(1/2)\,^+V^2_{xy}(1,m) -(1/2)\,^-V^2_{yx}(1,m+1) +(1/2)\,^+V^2_{yx}(1,m) \qquad (3.116)$$

Since

$$^+V^2_{yx}(1,m) = -1/2 \tag{3.117a}$$

$$^-V^2_{yx}(1,m+1) = 1/2 \tag{3.117b}$$

$$^+V^2_{xy}(1,m) = 1 \tag{3.117c}$$

we have

$$^-V^3_{xy}(1,m) = 0 \tag{3.118}$$

We next consider the $Z_{yx}(1,m)$ and $Z_{yx}(1,m+1)$ lines. $^+V^3_{yx}(1,m+1)$, for example, will consist of the wave scattered from the incident wave, $^+V^2_{xy}(1,m)$, the reflected wave of $^-V^2_{yx}(1,m+1)$, and the portion of the wave transferred from $^+V^2_{yx}1,m)$. Thus,

$$^+V^3_{yx}(1,m+1) = (-1/2)^+V^2_{xy}(1,m) + (1/2)^-V^2_{yx}(1,m+1) + (1/2)\,^+V^2_{yx}(1,m) \tag{3.119}$$

Since we know the wave amplitudes at $t = 2\Delta t$, from Eqs.(3.114) -(3.115), we see that the last two terms in Eq.(3.119) cancel , and we have

$$^+V^3_{yx}(1,m+1) = -1/2 \tag{3.120}$$

By symmetry,

$$^+V^3_{yx}(1,m) = -1/2 \tag{3.121}$$

Similar arguments, starting with $^-V^3_{yx}(1,m)$ lead to

$$^-V^3_{yx}(1,m) = 1/2 \tag{3.122a}$$

$$^-V^3_{yx}(1,m+1) = 1/2 \tag{3.122b}$$

Next we consider the waves in the $Z_{xy}(2,m)$ line. The forward wave will receive contributions from the $Z_{xy}(1, m)$, $Z_{yx}(1,m)$, and $Z_{yx}(1,m+1)$ lines, or,

$$^+V^3_{xy}(2,m) = (1/2)^+V^2_{xy}(1,m) + (1/2)\,^-V^2_{yx}(1,m+1) - (1/2)^+V^2_{yx}(1,m) \quad (3.123)$$

Using the wave amplitudes of the previous time step,

$$^+V^3_{xy}(2,m) = 1 \quad (3.124)$$

Thus, the plane wave amplitude is finally transferred from $Z(1,m)$ to $Z(2,m)$; the wave, however, experiences a delay of an additional time step. The backward wave in $Z_{xy}(2,m)$ is simply

$$^-V^3_{xy}(2,m) = (1/2)\,^+V^2_{xy}(1,m) = 1/4 \quad (3.125)$$

The scattering of $^+V^2_{xy}(2,m)$ determines the waves at $k=3$ in the $Z_{xy}(3,m)$, $Z_{yx}(2,m+1)$, and $Z_{yx}(2,m)$ lines. Thus $^+V^3_{yx}(2,m+1) = -(1/2)\,^+V^2_{xy}(2,m)$, $^-V^3_{yx}(2,m) = (1/2)\,^+V^2_{xy}(2,m)$, and $^+V^3_{xy}(3,m) = (1/2)\,^+V^2_{xy}(2,m)$. Using the wave amplitudes of the prior step, as well as the usual symmetry arguments, we obtain

$$^+V^3_{yx}(2,m) = {}^+V^3_{yx}(2,m+1) = -1/4 \quad (3..126)$$

$$^-V^3_{yx}(2,m+1) = {}^-V^3_{yx}(2,m) = 1/4 \quad (3.127)$$

$$^+V^3_{xy}(3,m) = 1/4 \quad (3.128)$$

Eqs.(3.116) and (3.118)-(3.128)constitute the waves present at the end of the third time step. The same process then is used to obtain the waves for the subsequent time steps. By the end of the seventh step, the fields, given in Figs.3.12, show certain trends which clearly indicate the need to modify the iteration. Fig. 3.13 shows the horizontal field $^+V^7_{xy}(n,m)$ as a function of n at the end of the seventh step. As expected $^+V^7_{xy}(n,m)$ has close to unity amplitude, at least for the first few cells. After n=4, however, the amplitude in $^+V^7_{xy}(n,m)$ begins to

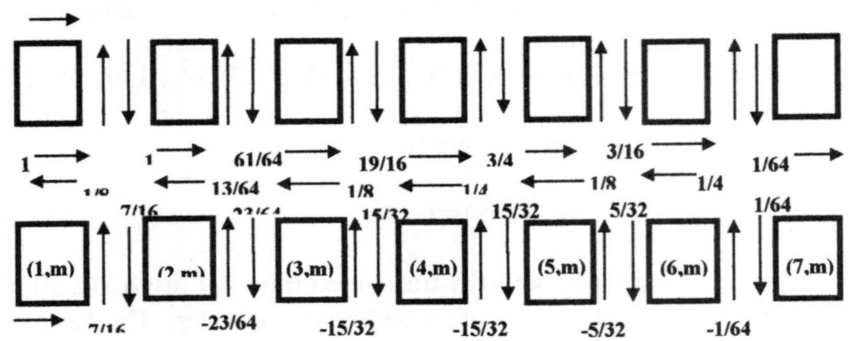

FIG.3.12 FIELD EVOLUTION IN 2D MATRIX AFTER SEVEN TIME STEPS, DUE TO INCIDENT PLANE WAVE FRONT. TRANSVERSE FIELDS FOR A GIVEN n ARE IDENTICAL.

VARIATION OF FORWARD WAVE AMPLITUDE WITH n AFTER K=7 TIME STEP

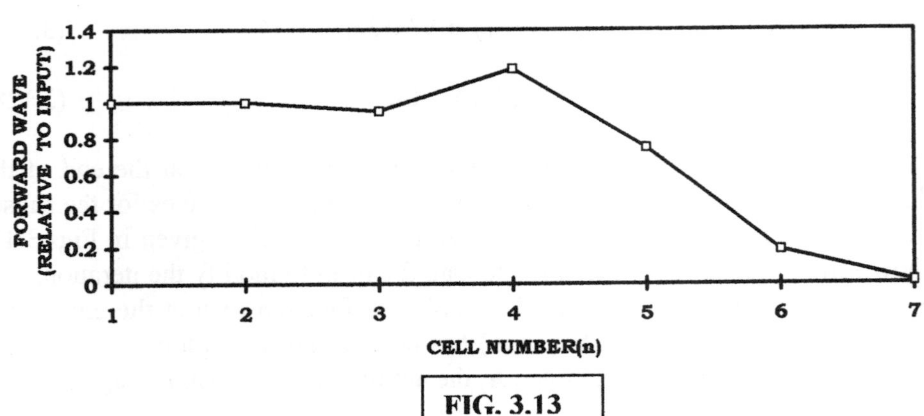

FIG. 3.13

diminish to values significantly less than one. The same amplitude diminution occurs in the transverse lines, $Z_{yx}(n,m)$ and $Z_{yx}(n,m+1)$, wherein the amplitudes approximates 1/2 for n=1 to 4, but drops off afterwards.

The computerized version of the previous graphical study confirms certain suspicions based on the observations of the graphical analysis carried out from n=1 to n=7. We find that for large n, the forward plane wave amplitude, $^{+}V^{k}_{xy}(n,m)$, is approximately unity, provided $n < k/2$, where k is of course the number of time steps. Beyond $n = k/2$, the amplitude begins to decline, finally approaching zero at n=k, which is the edge of the wave front. In fact this behavior is not surprising, based on the fact that the wave energy is traveling along a grid , and not entirely on a straight line path. Energy is therefore diverted into the transverse lines($Z_{yx}(n,m)$ and $Z_{yx}(n,m+1)$) and until these lines are "filled up" with their share of wave energy, the progress of the wave is slowed down. The effective velocity appears to be slowed down by about a factor of two, which again is not surprising if one considers the possible transverse paths taken by the waves, rather than the straight line path.

Other trends also are observed. As the number of time steps is increased, the reflected wave at the input, $^{-}V^{k}_{xy}(1,m)$, does not go to zero, as one would expect of a normal plane wave. In the interval k=4 to 9, for example, $^{-}V^{k}_{xy}(1,m)$ fluctuated between 0.125 and 0.250. Another observation has to do with the transverse waves. For $n < k/2$, these waves roughly satisfy

$$^{+}V^{k}_{yx}(n,m) = {}^{+}V^{k}_{yx}(n,m+1) \approx -1/2 \qquad (3.129a)$$

$$^{-}V^{k}_{yx}(n,m) = {}^{-}V^{k}_{yx}(n,m+1) \approx 1/2 \qquad (3.129.b)$$

and beyond n= k/2 the amplitudes diminish to ~ 0. Within each transverse line, for $n < k/2$, therefore, the energy may be classified as purely magnetic since the forward and backward waves are equal in amplitude and opposite in sign. In the case of the forward horizontal waves, $^{+}V^{k}_{xy}(n,m)$, the fields may be regarded as static in time (at least for $n < k/2$). We may then regard the energy in

$Z_{xy}(n,m)$ as purely electrical in nature and equal in magnitude to the magnetic energy, as expected for a pure transient plane wave.

3.15 Response of 2D Cell Matrix to Input Waves With Arbitrary Amplitudes

In this Section we determine the conditions under which a *limited* portion of a non-uniform input wave behaves simulates the behavior of a plane wave, as seen by the TLM matrix. Unlike the previous plane wave analysis, suppose we now allow the three forward waves, $^+V^1_{xy}(1,m-1)$, $^+V^1_{xy}(1,m)$, and $^+V^1_{xy}(1,m+1)$ to have arbitrary amplitudes, i.e., we assume the wave is in general non-uniform. As noted previously during the second time step the input waves fill up the longitudinal lines and partially transfer energy to $Z_{xy}(2,m)$. At the end of the third time step, the transfer of forward wave energy to $Z_{xy}(2,m)$ was more or less complete, as noted by the fact that $^+V^3_{xy}(2,m)$ was unity and approximately remained so for the subsequent steps. We apply the same criterion to the case when the three input waves have arbitrary amplitudes, carrying out the analysis to the third time step by which time the trend is observable. We then calculate $^+V^3_{xy}(2,m)$, using the same techniques described earlier, except for the differing input amplitudes. We then obtain the forward wave, in terms of the amplitudes during the first step,

$$^+V^3_{xy}(2,m)=(1/2)^+V^1_{xy}(1,m) + (1/4)[\,^+V^1_{xy}(1,m-1)+^+V^1_{xy}(1,m+1)] \qquad (3.130)$$

We now determine what constraints are imposed on the inputs if we require that the output during the third step satisfies

$$^+V^3_{xy}(2,m) \approx\, ^{=+}V^1_{xy}(1,m) \qquad (3.131)$$

If we combine Eqs.(3.130) and (3.131) we obtain the result

$$^+V^1_{xy}(1,m) = (1/2)[\,^+V^1_{xy}(1,m+1) + ^+V^1_{xy}(1,m-1) \qquad (3.132)$$

or, using more general notation,

$$^{+}V^{k}_{xy}(n,m) = (1/2)[\ ^{+}V^{k}_{xy}(n,m-1) + ^{+}V^{k}_{xy}(n,m+1)]\qquad(3.133)$$

We have the result, therefore, that the wave in the TLM matrix 'appears' to respond as though it were a plane wave, when Eq.(3.133) is satisfied. Eq.(3.133) states of course that $^{+}V^{k}_{xy}(n,m)$ is the average of the two neighboring forward waves at m-1 and m+1. We emphasize that the wave I the TLM matrix is only a "quasi plane wave" because of the unavoidable delay of the additional time step needed to fill up the transverse lines. The TLM formulation is corrected for this delay in the next Chapter.

3.16 Response of 3D Cell Matrix to Input Plane Wave

To explore the plane wave limit we consider a plane wave front, $^{+}V^{k}_{xy}(n,m,q)$, moving in the x direction with the front uniform throughout the yz plane. We assume that at k=1, i.e., t = Δt, the wave front occupies the first cell of the matrix (n,m,q). Just as in the 2D front, the subsequent time step is involved with "filling up" the transverse lines which are subsequently used to feed into the $Z_{xy}(n+1,m,q)$ line. At the end of the third step the full forward wave energy is delivered to the $Z_{xy}(n+1,m,q)$ line, just as in the 2D plane wave. With the help of Fig.3.14, we use a similar graphical to determine the origin of the wave distribution, again assuming that the incoming wave has unity amplitude.

We make several preparatory remarks regarding the scattering in the 3D matrix. In the event all the transmission lines are uniform, as is the case here, the quasi type aplanar coupling does not apply and the equivalent circuit simplifies, as seen in Fig.3.5. Based on this circuit, with negligible losses, an input wave at A, Fig.3.15, will see a load impedance of (3/7)Z. Thus the transfer coefficient to each of the two perpendicular lines, designated by $T_{A\rightarrow B}$ and $T_{A\rightarrow D}$,(or in cellular notation, $T_{xy}(n,m,q,53)$) and $T_{xy}(n,m,q,46)$ will be

$$T_{A\rightarrow B} = T_{A\rightarrow D} = 2(3/7)Z/[Z+(3/7)Z] = 3/5\qquad(3.134)$$

In the case of the coplanar coupling, there are three lines in series and thus the transfer to each one of the lines, designated by $T_{A\rightarrow C1}$, $T_{A\rightarrow C2}$, or $T_{A\rightarrow C3}$, is (in

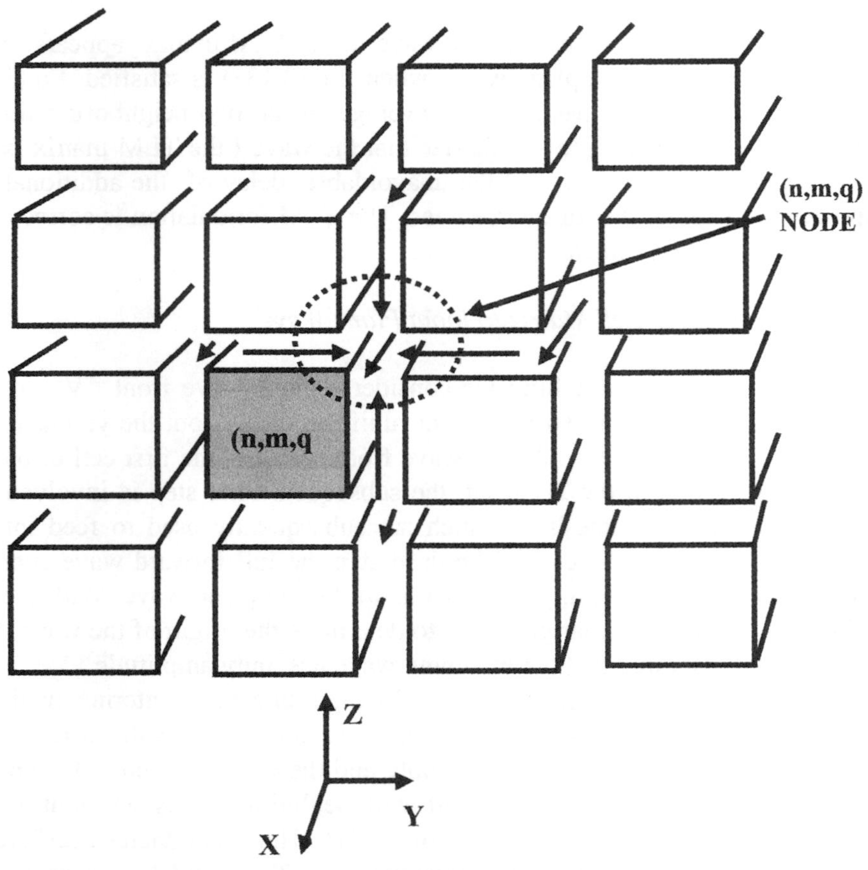

FIG 3.14 GRAPHICAL DEPICTION OF WAVES (APLANAR AND COPLANAR) WHICH CONTRIBUTE TO WAVE $^{+}V_{XY}(n+1,m,q)$.

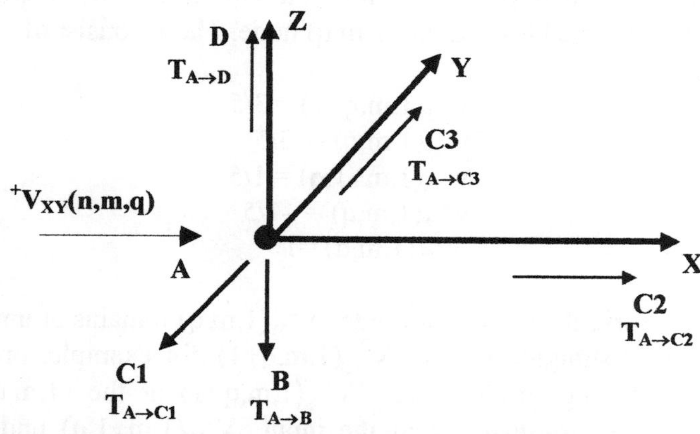

FIG. 3.15 3D TRANSFER OF WAVE $^{+}V_{XY}(n,m,q)$ INCIDENT AT (n,m,q) NODE . IN TERMS OF CELLULAR NOTATION:

$$T_{A \to D} = T_{xy}(n,m,q,46) \;\to\; {}^{+}V_{ZY}(n,m,q+1)$$
$$T_{A \to B} = T_{xy}(n,m,q,53) \;\to\; {}^{-}V_{ZY}(n,m,q)$$
$$T_{A \to C3} = T_{xy}(n,m,q,15) \;\to\; {}^{+}V_{YX}(n,m+1,q)$$
$$T_{A \to C2} = T_{xy}(n,m,q,1) \;\to\; {}^{+}V_{XY}(n+1,m,q)$$
$$T_{A \to C1} = T_{xy}(n,m,q,22) \;\to\; {}^{-}V_{YX}(n,m,q)$$

cellular notation the coefficients are $T_{xy}(n,m,q,22)$, $T_{xy}(n,m,q,15)$ and $T_{xy}(n,m,q,1)$)

$$T_{A \to C1} = T_{A \to C2} = T_{A \to C3} = 1/5 \qquad (3.135)$$

The reflection coefficient is $[(3/7)Z-Z]/[(3/7)Z+Z] = -2/5$. We now return to treat the plane wave impinging on the 3D matrix, making use of these coefficients.

For simplicity we assume that.at the end of the first step, k=1, all the Z_{xy} (1,m,q) lines contain the forward wave $^+V^1_{xy}(1,m,q)$ with unity amplitude. At the end of the second step we determine the amplitudes of the waves in the transverse lines, directed toward the (1,m,q) node. These consist of

$$^+V^2_{zy}(1,m,q+1) = 3/5 \qquad (3.136a)$$
$$^-V^2_{zy}(1,m,q) = 3/5 \qquad (3.136b)$$
$$^-V^2_{yx}(1,m+1,q) = 1/5 \qquad (3.136c)$$
$$^+V^2_{yx}(1,m,q) = -1/5 \qquad (3.136d)$$
$$^+V^2_{zy}(1,m,q) = 1 \qquad (3.136e)$$

where of course the input amplitude $^+V^2_{xy}(1,m,q)$ remains at unity. The origin of Eq.(3.136) is straightforward; $^+V^2_{zy}(1,m,q+1)$, for example, originates from the aplanar scattering of the input $^+V^1_{xy}(1,m,q+1)$ at the (1,m,q+1) node while $^+V^2_{yx}(1,m+1,q)$ originates from the input $^+V^1_{xy}(1,m+1,q)$ undergoing coplanar scattering at the (1,m+1,q) node, and similarly for the other incident waves. We use the scattering coefficients obtained in the previous paragraph. Knowing the waves at k=2 we can obtain the forward wave in $Z_{xy}(2,m,q)$ at k=3 , or

$$^+V^3_{xy}(2,m,q) = (1/5)\,^+V^2_{xy}(1,m,q) +(3/5)\,^-V^2_{zy}(1,m,q+1) + (3/5)\,^+V^2_{zy}(1,m,q)$$
$$+ (1/5)\,^-V^2_{yx}(1,m+1,q) - (1/5)\,^+V^2_{yx}(1,m,q) \qquad (3.137)$$

Substituting the amplitudes at t= $2\Delta t$, we then obtain the forward wave amplitude at $3\Delta t$, or

$$^+V^3_{xy}(2,m,q) = 1 \qquad (3.138)$$

which is the same as the result for the 2D matrix. As with the 2D matrix, the unity amplitude plane wave propagates into the matrix, but requires twice the normal amount of time.

3.17 Response of 3D Cell Matrix to Input Waves With Arbitrary Amplitudes

Following the 2D arguments we allow the input waves, $^+V^1_{xy}(1,m,q)$, to depart from uniformity and then inquire under what conditions, a limited portion of the input wave front behaves as though it were a plane wave. We first note that the waves incident on the (n,m,q) node during k=2 are related to the arbitrary inputs by

$$^-V^2_{zy}(1,m,q+1) = (3/5)\ ^+V^1_{xy}(1,m,q+1) \tag{3.139a}$$

$$^+V^2_{zy}(1,m,q) = (3/5)\ ^+V^1_{xy}(1,m,q-1) \tag{3.139b}$$

$$^-V^2_{yx}(1,m+1,q) = (1/5)\ ^+V^1_{xy}(1,m+1,q) \tag{3.139c}$$

$$^+V^2_{yx}(1,m,q) = -(1/5)\ ^+V^1_{xy}(1,m,q) \tag{3.139d}$$

$$^+V^2_{xy}(1,m,q) = 1 \tag{3.139e}$$

We then substitute Eq.(3.139) into (3.137) , and impose the following requirement

$$^+V^3_{xy}(2,m,q) \approx\ ^+V^1_{xy}(1,m,q) \tag{3.140}$$

As in the 2D case, by combining the previous equations, we can determine what constraints on the input will result in the continued propagation of the wave $^+V^1_{xy}(1,m,q)$ with the same amplitude without scattering. Generalizing the results to arbitrary n and k, $^+V^k_{xy}(n,m,q)$ must satisfy

$$^+V^k_{xy}(n,m,q) = (9/20)[\ ^+V^k_{xy}(n,m,q+1) +\ ^+V^k_{xy}(n,m,q-1)]$$
$$+(1/20)[\ ^+V^k_{xy}(n,m+1,q) +\ ^+V^k_{xy}(n,m-1,q)] \tag{3.141}$$

The previous Sections provide hints as to the conditions under which non-uniform input waves behave , to some degree, as plane waves , as they advance in the TLM matrix.   We saw that the forward wave in a particular cell advances as an approximate plane wave manner provided there is close "correlation" with its neighbors; in this case the correlation is provided by he closeness of the forward wave amplitude to the average, which takes into account the waves. in the neighboring cells.

However a serious problem still remains, even when the input is uniform. A plane wave launched in a TLM matrix, as it progresses , does not satisfy the basic criterion for a plane wave because of the diversion of wave energy into the transverse lines, which results in a propagation delay. This diversion is nothing more than an artifact of the TLM matrix , as it is presently formulated. What is needed is a systematic means for correcting the TLM method to account for plane wave effects. In addition, anisotropy effects have not been accounted for in the TLM method. These matters are taken up in the following Chapter.

APP. 3A.1 3D SCATTERING EQUATIONS: WITH BOTH COPLANAR AND APLANAR CONTRIBUTIONS INTO UNIT CELL LINES $Z_{YZ}(n,m,q)$, $Z_{ZY}(n,m,q)$ (yz PLANE).

$$
\begin{aligned}
{}^{+}V^{k+1}_{xy}(n,m,q) =\ & T_{yz}{}^{k}(n,m-1,q,29)\,{}^{+}V^{k}_{yz}(n,m-1,q) - T_{zy}{}^{k}(n,m-1,q,30)\,{}^{+}V^{k}_{zy}(n,m-1,q) \\
& + T_{zy}{}^{k}(n,m-1,q,31)\ {}^{-}V^{k}_{zy}(n,m-1,q+1) + B_{yz}{}^{k}(n,m-1,q,5)\ {}^{-}V^{k}yz(n,m,q) \\
& + T_{xz}{}^{k}(n,m-1,q,32)\,{}^{+}V^{k}_{xz}(n,m-1,q) + T_{xz}{}^{k}(n,m-1,q,33)\,{}^{+}V^{k}_{zy}(n+1,m-1,q) \\
& + T_{xyQ}{}^{k}(n,m-1,q,34)\,{}^{+}V^{k}_{xy}(n,m-1,q) + T_{xyQ}{}^{k}(n,m-1,q,35)\,{}^{-}V^{k}_{xy}(n+1,m-1,q)
\end{aligned}
$$

$$
\begin{aligned}
{}^{-}V^{k+1}_{xy}(n,m,q) =\ & T_{yz}{}^{k}(n,m,q,36)\ {}^{-}V^{k}_{yz}(n,m+1,q) - T_{zy}{}^{k}(n,m,q,37)\ {}^{-}V^{k}_{zy}(n,m,q+1) \\
& + T_{zy}{}^{k}(n,m,q,38)\ {}^{+}V^{k}_{zy}(n,m,q) + B_{yz}{}^{k}(n,m,q,6)\ {}^{-}V^{k}_{yz}(n,m,q) \\
& + T_{xz}{}^{k}(n,m,q,39)\ {}^{+}V^{k}_{zx}(n,m,q) + T_{xz}{}^{k}(n+1,m,q,40)\ {}^{-}V^{k}_{xz}(n+1,m,q) \\
& + T_{xyQ}{}^{k}(n,m,q,41)\,{}^{+}V^{k}_{xy}(n,m,q) + T_{xyQ}{}^{k}(n+1,m,q,42)\ {}^{-}V^{k}_{xy}(n+1,m,q)
\end{aligned}
$$

$$
\begin{aligned}
{}^{+}V^{k+1}_{zy}(n,m,q) =\ & -T_{yz}{}^{k}(n,m,q-1,43)\,{}^{+}V^{k}_{yz}(n,m,q-1) T_{zy}{}^{k}(n,m,q-1,44)\,{}^{+}V^{k}_{zy}(n,m,q-1) \\
& + T_{yz}{}^{k}(n,m,q-1,45)\ {}^{-}V^{k}_{yz}(n,m,q-1) + B_{zy}{}^{k}(n,m,q-1,7)\ {}^{-}V^{k}_{xy}(n,m,q) \\
& + T_{xy}{}^{k}(n,m,q-1,46)\,{}^{+}V^{k}_{xy}(n,m,q-1) + T_{xy}{}^{k}(n,m,q-1,47)\ {}^{-}V^{k}_{xy}(n+1,m,q-1) \\
& + T_{xzQ}{}^{k}(n,m,q-1,48)\ {}^{+}V^{k}_{xz}(n,m,q-1) + T_{xzQ}{}^{k}(n,m,q-1,49)\,{}^{-}V^{k}_{xz}(n,m,q-1)
\end{aligned}
$$

$$
\begin{aligned}
{}^{-}V^{k+1}_{zy}(n,m,q) =\ & T_{yz}{}^{k}(n,m,q,50)\ {}^{+}V^{k}_{yz}(n,m,q) + T_{zy}{}^{k}(n,m,q+1,51)\ {}^{-}V^{k}_{zy}(n,m,q+1) \\
& - T_{yz}{}^{k}(n,m,q,52)\ {}^{-}V^{k}_{yz}(n,m+1,q) + B_{zy}{}^{k}(n,m,q,8)\ {}^{+}V^{k}_{zy}(n,m,q) \\
& + T_{xy}{}^{k}(n,m,q,53)\ {}^{+}V^{k}_{xy}(n,m,q) + T_{xy}{}^{k}(n,m,q,54)\ {}^{-}V^{k}_{xy}(n+1,m,q) \\
& + T_{xzQ}{}^{k}(n,m,q,55)\ {}^{+}V^{k}_{zx}(n,m,q) + T_{xzQ}{}^{k}(n,m,q,56)\ {}^{-}V^{k}_{xz}(n+1,m,q)
\end{aligned}
$$

APP. 3A.1(CONT). 3D SCATTERING EQUATIONS: WITH BOTH COPLANAR AND APLANAR CONTRIBUTIONS INTO UNIT CELL LINES $Z_{ZX}(n,m,q)$, $Z_{XZ}(n,m,q)$ (zx PLANE).

$$^+V^{k+1}_{zx}(n,m,q)=T_{zx}{}^k(n,m,q-1,57)^+V^k_{zx}(n,m,q-1)-T_{xz}{}^k(n,m,q-1,58)^+V^k_{xz}(n,m,q-1)$$
$$+T_{xz}{}^k(n,m,q-1,59)\ ^-V^k_{xz}(n+1,m,q-1)\ +B_{zx}{}^k(n,m,q-1,9)\ ^-V^k_{zx}(n,m,q)$$
$$+T_{yx}{}^k(n,m,q-1,60)\ ^+V^k_{yx}(n,m,q-1)\ +T_{yx}{}^k(n,m,q-1,61)\ ^-V^k_{yx}(n,m+1,q-1)$$
$$+T_{yzQ}{}^k(n,m,q-1,62)\ ^+V^k_{yz}(n,m,q-1)+T_{yzQ}{}^k(n,m,q-1,63)^-V^k_{yz}(n,m+1,q-1)$$

$$^-V^{k+1}_{zx}(n,m,q-1)=T_{zx}{}^k(n,m,q,64)^-V^k_{zx}(n,m,q+1)\ -T_{xz}{}^k(n,m,q,65)\ ^-V^k_{xz}(n+1,m,q)$$
$$+T_{xz}{}^k(n,m,q,66)\ ^+V^k_{xz}(n,m,q)\ +B_{zx}{}^k(n,m,q,10)\ ^+V^k_{zx}(n,m,q)$$
$$+T_{yx}{}^k(n,m,q,67)\ ^+V^k_{yx}(n,m,q)\ +T_{yx}{}^k(n,m+1,q,68)\ ^-V^k_{yx}(n,m+1,q)$$
$$+T_{yzQ}{}^k(n,m,q,69)\ ^+V^k_{yz}(n,m,q)\ +T_{yzQ}{}^k(n,m+1,q,70)\ ^-V^k_{yz}(n,m+1,q)$$

$$^+V^{k+1}_{xz}(n,m,q)=-T_{zx}{}^k(n-1,m,q,71)^+V^k_{zx}(n-1,m,q)+T_{xz}{}^k(n-1,m,q,72)^+V^k_{xz}(n-1,m,q)$$
$$+T_{zx}{}^k(n-1,m,q,73)\ ^-V^k_{zx}(n-1,m,q+1)\ +B_{xz}{}^k(n-1,m,q,11)\ ^-V^k_{xz}(n,m,q)$$
$$+T_{yz}{}^k(n-1,m,q,74)^+V^k_{yz}(n-1,m,q)\ +T_{yz}{}^k(n-1,m,q,75)^-V^k_{yz}(n-1,m+1,q)$$
$$+T_{yxQ}{}^k(n-1,m,q,76)^+V^k_{yx}(n-1,m,q)+T_{yxQ}{}^k(n-1,m,q,77)^-V^k_{yx}(n-1,m+1,q)$$

$$^-V^{k+1}_{xz}(n,m,q)=T_{zx}{}^k(n,m,q,78)^+V^k_{zx}(n,m,q)+T_{xz}{}^k(n+1,m,q,79)\ ^-V^k_{zx}(n+1,m,q)$$
$$-T_{zx}{}^k(n,m,q,80)\ ^-V^k_{zx}(n,m,q+1)\ +B_{xz}{}^k(n,m,q,12)\ ^+V^k_{xz}(n,m,q)$$
$$+T_{yz}{}^k(n,m,q-1,81)\ ^+V^k_{yz}(n,m,q)+T_{yz}{}^k(n,m,q,82)\ ^-V^k_{yz}(n,m+1,q)$$
$$+T_{yxQ}{}^k(n,m,q,83)\ ^+V^k_{yx}(n,m,q)+T_{yxQ}{}^k(n,m,q,84)\ ^-V^k_{yx}(n,m+1,q)$$

APP.3A.2 SCATTERING COEFICIENTS WITH COPLANAR AND APLANAR CONTRIBUTIONS. <u>INTO</u> UNIT CELL LINES $Z_{yz}(n,m,q)$ AND $Z_{zy}(n,m,q)$ (yz PLANE).

TRANSFER TYPE

COEFFICIENT	FROM	TO
$T_{yz}(n,m-1,q,29)$	$Z_{yz}(n,m-1,q)$	$Z_{yz}(n,m,q)$
$(-)T_{zy}(n,m-1,q,30)$	$Z_{zy}(n,m-1,q)$	$Z_{yz}(n,m,q)$
$T_{zy}(n,m-1,q,31)$	$Z_{zy}(n,m-1,q)$	$Z_{yz}(n,m,q)$
$T_{xz}(n,m-1,q,32)$	$Z_{xz}(n,m-1,q)$	$Z_{yz}(n,m,q)$
$T_{xz}(n,m-1,q,33)$	$Z_{xz}(n+1,m-1,q)$	$Z_{yz}(n,m,q)$
$T_{xyQ}(n,m-1,q,34)$	$Z_{xy}(n,m-1,q)$	$Z_{yz}(n,m,q)$
$T_{xyQ}(n,m-1,q,35)$	$Z_{xy}(n+1,m-1,q)$	$Z_{yz}(n,m,q)$
$T_{yz}(n,m,q,36)$	$Z_{yz}(n,m+1,q)$	$Z_{yz}(n,m,q)$
$(-)T_{zy}(n,m,q,37)$	$Z_{zy}(n,m,q+1)$	$Z_{yz}(n,m,q)$
$T_{zy}(n,m,q,38)$	$Z_{zy}(n,m,q)$	$Z_{yz}(n,m,q)$
$T_{xz}(n,m,q,39)$	$Z_{xz}(n,m,q)$	$Z_{yz}(n,m,q)$
$T_{xz}(n+1,m,q,40)$	$Z_{xz}(n+1,m,q)$	$Z_{yz}(n,m,q)$
$T_{xyQ}(n,m,q,41)$	$Z_{xy}(n,m,q)$	$Z_{yz}(n,m,q)$
$T_{xyQ}(n,m,q,42)$	$Z_{xy}(n+1,m,q)$	$Z_{yz}(n,m,q)$
$(-)T_{yz}(n,m,q-1,43)$	$Z_{yz}(n,m,q-1)$	$Z_{zy}(n,m,q)$
$T_{zy}(n,m,q-1,44)$	$Z_{zy}(n,m,q-1)$	$Z_{zy}(n,m,q)$
$T_{yz}(n,m,q-1,45)$	$Z_{yz}(n,m+1,q-1)$	$Z_{zy}(n,m,q)$
$T_{xy}(n,m,q-1,46)$	$Z_{xy}(n,m,q-1)$	$Z_{zy}(n,m,q)$
$T_{xy}(n,m,q-1,47)$	$Z_{xy}(n+1,m,q-1)$	$Z_{zy}(n,m,q)$
$T_{xzQ}(n,m,q-1,48)$	$Z_{xz}(n,m,q-1)$	$Z_{zy}(n,m,q)$
$T_{xzQ}(n,m,q-1,49)$	$Z_{xz}(n+1,m,q-1)$	$Z_{zy}(n,m,q)$
$T_{yz}(n,m,q,50)$	$Z_{yz}(n,m,q)$	$Z_{zy}(n,m,q)$
$T_{zy}(n,m,q+1,51)$	$Z_{zy}(n,m,q+1)$	$Z_{zy}(n,m,q)$
$(-)T_{yz}(n,m,q,52)$	$Z_{yz}(n,m+1,q)$	$Z_{zy}(n,m,q)$
$T_{xy}(n,m,q,53)$	$Z_{xy}(n,m,q)$	$Z_{zy}(n,m,q)$
$T_{xy}(n,m,q,54)$	$Z_{xy}(n+1,m,q)$	$Z_{zy}(n,m,q)$
$T_{xzQ}(n,m,q,55)$	$Z_{xz}(n,m,q)$	$Z_{zy}(n,m,q)$
$T_{xzQ}(n,m,q,56)$	$Z_{xz}(n+1,m,q)$	$Z_{zy}(n,m,q)$

REFLECTION TYPE

$B_{yz}(n,m-1,q,5)$	$Z_{yz}(n,m,q)$	$Z_{yz}(n,m,q)$
$B_{yz}(n,m,q,6)$	$Z_{yz}(n,m,q)$	$Z_{yz}(n,m,q)$
$B_{zy}(n,m,q-1,7)$	$Z_{zy}(n,m,q)$	$Z_{yx}(n,m,q)$
$B_{zy}(n,m,q,8)$	$Z_{zy}(n,m,q)$	$Z_{zy}(n,m,q)$

APP. 3A.2 (CONT). 3D SCATTERING COEFICIENTS WITH COPLANAR AND APLANAR CONTRIBUTIONS. <u>INTO</u> UNIT CELL LINES $Z_{zx}(n,m,q)$ AND $Z_{zz}(n,m,q)$ (zx PLANE).

TRANSFER TYPE

COEFFICIENT	FROM	TO
$T_{zx}(n,m,q-1,57)$	$Z_{zx}(n,m,q-1)$	$Z_{zx}(n,m,q)$
$(-)T_{xz}(n,m,q-1,58)$	$Z_{xz}(n,m,q-1)$	$Z_{zx}(n,m,q)$
$T_{xz}(n,m,q-1,59)$	$Z_{xz}(n,m,q-1)$	$Z_{zx}(n,m,q)$
$T_{yx}(n,m,q-1,60)$	$Z_{yx}(n,m,q-1)$	$Z_{zx}(n,m,q)$
$T_{yx}(n,m,q-1,61)$	$Z_{yx}(n,mzx,q-1)$	$Z_{zx}(n,m,q)$
$T_{yzQ}(n,m,q-1,62)$	$Z_{yz}(n,m,q-1)$	$Z_{zx}(n,m,q)$
$T_{yzQ}(n,m,q-1,63)$	$Z_{yz}(n,m+1,q-1)$	$Z_{zx}(n,m,q)$
$T_{zx}(n,m,q,64)$	$Z_{zx}(n,m,q+1)$	$Z_{zx}(n,m,q)$
$(-)T_{xz}(n,m,q,65)$	$Z_{xz}(n+1,m,q)$	$Z_{zx}(n,m,q)$
$T_{xz}(n,m,q,66)$	$Z_{xz}(n,m,q)$	$Z_{zx}(n,m,q)$
$T_{yx}(n,m,q,67)$	$Z_{yx}(n,m,q)$	$Z_{zx}(n,m,q)$
$T_{yx}(n,m+1,q,68)$	$Z_{yx}(n,m+1,q)$	$Z_{zx}(n,m,q)$
$T_{yzQ}(n,m,q,69)$	$Z_{yz}(n,m,q)$	$Z_{zx}(n,m,q)$
$T_{yzQ}(n,m,q,70)$	$Z_{yz}(n,m+1,q)$	$Z_{zx}(n,m,q)$
$(-)T_{zx}(n-1,m,q,71)$	$Z_{xyzx}(n-1,m,q)$	$Z_{xz}(n,m,q)$
$T_{xz}(n-1,m,q,72)$	$Z_{xz}(n-1,m,q)$	$Z_{xz}(n,m,q)$
$T_{zx}(n-1,m,q,73)$	$Z_{zx}(n-1,m,q+1)$	$Z_{xz}(n,m,q)$
$T_{yz}(n-1,m,q,74)$	$Z_{yz}(n-1,m,q)$	$Z_{xz}(n,m,q)$
$T_{yz}(n-1,m,q,75)$	$Z_{yz}(n-1,m+1,q)$	$Z_{xz}(n,m,q)$
$T_{yxQ}(n-1,m,q,76)$	$Z_{yx}(n-1,m,q)$	$Z_{xz}(n,m,q)$
$T_{yxQ}(n-1,m,q,77)$	$Z_{yx}(n-1,m+1,q)$	$Z_{xz}(n,m,q)$
$T_{zx}(n,m,q,78)$	$Z_{zx}(n,m,q)$	$Z_{xz}(n,m,q)$
$T_{xz}(n+1,m,q,79)$	$Z_{xz}(n+1,m,q)$	$Z_{xz}(n,m,q)$
$(-)T_{zx}(n,m,q,80)$	$Z_{zx}(n,m,q+1)$	$Z_{xz}(n,m,q)$
$T_{yz}(n,m,q,81)$	$Z_{yz}(n,m,q)$	$Z_{xz}(n,m,q)$
$T_{yz}(n,m,q,82)$	$Z_{yz}(n,m+1,q)$	$Z_{xz}(n,m,q)$
$T_{yxQ}(n,m,q,83)$	$Z_{yx}(n,m,q)$	$Z_{xz}(n,m,q)$
$T_{yxQ}(n,m,q,84)$	$Z_{yx}(n,m+1,q)$	$Z_{xz}(n,m,q)$

REFLECTION TYPE

$B_{zx}(n,m,q-1,9)$	$Z_{zx}(n,m,q)$	$Z_{zx}(n,m,q)$
$B_{zx}(n,m,q,10)$	$Z_{zx}(n,m,q)$	$Z_{zx}(n,m,q)$
$B_{xz}(n-1,m,q,11)$	$Z_{xz}(n,m,q)$	$Z_{xz}(n,m,q)$
$B_{xz}(n,m,q,12)$	$Z_{xz}(n,m,q)$	$Z_{xz}(n,m,q)$

**APP.3A.3 3D SCATTERING COEFFICIENTS, WITHOUT QUASI-COUPLING ,
IN TERMS OF CIRCUIT PARAMETERS. FOR CO-PLANAR AND APLANAR
SCATTERING INTO XY PLANE ABOUT (n,m,q) NODE. COEFFICIENTS
FOR YZ AND ZX PLANES ARE OBTAINED FROM TABLE 3.8.**

TRANSFER COEFFICIENTS

$$T_{XY}(n,m,q,1)= (2RL1_{XY}(n,m,q)/[RL1_{XY}(n,m,q)+Z_{XY}(n,m,q)])*A1$$
$$A1=R3_{XY}(n,m,q)/[R2_{YX}(n,m,q)+R3_{XY}(n,m,q)+R4_{YX}(n,m,q)]$$

$$T_{YX}(n,m,q,2)= (2RL2_{YX}(n,m,q)/[RL2_{YX}(n,m,q)+Z_{YX}(n,m,q)])*A2$$
$$A2=R3_{XY}(n,m,q)/[R1_{XY}(n,m,q)+R3_{XY}(n,m,q)+R4_{YX}(n,m,q)]$$

$$T_{YX}(n,m,q,3)= (2RL4_{YX}(n,m,q)/[RL4_{YX}(n,m,q)+Z_{YX}(n,m+1,q)])*A3$$
$$A3=R3_{XY}(n,m,q)/[R1_{XY}(n,m,q)+R3_{XY}(n,m,q)+R2_{YX}(n,m,q)]$$

$$T_{ZY}(n,m,q,4) = 2RL2_{ZY}(n,m,q)/[RL2_{ZY}(n,m,q)+Z_{ZY}(n,m,q)]$$

$$T_{ZY}(n,m,q,5) = 2RL4_{ZY}(n,m,q)/[RL4_{ZY}(n,m,q)+Z_{ZY}(n,m,q+1)]$$

$$T_{ZXQ}(n,m,q,6)=0$$

$$T_{ZXQ}(n,m,q,7)=0$$

$$T_{XY}(n,m,q,8)= (2RL3_{XY}(n,m,q)/[RL3_{XY}(n,m,q)+Z_{XY}(n+1,m,q)])*A8$$
$$A8=R1_{XY}(n,m,q)/[R2_{YX}(n,m,q)+R1_{XY}(n,m,q)+R4_{YX}(n,m,q)]$$

$$T_{YX}(n,m,q,9)= (2RL4_{YX}(n,m,q)/[RL4_{YX}(n,m,q)+Z_{YX}(n,m+1,q)])*A9$$
$$A9=R1_{XY}(n,m,q)/[R2_{YX}(n,m,q)+R1_{XY}(n,m,q)+R3_{XY}(n,m,q)]$$

$$T_{YX}(n,m,q,10)= (2RL2_{YX}(n,m,q)/[RL2_{YX}(n,m,q)+Z_{YX}(n,m,q)])*A10$$
$$A10=R1_{XY}(n,m,q)/[R4_{YX}(n,m,q)+R1_{XY}(n,m,q)+R3_{XY}(n,m,q)]$$

$T_{ZY}(n,m,q,11) = 2RL2_{ZY}(n,m,q)/[RL2_{ZY}(n,m,q)+Z_{ZY}(n,m,q)]$

$T_{ZY}(n,m,q,12) = 2RL4_{ZY}(n,m,q)/[RL4_{ZY}(n,m,q)+Z_{ZY}(n,m,q)]$

$T_{ZXQ}(n,m,q,13)=0$

$T_{ZXQ}(n,m,q,14)=0$

$T_{XY}(n,m,q,15)= (2RL1_{XY}(n,m,q)/[RL1_{XY}(n,m1,q)+Z_{XY}(n,m,q)])*A15$

$\qquad A15=R4_{YX}(n,m,q)/[R2_{YX}(n,m,q)+R3_{XY}(n,m,q)+R4_{YX}(n,m,q)]$

$T_{YX}(n,m,q,16)= (2RL2_{YX}(n,m,q)/[RL2_{YX}(n,m,q)+Z_{YX}(n,m,q)])*A16$

$\qquad A16=R4_{YX}(n,m,q)/[R4_{YX}(n,m,q)+R1_{XY}(n,m,q)+R3_{XY}(n,m,q)]$

$T_{XY}(n,m,q,17)= (2RL3_{XY}(n,m,q)/[RL3_{XY}(n,m,q)+Z_{XY}(n+1,m,q)])*A17$

$\qquad A17=R4_{YX}(n,m,q)/[R2_{YX}(n,m,q)+R1_{XY}(n,m,q)+R4_{YX}(n,m,q)]$

$T_{ZX}(n,m,q,18) = 2RLI_{ZX}(n,m,q)/[RL1_{ZX}(n,m,q)+Z_{ZX}(n,m,q)]$

$T_{ZX}(n,m,q,19) = 2RL3_{ZX}(n,m,q+1)/[RL3_{ZX}(n,m,q+1)+Z_{ZX}(n,m,q+1)]$

$T_{ZYQ}(n,m,q,20)=0$

$T_{ZYQ}(n,m,q,21)=0$

$T_{XY}(n,m,q,22)= (2RL1_{XY}(n,m,q)/[RL1_{XY}(n,m,q)+Z_{XY}(n,m,q)])*A22$

$\qquad A22=R2_{YX}(n,m,q)/[R2_{YX}(n,m,q)+R3_{XY}(n,m,q)+R4_{YX}(n,m,q)]$

$T_{YX}(n,m,q,23)= (2RL4_{YX}(n,m,q)/[RL4_{YX}(n,m,q)+Z_{YX}(n,m+1,q)])*A23$

$\qquad A23=R2_{YX}(n,m,q)/[R2_{YX}(n,m,q)+R3_{XY}(n,m,q)+R1_{XY}(n,m,q)]$

$T_{XY}(n,m,q,24)= (2RL3_{XY}(n,m,q)/[RL3_{XY}(n,m,q)+Z_{XY}(n+1,m,q)])*A24$

$\qquad A24=R2_{YX}(n,m,q)/[R2_{YX}(n,m,q)+R1_{XY}(n,m,q)+R4_{YX}(n,m,q)]$

$$T_{ZX}(n,m,q,25) = 2RLI_{ZX}(n,m,q)/[RL1_{ZX}(n,m,q)+Z_{ZX}(n,m,q)]$$

$$T_{ZX}(n,m,q,26) = 2RL3_{ZX}(n,m,q)/[RL3_{ZX}(n,m,q)+Z_{ZX}(n,m,q+1)]$$

$$T_{ZYQ}(n,m,q,27)=0$$

$$T_{ZYQ}(n,m,q,28)=0$$

REFLECTION COEFFICIENTS(CONT)

$$B_{XY}(n,m,q,1) = [RL3_{XY}(n,m,q)-Z_{XY}(n+1,m,q)] \: / \: [RL3_{XY}(n,m,q)+Z_{XY}(n+1,m,q)]$$
$$B_{XY}(n,m,q,2) = [RL1_{XY}(n,m,q)-Z_{XY}(n,m,q)] \: / \: [RL1_{XY}(n,m,q)+Z_{XY}(n,m,q)]$$
$$B_{YX}(n,m,q,3) = [RL4_{YX}(n,m,q)-Z_{YX}(n,m-1,q)] \: / \: [RL4_{YX}(n,m,q)+Z_{YX}(n,m-1,q)]$$
$$B_{YX}(n,m,q,4) = [RL2_{YX}(n,m,q)-Z_{YX}(n,m,q)] \: / \: [RL2_{YX}(n,m,q)+Z_{YX}(n,m,q)]$$

APP. 3A.3(CONT) NODE PARAMETERS (ALL THREE PLANES) WITHOUT QUASI-COUPLING.THE NODE PARAMETERS ARE REFERENCED TO THE (n,m,q) NODE. REPLACE n WITH n+1, n-1, ETC.. , AS APPROPRIATE, IN THE SCATTERING EQUATIONS. TO SAVE SPACE WE HAVE OMITTED THE (n,m,q) ARGUMENT IN THE LOAD RESISTANCE PARAMETERS.

LOAD RESISTANCE PARAMETERS

$$RL1_{XY}=[R4_{ZY}*R2_{ZY}*R*D1]/\{R4_{ZY}*R2_{ZY}*R+D1*R4_{ZY}*R2_{ZY}+D1*R*R4_{ZY}+D1*R*R2_{ZY}\}$$
$$D1 = R2_{YX} +R3_{XY} +R4_{YX}$$

$$RL2_{YX}=[R1_{ZX}*R3_{ZX}*R*D2]/\{R1_{ZX}*R3_{ZX}*R+D2*R1_{ZX}*R3_{ZX}+D2*R*R1_{ZX}+D2*R*R3_{ZX}\}$$
$$D2 = R1_{XY} +R4_{YX} +R3_{XY}$$

$$RL3_{XY}= [R4_{ZY}*R2_{ZY}*R*D3]/ \{R4_{ZY}*R2_{ZY}*R+D3*R4_{ZY}R2_{ZY}+D3*R*R2_{ZY}+D3*R*R4_{ZY}\}$$
$$D3 = R2_{YX} +R1_{XY}+R4_{YX}$$

$$RL4_{YX}=[R1_{ZX}*R3_{ZX}*R*D4]/\{R1_{ZX}*R3_{ZX}*R+D4*R1_{ZX}*R3_{ZX}+D4*R*R1_{ZX}+D4*R*R3_{ZX}\}$$
$$D4 = R1_{XY} +R2_{YX} +R3_{XY}$$

$$RL1_{YZ}=[R4_{XZ}*R2_{XZ}*R*D5]/\{R4_{XZ}*R2_{XZ}*R+D5*R4_{XZ}*R2_{XZ}+D5*R*R4_{XZ}+D5*R*R2_{XZ}\}$$
$$D5 = R2_{ZY} +R3_{YZ} +R4_{ZY}$$

$$RL2_{ZY}=[R1_{XY}*R3_{XY}*R*D6]/\{R1_{XY}*R3_{XY}*R+D6*R1_{XY}*R3_{XY}+D6*R*R1_{XY}+D6*R*R3_{XY}$$
$$D6 = R1_{YZ} +R4_{ZY} +R3_{YZ}$$

$$RL3_{YZ}= [R4_{XZ}*R2_{XZ}*R*D7]/ \{R4_{XZ}*R2_{XZ}*R+D7*R4_{XZ}R2_{XZ}+D7*R*R2_{XZ}+D7*R*R4_{XZ}\}$$
$$D7 = R2_{ZY} +R1_{YZ}+R4_{ZY}$$

$$RL4_{ZY}=[R1_{XY}*R3_{XY}*R*D8]/\{R1_{XY}*R3_{XY}*R+D8*R1_{XY}*R3_{XY}+D8*R*R1_{XY}+D8*R*R3_{XY}\}$$
$$D8 = R1_{YZ} +R2_{ZY} +R3_{YZ}$$

$$RL1_{ZX}=[R4_{YX}*R2_{YX}*R*D9]/\{R4_{YX}*R2_{YX}*R+D9*R4_{YX}*R2_{YX}+D9*R*R4_{YX}+D9*R*R2_{YX}\}$$
$$D9 = R2_{XZ} +R3_{ZX} +R4_{XZ}$$

$$RL2_{XZ}=[R1_{YZ}*R3_{YZ}*R*D10]/\{R1_{YZ}*R3_{XY}*R+D10*R1_{YZ}*R3_{YZ}+D10*R*R1_{YZ}$$
$$+D10*R*R3_{YZ}\}$$
$$D10 = R1_{ZX} +R4_{XZ} +R3_{ZX}$$

$$RL3_{ZX}=[R4_{YX}*R2_{YX}*R*D11]/\{R4_{YX}*R2_{YX}*R+D11*R4_{YX}R2_{YX}+D11*R*R2_{YX}$$
$$+D11*R*R4_{YX}\}$$
$$D11 = R2_{XZ} +R1_{ZX}+R4_{XZ}$$

$$RL4_{XZ}=[R1_{YZ}*R3_{YZ}*R*D12]/\{R1_{YZ}*R3_{YZ}*R+D12*R1_{YZ}*R3_{YZ}+D12*R*R1$$
$$+D12*R*R3_{YZ}\}$$
$$D12 = R1_{ZX} +R2_{XZ} +R3_{ZX}$$

PARALLEL RESISTANCE NODE PARAMETERS(ALL THREE PLANES).

$$R1_{XY} = R(n,m,q) *Z_{XY}(n,m,q)/[R(n,m,q)+Z_{XY}(n,m,q)]$$

$$R2_{YX} = R(n,m,q) *Z_{YX}(n,m,q)/[R(n,m,q)+Z_{YX}(n,m,q)]$$

$$R3_{XY} = R(n,m,q) *Z_{XY}(n+1,m,q)/[R(n,m,q)+Z_{XY}(n+1,m,q)]$$

$$R4_{YX} = R(n,m,q) *Z_{YX}(n,m+1,q)/[R(n,m,q)+Z_{YX}(n,m+1,q)]$$

$$R1_{YZ} = R(n,m,q) *Z_{YZ}(n,m,q)/[R(n,m,q)+Z_{YZ}(n,m,q)]$$

$$R2_{ZY} = R(n,m,q) *Z_{ZY}(n,m,q)/[R(n,m,q)+Z_{ZY}(n,m,q)]$$

$$R3_{YZ} = R(n,m,q) *Z_{YZ}(n,m+1,q)/[R(n,m,q)+Z_{YZ}(n,m+1,q)]$$

$$R4_{ZY} = R(n,m,q) *Z_{ZY}(n,m,q+1)/[R(n,m,q)+Z_{ZY}(n,m,q+1)]$$

$$R1_{ZX} = R(n,m,q) *Z_{ZX}(n,m,q)/[R(n,m,q)+Z_{ZX}(n,m,q)]$$

$$R2_{XZ} = R(n,m,q) *Z_{XZ}(n,m,q)/[R(n,m,q)+Z_{XZ}(n,m,q)]$$

$$R3_{ZX} = R(n,m,q)*Z_{ZX}(n,m,q+1)/[R(n,m,q)+Z_{ZX}(n,m,q+1)]$$

$$R4_{XZ} = R(n,m,q))*Z_{XZ}(n+1,m,q)/[R(n,m,q)+Z_{XZ}(n+1,m,q)]$$

IV. Corrections for Plane Wave and Anisotropy Effects

The previous Chapter revealed certain flaws in the TLM method. In particular, the TLM method, as it now stands, does not properly account for plane wave behavior in the electromagnetic field. In addition, the artificial anisotropy introduced by the cell matrix has not been resolved. This Chapter addresses these issues. In preparation, however, we first require discussion of the partitioning of the TLM wave into two or more component waves, described in the following Section.

4.1 Partition of TLM Waves into Component Waves

Before proceeding further we must outline a general means for dividing a wave into two parts. Once we acquire this technique, we will then be able to partition a wave into a plane wave part, which does not scatter normal to the propagation direction , and the usual "symmetric" part which scatters to all the TLM lines.

We inquire as to the possibility of replacing a wave $^+V_{xy}(n,m)$ with a pair of separate, independent waves, given by $^+V_{xyA}(n,m)$ and $^+V_{xyB}(n,m)$, such that their effect is exactly equivalent to the original wave. We   assume the wave amplitudes satisfy

$$^+V_{xy}(n,m) = {}^+V_{xA}(n,m) + {}^+V_{xyB}(n,m) \qquad (4.1)$$

Our initial impulse is to employ two separate waves; $^+V_{xyA}(n,m)$ and $^+V_{xyB}(n,m)$ without any modification of their wave amplitudes; this inclination is incorrect, however, since $^+V_{xy}(n,m)$ and the component waves are not referenced to the same line impedance. The modified "effective" wave amplitudes may be obtained with the help of Fig.4.1, where we have suppressed unnecessary subscripts and arguments for the present discussion. We imagine the line impedance, Z, to be partitioned into two series connected lines, $Z = Z_A + Z_B$, where

$$Z_A = [\,^+V_A/\,^+V]\, Z \qquad\qquad (4.2a)$$

$$Z_B = [\,^+V_B/\,^+V]\, Z \qquad\qquad (4.2b)$$

and $^+V = \,^+V_A + \,^+V_B$. The power in the line Z is equal to the sum of the powers in lines Z_A and Z_B, or

$$(^+V)^2\,/Z = (^+V_A)^2\,/\,Z_A + (^+V_B)^2\,/Z_B = (^+V_A\,^+V)/Z + (^+V_B\,^+V)/Z \qquad (4.3)$$

The effective wave amplitudes, associated with ^+V_A and ^+V_B, are thus

$$^+V_{A,Eff} = (^+V_A\,^+V)^{1/2} \qquad\qquad (4.4a)$$
$$^+V_{B,Eff} = (^+V_B\,^+V)^{1/2} \qquad\qquad (4.4b)$$

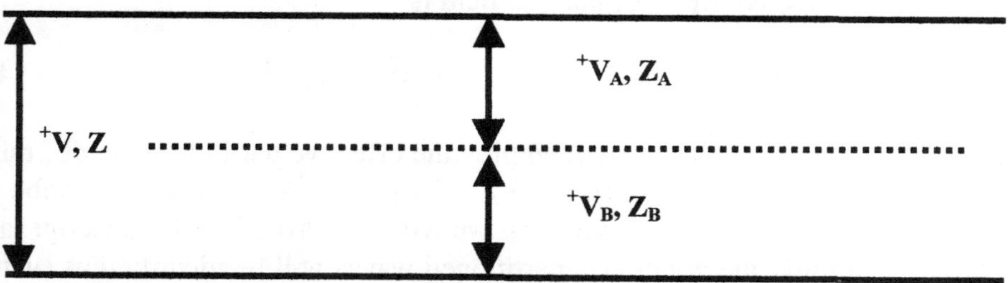

FIG. 4.1 WAVE PARTITION INTO COMPONENT WAVES. THE EFFECTIVE WAVE AMPLITUDES OF ^+V_A, ^+V_B, REFERENCED TO LINE IMPEDANCE Z, ARE : $^+V_{A,EFF} = \{^+V_A\,^+V\}^{1/2}$, $^+V_{B,EFF} = \{^+V_B\,^+V\}^{1/2}$

where $^+V_{A,Eff}$ and $^+V_{B,Eff}$ are each referenced to the full line impedance Z, rather than a portion of the impedance. The fact that the effective fields are referenced to the same original impedance provides for a much simpler interpretation of the wave picture. The effective amplitude of ^+V_A is the mean square of the full voltage and ^+V_A and similarly for ^+V_B. We should note that the above relationships satisfy

$$(^+V_{A,Eff})^2 + (^+V_{B,Eff})^2 = (^+V)^2 \qquad (4.4c)$$

which should not come as a surprise; the sum of the partioned wave energies, using the effective amplitudes, is the same as the original energy. The two waves then are said to add in quadrature. We have therefore divided the original wave V into two independent waves, $^+V_{A,Eff}$ and $^+V_{B,Eff}$, each referenced to the full line impedance .

For more than two components, the same root mean square relationship holds. Suppose $^+V_{xy}(n,m)$ is divided into N_t components,

$$^+V_{xy}(n,m) = \sum_{N=1}^{Nt} {}^+V_{xyN}(n,m) \qquad (4.5)$$

where N denotes the Nth component. Using the same arguments as before we can show that the effective field for the Nth field is

$$(^+V_{xyN}(n,m))_{Eff} = [^+V_{xy}(n,m) \, ^+V_{xyN}(n,m)]^{1/2} \qquad (4.6)$$

In the following discussion we will employ the effective partitioned fields , rather than the component fields (such as ^+V_A , ^+V_B or $^+V_{xy}(n,m)$) to describe the divided fields. In subsequent discussion we will also drop the Eff subscript label in order to simplify the notation ; partitioned waves will be identified as such, it being understood that any such waves are to be added in quadrature.

4.2 Scattering Corrections for 2D Plane Waves :Plane Wave Correlations Between Cells

In Chapter III, Sections 3.15 to 3.18, we saw that unless some correction is introduced, the TLM matrix technique will distort the time of arrival and the strength of an input plane wave electromagnetic signal propagating in a medium. By the very nature of the TLM formulation, a substantial portion of an input plane wave will divert significant energy to the transverse lines, effectively attenuating the leading edge of the signal and slowing down the effective propagation velocity of the wave *amplitude*. This effect of the TLM matrix on the plane waves will manifest itself not only in a plane waves of infinite extent, but also in much more limited wave fronts, extending only a few cells in the transverse direction. There is therefore a definite need to correct for plane wave effects. How do we begin to implement these corrections? First consider a plane wave of infinite extent in the transverse direction. The TLM formulation cannot be brought into accord with the plane wave properties unless we exclude scattering normal to the propagation direction; we postulate that the plane wave behavior is due to "correlation" effects between waves in adjacent TLM lines. The correlation effects are maximized when the fields in the neighboring lines have identical amplitudes (as well as sign). For a perfect plane wave the waves in adjacent TLM lines are identical , and the correlation between such waves is maximized. We now assert we can apply the same correlation to wave fronts which are not perfect plane waves, i.e, to arbitrary wave fronts. Neighboring waves will therefore exhibit plane wave correlation with one another, arising from the degree to which the waves have equal amplitudes (assuming the same direction and sign; opposing waves or waves with unequal sign are assumed to have zero correlation). Conversely, neighboring waves with highly disparate amplitudes will exhibit normal (i.e., symmetric) scattering. Implicit in the correation process is the assumption that neighboring TLM plane waves are phase coherence, having originated from a common source. In this regard, we remark that although we are employing strictly classical concepts to describe the correlation, one may also employ quantum mechanical considerations to obtain the equivalent correlation between neighboring photons. App.4A.2 presents simple quantum mechanical arguments which support the assumption of plane wave correlations.

As a first step we adapt a technique in which the field in each cell is divided into two parts: a part exhibiting plane wave effects and a part exhibiting the "normal" TLM scattering into all the available lines. In this Section we outline a technique for partitioning the wave into two such components. The first component wave behaves as a plane wave, with the wave progressing straight ahead(or reflected backward if the next TLM line has a differing dielectric constant) with no diversion to the transverse lines. In order to implement the plane wave progression, the scattering coefficients must be of course be modified. The second component behaves as a normal wave in a TLM matrix, and the scattering coefficients previously derived apply. The partitioning makes use of the results given in Section 4.1.

We examine the partitioning of the wave with the help of Fig.4.2 which shows the forward wave $^+V_{xy}(n,m)$ and its two nearest neighbors , assuming the same impedance, $^+V_{xy}(n,m+1)$ and $^+V_{xy}(n,m-1)$. The amplitudes of the three waves are arbitrary, but in any event they fall into one of the six categories in Fig.4.2. Thus if category I applies, for example, $^+V_{xy}(n,m+1)$ is greater or equal

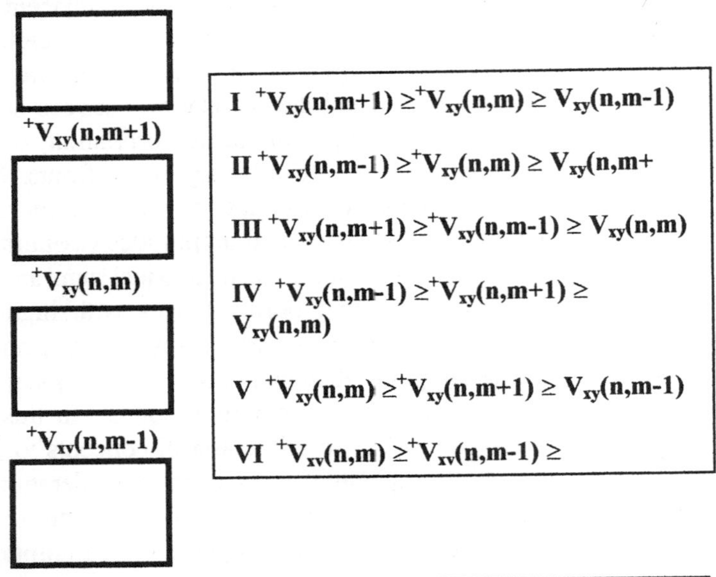

FIG. 4.2 VARIOUS CATEGORIES USED TO DETERMINE PLANE WAVE CORRELATION.

than $^+V_{xy}(n,m)$ while $^+V_{xy}(n,m)$ is greater or equal than $^+V_{xy}(n,m-1)$. With the hierarchies given in Fig.4.2, we assume in the ensuing discussion, that if $^+V_{xy}(n,m)$ is positive then any negative values of $^+V_{xy}(n,m-1)$ and $^+V_{xy}(n,m+1)$

are treated as having zero values for correlation purposes. Likewise, if $^+V_{xy}(n,m)$ is negative , any *positive* values of $^+V_{xy}(n,m-1)$ and $^+V_{xy}(n,m+1)$ are assigned zero values when calculating the correlation. The same Categories and conditions also apply of course to the backward waves and the transverse waves.

In the following we use Category I as an illustrative example, with the forward wave $^+V_{xy}(n,m)$ considered positive. The first step is to divide the wave in each line into two identical waves, the sum of whose energies is equal to the original energy. Using the partitioning given in Eqs.(4.4a)-(4.4b), the half energy effective wave is denoted by $^+V_{xy,D}(n,m)$ and

$$^+V_{xy,D}(n,m) = {}^+V_{xy}(n,m) /2^{1/2} \tag{4.7}$$

with similar relationships for $^+V_{xy,D}(n,m+1)$ and $^+V_{xy,D}(n,m-1)$. We now invoke a symmetry argument and state that one of the two waves, $^+V_{xy,D}(n,m)$, correlates with $^+V_{xy,D}(n,m+1)$, while the other $^+V_{xy,D}(n,m)$ correlates to $^+V_{xy}(n,m-1)$. Based on these correlations we must now determine whether each of the $^+V_{xy,D}(n,m)$ must be further partitioned. We first look at the correlation of $^+V_{xy,D}(n,m+1)$ to $^+V_{xy,D}(n,m)$. Since category I has been selected, $^+V_{xy,D}(n,m+1) \geq {}^+V_{xy,D}(n,m)$. There is no need, therefore, to partition the upper $^+V_{xy,D}(n,m)$ since the amplitude in $Z_{xy}(n,m+1)$ is more than sufficient to produce plane wave correlation. The upper $^+V_{xy,D}(n,m)$ is entirely of the plane wave type(note that if we were to start out with $Z_{xy}(n,m+1)$, however, a partitioning of $Z_{xy}(n,m+1)$ would then be necessary) .The "upper" plane wave component, designated by $^+V_{xyPU}(n,m)$, is therefore

$$^+V_{xyPU}(n,m) = {}^+V_{xy,D}(n,m) = {}^+V_{xy}(n,m) /2^{1/2} \tag{4.8a}$$

Thus there is no need for an "upper " symmetric component. Using a similar notation for the symmetric component,

$$^+V_{xySU}(n,m) = 0 \tag{4.8b}$$

Next we consider the correlation of $^+V_{xy,D}(n,m)$ to $^+V_{xy,D}(n,m-1)$, where $^+V_{xy,D}(n,m) \geq {}^+V_{xy,D}(n,m-1)$. Here a partitioning of $^+V_{xy,D}(n,m)$ is necessary since there is insufficient amplitude in $^+V_{xy,D}(n,m-1)$ to produce a complete plane wave correlation. We therefore set $^+V_{xy,D}(n,m)$ equal to

$$^+V_{xy,D}(n,m) = {}^+V_{xy,D}(n,m-1) + {}^+\Delta_{xy}(n,m) \qquad (4.9)$$

where $^+\Delta_{xy}(n,m)$ is defined by Eq.(4.9). We must not forget to use Eqs.(4.4a)-(4.4b) to insure proper scaling. The new effective components for the lower components, with similar notation(replacing subscript U with L) , are

$$^+V_{xyPL}(n,m) =[{}^+V_{xy,D}(n,m-1)\, {}^+V_{xy,D}(n,m)]^{1/2} \qquad (4.10a)$$

$$^+V_{xySL}(n,m) = [{}^+\Delta_{xy}(n,m)\, {}^+V_{xyD}(n,m)]^{1/2} \qquad (4.10b)$$

Eq.(4.10a) represents the plane wave contribution due to the correlation of the $^+V_{xy}(n,m)$ and $^+V_{xy}(n,m-1)$ waves. Eq.(4.10b) is the "normal' wave which is scattered in all directions.(lines). In addition to Eqs.(4.10a)-(4.10b) we must also add, in quadrature, the plane wave contribution resulting from the correlation with $^+V_{xy}(n,m+1)$, which we have shown to be equal to $^+V_{xyD}(n,m)$. The total plane wave, designated by $^+V_{xyP}(n,m)$, relates in quadrature to $^+V_{xyPL}(n,m)$ and $^+V_{xyD}(n,m)$ by

$$[{}^+V_{xyP}(n,m)]^2 = [{}^+V_{xyPL}(n,m)]^2 + [{}^+V_{xyD}(n,m)]^2 \qquad (4.11)$$

The portion of the wave which undergoes normal, symmetric, scattering is designated by $^+V_{xyS}(n,m)$, and is simply

$$^+V_{xyS}(n,m) = {}^+V_{xySL}(n,m) \qquad (4.12)$$

It is more useful to express $^+V_{xyP}(n,m)$ and $^+V_{xyS}(n,m)$ in terms of the original wave amplitudes. Using Eqs.(4.7)-(4.10), $^+V_{xyP}(n,m)$ and $^+V_{xyS}(n,m)$ then become

$$^+V_{xyP}(n,m) = (1/2)^{1/2}[{}^+V_{xy}(n,m-1)\, {}^+V_{xy}(n,m) + \{{}^+V_{xy}(n,m)\}^2\,]^{1/2} \qquad (4.13)$$

$$^+V_{xyS}(n,m) = (1/2)^{1/2} \left[\{^+V_{xy}(n,m)\}^2 - \, ^+V_{xy}(n,m) \, ^+V_{xy}(n,m-1) \right]^{1/2} \qquad (4.14)$$

$$\{ \, ^+V_{xy}(n,m) \geq 0 \; ; \; \text{if} \, ^+V_{xy}(n,m-1) < 0 \text{ then } ^+V_{xy}(n,m-1) = 0 \}$$

Eqs.(4.13) -(4.14) are the sought after plane wave and normal components of the original wave. We note certain simple but important features of the partition. If $^+V_{xy}(n,m-1) = 0$ then $^+V_{xyP}(n,m) = \, ^+V_{xyS}(n,m) = (1/2)^{1/2} \, ^+V_{xy}(n,m)$, i.e., the planar and symmetric components are equal. Also note that when $^+V_{xy}(n,m-1)$ $= \, ^+V_{xy}(n,m)$ then $^+V_{xyS}(n,m) = 0$ and $^+V_{xyP}(n,m) = \, ^+V_{xy}(n,m)$, as expected.

We should also observe that $^+V_{xyP}(n,m)$ and $^+V_{xyS}(n,m)$ are the *effective* amplitudes of the planar and symmetric fields, and thus $^+V_{xy}(n,m) \neq \, ^+V_{xyP}(n,m)$ $+ \, ^+V_{xyS}(n,m)$, but rather $[^+V_{xy}(n,m)]^2 = [^+V_{xyP}(n,m)]^2 + [^+V_{xyS}(n,m)]^2$, as may be verified from Eqs.(4.13) and (4.14). From Eqs.(4.4a) and (4.4b) the partitioned waves , denoted by $^+V_{xyPM}(n,m)$ and $^+V_{xySM}(n,m)$, are related to the effective amplitudes by

$$^+V_{xyPM}(n,m) = [^+V_{xyP}(n,m)]^2 \, / \, ^+V_{xy}(n,m) \qquad (4.15)$$

$$^+V_{xySM}(n,m) = [^+V_{xyS}(n,m)]^2 \, / \, ^+V_{xy}(n,m) \qquad (4.16)$$

The above equations satisfy $^+V_{xy}(n,m) = \, ^+V_{yxPM}(n,m) + \, ^+V_{yxSM}(n,m)$. The relationships for the transverse components $^+V_{xyP}(n,m)$ and $^+V_{xyS}(n,m)$ are of course similar.

We also emphasize that the above partition applies only to category I. Category II is similar except for the replacement of $^+V_{xy}(n,m-1)$ with $^+V_{xy}(n,m+1)$. Thus

$$^+V_{xyP}(n,m) = (1/2)^{1/2} [^+V_{xy}(n,m+1) \, ^+V_{xy}(n,m) + [^+V_{xy}(n,m)]^2]^{1/2} \qquad (4.17)$$

$$^+V_{xyS}(n,m) = (1/2)^{1/2} \left[\{^+V_{xy}(n,m)\}^2 - \, ^+V_{xy}(n,m) \, ^+V_{xy}(n,m+1) \right]^{1/2} \qquad (4.18)$$

$$\{ \, ^+V_{xy}(n,m) \geq 0; \; \text{if} \, ^+V_{xy}(n,m+1) < 0 \text{ then } ^+V_{xy}(n,m+1) = 0 \}$$

In Categories III and IV, $^+V_{xy}(n,m)$ is the minimum among its neighbors. $^+V_{xy}(n,m)$ then correlates perfectly with its neighbors with respect to its plane wave properties. The normal scattering then vanishes. Thus

$$^+V_{xyP}(n,m) = {}^+V_{xy}(n,m) \qquad\qquad (4.19)$$

$$^+V_{xyS}(n,m) = 0 \qquad\qquad (4.20)$$

$$\{{}^+V_{xy}(n,m) \geq 0\}$$

In Categories V and VI, $^+V_{xy}(n,m)$ has the highest amplitude among its neighbors and the effect is somewhat the opposite of categories III and IV, but not quite. The planar and normal fields are

$$^+V_{xyP}(n,m) = (1/2)^{1/2}[{}^+V_{xy}(n,m)\ ^+V_{xy}(n,m+1) + {}^+V_{xy}(n,m)\ ^+V_{xy}(n,m-1)\,]^{1/2} \quad (4.21)$$

$$^+V_{xyS}(n,m)=(1/2)^{1/2}[2({}^+V_{xy}(n,m))^2-{}^+V_{xy}(n,m)^+V_{xy}(n,m+1)-{}^+V_{xy}(n,m)^+V_{xy}(n,m1)]^{1/2}$$
$$(4.22)$$

$\{{}^+V_{xy}(n,m) \geq 0$; if $^+V_{xy}(n,m-1) < 0$ then $^+V_{xy}(n,m-1) = 0$; same for $^+V_{xy}(n,m+1)$ $\}$

From Eqs.(4.21)-(4.22) we see that the planar field vanishes only when $^+V_{xy}(n,m+1)$ and $^+V_{xy}(n,m-1)$ are equal to or less than zero and under these conditions we have $^+V_{xyS}(n,m) = {}^+V_{xy}(n,m)$. With these conditions we have an isolated(from its immediate neighbors) wave from which plane wave effects are impossible and we should expect scattering in all the available TLM lines.

    Care should be exercised in the expressions for $^+V_{xyP}(n,m)$ and $^+V_{xyS}(n,m)$, concerning the signs of $^+V_{xy}(n,m+1)$ and $^+V_{xy}(n,m-1)$. As noted beneath the expressions, when $^+V_{xy}(n,m)$ is positive and $^+V_{xy}(n,m+1)<0$ then we set $^+V_{xy}(n,m+1) = 0$. in the above expressions. A similar statement holds for $^+V_{xy}(n,m-1)$. We are therefore asserting that there is no plane wave correlation between neighbors   exists when the polarities of the waves are opposite.

Similarly when $^+V_{xy}(n,m)$ is *negative*, $^+V_{xy}(n,m+1)$ and $^+V_{xy}(n,m-1)$ will vanish when they are *positive*. We should also note that when $^+V_{xy}(n,m)$ is negative the roles of $^+V_{xy}(n,m-1)$ and $^+V_{xy}(n,m+1)$ are reversed so that Eqs.(4.13) and (4.14) apply to Category II (instead of I) while Eqs.(4.17) and (4.18) apply to Category I, and so forth. App.4.1 goes through the steps , parallel to the 2D development, for deriving the 3D planar and normal components for the various Categories.

Throughout the previous discussion, we postulated that a plane wave correlates with an adjacent wave, with the correlation strength roughly proportional to the product of the waves; however, any excess in the wave amplitude relative to the adjacent wave, causes that portion of the wave to scatter in all directions. As mentioned before, this assumption is supported by elementary quantum mechanical arguments, described in App.4A.2, where we show that the quantum cross-coupling between adjacent regions is consistent with plane wave correlations.

One may have noticed that throughout the discussion we have avoided the following question concerning the plane wave analysis. The plane wave correlation process can only occur instantaneously if the wave in a particular line "knows" what the status is of the wave in the adjoining line. If we assume the two lines somehow "communicate" with each other, however, we return to our original problem, in which a signal delay occurs because of signals propagating in the transverse lines. The only way to "resolve" this dilemma is through intensive quantum mechanical considerations, where such questions are posed in a different manner. The issue is a matter for debate even to this day. Indeed, our treatment of wave correlations represents a classical description of phenomena more appropriately described by quantum mechanics.   Needless to say, the issues just discussed become more important as the cell size is reduced.   Further examination of this topic will take us far beyond the scope of the present subject.

4.3 *Changes to 2D Scattering Coefficients*

Having determined the effective wave amplitudes for a partitioned wave, our task is now to modify the scattering coefficients. In particular, we must modify the scattering equations to account for the plane wave component of the wave. Thus we need to change Eqs.(3.33)-(3.36) for coplanar scattering or Eqs.(3.97)-

(3.100) for the general aplanar case. We now have a more complicated situation in which each wave incident on a node contains two components with differing scattering components. In effect this doubles the number of 3D scattering coefficients, from 96 to 192, with the additional 96 elements arising from the planar wave component in each transmission line.

The remaining issue is what changes should be introduced in the scattering coefficients for the planar wave component of each wave. The goal is to eliminate the indirect path of the wave, which effectively distorts the signal, slowing down the time of arrival of the full amplitude to its destination. The remedy adapted is fairly straightforward. If, for example, $^+V_{xy,P}(n,m)$ is incident on the (n,m) node then the 2D scattering coefficients are modified. For purposes of calculating the coefficients, we simply set $Z_{yx}(n,m)$ and $Z_{yx}(n,m+1)$ equal to zero, essentially shorting out the transverse lines. In the absence of any node resistance, therefore, the wave will continue unimpeded to the next line, $Z_{xy}(n+1,m)$, assuming $Z_{xy}(n+1,m) = Z_{zy}(n,m)$. In formal language, the scattering coefficient for the coupling of $^+V_{xy}(n,m)$ into the $Z_{xy}(n+1,m)$ line is

$$T_{xy,P}(n,m,1) = \ 1: \{Z_{yx}(n,m) = Z_{yx}(n,m+1) = 0\} \tag{4.23}$$

while the reflection coefficient is zero. In the event that $Z_{xy}(n+1,m) \neq Z_{xy}(n,m)$ then the modified transfer and reflection coefficients are

$$T_{xyP}(n,m,1) = 2Z_{xy}(n+1,m)/[\ Z_{yx}(n,m+1) + Z_{yx}(n,m)] \tag{4.24a}$$

$$B_{xyP}(n,m,2) = [Z_{yx}(n,m+1) - Z_{yx}(n,m)]\ /\ [Z_{yx}(n,m+1) + Z_{yx}(n,m)] \tag{4.24b}$$

The modification for the presence of node conductivity is straightforward and follows the discussion for the 1D scattering In the event $Z_{xy}(n+1,m) \neq Z_{xy}(n,m)$ or if node resistance is present, part of the wave energy is reflected back into $Z_{xy}(n,m)$, but in any case, there is still no energy transfer to the transverse lines. As a second example, consider $T_{yx}(n,m,2)$, which couples energy from the $Z_{yx}(n,m)$ to the $Z_{xy}(n+1,m)$ lines. For the plane wave component, $^+V_{yxP}(n,m)$,

$$T_{yx,P}(n,m,2) = 0 : \qquad (4.25)$$

$$\{Z_{xy}(n,m) = Z_{xy}(n+1,m)=0\}$$

since $Z_{xy}(n+1,m)$ and $Z_{xy}(n,m)$ are shorted out .

In the case of 3D assume the wave $^{+}V_{xy,P}(n,m,q)$ is incident on the (n,m,q) node. A total of six transverse lines must be decoupled by setting $Z_{yx}(n,m,q)$, $Z_{yx}(n,m+1,q)$, $Z_{zy}(n,m,q)$, $Z_{zy}(n,m,q+1)$, $Z_{yz}(n,m,q)$, and $Z_{yz}(n,m+1,q)$ equal to zero. These zero values for the impedance lines are then substituted in the 3D scattering coefficients.

Once the scattering is completed each line will contain forward and backward waves, each comprised of plane wave and symmetric components. Provided that the net plane wave and net symmetric components have the same sign, the two components will add in quadrature, resulting in the final field. In the event of a sign disparity between the two components, a "decorrellation" process must be implemented , to be discussed later in the Chapter.

A computer simulation of plane wave correlations in a light activated, semiconductor switch , with a parallel plate geometry, is given in Chapter VII. As we shall see, there are very substantial differences, compared to the situation in which only symmetric scattering prevails.

Finally, one should allow for the possibility that the node resistance (or any other node parameter) is dependent on the wave amplitude. If so, is the node dependence changed when the wave is partitioned into plane wave and symmetric parts? Although second order effects are always possible, in this study we assume the dependence of the node on the full wave amplitude is unchanged when the wave is partitioned. Thus, for example, the avalanche threshold at a node is assumed the same, regardless of the degree of partitioning.

Corrections to Plane Wave Correlation

The plane wave correlation described in the previous Sections is satisfactory provided the neighboring TLM lines are uniform. When the bordering TLM lines differ, however, the previous correlation is not complete and we must consider augmenting the   correlation appropriately. First we consider changes to the correlation when the adjoining cells have different dielectric constants. Correlation changes should not be surprising since the dielectric constant affects the wave energy and this in turn will influence the correlation. We then go on to discuss the modifications to the wave correlation required when the TLM wave is adjacent a conductor.   This is an important modification because of the omnipresence of conductors, as well as the frequent application of guided waves adjacent to a conductor(or for that matter, a dielectric). The topic of *de-correlation* of the waves, i.e., the conversion of plane waves into symmetric waves, is discussed later in the Chapter.

4.4   Correlation of Waves in Adjoining Media With Differing Dielectric Constants

In the previous discussion we have tacitly assumed that the correlation process takes place between the amplitudes of adjoining waves propagating in the same dielectric media. With differing dielectric media , however, proper weight must be given to the energy residing in each wave, taking into account the dielectric constant. A simple example will illustrate the point. If a wave in $Z_{xy}(n,m)$ correlates with another wave in $Z_{xy}(n,m+1)$, situated in a very high impedance region , the correlation ignores the fact that the wave in $Z_{xy}(n,m+1)$ carries little energy and should therefore carry less weight. The simplest way to correct for this oversight is to replace every amplitude $^+V_{xy}(n,m)$, located in $Z_{xy}(n,m)$, by $^+V_{xy}(n,m) \, /(Z_{xy}(n,m))^{1/2}$, which takes into account the wave impedance. The previous correllation equations, such as Eqs.(4.13)-(4.14) and Eqs.(4.17)-(4.22), should be modified by the following replacements

$$^+V_{xy}(n,m) \rightarrow {}^+V_{xy}(n,m) \, /(Z_{xy}(n,m))^{1/2} \qquad (4.26)$$

$$^+V_{xy}(n,m+1) \rightarrow {}^+V_{xy}(n,m+1) \, /(Z_{xy}(n,m+1))^{1/2} \qquad (4.27)$$

$$^+V_{xy}(n,m-1) \rightarrow {}^+V_{xy}(n,m-1) \, /(Z_{xy}(n,m-1))^{1/2} \qquad (4.28)$$

as well as a similar set of transformations for ${}^{+}V_{xyP}(n,m)$, ${}^{+}V_{xyP}(n,m-1)$, ${}^{+}V_{xyP}(n,m+1)$ and ${}^{+}V_{xyS}(n,m)$, ${}^{+}V_{xyS}(n,m-1)$, ${}^{+}V_{xyS}(n,m+1)$. The criteria for the six Categories of wave correlation, listed in Fig.4.2, are also modified accordingly. Thus Category I, for example, is: $[{}^{+}V_{xy}(n,m+1)\ /(Z_{xy}(n,m+1))^{1/2}\ ] \geq [{}^{+}V_{xy}(n,m)\ /(Z_{xy}(n,m))^{1/2}] \geq [{}^{+}V_{xy}(n,m-1)\ /(Z_{xy}(n,m-1))^{1/2}]$.

Another issue arises when we examine the correlation results at the dielectric-dielectric interface. Suppose we wish to obtain the correlation for the TLM wave belonging to the lower dielectric region (the larger cell) and propagating parallel the interface. Normally this wave will be correlated with a single wave, in the high dielecric region, corresponding to the "nearest node" at the interface(see Sections 5.3 and 5.4 in the next Chapter). Rather than consider only a single cell, a better approximation takes into consideration the multitude of high dielectric cells sharing the border with the larger low dielectric cell.

For example, suppose at the interface the ratio of dielectric constants is nine. The cells in the low dielectric region will then be 3X larger than the their counterparts in the high dielectric region, and therefore each large cell will share a border with three of the smaller cells. We can improve the accuracy of the correlation of the wave in the larger cell if we perform the correlation with all three waves, belonging to the smaller cells, rather than with only a single wave(belonging to the nearest node). In performing the correlation we utilize the *average* of the fields in the three cells, bordering the large cell. During the correlation process we still utilize the modified relationships, Eqs.(4.26)-(4.28), discussed previously. The same averaging technique in the correlation may of course be extended to arbitrary ratios for the dielectric constant.

4.5 Modification of Wave Correlation Adjacent a Conducting Boundary

The need for this modification is most easily understood if we consider a plane wave next to a conducting plane. If we *assume* the plane wave ${}^{+}V_{xy}(n,m)$ can propagate next to the conducting zx plane, without loss of its plane wave properties, then the previous wave correlation must be modified, as we shall see. Suppose the zx conducting plane is just beneath the $Z_{xy}(n,m)$ line. Since the wave in the $Z_{xy}(n,m-1)$ vanishes there is no plane wave correlation with that line, and a symmetric component equal to $^+V_{xy}(n,m)/(1/2)^{1/2}$ exists in the line. In the light of Eq.(4.14), e.g., this component will then scatter to the transverse line $Z_{yx}(n,m+1)$, thus creating a situation which eventually   destroys the plane wave correlation. One way to remedy this predicament is to simply postulate that the wave $^+V_{xy}(n,m)$, when adjacent a *perfect* conductor, *fully* correlates with $^+V_{xy}(n,m+1)$. Using the same notation as before *both* the "upper" and " lower" $^+V_{xyD}(n,m)$ then correlate with $^+V_{xy}(n,m+1)$. The plane wave property of $^+V_{xy}(n,m)$ is therefore maintained. An alternative view, which is   physically motivated and also produces the same result, is given by the following. Assume the line $Z_{xy}(n,m)$ is next to the conductor. This conductor then may be replaced by an image line with the same field,   located at $Z_{xy}(n,m-1)$. This is nothing more than the invocation of well known concepts from potential theory   which replaces a conducting plane with opposite charge at the image location . To see this more clearly, we first simulate the TLM field in $Z_{xy}(n,m)$ with + and - charges residing on the opposite TLM conductors. The conducting boundary plane is then replaced by inserting the image charges(after changing the sign of the charge) to their respective image locations. The image charges produce a replica of the field in $Z_{xy}(n,m)$, but now located at $Z_{xy}(n,m-1)$. $^+V_{xyD}(n,m)$ is then allowed to correlate with $^+V_{xyD}(n,m-1)$, i.e. , the lower $^+V_{xyD}(n,m)$ correlates with its own field. From this we may immediately deduce that when a plane wave propagates adjacent a perfect conducting plane the plane wave property in $^+V_{xy}(n,m)$ is maintained since the lower $^+V_{xyD}(n,m)$ correlates perfectly with its image field while the upper $^+V_{xyD}(n,m)$ correlates perfectly with its plane wave neighbor in $^+V_{xy}(n,m+1)$.

Having demonstrated the plane wave properties next to a conductor, we should then go on to treat the general case when $^+V_{xy}(n,m)$ is not part of a plane wave. In this case the number of Categories simplifies, reducing from six to two possibilities. This is because $^+V_{xy}(n,m)$ is always equal to its image field at $^+V_{xy}(n,m-1)$. In Category I $^+V_{xy}(n,m+1) \geq {}^+V_{xy}(n,m)$. From Eqs.(4.13)-(4.14), setting $^+V_{xy}(n,m-1) = {}^+V_{xy}(n,m)$,

$$^+V_{xyP}(n,m) = {}^+V_{xy}(n,m) \tag{4.29}$$

$$^{+}V_{xyS}(n,m) = 0 \qquad (4.30)$$

In Category II , $^{+}V_{xy}(n,m) \geq {}^{+}V_{xy}(n,m+1)$. In this case there is no change in the expressions for the two components, given in Eqs.(4.17)-(4.18) and repeated here for completeness,

$$^{+}V_{xyP}(n,m) = (1/2)^{1/2}[^{+}V_{xy}(n,m+1)\ {}^{+}V_{xy}(n,m) +[^{+}V_{xy}(n,m)\ ]^{2}]^{1/2} \qquad (4.31)$$

$$^{+}V_{xyS}(n,m) = (1/2)^{1/2}\ [[^{+}V_{xy}(n,m)\ ]^{2} - {}^{+}V_{xy}(n,m)\ {}^{+}V_{xy}(n,m+1)\ ]^{1/2} \qquad (4.32)$$

In the previous discussion, we did not inquire as to how the plane wave was launched in the region next to the conductor. One should expect the plane wave conditions to be upset when the wave encounters the conducting plane. One possible way to avoid such a disturbance is to employ an "infinitely" long , conducting plane in which the conductor is tapered, with the grazing angle slowly varying from 0 to 90 degrees. Such tapers must of course be approximated. An alternative is to employ the TLM matrix, including plane wave correlations, to analyze the changes to the plane wave conditions, even in cases where the transition to the conductor is abrupt. Fig.4.3 shows such a situation for a half-infinite conducting plane. For the most part the plane wave, denoted by ^{+}V, is maintained in regions far away from the corner of the half plane, designated by O. In the region near O, however, a definite departure from plane wave conditions is encountered , due to the limited front. This takes the form of non-uniformity of the waves in adjacent cells, as well as the initiation of waves in the transverse direction (i.e., waves propagating in the vertical direction). As the waves proceed further and further away from O, however, the waves return to their plane wave properties. These properties may be deduced in detail, using the iterative equations with plane wave correlations. One may also employ the semi-graphical, step by step technique, similar to that used in Section 3.4.

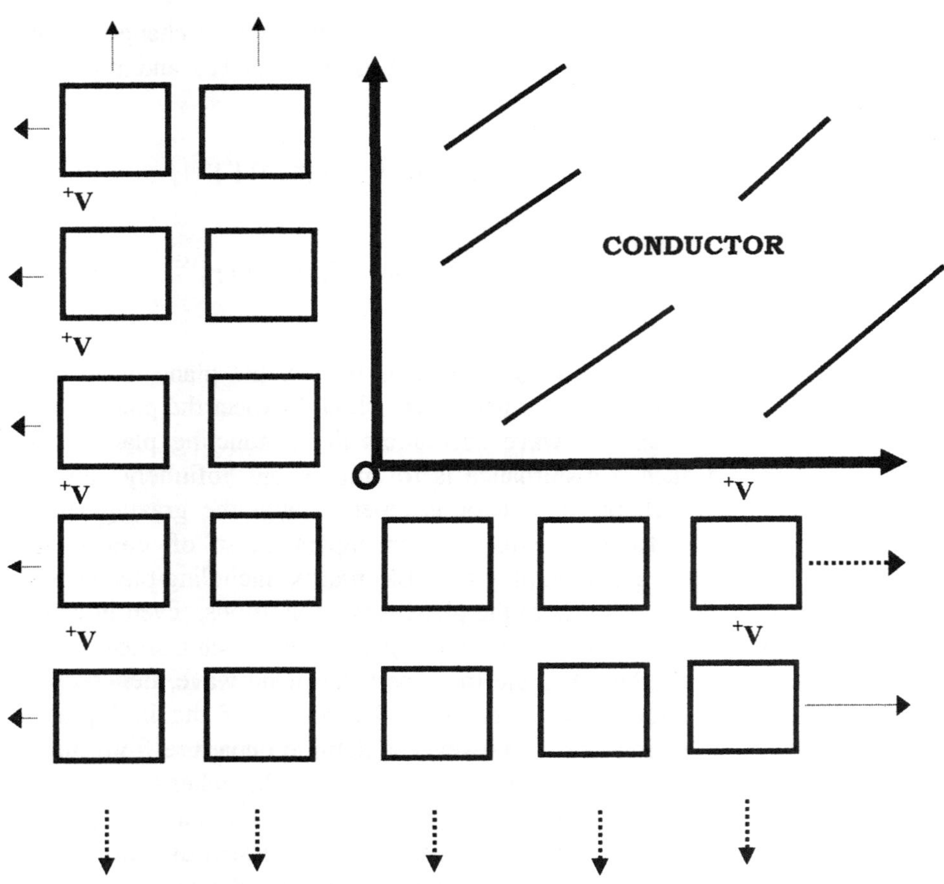

FIG. 4.3 PLANE WAVE INCIDENT ON HALF INFINITE CONDUCTOR PLANE. PLANE WAVE CONDITIONS ARE UPSET NEAR THE ORIGIN O.

4.5(a) Non-Perfect Conducting Plane

What effect does an imperfect conducting plane have on a plane wave which borders the conductor? It would seem that even with very small imperfections, the plane wave properties will eventually be destroyed. However, if the conducting surface has random irregularities with the usual usual peaks and valleys, then one may apply very small TLM cells to demonstrate the overall effect. With a random surface, the decorrelation effects are essentially confined to the surface region(defined by the size of the imperfections) and thus the destruction of the plane wave does not spread beyond the region of the imperfections. Stated another way, the symmetric waves are cancel out and are extinguished once they venture beyond the imperfection region. The effect is to slow down the wave propagation(in the directions parallel to the surface) near the conducting surface, in much the same manner as a pure symmetric matrix. To be sure this is an intuitive argument , which may be verified by applying the TLM matrix to various models of imperfections on the conducting surface, using a very high resolution TLM matrix. An alternate (but cruder) way of looking at the problem is to apply a large cell matrix, whose cell size is much larger than the imperfections. One can then estimate that the imperfections average out to zero, and thus the imperfect surface may be replaced by a perfectly smooth surface.

Decorrelation Processes

4.6 De-Correlation Due to Sign Disparity of Plane and Symmetric Waves

In earlier discussion we described field correlation of a wave $^+V_{xy}(n,m)$ with its neighbors $^+V_{xy}(n,m+1)$ and $^+V_{xy}(n,m-1)$. Is there a simultaneous process which brings about the decorrelation of the field(i.e., the decrease of the plane wave component)? In the following we shall see that indeed such a decorrelation does occur , depending on the four waves incident on the (n,m) node.

We start with a simple example. Suppose a plane wave is advancing along the TLM matrix in the x direction and that the wave front is located at the nth element. Thus for the (n,m) cell the field in the $Z_{xy}(n,m)$ line is $^+V_{xy}(n,m)$ and since we assume the wave is a purely planar type, we have $^+V_{xy}(n,m) =$

$^+V_{xyP}(n,m)$. At the same time that we have the plane wave, suppose in addition we have a *solitary* transverse symmetric wave $^+V_{yxS}(n,m)$ (i.e., there is no plane wave component contained in $^+V_{yxS}(n,m)$).   What happens during the ensuing time step. ? The planar wave $^+V_{xyP}(n,m)$ will advance to the $Z_{xy}(n+1,m)$ line. The $^+V_{yxS}(n,m)$ wave, however, will scatter in symmetric fashion about the (n,m) node. The significant point is that the symmetric component, scattered into the $Z_{xy}(n+1,m)$ line, will be opposite in sign to the planar component $^+V_{xyP}(n,m)$. In order to simultaneously account for this disparity in sign, *and* the conservation of scattered energy about the node, we must surrender the existence of a plane wave in the $Z_{xy}(n,m)$ line and assume instead that the wave is composed of both plane wave and symmetric components(despite the correlation with the waves in the $Z_{xy}(n,m+1)$ and $Z_{xy}(n-1,m)$ lines). The sign disparity immediately signifies that a decorrelation of the incident wave field must take place. We should add that the symmetric wave in the $Z_{yx}(n,m)$ line, in some ways, may be regarded as a physical obstacle which undoes   the plane wave properties in the $Z_{xy}(n,m)$ line.

4.6(a). Simplified De-correlation Process For Removal of Sign Disparity

We consider the case in which the incoming waves, $^+V_{xy}(n,m)$, $^-V_{xy}(n+1,m)$, $^+V_{yx}(n,m)$, and $^-V_{yx}(n,m+1)$ converge on the (n,m) node and we allow each wave the possibility of having plane wave as well as symmetric components. We of course assume there is no sign disparity in any of he incoming waves. Now suppose a sign disparity develops in one or more of the *outgoing* waves, $^-V_{xy}(n,m)$, $^+V_{xy}(n+1,m)$, $^-V_{yx}(n,m)$, and $^+V_{yx}(n,m+1)$, during the ensuing $(k+1)$ time step . The sign disparity means that the plane wave and symmetric waves are opposite to one another, an unacceptable condition requiring some type of decorrelation of the *incoming* waves.

The simplest and most direct means for removing the disparity is to convert all the incoming waves to purely symmetric waves. In this case $^+V_{xyS}(n,m)$ $=^+V_{xy}(n,m)$ and $^+V_{xyP}(n,m)$ $=0$, etc.. This obviously insure that no sign disparity will occur since we have removed all the plane wave components heading toward the (n,m) node. A less crude version of this approach is to convert only those co-linear incoming waves for which a sign disparity exists in the outgoing waves of the same co-linear TLM lines. Thus for example, if the only outgoing waves which have a sign disparity are either $^-V_{xy}(n,m)$ or $^+V_{xy}(n+1,m)$ (or both), but there is no disparity in the other outgoing waves, then obviously we can eliminate the disparity by removing the planar components from the incoming co-linear waves $^+V_{xy}(n,m)$ and $^-V_{xy}(n+1,m)$ and nothing need be done for the other incoming waves $^+V_{yx}(n,m)$ and $^-V_{xy}(n,m+1)$.

Before discussing the general case, it is helpful to first cite the earlier example of a pure plane wave $^+V^k_{xyP}(n,m) = {}^+V^k_{xy}(n,m)$, advancing toward the (n,m) node and, in addition, a pure symmetric wave, $^+V^k_{yxS}(n,m) = {}^+V^k_{xy}(n,m)$, also advancing toward the (n,m) node (here we include the time step superscript for clarity). For simplicity assume for the time being that the TLM lines are identical. The wave scattered from $Z_{yx}(n,m)$ to line $Z_{xy}(n+1,m)$, is equal to $-{}^+V^k_{yxS}(n,m)/2$, where the minus sign indicates the negative polarity. Assume $^+V^k_{xyP}(n,m)$ and $^+V^k_{yxS}(n,m)$ both have the same sign. If the *magnitude* of $^+V^k_{yxS}(n,m)$ is greater than or equal to $2{}^+V^k_{xyP}(n,m)$, then in fact $^+V^k_{xyP}(n,m)$ must be replaced by $^+V^k_{xyS}(n,m)$, i,e., the plane wave must be completely decorrelated. Suppose the magnitude of $^+V^k_{yxS}(n,m)$ is less than $2{}^+V^k_{xyP}(n,m)$. In this case $^+V^k_{xyP}(n,m)$ is replaced by a new pair of components , $^+V'^k_{xyP}(n,m)$ and $^+V'^k_{xyS}(n,m)$, where $[{}^+V^k_{xyP}(n,m)]^2 = ({}^+V'^k_{xyP}(n,m))^2 + ({}^+V'^k_{xyS}(n,m))^2$ and $^+V'^k_{xyS}(n,m)$ is equal in magnitude to $^+V^k_{yxS}(n,m)$. (Here and in the remainder of the Chapter, the prime may signify transformed waves resulting from either decorrelation or grid change). Their contributions to $Z_{xy}(n+1,m)$ then cancel one another and we are left with a smaller plane wave component $^+V^{k+1}_{xyP}(n+1,m)$ (compared to $^+V^k_{xyP}(n,m)$).

In the previous discussion the sign disparity between the plane and symmetric components was resolved for a particular situation in which the incident wave , for the horizontal line, was completely planar and the incident wave for the transverse line was completely symmetric. The same sign disparity can also arise when all the lines, surrounding the node, have both plane wave and symmetric components. The sign disparity between the resultant planar and symmetric waves may then be removed using the same *decorrelation* techniques as before, but we use a more general format. This is done by first examining the waves emanating from the (n,m) node in each of the four lines surrounding the node. Assume, as a matter of generality that in each line the emerging line has both symmetric and plane wave components. If we find no sign disparity in any of the four lines then no alteration is necessary and we allow the iteration to continue on to the next step. On the other hand, suppose a sign disparity does exist in one or more of the lines. This disparity may then be removed by appropriately reducing the planar components of the *incident co-linear* waves, and similarly for any other disparity present in any of the other lines. Once we dispose of the sign disparity we allow the iteration to proceed for that step, combining the symmetric and plane waves in each line and thus providing the total field.   With the new total fields, the process starts over again, beginning with the *correlation* process(which compares the fields in the two parallel neighbors)immediately followed by the *de-correlation* process.

In the previous discussion we considered only the simple case in which the sign disparity is removed by decorrelation of the co-linear waves.   Such a solution will in general, however, not be unique. For example, since it may also be possible to remove the disparity by eliminating the plane wave component in the transverse lines, without changing the co-linear waves. Indeed we may have all four incoming waves participating in the decorrelation process. This raises the issue of exactly what is required of the decorrelation process, given the fact that more than one solution may be possible.

In order to proceed, we postulate that the decorrelation represents a *minimal path* solution., i.e., a solution in which the least amount of plane wave energy undergoes decorrelation. This has two desirable properties. The first is that the decorrelated field solution is unique. The second is that the new decorrelated waves will "least disturb" the old correlated fields. This helps to insure stability and prevents the occurrence of catastrophic changes in the plane waves, brought about by small sign disparities. A corollary of the minimal solution is that it allows for an elegant mathematical formulation, namely, the calculation of variations, which often leads to new insights. With this approach, the voltage integral of the decorrelation energy is assumed to be an extremum.

Suppose the plane wave energy converging on a node is E_T, and there are sign disparities in the subsequent step. We designate the decorrelation energy by ΔE_T, remembering that ΔE_T is one of many possible solutions. Using standard minimization methods(including the calculus of variations) requires that $\partial(\Delta E_T)/\partial V_i = 0$, where V_i is any one of the four TLM waves incident on the node. In this case the only allowable extremum is a minimum since the maximum calls for the elimination of the entire incident plane wave energy - a generally unstable condition. The minimal solution may be determined, in principle, by analytic methods, but it is often more convenient to obtain such a solution by means of computerized, iterative techniques. An outline of one such approach is given later in the Chapter.

Before outlining the techniques for the minimal solution, it is worthwhile to devote attention to several general decorrelation concepts and also to solutions which are not minimal. The approximate solutions afford much insight into the decorrelation process, and will be useful in obtaining the minimal solutions. We reiterate that whatever process is employed, the decorrelation of a wave is allowed only if it results in the reduction of the sign disparity. Upon completing the decorrelation, the newly acquired waves are then incorporated into the overall computer iteration.

4.6(b) General Concepts Underlying Removal of the Sign Disparity

Before delving into the analytic and minimal solution methods, we first discuss several concepts pertaining to the decorrelation. We suppose the general case in which each of the four lines surrounding the (n,m) node has both symmetric and plane wave components, incident on the node, and that there exists a sign disparity in one or more of the four waves emerging from the node. In the general case we must consider all four input and output waves simultaneously, since merely "fixing" the problem in one wave, along the lines described in the previous paragraphs, may induce an additional disparity in another line. For convenience, we mostly focus attention on only one output line, $Z_{xy}(n+1,m)$, but the method is exactly the same for the other lines.

In order to proceed we utilize the iterative equations described in Chapter III, Eqs.(3.20)-(3.23), (We point out the difference that in the present discussion, the scattering is *about* the node, and not scattering into the TLM cell , so that for example $T_{xy}(n-1,m,1) \rightarrow T_{xy}(n,m,1)$, $^{+}V^{k+1}_{xy}(n,m) \rightarrow {}^{+}V^{k+1}_{xy}(n+1,m)$, etc...) . We first state the expression for $^{+}V^{k+1}_{xy}(n+1,m)$, which from Section 3.2 is

$$^+V^{k+1}{}_{xyS}(n+1,m) = T^k{}_{xyS}(n,m,1) \, ^+V^k{}_{xyS}(n,m) - T^k{}_{yxS}(n,m,2) \, ^+V^k{}_{yxS}(n,m)$$
$$+T^k{}_{yxS}(n,m,3) \, ^-V^k{}_{yxS}(n,m+1) + B^k{}_{xyS}(n,m,1) \, ^-V^k{}_{xyS}(n+1,m) \qquad (4.33)$$

We also have need of the expression for the companion planar component in the $Z_{xy}(n+1,m)$ line, given by

$$^+V^{k+1}{}_{xyP}(n+1,m) = T^k{}_{xyP}(n,m,1) \, ^+V^k{}_{xyP}(n,m) + B^k{}_{xyP}(n,m,1) \, ^-V^k{}_{xyP}(n+1,m) \qquad (4.34)$$

where the P subscript in the coefficients indicates that we short out the transverse lines, as is appropriate for the plane wave components.

We may utilize the previous expressions to estimate the condition for the elimination of any assumed sign disparity of the outgoing wave in $Z_{xy}(n+1,m)$. We require that

$$\text{SIGN}\{\, ^+V^{k+1}{}_{xyP}(n+1,m) \,\} = \text{SIGN}\{\, ^+V^{k+1}{}_{xyS}(n+1,m) \,\} \qquad (4.35a)$$

We proceed in similar fashion to obtain the components for the three other outgoing waves, again utilizing Eqs.(3.20)-(3.23) and the companion equations for the planar components. The required sign conditions are

$$\text{SIGN}\{\, ^-V^{k+1}{}_{xyP}(n,m) \,\} = \text{SIGN}\{\, ^-V^{k+1}{}_{xyS}(n,m) \,\} \qquad (4.35b)$$

$$\text{SIGN}\{\, ^-V^{k+1}{}_{yxP}(n,m) \,\} = \text{SIGN}\{\, ^-V^{k+1}{}_{yxS}(n,m) \,\} \qquad (4.35c)$$

$$\text{SIGN}\{\, ^+V^{k+1}{}_{yxP}(n,m+1) \,\} = \text{SIGN}\{\, ^+V^{k+1}{}_{yxS}(n,m+1) \,\} \qquad (4.35d)$$

In the event of a sign disparity in an outgoing line, it is useful to define a quantity which reflects the degree of sign disparity. Suppose, for example,

$^{+}V^{k+1}_{xyP}(n+1,m)$ and $^{+}V^{k+1}_{xyS}(n+1,m)$ have disparate signs and the absolute value of $^{+}V^{k+1}_{xyP}(n+1,m)$ exceeds that of $^{+}V^{k+1}_{xyS}(n+1,m)$. The degree of sign *identity* index, denoted by γ_D, is defined by

$$\gamma_D = [((^{+}V^{k+1}_{xyP}(n+1,m))^2 - (^{+}V^{k+1}_{xyS}(n+1,m))^2]^{1/2} / [(^{+}V^{k+1}_{xyP}(n+1,m))^2 + (^{+}V^{k+1}_{xyS}(n+1,m))^2]^{1/2}$$

(4.36)

$$SIGN(^{+}V^{k+1}_{xyP}(n+1,m)) \neq SIGN(^{+}V^{k+1}_{xyS}(n+1,m));$$
$$|^{+}V^{k+1}_{xyP}(n+1,m)| \geq |^{+}V^{k+1}_{xyS}(n+1,m)|$$

and the P and S subscripts are of course interchanged if $|^{+}V^{k+1}_{xyS}(n+1,m)|$ exceeds $|^{+}V^{k+1}_{xyP}(n+1,m)|$. Note that when the signs are opposite but the magnitudes of the two waves are equal, $\gamma_D = 0$, and the sign disparity is correspondingly a maximum with the plane and symmetric components having equal magnitudes. For the other extreme, when the magnitude of one component is much higher than the magnitude of the other, γ_D is approaches unity and there is very little disparity. The inverse of γ_D of course expresses the degree of sign disparity.

Looking at the scattering equations , we can immediately assert that there is at least one(non-minimal) solution, namely, the simplified decorrellation, discussed earlier, in which the sign disparities are removed by setting equal to zero the planar components in each co-linear direction. If there are sign disparities in both the x and y directions, then *all* the plane wave components are set equal to zero and the input waves satisfy, after decorrelation,

$$^{+}V^{k}_{xyS}(n,m) = {}^{+}V^{k}_{xy}(n,m) \ ; \ ^{-}V^{k}_{xyS}(n+1,m) = {}^{-}V^{k}_{xy}(n+1,m)$$ (4.37a,b)
$$^{+}V^{k}_{yxS}(n,m) = {}^{+}V^{k}_{yx}(n,m) \ {}^{-}V^{k}_{yxS}(n,m+1) = {}^{-}V^{k}_{yx}(n,m+1)$$ (4.37c,d)
$$^{+}V^{k}_{xyP}(n,m) = {}^{+}V^{k}_{yxP}(n,m) = {}^{-}V^{k}_{xyP}(n+1,m) = {}^{-}V^{k}_{yxP}(n,m+1) = 0$$ (4.37e,f)

The more refined procedure is to remove only that amount of planar field necessary to eliminate the sign disparity. The simplest strategy starts with the largest disparity in the four output lines, using the sign identity index γ_D as a measure. For example, suppose the largest disparity is in $Z_{xy}(n+1,m)$. There are then four possible sources of *incoming* plane wave components which we can seek to reduce in order to remove the sign disparity: the forward planar wave $^{+}V^{k}_{xyP}(n,m)$, the backward planar wave, $^{-}V^{k}_{xyP}(n+1,m)$, and the corresponding transverse planar waves , $^{+}V^{k}_{yxP}(n,m)$ and $^{-}V^{k}_{yxP}(n,m+1)$.

In proceeding with the decorrelation the first step is to determine the existence of any sign disparities in any of the outgoing waves and then to determine the sign identity index γ_D for each instance where such a disparity exists. Once we have calculated the these indices then the next step is to determine which of the four input planar fields must undergo reduction in order for the sign identity index to *increase* toward unity, and thus remove the sign disparity. How we set γ_D equal to unity depends on the relative magnitudes of the plane and symmetric components, and the elimination of one or the other.

4.6(c) Approximate Decorrelation Methods(Non-Minimal Solution)

In this Sub-Section we discuss approximate methods for finding *non-minimal* decorrelation solutions. As mentioned previously, although not a true solution , the method does provide insight into the decorrelation process. The method relies on the progressive increase of γ_D , seeking out solutions which, though not strictly minimal, are efficient in increasing γ_D.

How we approach the increase in γ_D , as we force it toward unity, requires some attention. Suppose $Z_{xy}(n+1,m)$ has a sign disparity between the two waves $^{+}V^{k+1}_{xyP}(n+1,m)$ and $^{+}V^{k+1}_{xyS}(n+1,m)$. There are *two* possible routes for obtaining $\gamma_D=1$, depending on the relative magnitudes of the components and the composition of the input waves. The two paths will in general differ as to the total amount of decorrelation needed to remove the sign disparity. The desired, of course, path is the one requiring the lesser amount of decorrelation , i.e., the one which results in a smaller disturbance of the prior fields, as mentioned before. The quantitative meaning of " lesser" is discussed subsequently . We classify the two routes as follows, using the output in the $Z_{xy}(n+1,m)$ line as an example.

PATH   COMMENT<u>S</u>

I : Eliminate $^+V^{k+1}_{xyP}(n+1,m)$

Guaranteed to Remove Sign Disparity. More likely to be Lesser Path when:
$$|{^+V^{k+1}_{xyS}}(n+1,m)| > |{^+V^{k+1}_{xyP}}(n+1,m)|$$

II: Eliminate $^+V^{k+1}_{xyS}(n+1,m)$

No guarantee to Remove Sign Disparity. More likely to be Lesser Path when:
$$|{^+V^{k+1}_{xyP}}(n+1,m)| > |{^+V^{k+1}_{xyS}}(n+1,m)|$$

If Path I applies, then the disparity is removed by the decorrelation of other input plane waves such that $^+V^{k+1}_{xyP}(n+1,m)$ is eliminated. In this case the input plane waves to be decorrelated are confined to the in-line type, namely, $^+V^k_{xyP}(n,m)$ and $^-V^k_{xyP}(n+1,m)$; the transverse waves then play no role. Using this path , although one is assured of eliminating $^+V^{k+1}_{xyP}(n+1,m)$, it is no guarantee that this is a lesser path, although this is more likely to happen when, as noted, $|{^+V^{k+1}_{xyS}}(n+1,m)| \geq |{^+V^{k+1}_{xyP}}(n+1,m)|$. We must also look into the possibility that the symmetric component , $^+V^{k+1}_{xyS}(n+1,m)$, goes to zero before the plane wave type, i.e., we must consider the possibility of Path II. Unlike Path I , however, there is no guarantee that $^+V^{k+1}_{xyS}(n+1,m)$ can be eliminated at all, much less qualify as a lesser path. As noted above Path II has greater chances of success when $|{^+V^{k+1}_{xyP}}(n+1,m)| \geq |{^+V^{k+1}_{xyS}}(n+1,m)|$. With Path II all four input waves may play a role in the removal of the sign disparity.

We now quantify and generalize several of the ideas described in the previous paragraphs using analytic means. Once we have determined which of the four incoming planar fields must be reduced then we attach a common decorrelation factor α_D to each of the applicable incoming fields. Under the most general conditions all four input planar waves will have to be reduced, in which case α_D will apply to all the input waves. The revised fields are denoted by a

prime superscript in order to help differentiate them from the original fields.
Thus,

$$(\, ^+V'^k_{xyS}(n,m) \,)^2 + (\, ^+V'^k_{xyP}(n,m))^2 = (\, ^+V^k_{xy}(n,m))^2 \tag{4.38a}$$
$$(\, ^-V'^k_{xyS}(n+1,m))^2 + (\, ^-V'^k_{xyP}(n+1,m))^2 = (\, ^-V^k_{xy}(n+1,m))^2 \tag{4.38b}$$
$$(\, ^+V'^k_{yxS}(n,m))^2 + (\, ^+V'^k_{yxP}(n,m))^2 = (\, ^+V^k_{yx}(n,m))^2 \tag{4.38c}$$
$$^-V'^k_{yxS}(n,m+1))^2 + (\, ^-V'^k_{yxP}(n,m+1))^2 = (\, ^-V^k_{yx}(n,m+1))^2 \tag{4.38d}$$

Now suppose we wish to break up $^+V^k_{xyP}(n,m)$ into two quadrature components, $\alpha_D \, ^+V^k_{xyP}(n,m)$ and $(1-\alpha_D^2)^{1/2} \, ^+V^k_{xyP}(n,m)$, with similar results for the other input waves. Using the partitioning process described at the start of the Chapter, the revised(primed) and the original fields are related by

$$^+V'^k_{xyP}(n,m)) = \alpha_D \, ^+V^k_{xyP}(n,m)) \tag{4.39a}$$
$$^-V'^k_{xyP}(n+1,m)) = \alpha_D \, ^-V^k_{xyP}(n+1,m)) \tag{4.39b}$$
$$^+V'^k_{xyP}(n,m)) = \alpha_D \, ^+V^k_{xyP}(n,m)) \tag{4.39c}$$
$$^-V'^k_{yxP}(n,m+1)) = \alpha_D \, ^-V^k_{yxP}(n,m+1)) \tag{4.39d}$$

Thus the revised and original fields for the symmetric component are related by

$$(\, ^+V'^k_{xyS}(n,m) \,)^2 = (\, ^+V^k_{xyS}(n,m) \,)^2 + (1-\alpha_D^2)((\, ^+V^k_{xyP}(n,m))^2 \tag{4.40a}$$
$$(\, ^-V'^k_{xyS}(n+1,m))^2 = (\, ^-V^k_{xyS}(n+1,m))^2 + (1-\alpha_D^2)(\, ^-V^k_{xyP}(n+1,m))^2 \tag{4.40b}$$
$$(\, ^+V'^k_{yxS}(n,m))^2 = (\, ^+V^k_{yxS}(n,m))^2 + (1-\alpha_D^2)(\, ^+V^k_{yxP}(n,m))^2 \tag{4.40c}$$
$$(\, ^-V'^k_{yxS}(n,m+1))^2 = (\, ^-V^k_{yxS}(n,m+1))^2 + (1-\alpha_D^2)(\, ^-V^k_{yxP}(n,m+1))^2 \tag{4.40d}$$

α_D is in the range $0 \leq \alpha_D \leq 1$ For $\alpha_D = 0$, the decorrelation is complete whereas for $\alpha_D = 1$, no decorrellation occurs. We again reiterate that the decorrelation is applied only to those planar fields whose reduction will result in a field increase in the sign identity index.

The decorrelation is achieved by combining Eqs(4.38)-(4.40) with the scattering equations, Eqs.(4.33)-(4.34) and setting either the symmetric or plane wave portion of $^+V^{k+1}_{xy}(n+1,m)$, for example, equal to zero. The solution for α_D

is then obtained for each of these two cases. The *larger* value of α_D is then selected, which corresponds to the lesser degree of decorrelation needed to remove the sign disparity, i.e, the minimum Path whether it be I or II. In certain cases the only solution will be $\alpha_D = 0$, in which case all the input plane wave fields will converted to symmetric fields.

A simple example is discussed in order to render the concepts more clearly. Suppose we wish to eliminate a sign disparity in $Zxy(n+1,m)$. The de-correlation factor α_D based on the planar scattering , must then be determined. We assume for the moment that no planar fields are present in the transverse lines and thus no decorrellation need be applied to these fields. Given these circumstances, how do we go about setting ${}^+V^{k+1}_{xyP}(n+1,m)$ or ${}^+V^{k+1}_{xyS}(n+1,m)$ equal to zero, assuming a sign disparity is present. Several possibilities exist. To simplify we first assume Path I prevails , and to make matters concrete, we also assume $B(n,m,1)$ is positive and ${}^-V^k_{xyP}(n+1,m)$, ${}^+V^k_{xyP}(n,m)$ are as well. We then take ${}^+V^{k+1}_{xyS}(n+1,m)$ to be negative , which thus makes ${}^+V^{k+1}_{xyP}(n+1,m)$ positive. Now since Path I applies, ${}^+V^{k+1}_{xyP}(n+1,m)$ is to be eliminated. Because both incoming plane waves contribute to the reduction, we attach α_D to *both* ${}^+V^k_{xyP}(n,m)$ and ${}^-V^k_{xyP}(n+1,m)$. In this case, in order to reduce the output plane wave component to zero, we must likewise force the two planar inputs to zero, i..e., $\alpha_D=0$. During this reduction ${}^+V^{k+1}_{xyS}(n+1,m)$ may change in value but will remain negative. Now consider a slightly different variation of the previous, in which all the field magnitudes and polarities are unchanged but $B(n,m,1)$ is negative. In this situation the decorrelation factor attaches to only one of the two inputs, namely, ${}^+V^k_{xyP}(n,m)$. In this instance it may be possible to analytically determine α_D from the scattering equation, Eq.(4.34)

$$\alpha_D = -B_{xyP}(n,m,1)\ {}^-V^k_{xyP}(n+1,m) / T_{xyP}(n,m,1)\ {}^+V^k_{xyP}(n,m) \qquad (4.41a)$$

Suppose the solution of the above does not satisfy $0 \leq \alpha \leq 1$? Then we proceed as before and completely decorrelate the two inputs, effectively forcing $\alpha_D = 0$.

We use the same example as before but now examine the possibility that Path II applies. $B(n,m,1)$ is once more negative and the field magnitudes and polarities are as before. The strategy followed is then to reduce ${}^+V^{k+1}_{xyS}(n+1,m)$

to zero. Since we assume in this example that there are no transverse plane waves, we need only attach the decorrelation factor to the input plane wave $^+V^k_{xyP}(n,m)$. In this simple case we can resort analytic means, using the scattering relationship,Eq.(4.33),to find α_D , remembering to set $^+V^{k+1}_{xyS}(n+1,m)$ equal to zero:

$$T_{xyS}(n,m,1)\ ^+V'^k_{xyS}(n,m) = -B_{xyS}(n,m,1)\ ^-V^k_{xyS}(n+1,m) +T_{yxS}(n,m,2)\ ^+V^k_{yxS}(n,m)$$
$$- T_{yxS}(n,m,3)\ ^-V^k_{yxS}(n,m+1) \tag{4.41b}$$

where we use Eq.(4.40a) to find $^+V'^k_{xyS}(n,m)$

$$(^+V'^k_{xyS}(n,m))^2 = (^+V^k_{xyS}(n,m))^2 +(1-\alpha_D^2)\ (^+V^k_{xyP}(n,m))^2 \tag{4.41c}$$

which then allows us to solve for α_D. As indicated before the prime superscript indicates that we are using the revised symmetric field. For clarity we retain the subscripts S in the transverse fields despite the assumed absence of plane wave fields. As this is the only field undergoing decorrelation, only this field requires revision. Finally we compare the α_D here with the α_D obtained for the Path I trial. The Path with the smaller α_D then prevails.

The previous example was for a specific wave having a sign disparity, but the non-minimal method may be generalized to all four output waves with or without sign disparities. Starting with the wave with the largest sign disparity , we employ both Paths I and II to remove the disparity We then select the Path, which corresponds to the largest α_D. We then re-examine the outputs and once more identify the largest disparity, followed once more by the same employment of Paths I and II, etc... The process ends when all the sign disparities are removed in all four output waves.

4.7 Minimal Solution Using Differing Decorrelation Factors to Remove Sign Disparities

Although useful, the simple solutions just discussed suffer from certain deficiencies. First of all there is no reason to assume that the *same* decorrelation factor α_D applies to all of the four input waves surrounding the node. For completeness, each wave should have its own decorrelation factor. More importantly, there is no guarantee that the previous methods will yield a minimal, solution, i.e., the decorrelation path which least disturbs the initial plane wave energy about the node. The minimal solution , employing differing decorrelation factors, should provide the necessary stability for the resultant waves. If we assume the stable solution embodies the correct solution, then the minimal solution will provide greater accuracy as well.

In seeking the minimal solution, we have to spell out exactly what quantity we are trying to minimize. This quantity, as one might expect, is the energy flow associated with the initial plane wave components directed at the particular node. The "inertia" of the plane waves is associated with the energy, and not the amplitudes. To obtain the plane wave energy flow in each line, we must also take into account the impedance value of each line.

To proceed with this method we first assign a decorrelation factor to each input wave about the (n,m) node. We use α_{D1}, α_{D2} , α_{D3}, and α_{D4} as the factors applied to the waves ${}^{+}V^{k}_{xyP}(n,m)$, ${}^{+}V^{k}_{yxP}(n,m)$, ${}^{-}V^{k}_{xyP}(n+1,m)$, and ${}^{-}V^{k}_{yxP}(n,m+1)$ respectively. Thus, for example, the *difference* in the energy flow, ΔE_1, for the ${}^{+}V^{k}_{xyP}(n,m)$ wave is $\Delta E_1 = ({}^{+}V^{k}_{xyP}(n,m))^2/Z_{xy}(n,m) - (\alpha_{D1})^2({}^{+}V^{k}_{xyP}(n,m))^2/Z_{xy}(n,m)$, or,

$$\Delta E_1 = (({}^{+}V^{k}_{xyP}(n,m))^2/Z_{xy}(n,m))(1-\alpha_{D1}^2) \qquad (4.42a)$$

When α_{D1} is equal to one then there is no decorrelation and $\Delta E_1 = 0$. When $\alpha_{D1} = 0$ then the decorrelation is complete and ΔE_1 is equal to the initial energy flow of the plane wave. The decrements of the energy flow for the other three lines are, in similar fashion,

$$\Delta E_2 = (({}^{+}V^{k}_{yxP}(n,m))^2/Z_{yx}(n,m))(1-\alpha_{D2}^2) \qquad (4.42b)$$
$$\Delta E_3 = (({}^{-}V^{k}_{xyP}(n+1,m))^2/Z_{xy}(n+1,m))(1-\alpha_{D3}^2) \qquad (4.42c)$$
$$\Delta E_4 = (({}^{-}V^{k}_{yxP}(n,m+1))^2/Z_{yx}(n,m+1))(1-\alpha_{D4}^2) \qquad (4.42d)$$

The general procedure for finding the minimal solution is fairly straightforward. Suppose only a single output wave from the (n,m) node exhibits a sign disparity. Then there are two possible ways, or paths, for removing the sign disparity- either eliminating the plane wave or symmetric components. When the sign disparity exists in two, three, or all four outputs, the number of paths is four, eight, or sixteen, respectively.

As an example, suppose two of the outputs , $^+V^{k+1}_{xy}(n+1,m)$ and $^-V^{k}_{yx}(n,m)$ exhibit a sign disparity. The following Table lists all the possible combinations or paths for removing the sign disparity.

POSSIBLE MINIMAL PATHS FOR REMOVAL OF SIGN DISPARITY
EXAMPLE: DISPARITIES IN $^+V^{k+1}_{xy}(n+1,m)$ and $^-V^{k+1}_{yx}(n,m)$

$$A : \quad ^+V^{k+1}_{xyP}(n+1,m) , ^-V^{k+1}_{yxP}(n,m)$$
$$B: \quad ^+V^{k+1}_{xyP}(n+1,m) , ^-V^{k+1}_{yxS}(n,m)$$
$$C: \quad ^+V^{k+1}_{xyS}(n+1,m) , ^-V^{k+1}_{yxP}(n,m)$$
$$D : \quad ^+V^{k+1}_{xyS}(n+1,m) , ^-V^{k+1}_{yxS}(n,m)$$

Thus , for example, in path B we reduce $^+V^{k+1}_{xyP}(n+1,m)$ and $^-V^{k+1}_{yxS}(n,m)$ to zero in order to remove the disparity. Continuing with Path B, we first require the total difference in energy flow ΔE_T to be minimized. The total is

$$\Delta E_T = \Delta E_1 + \Delta E_2 + \Delta E_3 + \Delta E_4 \tag{4.42e}$$

In the above expression, not all terms will be present, depending on whether plane waves are present or absent. Thus, if there are no plane waves in $^-V^{k}_{yx}(n,m+1)$ then $\Delta E_4 = 0$. The minimum of ΔE_T, with respect to the decorrelation factors, is then determined subject to several constraints. First of all we require that any solutions must satisfy $0 \leq \alpha_{DN} \leq 1$, where $N=1,2,3,4$. Secondly, the minimization is accompanied by two subsidiary relations which must be satisfied, namely, the two scattering equations which cause $^+V^{k+1}_{xyP}(n+1,m)$ and $^-V^{k+1}_{yxS}(n,m)$ to vanish. In the case of $^+V^{k+1}_{xyP}(n+1,m)$, the scattering equation is (see Eq.(4,34) set equal to 0)

$$T^{k}_{xyP}(n,m,1)\alpha_{D1}{}^+V^{k}_{xyP}(n,m)+B^{k}_{xyP}(n,m,1)\alpha_{D3}{}^-V^{k}_{xyP}(n+1,m)=0 \tag{4.42f}$$

Similarly, for $^-V^{k+1}{}_{yxS}(n,m)$ to vanish, we have(see Eq.(3.23)):

$$T^k{}_{xyS}(n,m,10)\ ^+V'^k{}_{xyS}(n,m) + T^k{}_{yxS}(n,m,11)\ ^-V'^k{}_{yxS}(n,m+1)$$
$$-T^k{}_{xyS}(n,m,12)\ ^-V'^k{}_{xyS}(n+1,m) +B^k{}_{yxS}(n,m,4)\ ^+V'^k{}_{yxS}(n,m) =0 \qquad (4.42g)$$

where the primes indicate we are using the decorrelated symmetric fields. These fields will depend on the decorrelation factors in a manner similar to Eq.(4.40) but with an important difference: each input wave will have its own decorrelation factor. Thus

$$(^+V'^k{}_{xyS}(n,m))^2 =(^+V^k{}_{xyS}(n,m))^2 +(1-\alpha_{D1}{}^2)\ (^+V^k{}_{xyP}(n,m))^2 \qquad (4.42h)$$

$$(^-V'^k{}_{yxS}(n,m+1))^2=(^-V^k{}_{yxS}(n,m+1))^2+(1-\alpha_{D4}{}^2)(^-V^k{}_{yxP}(n,m+1))^2 \qquad (4.42i)$$

$$(^-V'^k{}_{xyS}(n+1,m))^2 =(^-V^k{}_{xyS}(n+1,m))^2 +(1-\alpha_{D3}{}^2)\ (^-V^k{}_{xyP}(n+1,m))^2 \qquad (4.42j)$$

$$(^+V'^k{}_{yxS}(n,m))^2 =(^+V^k{}_{yxS}(n,m))^2 +(1-\alpha_{D2}{}^2)\ (^+V^k{}_{yxP}(n,m))^2 \qquad (4.42k)$$

The minimization is then carried out subject to the previously mentioned constraints on the decorrelation factors and the two scattering equations. In this case , for Path B, a solution is not guaranteed since one can never assume that $^-V^{k+1}{}_{yxS}(n,m)$ can be attain a zero value. Once we finish with Path B we follow the procedure for the other Paths(A, C, and D), obtaining ΔE_T and a set of decorrelation factors for each Path. We then select the smallest ΔE_T , among the four possibilities.

The process is then repeated for all the nodes until all the sign disparities throughout the space are eliminated. The result is the minimal solution for the entire matrix . We then go forward with the next iteration.

4.7(a). Iterative Method

The previously outlined   solution relies on the use of standard analytic methods for minimizing a function A purely numerical method may be employed provided the second derivative of ΔE_T is positive throughout the region of interest. This will be the case if ΔE_T is close to its minimum value. To implement this method we first define a *constant* decrement of the energy flow δE . Thus if we wish to decrement the energy of the initial plane wave field in $^+V^k_{xyP}(n,m)$, the old and the new fields are related by

$$[^+V'^k_{xyP}(n,m)]^2 = [^+V^k_{xyP}(n,m)]^2 - [\delta E]^2 \qquad (4.42l)$$

$$[^+V'^k_{xyS}(n,m)]^2 = [^+V^k_{xyS}(n,m)]^2 + [\delta E]^2 \qquad (4.42m)$$

As in our previous example suppose there exists sign disparities in $^+V^{k+1}_{xy}(n+1,m)$, $^-V^{k+1}_{yx}(n,m)$. We decorrelate the input of each input plane wave (a total of four possibilities), one at a time, by the amount δE examining the reduction in the output wave energy . If we are focusing on Path B then we examine the reduction in the energy flow $(^+V^{k+1}_{xyP}(n+1,m))^2/ Z_{xy}(n+1,m) + (^-V^{k+1}_{yxS}(n,m))^2/Z_{yx}(n,m)$ at each step. The particular decorrelation which yields the *largest* reduction is then selected(in the event that two decorrelations yield the same maximum reduction, then the energy decrement may be distributed equally among the two types of decorrelations). This process is continued until $^+V^{k+1}_{xyP}(n+1,m)$ and $^-V^{k+1}_{yxS}(n,m)$ are completely extinguished. We are then left with decorrelated fields obtained in a number of energy decrements, designated by N_B. The same process is then performed for the other three paths , yielding energy decrement numbers of N_A, N_C, and N_D and their corresponding decorrelated fields. Finally we select the minimum among N_A, N_B, N_C, and N_D. Small values of δE may be required in order to differentiate between paths. It may also occur(though unlikely) that a decorrelation may lead to an additional sign disparity, in which case additional paths must be considered.

4.8 Non-Essential De-Correlation Caused by Simultaneous Presence of Forward and Backward In-Line Plane Waves With Same Polarity

There is another source of de-correlation, which in some ways may be considered an adjunct of the previous sign disparity type, and having to do with the with the simultaneous presence of *forward and backward plane waves* with the same polarity in neighboring series TLM lines. This decorrelation is termed non-essential in the sense that it is used to simplify the decorrelation process, but is not needed to obtain the final decorrelation. For this decorrelation, we take a clue from equilibrium situations in which we know that at equilibrium, the forward and backward waves are equal in amplitude and sign. In the process of achieving the equilibrium state, we may regard the correlation between the two waves as indicative of the degree to which the system has achieved equilibrium. In order to insure that the system tends toward equilibrium, however, we require a corresponding increase in the symmetric scattering , at the expense of the plane wave scattering, i.e., we perform a decorrelation of the opposite waves As we will, see this type decorrelation applies not only to equilibrium states, but to non-equilibrium situations as well, i.e., time dependent situations in which plane wave components always appear.

We can best illustrate the previous assertion by considering two oppositely directed plane waves approaching one another with equal amplitudes, one in the positive x direction and the other in the negative x direction. Until the two waves converge , the waves proceed as normal plane waves. However, when the waves border one another, or even overlap, our interpretation of the two waves will change. In fact the two waves converging on the (n,m) node, $^+V_{xyP}(n,m)$ and $^-V_{xyP}(n+1,m)$ may just as well be regarded as symmetric waves $^+V_{xyS}(n,m)$ and $^-V^k_{xyS}(n+1,m)$. Indeed it is very easy to show that, for these oppositely directed symmetric waves, the net scattering to the transverse lines $Z_{yx}(n,m)$ and $Z_{yx}(n,m+1)$ vanishes. We should also take note that in arriving at the zero scattering to the transverse lines, we considered opposite waves in the $Z_{xy}(n,m)$ and $Z_{xy}(n+1,m)$ lines, rather than both waves in the $Z_{xy}(n,m)$. However, if the changes in the x direction vary slowly, and the TLM length is correspondingly small, then we can compare forward and backward waves in the same $Z_{xy}(n,m)$ line.

In the previous discussion we saw that a pair of opposite plane waves, of equal amplitude, are equivalent to symmetric waves so far as transverse scattering is concerned. What about the reflected and transmitted waves in $Z_{xy}(n,m)$ and $Z_{xy}(n+1,m)$? Indeed it is easy to show, for example, that the reflected plane wave differs from that of a symmetric wave (for the same input amplitude). The same is true of the transmitted wave. However, when we consider the *total* wave amplitude leaving the node, $^-V_{xy}(n,m)$ and $^+V_{xy}(n+1,m)$, we find that $^-V_{xy}(n,m) = {}^+V_{xy}(n+1,m) = {}^+V_{xy}(n,m) = {}^-V_{xy}(n+1,m)$, *irregardless* of whether the inputs are pure plane waves or symmetric waves. In summary, therefore, we can replace a pair of opposite plane waves of equal amplitude, with a pair of symmetric waves of the same amplitude.

Thus plane wave components, when they are part of equal and opposite directed uniform plane wave fronts, and also uniform in the direction in the propagation directions, are redundant and may be replaced by their symmetric counterparts. In such situations, the replacement of the plane wave components will *not* change the iteration, as we previously indicated. However, because we are converting plane waves to symmetric waves, we may be able to avoid the need for a follow-on sign disparity decorrelation, thus simplifying the process. We emphasize, however, that if there is no sign disparity to begin with, then no decorrelation of this type (i.e., due to the presence of forward and backward plane waves)need be performed and we may proceed directly to the scattering equations. An exception to this rule occurs, however, when we do not perform a correlation after each time step, allowing the plane wave and symmetric components to proceed for a number of steps before intervening with a correlation.

We now describe the general decorrelation in which the forward and backward plane wave components , in adjoining series TLM lines, are not part of plane wave fronts and there is non-uniformity of the plane wave components. In other words, what happens when the two opposite waves have the same polarity but differ in magnitude. Also, what occurs if the opposite wavesboth contain symmetric as well as plane components? These issues are discussed in the following.

As usual, we can best explain this type decorrelation by means of an example. Suppose the forward plane wave in line $Z_{xy}(n,m)$ is $^+V_{xyP}(n,m)$ and the backward plane wave, in line $Z_{xy}(n+1,m)$, $^-V_{xyP}(n+1,m)$. Assume for the moment that both are positive and that $^+V_{xyP}(n,m) \geq {}^-V_{xyP}(n+1,m)$, and that there are *no* symmetric components present. Under these circumstances we assume a decorrelation of $^+V_{xyP}(n,m)$ occurs , converting into plane wave and symmetric parts, $^+V'_{xyP}(n,m)$ and $^+V'_{xyS}(n,m)$, where we append a prime superscript to differentiate the revised fields from the original ones. At the same time the backward wave $^-V'_{xyP}(n+1,m)$ is converted to $^-V'_{xyS}(n+1,m)$. The new and old fields are related by

$$(^+V'_{xyP}(n,m))^2 + (^+V'_{xyS}(n,m))^2 = (^+V_{xyP}(n,m))^2 \tag{4.43a}$$

$$^+V'_{xyS}(n,m) = {^-V^k_{xyP}(n+1,m)} \tag{4.43b}$$

and

$$^-V'_{xyS}(n+1,m) = {^-V_{xyP}(n+1,m)} \tag{4.43c}$$

and of course $^+V'_{xyP}(n,m)$ is explicitly determined by combining Eqs.(4.43a) and (4.43b). We again point out that $^+V'_{xyP}(n,m)$ is determined by the degree of *decorrelation* between the forward and backward plane waves in the adjoining series TLM lines. As implied by the existence of $^+V_{xyP}(n,m)$ and $^-V^k_{xyP}(n+1,m)$, this type de- correlation follows the correlation process with the adjacent horizontal waves $^+V_{xy}(n,m-1)$ and $^+V_{xy}(n,m+1)$ as well as $^-V_{xy}(n+1,m-1)$ and $^-V_{xy}(n+1,m+1)$.

In the previous example we assumed that no symmetric fields were present in the original fields. If symmetric fields are present, the repair of the outlined decorrelation method is straightforward; we simply add , in *quadrature* , the new symmetric fields, $^+V'_{xyS}(n,m)$ and $^-V'_{xyS}(n+1,m)$, to the old symmetric fields $^+V_{xyS}(n,m)$ and $^-V_{xyS}(n+1,m)$.

The consideration of other examples is straightforward. Suppose the amplitude of the backward plane wave $^-V_{xyP}(n+1,m)$ is larger than that of $^+V_{xyP}(n,m)$, although both are considered to have the same sign. In this case the situation is reversed from that of the previous example and we are left with the resultant backward plane wave.

As another example suppose that $^+V_{xyP}(n,m)$ is positive while the backward plane wave $^-V_{xyP}(n+1,m)$ is negative. In this case $^+V_{xyP}(n,m)$ and $^-V_{xyP}(n+1,m)$

remain intact as plane wave components and no further decorrelation occurs. Finally, although all our comments have focused on the components in the x direction, naturally the same technique applies to the y direction, i.e., to the $^+V_{yxP}(n,m)$ and $^-V_{yxP}(n,m+1)$.

Although the examples outlined certainly do not exhaust all the possibilities they are sufficient to completely make the method very clear. We should note the similarity with the plane wave correlation in adjacent parallel lines with that of the forward and backward plane waves in the series lines. The difference is that in the latter the plane wave remains intact to the degree that it differs in sign and magnitude with its oppositely directed counterpart; in the former case the correlation is enhanced by adjacent waves with the same sign and magnitude.

4.9 Decorrelation Treatment of Plane Waves Incident on a Dielectric Interface

Another type of decorrelation to be discussed requires us to anticipate results discussed fully in Chapter V, having to do with the boundary between two regions with differing   dielectric constants. The differing values in ε imply differences in the propagation velocity, which in turn signifies differing cell sizes in each region. Fig.4.4 depicts the interface separating regions ε_1 and ε_2, and we of course note the larger cells in region ε_2 corresponding to the presumed smaller value of ε_2 compared to ε_1 .

4.9(a) Decorrelation of Partial Node When Using Nearest Node Method

We focus our attention first on the line $Z_{yx}(n,m)$ and also note that the wave $^+V_{yx}(n,m)$ does not penetrate directly into the ε_2 region , but must instead first enter lines $Z_{xy}(n,m)$ and $Z_{xy}(n+1,m)$. This type of node is termed a partial node , whose designation arises because the larger cell essentially blocks the direct entry of incident waves at certain nodes, as in Fig.4.4a. Note from Fig.4.4a that we

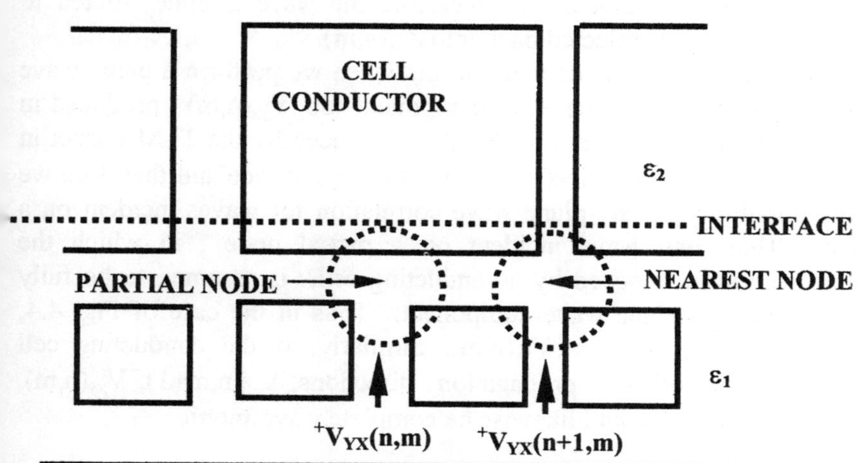

FIG. 4.4(a) WITH NEAREST NODE COUPLING, THE INCIDENT WAVE $^+V(n,m)$ IS PURELY SYMMETRIC AT A PARTIAL NODE.

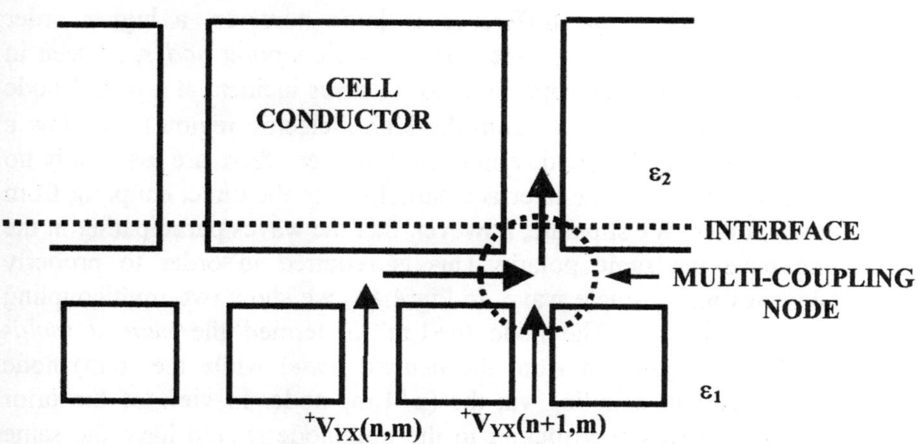

FIG. 4.4(b) DEMONSTRATION OF MULTICOUPLING NODE WITH NODES (n,m) AND (n+1,m) PARTICIPATING. FOR WAVES INCIDENT ON ε_2 WAVES $^+V((n,m)$ AND $^+V(n+1,m)$ ARE COMBINED.

assume the partial node is *not* a "nearest node", discussed extensively in Chapter V. The blocking of the wave at the partial node occurs of course since the large cell in ε_2 appears as a conductor and therefore the wave is either routed to $Z_{xy}(n,m)$ or $Z_{xy}(n+1,m)$, or reflected back into $Z_{yx}(n,m)$.

The problem with this type of node occurs when we perform a plane wave correlation in $Z_{yx}(n,m)$. Any plane wave component of $^+V_{yx}(n,m)$, produced in $Z_{yx}(n,m)$, runs into a short circuit , artificially produced by the TLM matrix in the ε_2 region. There is no physical reason for this occurrence and therefore we eliminate the possibility of any plane wave correlation for waves incident on a partial node.   Thus, any wave incident on a partial node , in which the propagation direction is blocked by a conducting cell, is assumed to be fully symmetric and with no plane wave component. Thus in the case of Fig. 4.4, $^+V_{yxP}(n,m) =0$ and $^+V_{yxS}(n,m) =\,^+V_{yx}(n,m)$. Similarly, if the conducting cell blocked any of the other propagation directions, $^-V_{yx}(n,m+1),\,^+V_{xy}(n,m)$, $^-V_{yx}(n+1,m)$, these waves would likewise be completely symmetric.

4.9(b) Decorrelation at Dielectric Interface Using Multi-Coupling Nodes

Usually the nearest node coupling approximation is quite adequate, particularly when the waves are incident on a dielectric interface whose dielectric constants are close in value. When this difference is large however, a higher order approximation may be warranted, one using *multi-coupling nodes*, as seen in Fig.4.4b. With the nearest node approximation, waves incident at a partial node do not have do not couple *directly* from the high dielectric region to the low ε region With the multi-coupling node approach, however, there are essentially no partial nodes and all nodes at the interface participate in the direct coupling from one region to the other. We stipulate, however, that the waves participating in the multicoupling have the same polarity.This is required in order to properly combine the waves into a single wave. In Fig(4.4b) we show two multicoupling nodes, (n,m) and (n+1,m). The node (n+1,m) is termed the *nearest multi-coupling node*(to distinguish it from the *nearest node*) while the (n,m) node participates in the direct coupling via the (n+1,m) node. In view of the prior comments, the two waves contributing to the multinode should have the same polarity.   In the prior nearest node approximation,   (n,m) and the (n+1,m) represented the partial and nearest nodes respectively. The procedures for locating in cell notation the nearest multicoupling and nearest nodes, as well as the participating nodes, are discussed in Chapter V.

Having identified the nearest multicoupling and the participating nodes, the next step is to outline the procedure for combining such waves. We illustrate with the two nodes in Fig.4.4b, which assumes the (n,m) and (n+1,m) are the only nodes participating in the multicoupling, via the nearest multicoupling node (n+1,m). We then consider the incident waves $^+V_{yx}(n,m)$ and $^+V_{yx}(n+1,m)$. In order for the multi-coupling to be allowed, the polarity of $^+V_{yx}(n,m)$ and $^+V_{yx}(n+1,m)$ must be the same; else we fall back on the nearest node approximation. Assuming the polarities are the same the procedure is then to "remove " $^+V_{yx}(n,m)$ and add it in quadrature to $^+V_{yx}(n+1,m)$, i.e., the plane wave and symmetric components are each added resulting in a new pair of $^+V_{yxP}(n+1,m)$ and $^+V_{yxS}(n+1,m)$. The usual scattering of these waves is then allowed to take place. If a decorrelation of the new wave is required , we use the minimal procedure outlined previously. What about a vertical wave emanating from the low dielectric side and incident on the (n,m+1) node. In this case the outputs among the lines surrounding the (n+1,m) node in the ε_1 region, are re-distributed in equal fashion among the lines about (n,m) and (n+1,m). For example, if after scattering about (n+1,m), suppose the output wave in the line $Z_{yx}(n+1,m)$ is $^-V_{yx}(n+1,m)$.and has both plane wave and symmetric components. Each component is then divided in quadratuere between $Z_{yx}(n,m)$ and $Z_{yx}(n+1,m)$.

When combining waves in the multicoupling approach, we should take note of the fact that we have artificially inflated the amplitude in a particular TLM line. This is not a problem unless the node function(e.g, the node resistance) is amplitude dependent, as is often the case. If such is the case then we must use the average of the *full* wave TLM wave amplitudes participating in the multicoupling process.

We point out that the multicoupling node approach should not result in any new results, compared to the nearest node method. With the nearest node method, the plane waves incident on partial nodes eventually make their way into the low dielectric region where, after some delay they participate in the plane wave process. Thus, the multicoupling should speed up the iteration, since it will reduce the amount of delay at the interface., for a given cell size. The main disadvantage is the additional iterative complexity of the multicoupling.

4.10   *Comments on Interaction of a Plane Wave Front and a Source Emitting Both Plane Wave and Symmetric Components*

The interpretation of plane wave correlations/decorrelations may be aided by considering the interaction of a plane wave front in which a source is immersed and emitting both plane and symmetric wave components. What effect should we observe in the behavior of the plane wave, in the region of the source? We assume of course that the surface is compatible with the TLM matrix so that the source roughly coincides with a TLM line at the interface. Suppose the source is immersed in the plane wave, and the source emits a constant amplitude. A TLM analysis in the region of the source then reveals a definite perturbation of the plane wave in the region of the source. Now suppose the frequency of the source is changed, from say "dc" to an alternating signal. In other words we alternate the sign of the signal amplitude, during successive transit times, emanating from the source(Section 7.16 discusses a simulation of a signal with an alternating sign). In this case also, the effect on the plane wave is in the form of low amplitude ripples superimposed on the plane wave, which decay also in amplitude as one goes farther and farther away from the source. We emphasize, however, that both these perturbations occur because of the tacit assumption that the plane wave and source signal are locked in time and therefore the two signals are "coherent" and therefore an interaction is to be expected. Suppose, however, the source signal is random in nature in either phase or frequency(or both). Averaged over many cells the effect of the source on the plane wave vanishes.

The situation also applies to two interacting laser signals which cross paths. Each laser signal is assumed to be made up of a dominant plane wave component, except for the laser boundary region, where symmetric components are assumed to be present.   When the lasers are locked in phase, the laser signals will experience coherent interaction due to the correlation of the plane wave components. If the lasers are not locked together, however, the interaction will be very weak.

The features of all the interactions may be examined more closely using the graphical TLM analysis as in Sections 3.14 to 3.17, or by using the TLM simulations of Chapter VII.

4.11 Summary of Correlation/Decorrelation Processes

Fig.4.5 shows a flow diagram of the wave decomposition, followed by the de-correlation processes. Although not specifically noted the correlation is assumed to include correlation effects due to the presence of conductors and differing dielectric constants, as discussed earlier. Note that the decorrelation due to the simultaneous presence of forward and backward waves, in series TLM lines, is performed first, before that caused by the sign disparity. This type decorrelation is not essential, but can shorten the follow-on decorrelation processes. The second decorrelation is that due to either a nearest node, or a multicoupling node, taking place at a dielectric interface. Finally, if a sign disparity is predicted, a further decorrelation is performed. Once the waves are correctly decorrelated they are delivered to the scattering equations. Following scattering, we then perform the process of wave reconstruction. The process is repeated for each node and for each time step.

Grid Orientation Effects

4.12 Dependence of Wave Energy Dispersal on Grid Orientation

Since the electromagnetic energy is guided along the edges of either a square (for 2D) or a cube(for 3D), these edge directions are preferred and energy will be transmitted more rapidly along these directions compared to any other directions, as indicated earlier in Fig.1.17. The earliest arrival of an electromagnetic disturbance initiated at the region O will reach regions A and B more quickly than the earliest arrival at region B. This stems from the fact that A and C may be reached via a straight line path whereas region B may be accessed only through a tortuous zigzag path through both transverse and longitudinal lines.

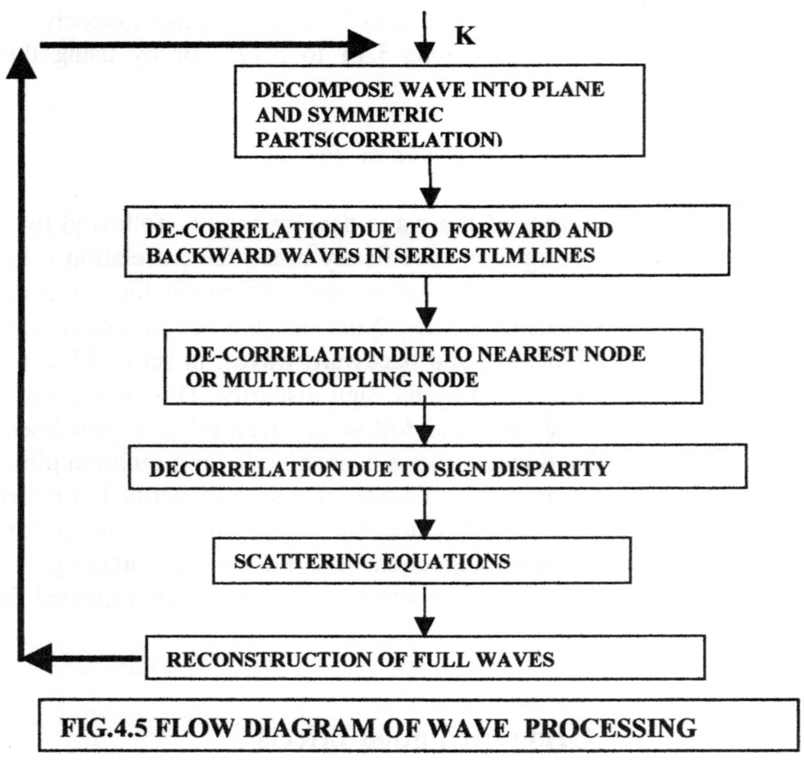

FIG.4.5 FLOW DIAGRAM OF WAVE PROCESSING

We generalize the concept by comparing the earliest arrival times from the origin to points on a circle of radius L_0, as in Fig.4.6. Obviously, the shortest path is along a grid axis, as with point A. Next consider an arbitrary point B in which the radius vector makes an angle θ with the x axis. The shortest path, L, from the origin to B is not unique, but nevertheless may be expressed by

$$L = L_0[\cos\theta + \sin\theta] \tag{4.44}$$

Note that there may be many paths , such as L', which have the same total length as L; so long as one does not go backwards in x or y, the path lengths woll be

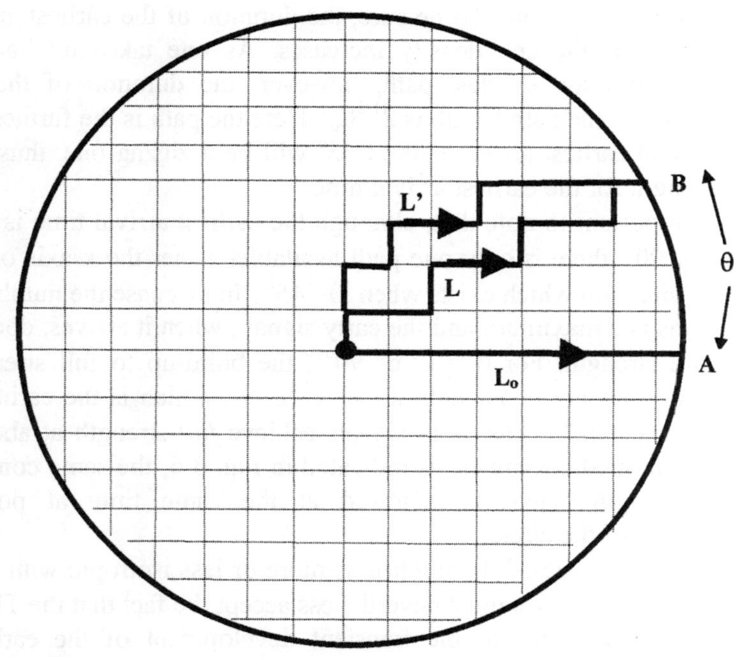

FIG. 4.6 VARIOUS TLM PATHS FROM ORIGIN TO POINTS ON CIRCUMFERENCE. PATH L_o IS PARALLEL TO X AXIS AND IS SHORTEST POSSIBLE PATH LENGTH, EQUAL TO RADIUS. PATHS L AND L' TO POINT B ARE INDIRECT AND LONGER, EQUAL TO $L=L'= [COS\theta +SIN\theta]L_o$

identical. In any event, the earliest arrival time, to point B, will be dependent on the orientation of the TLM matrix. An inherent anisotropy therefore exists in the distribution of electromagnetic energy, at least during the initial transient phase, resulting from the particular symmetry elements employed (such as the 2D square or the 3D cube) throughout the space. From Eq.(4.44), the maximum occurs at $\theta = 45^\circ$ while the minimum occurs at either 0° or 90°. These results are not surprising. At $\theta =0$ or 90° , the path length is L_o and the signal has a straight line path to its destination and therefore the earliest arrival time will be the shortest time possible. To be sure, the duration of the earliest arriving signal will diminish as the cell density increases. As one takes into account slight excursions from the shortest path, however, the duration of the signal will increase. At $45°$ the path length is $2^{1/2}L_o$. Here the path is the furthest away from both axes and earliest arrival path at $45°$ will be a zigzag one, thus maximizing the path length for the earliest arrival time.

It is important to note that, although the earliest arrival time is a minimum at $\theta = 0°$ or $90°$, there is only one path available(either the x axis or the y axis), unlike the situation which exists when $\theta = 45°$. In this case the number of earliest arrival paths is a maximum and the early signal , when it arrives, does so at more or less full strength. For $\theta = 0°$ or $90°$, the build-up to full strength is more gradual and does not occur until about $2^{1/2}L_o/v$, which is the earliest arrival of the signal at $\theta = 45°$. Thus, both signals achieve *full* strength at about the same time. For intermediate angles, as indicated in Fig. 3.6, the same concept applies; the full strength signal is achieved at the same time at points on the circumference of the circle.

Although the signal distribution is more or less isotropic with regard to the full amplitude signal, we must nevertheless accept the fact that the TLM matrix is anisotropic with regard to the transient development of the earliest arriving signal. This must be regarded as an unacceptable artifact of the TLM method. The issue to be addressed, therefore , is how the TLM matrix may be modified so as to eliminate or at least minimize this anisotropic effect, which is built into the matrix. One crude approach is to reorient the axes $45°$, obtain the solution in the new coordinates, and average the new solution with that of the old one. If the disturbance source is spherically symmetric, the new solution will be exactly the same as the old one, but rotated $45°$. Setting $L_o = 1$, the two paths are $L_1 = \cos\theta + \sin\theta$ and $L_2 = \cos(\theta + 45°) + \sin(\theta + 45°)$, and the average of the two paths is

$$AvL = (L_1 + L_2)/2 = [(1 + 2^{1/2})\cos\theta + \sin\theta]/2 \qquad (4.45)$$

Because we are averaging two of these functions, which tend to dilute the extreme path values, the earliest arrival time will exhibit more spherical symmetry, with the path length varying from 1.2~1.3, compared to that in Fig. 4.7 where the variation is from 1 to $2^{1/2}$. The averaging process may be continued by dividing into smaller angles followed by an averaging over the various angles. This allows the angular dependence of the earliest arrival to be further reduced. Repeated averaging leads to the elimination of the angular dependence, and a path length of ~ 1.25. The risetime of the signal also becomes considerably reduced, compared to the on-axis path.

Despite the removal of the anisotropy there still remains the delay factor (of about 1.25). This delay is illusory, however, since we have not incorporated the plane wave correlation effects, or directivity, discussed earlier, into the description. If we look at a limited portion of a spherical wave front, the directivity will tend to keep that portion of the front moving in the same direction along the longitudinal lines, particularly for those grids in which the axes are aligned in the same direction as the direction of the front. As discussed previously, the plane wave correlation property prevents energy from being diverted into the transverse lines; such diversion effectively slows down the front velocity.

4.13 Transformation Properties Between Grids

As we have seen, we may remove the anisotropy inherent in the TLM matrix , by considering an array of TLM grids, oriented at various angles In order to proceed, however, we must determine how the various properties change under orthogonal transformations. Such transformations are well known; our main goal will be to recast such changes using cellular notation. This is to be followed by averaging the properties among the various grids.

We begin with 2D orthogonal transformations arising from a rotation about the origin. Only a single parameter, the angle θ is required. Using primes to specify the transformed coordinates, the coordinates in the new system are

$$x' = x\cos\theta \ + y\sin\theta \qquad\qquad (4.46a)$$
$$y' = -x\sin\theta \ + y\cos\theta \qquad\qquad (4.46b)$$

Conversely, the old coordinates may be related to the new ones by

$$x = x'\cos\theta \ - y'\sin\theta \qquad\qquad (4.47a)$$

$$y = x'\sin\theta + y'\cos\theta \qquad (4.47b)$$

We now determine the node resistors and voltage waves in the transformed grid. The approach taken is to first find the node (n',m') closest to the (n,m) node. For this we use Eqs.(4.46a)-(4.46b), setting $x=n\Delta l$, $y= m\Delta l$, $x'=\gamma\Delta l$, and $y'=\eta\Delta l$. These equations then become

$$\gamma = n\cos\theta + m\sin\theta \qquad (4.48)$$

$$\eta = -n\sin\theta + m\cos\theta \qquad (4.49)$$

Since we have not placed any constraints on , x and y, γ and η in Eqs.(4.48)-(4.49) will in general not be integral . We then state

$$n' = INT(\gamma) + Rem(\gamma) \qquad (4.50)$$

$$m' = INT(\eta) + Rem(\eta) \qquad (4.51)$$

where $INT(\gamma)$ and $INT(\eta)$ are the largest integers for γ and η respectively and the Rem( ) functions are the *fractional* remainders of γ and η. Applying the nearest node approximation,

$$Rem(\gamma) = 0 \text{ if } Rem(\gamma) < 0.5 \qquad (4.52)$$
$$Rem(\gamma) = 1 \text{ if } Rem(\gamma) \geq 0.5 \qquad (4.53)$$
$$Rem(\eta) = 0 \text{ if } Rem(\eta) < 0.5 \qquad (4.54)$$
$$Rem(\eta) = 1 \text{ if } Rem(\eta) \geq 0.5 \qquad (4.55)$$

From Eqs.(4.48)-(4.55) we now can identify the node (n',m') closest to (n,m), for which we will use the designation ,

$$(n',m') \leftrightarrow (n,m) \qquad (4.56)$$

Once we know the corresponding nodes in the oriented grid, we can readily obtain the node resistors and voltage waves in the same grid. Since $\rho(n,m)$ is a scalar, $\rho(n',m') \cong \rho(n,m)$ for the transformation $(n,m) \leftrightarrow (n',m')$, and thus

$$R(n',m') \cong R(n,m) \ ; \ \ (n,m) \leftrightarrow (n',m') \tag{4.57}$$

As the cell density is increased, $R(n',m')$ approaches $R(n,m)$ with greater and greater accuracy(this neglects grid dependence due to nonlinear effects, to be discussed later). Next we find the voltage waves in the new grid. Since we know the fields $V_{xy}(n,m)$ and $V_{yx}(n,m)$ associated with the (n,m) node , we may find $V'_{x'y'}(n',m')$ and $V'_{y'x'}(n',m')$, by simply applying the transformations , Eqs.(4.46a)-(4.46b). In so doing we regard $V_{xy}(n,m)$ and $V_{yx}(n,m)$ as the components of the field vector $\mathbf{V}(n,m)$, undergoing an orthogonal transformation. Thus

$$V'_{y'x'}(n',m') = V_{yx}(n,m)\cos\theta + V_{xy}(n,m)\sin\theta \tag{4.58}$$

$$V'_{x'y'}(n',m') = -V_{yx}(n,m)\sin\theta + V_{xy}(n,m)\cos\theta \tag{4.59}$$

At the same time that the fields transform by the above, the node indices also transform, $(n',m') \leftrightarrow (n,m)$, according to Eqs.(4.48)-(4.51). Each of the voltage amplitudes in Eqs.(4.58)-(4.59) represent the sum of forward and backward waves. Exactly the same equations apply individually to the forward and backward voltage waves. Thus, for example, Eqs.(4.58)-(4.59) become $^{+}V'_{y'x'}(n',m')$, for the forward wave,

$$^{+}V'_{y'x'}(n',m') = {}^{+}V_{yx}(n,m)\cos\theta + {}^{+}V_{xy}(n,m)\sin\theta \tag{4.60}$$

$$^{+}V'_{x'y'}(n',m') = -{}^{+}V_{yx}(n,m)\sin\theta + {}^{+}V_{xy}(n,m)\cos\theta \tag{4.61}$$

and a similar pair of equations for the backward ones. We will also have need of the inverse transformation,

$$V_{yx}(n,m) = V'_{y'x'}(n'm')\cos\theta - V'_{x'y'}(n',m')\sin\theta \qquad (4.62)$$

$$V_{xy}(n,m) = V'_{y'x'}(n',m')\sin\theta + V'_{x'y'}(n',m')\cos\theta \qquad (4.63)$$

where as before the above represent the forward and backward waves as well.

In the previous mapping equations for the wave amplitudes and node resistance, we *assumed* a knowledge of the system at a particular instant in time. Such information might be available, for example, during static or steady state conditions, or possibly by measurement of the properties at a particular instant, even during the transient phase. However, we stress that, if, from that particular point in time, we allow the iteration to proceed in different grids, it is quite possible that a prediction of different distributions of waves and node resistors will evolve, depending on the particular grid selected. Indeed such differences are inevitable, if we recall the phenomenon of the earliest arrival of a signal, which is entirely based on the particular grid employed. In addition, the interplay of the wave amplitudes and the node resistance, through some non-linear process, such as avalanching , will lead to further differences. Thus, after the iteration proceeds for some number of steps(or even one step) we need to adopt an averaging procedure, based on the various grids , so as to determine the properties in the medium. We use these averages, with respect to a reference grid, to then perform a mapping back onto the various grids, thus continuing the process. In the following we describe the averaging procedure.

Averaging Procedure Among Grids

4.14 General Procedure and Grid Specification

At a given instant, which includes the initial conditions, we assume a knowledge of the fields and node resistors throughout the medium, referenced to a particular grid. We can then determine the fields and node resistors in the various grids, oriented at different angles. The idea is to first express the existing conditions in the various grids, using the corresponding transformations previously discussed. We then allow the solutions to evolve within each grid. At the completion of the final time step, or at some intermediate number of steps, the various solutions are averaged together, referencing them to the particular grid. The process is then repeated as before if the state is an intermediate one.

Although it is relatively easy to see why the fields will depend on the orientation of a particular grid, it is less evident why the node resistors should change with grid orientation and, therefore, why it is necessary to obtain averages (from the various grid orientations)for the node resistors. As mentioned previously, the reason has to do with the possible dependence of the node resistance on the field which may, for example, trigger an avalanche. If the field depends on the particular grid, therefore, the node resistance also will depend on the grid. This in turn will further impact the wave amplitude in the grid. It is possible, therefore, each grid will have its own version of electromagnetic events different from those in another grid. The averaging of the grids is therefore important.

In preparation for the averaging we first specify the type and number of oriented grids, with the orientation angle ranging from -45° to +45°. We assume the angular spacing between grids is uniform. If N_T is the total number of grids, and θ_N represents the orientation of the Nth grid, then

$$\theta_N = 90^\circ\{N/N_T\} - 45^\circ \tag{4.64}$$

where N= 0,1,2,.......N_T. Note that the angular spacing between grids is 90/ N_T. and that the our customary grid, $\theta_N = 0$, occurs when N = N_T /2. We have arbitrarily selected the range for θ_N to be from -45° to +45°. We could just as easily have chosen the range to be from 0° to 90°, since this also provides the same distribution of grids.

4.15 Vector Description of Plane and Symmetric Waves

It will be convenient to recast the plane and symmetric fields associated with each cell in terms of vector notation. Each grid will have its version of the wave distribution, and will contribute accordingly to the averaging process. Suppose that at node (n,m), the components for the forward plane wave fields associated with the node(or cell) are ${}^{+}V_{xyP}(n,m)$ and ${}^{+}V_{yxP}(n,m)$. The magnitude and direction of the mini- front field, $V_{F,P}(n,m)$, existing at the (n,m) node is then found from the vector addition of the transmission line fields. First considering only the forward waves in $Z_{xy}(n,m,)$ and $Z_{yx}(n,m)$,

$$V_{F,P}(n,m) = {}^{+}V_{yxP}(n,m)\ \mathbf{i}\ +\ {}^{+}V_{xyP}(n,m)\ \mathbf{j} \qquad (4.65)$$

$V_{F,P}(n,m)$, however does not constitute the total vector field at the (n,m) node. There is also another portion of the field, the symmetric fields, which have no directivity, designated by $V_{FS}(n,m)$, and satisfying

$$V_{F,S}(n,m) = {}^{+}V_{yxS}(n,m)\ \mathbf{i}\ +\ {}^{+}V_{xyS}(n,m)\ \mathbf{j} \qquad (4.66)$$

Unlike $V_{FP}(n,m)$, $V_{FS}(n,m)$ scatters in all directions when it encounters a node. For completeness, we state the total forward wave associated with (n,m) node :

$$V_{F}(n,m) = V_{FP}(n,m) + V_{FS}(n,m) \qquad (4.67)$$

Care should be exercised in the interpretation of Eq.(4.67), however. The vector components for $V_{FP}(n,m)$ and $V_{FS}(n,m)$ are added in quadrature, as outlined Section 4.1 and in subsequent Sections.

We emphasize that Eqs.(4.65)-(4.67) are cast in terms of the vector fields and *not* the propagation direction. We may recast these equations in terms of the propagation direction, in which case Eqs.(4.65)-(4.66) become

$$V_{FP\perp}(n,m) = {}^{+}V_{xyP}(n,m)\ \mathbf{i}\ -\ {}^{+}V_{yxP}(n,m)\ \mathbf{j} \qquad (4.67b)$$

$$V_{FS\perp}(n,m) = {}^{+}V_{xyS}(n,m)\ \mathbf{i}\ -\ {}^{+}V_{yxS}(n,m)\ \mathbf{j} \qquad (4.67b)$$

Note that we appended the subscript $\perp$ to indicate the propagation direction; also note the minus sign in front of the $\mathbf{j}$ component. This is needed to insure the orthogonality condition, $V_{F,P\perp}(n,m) \cdot V_{F,P}(n,m) = 0$. Although we employ $V_{F,P}(n,m)$ in the following development, one should keep in mind the associated $V_{F,P\perp}(n,m)$ (as well as $V_{F,S\perp}(n,m)$); indeed the propagation type vectors help to visualize the plane wave propagation.

Looking at Eqs.(4.65)-(4.67) we have only considered forward waves whereas any combination of forward and backward waves is possible. Eq.(4.65), for example, is only one of four possibilities, depending on the particular combination of forward and backward waves in $Z_{xy}(n,m)$ and $Z_{yx}(n,m)$. The four possibilities, denoted by $V_{F,P,O}(n,m)$ with O=1,2,3,4, are

$$V_{FP1}(n,m) = {}^+V_{yxP1}(n,m)\, \mathbf{i} + {}^+V_{xyP1}(n,m)\, \mathbf{j} \qquad (4.68)$$

$$V_{FP2}(n,m) = {}^-V_{yxP2}(n,m)\, \mathbf{i} + {}^-V_{xyP2}(n,m)\, \mathbf{j} \qquad (4.69)$$

$$V_{FP3}(n,m) = {}^+V_{yxP3}(n,m)\, \mathbf{i} + {}^-V_{xyP3}(n,m)\, \mathbf{j} \qquad (4.70)$$

$$V_{FP4}(n,m) = {}^-V_{yxP4}(n,m)\, \mathbf{i} + {}^+V_{xyP4}(n,m)\, \mathbf{j} \qquad (4.71)$$

The above four equations are redundant and it is sufficient to specify either the first pair, Eqs.(4.68)-(4.69), or the second pair, Eqs.(4.70)-(4.71), since either pair comprises all possible individual waves in the TLM lines. Note that in Eqs.(4.68)-(4.71), although redundant, we retain the subscript for the O index in the interest of clarity. We also construct the symmetric wave counterparts to the above, given by

$$V_{FS1}(n,m) = {}^+V_{yxS1}(n,m)\, \mathbf{i} + {}^+V_{xyS1}(n,m)\, \mathbf{j} \qquad (4.72)$$
$$V_{FS2}(n,m) = {}^-V_{yxS2}(n,m)\, \mathbf{i} + {}^-V_{xyS2}(n,m)\, \mathbf{j} \qquad (4.73)$$
$$V_{FS3}(n,m) = {}^+V_{yxS3}(n,m)\, \mathbf{i} + {}^-V_{xyS3}(n,m)\, \mathbf{j} \qquad (4.74)$$
$$V_{FS4}(n,m) = {}^-V_{yxS4}(n,m)\, \mathbf{i} + {}^+V_{xyS4}(n,m)\, \mathbf{j} \qquad (4.75)$$

As with the planar waves we need only concern ourselves with either $V_{FS1}(n,m)$ and $V_{FS2}(n,m)$, or $V_{FS3}(n,m)$ and $V_{FS4}(n,m)$. Also, as we shall see in later Sections, the plane wave vectors given by Eqs.(4.68) and (4.69) will be the most useful pair in formulating the iteration.

4.16 Energy Content of Plane Waves and Symmetric Waves

It will prove useful to determine the energy content of $\mathbf{V}_{FP,O}$ and $\mathbf{V}_{FS,O}$ where O $=1,2,3,$ or 4. Using O$=1$ as an example, we know that the amplitudes of $^{+}V_{xyP1}(n,m)$, $^{+}V_{xyS1}(n,m)$, $^{+}V_{yxP1}(n,m)$, and $^{+}V_{yxS1}(n,m)$, which comprise $\mathbf{V}_{FP,1}$, $\mathbf{V}_{FS,1}$, are effective amplitudes, and therefore the wave energies are proportional to the square of the amplitudes of the components. The energy of the forward and backward plane waves for the cell, denoted by $U_{FP1}(n,m)$ and $U_{FP2}(n,m)$, respectively, are given by

$$U_{FP1}(n,m) = \left(^{+}V_{yxP1}(n,m)\right)^{2}/Z_{o} \; + \left(^{+}V_{xyP1}(n,m)\right)^{2}/Z_{o} \qquad (4.76a)$$

$$U_{FP2}(n,m) = \left(^{-}V_{xyP2}(n,m)\right)^{2}/Z_{o} \; + \left(^{-}V_{yxP2}(n,m)\right)^{2}/Z_{o} \qquad (4.76b)$$

where we assume for simplicity that the TLM lines surrounding the node are identical, Z_{o}. Similarly, the corresponding energy relationships for the symmetric terms are

$$U_{FS1}(n,m) = \left(^{+}V_{yxS1}(n,m)\right)^{2}/Z_{o} \; + \left(^{+}V_{xyS1}(n,m)\right)^{2} /Z_{o} \qquad (4.77a)$$

$$U_{FS2}(n,m) = \left(^{-}V_{xyS2}(n,m)\right)^{2}/Z_{o} \; + \left(^{-}V_{yxS2}(n,m)\right)^{2}/Z_{o} \qquad (4.77b)$$

The total forward wave energy of $U_{F1}(n,m)$, belonging to the (n,m) cell, consists of both types of energy(plane wave and symmetric), or

$$U_{F1}(n,m) = U_{FP1}(n,m) +U_{FS1}(n,m) = \left(^{+}V_{xy}(n,m)\right)^{2}/Z_{o} +\left(^{+}V_{yx}(n,m)\right)^{2}/Z_{o} \quad (4.78)$$

and a similar relationship for $U_{F2}(n,m)$

$$U_{F2}(n,m)= U_{FP2}(n,m)+U_{FS2}(n,m)=\left(^{-}V_{xy}(n,m)\right)^{2}/Z_{o}+\left(^{-}V_{yx}(n,m)\right)^{2}/Z_{o} \qquad (4.79)$$

It is worthwhile to recall that we are using effective amplitudes for $^+V_{xyP1}(n,m)$, $^+V_{xyS1}(n,m)$, etc... Thus, for example, $^+V_{xy}(n,m)$ is related to the symmetric and plane wave components by the energy relations, $(^+V_{xy}(n,m))^2 = (^+V_{xyP1}(n,m))^2 + (^+V_{xyS1}(n,m))^2$ with a similar relationship for the backward wave. As we have alluded to before, the scattering coefficients of the two components will differ, with the symmetric wave scattering in all directions while the plane wave component avoids scattering in the transverse directions.

4.17 Principal Axis Grid

It will be useful, as indicated in the ensuing discussion, to obtain the grid with one of the principal axes in the same direction of $V_{F,P1}$, at a particular cell site. This is usually done at some time step prior to the start of the averaging process. For example, with the grid axis x' along $V_{F,P1}$, the x' component of the field disappears and the following condition is satisfied, using Eqs.(4.60):

$$^+V_{yxP1}(n,m) \cos\theta + {}^+V_{xyP1}(n,m) \sin\theta = 0 \qquad (4.80)$$

At the same time the companion transformation, Eq.(4,61), provides the actual wave

$$^+V_{x'y'P1}(n',m') = V_{F,P1}(n,m) = -{}^+V_{yxP1}(n,m) \sin\theta + {}^+V_{xyP1}(n,m) \cos\theta \qquad (4.81a)$$

$$|V_{F,P1}(n,m)| = [(^+V_{yxP1}(n,m))^2 + (^+V_{xyP1}(n,m))^2]^{1/2} \qquad (4.81b)$$

From Eq.(4.80) the solution for θ is designated as the principal angle θ_x,

$$\tan\theta_x = -[^+V_{yxP1}(n,m) / {}^+V_{xyP1}(n,m)] \qquad (4.82)$$

A similar calculation for the y' axis gives the angle θ_y, or

$$\tan\theta_y = [{}^+V_{xyP1}(n,m) / {}^+V_{yxP1}(n,m)] \tag{4.83}$$

Note that the right hand side of Eq.(4.83) is the negative reciprocal of that for Eq.(4.82). Between θ_x and θ_y we choose the angle having the smaller absolute value. Thus if $\tan\theta_x$ is less than or equal to one than we choose θ_x and if it is greater than one we choose θ_y. In any event it is only necessary to orient the grid somewhere between $-45°$ to $+45°$ in order for either the x' or y' axis to be aligned with $V_{F,P1}$. This choice then dictates the particular θ_N grid , whichever is closer to the smaller absolute value of θ_x or θ_y .

4.18 Simple Averaging Example Without Plane Wave Effects

We first illustrate the averaging procedure for the fields and conductivity using a very simple example, consisting of the original grid and an additional one rotated 45 degrees about the origin, applicable to each cell.   Initially we know the conductivity and wave amplitudes in the original grid. Plane wave correlations are ignored for the moment. We obtain the corresponding waves in the oriented system, using the transformation results of the   discussion in Section 4.6, and setting $\cos45°$ $=\sin45°$ $=2^{-1/2}$. We then allow the iteration to proceed in both systems for the same number of time steps. At the conclusion we compare the results in regard to the node resistance and wave amplitudes. The comparison is more easily interpreted if, after the iteration, we transform the results of the oriented grid to that of the original grid, which we use as the basis for the comparison. We then   average   the results for the two grids, designated by N_1(for the original) and N_2(for the orientated grid). Thus, for example, for the voltage component ${}^+V_{xy}(n,m)$

$$ {}^+V_{xy}(n,m) = [{}^+V_{xy}(n,m)_{N1} + {}^+V_{xy}(n,m)_{N2}]/ 2 \tag{4.84}$$

and for the conductivity

$$\sigma(n,m) = [\sigma(n,m)_{N1} + \sigma(n,m)_{N2}] / 2 \qquad (4.85)$$

where the results are always referenced to the original N_1 grid. Under ordinary circumstances $^+V_{xy}(n,m)$ evaluated in grid I will be the same as $^+V_{xy}(n,m)$ traveling in grid II (following the transformation back to grid I). In the event of nonlinear effects, however, this will no longer be true and the averaging process is required. The same is true of the conductivity. Note that we use the inverse of the node resistance in taking the average. Also, suppose there is a substantial discrepancy between the results of the N_I and N_2 grids. In this case, higher order averaging is required, i.e., additional more closely spaced grids are needed in the averaging. The generalization to a larger number of grids is straightforward. Because of symmetry, we choose N_T grids(N_T a positive integer) with the angular spacing equal to $90/N_T$. Until this point, we have said nothing about attaching any weight to a particular grid; each grid is assumed to have the same weight as any other grid. This topic is discussed later in regard to the average node resistance. In addition, the plane wave correlations(ignored here) are discussed in the following Sections.

4.19 General Averaging Procedure For Cell Waves, Including Plane Wave Correlations

We summarize the averaging process as follows. Preparatory to the averaging we create N_T grids, with uniform angular spacing given by $90°/N_T$. At some time step we know the fields and node resistance(whether from static, steady state, experiment, or a prior averaging). We examine the fields associated with the (n,m) cell in the reference grid, or "original" grid, and split the fields in the $Z_{xy}(n,m)$ and $Z_{yx}(n,m)$ lines into plane wave and symmetric parts, using the correlation technique outlined earlier in the Chapter. When we use the term correlation, we always mean the net correlation in which decorrelation factors have been taken into account as well. Further, we always return to the reference grid since this is the grid shared by all the cells, and therefore forms a convenient basis for comparison. In general, however, keep in mind that the reference grid is not the grid in which the waves actually undergo their motion. We first construct the vectorial plane wave vectors $\mathbf{V}_{F,P1}(n,m)$ and $\mathbf{V}_{FP2}(n,m)$ as well as the symmetric vectors $\mathbf{V}_{F,S1}(n,m)$ and $\mathbf{V}_{FS2}(n,m)$, which will prepare us for the averaging process of the various grids. From the two planar fields, $\mathbf{V}_{F,P1}(n,m)$ and $\mathbf{V}_{FP2}(n,m)$, we select the two principal axis grids *most closely aligned* with the directions of the two fields; we designate these two grids by N_{P1} and N_{P2}. By selecting only these two (and no other) grids to represent the plane waves, from among the array of N_T grids, we assign proper weight to the plane waves .The utilization of the principle axes for the plane waves helps to avoid any artificial effects due to the grid. We then track the plane waves during the ensuing time step(or time steps). A new set of plane waves will then enter the (n,m) cell although they may not be necessarily restricted to the N_{P1} and N_{P2} grids, Thus, for example, a plane wave entering the (n,m) cell from the adjacent (n-1,m) cell may belong to a grid which is slightly different from N_{P1} or N_{P2} .

With regard to the symmetric fields, $\mathbf{V}_{FS1}(n, m)$ and $\mathbf{V}_{FS2}(n,m)$, we use all N_T grids available, unlike the case for the plane waves. Initially, before the time step iteration, the symmetric waves are transformed to each of the N_T grids, and we assume each of these grids has an equal weight. We then allow the waves, which include both the plane and symmetric waves, to proceed one or more time steps, taking into account the differing scattering coefficients of the plane wave and symmetric waves. We track the wave motion for each grid. The contribution from each grid is needed in order to obtain the proper averaging.

Once the time step(or time steps) is completed we are then ready to obtain the total plane wave as well obtain the symmetric wave, obtained by averaging over the various grids. First we obtain the various contributions separately for the plane and symmetric wave parts in each grid line, either emanating from adjacent cells or reflected within the cell. This is treated as a normal scattering process. To simplify the interpretation, we then proceed back to the reference (n,m,q) grid(for both plane and symmetric waves). The averaging of the symmetric portion occurs in the reference grid.

Starting with the plane wave case, let us consider first the components contributing to ${}^{+}V_{xyP1}(n,m)$. The field ${}^{+}V_{xyP1}(n,m)$ emanates partly from the motion of the waves along the principal grids determined by $V_{F,P1}(n-1,m)$ and $V_{FP2}(n,m)$ of the prior step, after which the fields are *transformed* back to the reference grid. This is followed by the correlation treatment (for which we also require the symmetric waves, to be discussed) to get an updated $^+V_{xyP1}(n,m)$, We do the same for $^+V_{yxP1}(n,m)$, from which we form a new $V_{F,P1}(n,m)$ and a principal grid N_{P1} Making use of the results in Section 4.14, the plane wave in the principal grid is given by

$$^+V'_{x'y'P1}(n',m') = |V_{F,P1}(n,m)| = -^+V_{yxP1}(n,m)\,\sin\theta_{P1} + {}^+V_{xyP1}(n,m)\,\cos\theta_{P1} \qquad (4.86)$$

while the $^+V'_{y'x'P1}(n',m')$ component vanishes. Here the P1 grid is assumed to be along the x' axis; in the event that the y' axis applies then Eq.(4.86a) is replaced by

$$^+V'_{y'x'P1}(n',m') = |V_{F,P1}(n,m)| = {}^+V_{yxP1}(n,m)\,\cos\theta_{P1} + {}^+V_{xyP1}(n,m)\,\sin\theta_{P1} \qquad (4.87)$$

A similar treatment applies to the backward wave and the selection of the grid based on $V_{F,P2}(n,m)$. The plane waves are then all allowed to scatter , creating the next iterative step.

We then turn our attention to the symmetric waves. The situation with the symmetric waves is different since most if not all N_T grids participate, and the process involves an averaging step. This should not come as a surprise since initially we transformed the symmetric field to the N_T grids using the same amplitude in each grid. As mentioned previously, the number of grids contributing to the symmetric averaging will therefore far exceed the plane waves grids. We pick up the description where the have completed their motion in each of N_T grids ; we then have to transform these waves back to our reference grid and perform an averaging. This given by as

$$AV[^+V_{yxS1}(n,m)] = (1/N_T)\,\sum_{N=1}^{N_T}[^+V_{yxS1}(n,m,\theta_N)] \qquad (4.88)$$

$$AV[^+V_{xyS1}(n,m)] = (1/N_T)\,\sum_{1}^{N_T}[^+V_{xyS1}(n,m,\theta_N)] \qquad (4.89)$$

and with similar equations for $^-V_{yxS2}(n,m)$ and $^-V_{xyS2}(n,m)$. Note that $^+V_{yxS1}(n,m,\theta_N)$ and $^+V_{xyS1}(n,m,\theta_N)$ represent the fields transformed to the reference grid. θ_N in the above represents the Nth grid (which is a "primed" grid). The actual transformation is implemented in the usual way; see Eqs.(4.60)-(4.61). Once we have these fields, in addition the plane wave fields, we subject the fields to a correlation process, providing us with a new set of waves from which we begin the next iterative step. For the symmetric field this means assigning the same symmetric field to each of N_T grids and allowing the iteration to proceed.

It may be worthwhile to summarize results for the plane and symmetric waves in the (n,m) cell, explicitly citing the transformation to the reference grid, but not yet performing the correlation.

$$^{+}V_{yxP1}(n,m) = [^{+}V_{yxP1}(n,m)]_{\theta P1,\, \theta P2}\, ,\ ^{+}V_{yxS1}(n,m) = AV[^{+}V_{yxS1}(n,m)\,] \qquad (4.90)$$

$$^{+}V_{xyP1}(n,m) = [^{+}V_{xyP1}(n,m)]_{\theta P1,\, \theta P2}\, ,\ ^{+}V_{xyS1}(n,m) = AV[^{+}V_{xyS1}(n,m)\,] \qquad (4.91)$$

$$^{-}V_{yxP2}(n,m) = [^{-}V_{yxP2}(n,m)]_{\theta P1,\, \theta P2}\, ,\ ^{-}V_{yxS2}(n,m) = AV[\,^{-}V_{yxS2}(n,m)\,] \qquad (4.92)$$

$$^{-}V_{xyP2}(n,m) = [^{-}V_{xyP2}(n,m)]_{\theta P1,\, \theta P2}\, ,\ ^{-}V_{xyS2}(n,m) = AV[^{-}V_{xyS2}(n,m)] \qquad (4.93)$$

As noted, the planar fields arise from transformations from principal grids while the symmetric fields are an average of the transformed fields of N_T grids, as indicated in Eqs.(4,88)-(4.89).We then have necessary information so that the total forward and backward wave in the TLM line may be obtained, which serves as the starting point for the correlation process. With the usual notation, the forward waves are:

$$^{+}V_{xy}(n,m) = [[^{+}V_{xyP1}(n,m)]^2 + [^{+}V_{xyS1}(n,m)]^2]^{1/2} \qquad (4.94)$$

$$^{+}V_{yx}(n,m) = [[^{+}V_{yxP1}(n,m)]^2 + [^{+}V_{yxS1}(n,m)]^2]^{1/2} \qquad (4.95)$$

A parallel expression may also be done for the backward waves . The results are

$$^{-}V_{xy}(n,m) = [[^{-}V_{xyP2}(n,m)]^2 + [^{-}V_{xyS2}(n,m)]^2]^{1/2} \qquad (4.96a)$$

$$^-V_{yx}(n,m) = [[^-V_{yxP2}(n,m)]^2 + [^-V_{yxS2}(n,m)]^2]^{1/2} \qquad (4.96b)$$

This process is done of course for all the cells in the region of interest. This completes the description of the iteration step. To continue the process we go back to the beginning of this Section in which we first examine for field correlation in order to break up the field into *new* symmetric and planar parts, followed by the determination of $V_{FP1}(n,m)$ and $V_{FP2}(n,m)$ to obtain the two "favored" grids, N_{P1} and N_{P2} . Note that we have concentrated on a single cell (n,m) . During the iteration process, of course, the same preparation process is repeated for all the cells, before the waves are "released".

The question of whether to allow the iteration to continue for a single time step or for many time steps is largely an issue of the computer capability and solution accuracy. If the medium properties change only a small amount, over the length of the cell size, then it probably is not necessary to re-examine the fields, at every step, in regard to the composition of the symmetric and planar fields.

In the previous discussion, we transformed the symmetric fields to the array of grids, allowed the iteration to proceed, and then transformed back to the original grid, taking a suitable average of the symmetric fields. There is a potential defect in using this procedure, however. The problem is that the amplitude of the symmetric wave will by definition be smaller than the total wave amplitude, which is the sum(in quadrature) of the symmetric and plane wave parts. The simulation in general will depend on the signal amplitude and therefore the voltage amplitude in each grid line should equal the original amplitude. This does not represent a problem for grids N_{P1} and N_{P2}, which contain both the planar and symmetric fields, and whose sum is equal to the original amplitude. But for other grids, the field(purely symmetric part) will be less than the total field.

A possible way to repair this defect is to replace the symmetric field in each grid with the full amplitude waves, and to solely ascribe symmetric properties to such waves. Thus, if $V_{F1}(n,m)$ and $V_{F2}(n,m)$ are the total fields(both symmetric and planar) associated with the (n,m) cell, then we designate the "full "symmetric waves, as $V_{FST1}(n,m)$ and $V_{FST2}(n,m)$, which have the same amplitudes as $V_{F1}(n,m)$ and $V_{F2}(n,m)$. We then allow the iteration to run separately, i.e. , in parallel, one for the full symmetric and the other for the combination of planar and symmetric fields in grids N_{P1} and N_{P2}. Once the iteration step is completed, we remove the added symmetric portion (equal in magnitude to the plane wave part) appearing in any grids. It is relatively easy to keep track of the two symmetric portions of the wave, using the by now familiar quadrature method for adding two waves. Alternatively, we may wish to retain the identity of the two symmetric components, since, the process of disentangling the components is made easier after the scattering is completed. Also, from a conceptual viewpoint, we may simplify matters, perhaps, by applying the full symmetric amplitude to all the N_T grids. This gives more weight to the symmetric field than is deserved , but for large N_T the differences between the two approaches become small.

Formally the procedure for using full symmetric waves is similar to that of the previous except that we now have two separate, parallel iterations, one for the full symmetric waves $V_{FST1}(n,m)$ and $V_{FST2}(n,m)$, which apply to all the grids, and the other for $V_{FP1}(n,m)$, $V_{FP2}(n,m)$ with grids N_{P1} and N_{P2} but which also includes the full symmetric fields as well. As an example, suppose that in line $Z_{xy}(n,m)$. the full symmetric component is $^+V'_{xyST}(n,m)$ where the prime indicates that the fields have not yet been transformed to the reference grid . $^+V'_{xyST}(n,m)$ will consist will consist of the normal symmetric part, $^+V'_{xyS}(n,m)$ and the complementary part, designated by $^+V'_{xySC}(n,m)$, where the complementary part may be determined from the fact that $^+V'_{xyS}(n,m)$ and $^+V'_{xySC}(n,m)$ add in quadrature. To obtain the full symmetric field $^+V'_{xyS}(n,m)$ the quadrature sum is

$$[^+V'_{xyS}(n,m)]^2 + [^+V'_{xySC}(n,m)]^2 = [^+V'_{xyST}(n,m)]^2 = [V_{F1}(n,m)]^2 \qquad (4.97)$$

which allows one to solve for $^+V'_{xySC}(n,m)$. Thus for both components, prior to transformation, the complementary part is deleted and

$$^{+}V'_{xyST}(n,m) \rightarrow \ ^{+}V'_{xyS}(n,m) \tag{4.98}$$

$$^{+}V'_{yxST}(n,m) \rightarrow \ ^{+}V'_{yxS}(n,m) \tag{4.99}$$

with similar relations for the backward waves. We then perform the averaging process, transforming the above to the reference grid, together with the symmetric fields in the other grids, including any symmetric fields associated with planar fields . Once the transformation is completed the final fields are the found by summing the component planar and symmetric fields in quadrature.

4.20 Summary of Field Averaging Procedure

Unfortunately the conceptualization of the averaging processes, described in the preceding Sections, are obscured somewhat by the fact that we have to take into account several processes: the usual time step iteration of a 2D (or 3D) cell matrix, field correlation , the break-up of the fields into plane and symmetric parts, the establishment of the grid array, and the averaging of the symmetric fields over various grids. The proper *sequencing* of these processes is important but not readily apparent. Fig.4.7 addresses these issues, where we use the format of a flow diagram , which should help remove any uncertainties and also render the processes more amenable to computer iteration. The deletion of the complementary portion of the symmetric wave is indicated as an option, depending on the particular application or the desired accuracy.

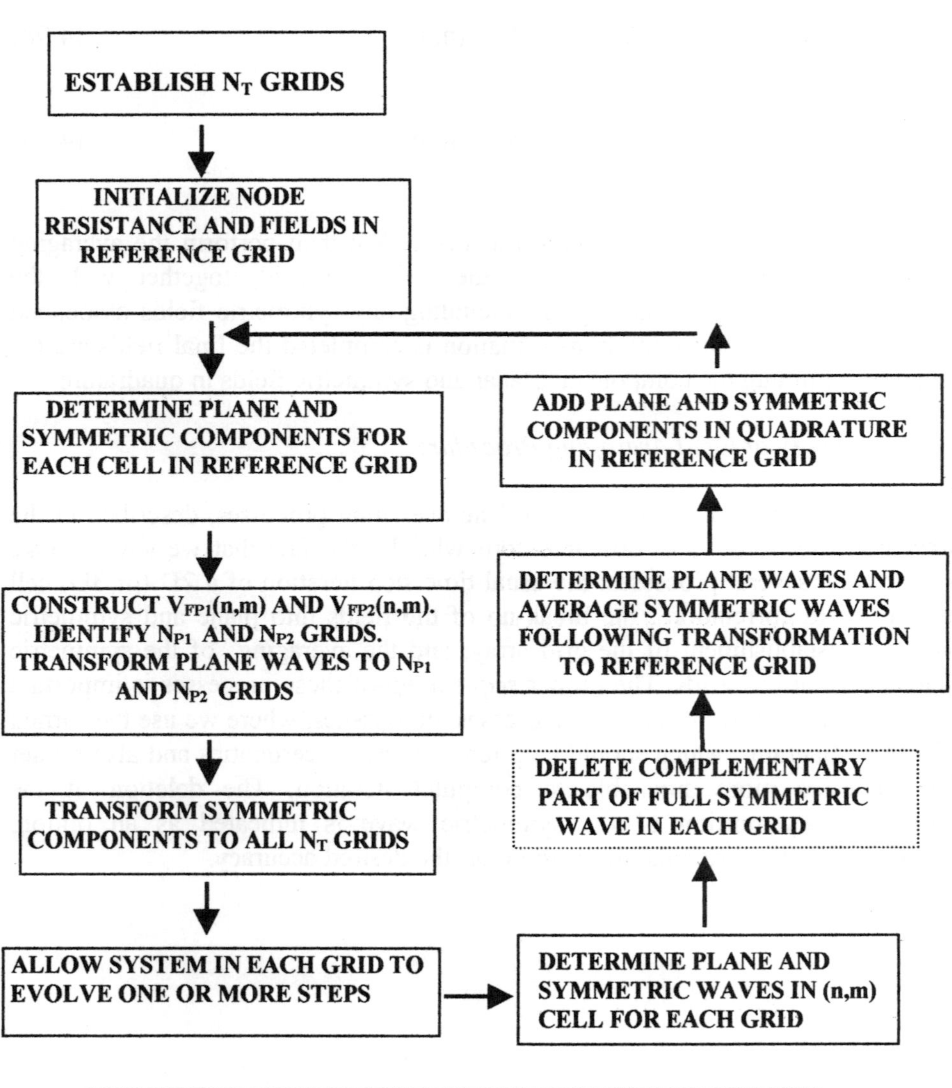

FIG. 4.7 ITERATIVE STEPS FOR FIELD AVERAGING

4.21 Averaging Procedure For Node Resistance

As mentioned previously, the node resistance may exhibit differing resistance values,   depending on the particular grid. This has to do with the possible nonlinear nature of the resistance in which the resistance is a function of the signal amplitude(as, for example, avalanching). We must therefore resort to an averaging of the node resistance over the various grids. For equal grid weight, the node resistance is thus expressed as

$$AV(1/R(n,m)) = (1/N_T)_1 \ {}^{N_T}\Sigma \ (1/ R(n,m,\theta_N)) \qquad\qquad (4.100)$$

where $R(n,m,\theta_N)$ is the node resistance of the Nth grid oriented at angle θ_N , and we have assumed each grid has equal weight. Note that the actual quantity being averaged is the inverse of the node resistance, $(1/R(n,m)$. We know this quantity is proportional to the conductivity, which is the appropriate parameter for averaging.

In the previous discussion on node resistance averaging, we assumed the weight given to each grid was equal to that of any other grid. Although very simple to interpret this is not the best choice for the grid weight, however. The brunt of the plane wave effects is represented by only two grids, N_{P1} and N_{P2} , while the symmetric waves are represented by N_T grids. If we assign an equal weight to each of the N_T grids, however, then the symmetric waves will swamp the plane waves and their effect on the node resistance will be skewed accordingly. The grid weights must be changed such that the effect of the plane waves are properly taken into account. One method is to separately average the node resistance over the two plane wave grids and the node resistance over the N_T symmetric grids; this is then followed by a final average of the plane wave and symmetric averages. A second method is to assign grid weights based on the relative content of the plane wave and symmetric wave energies.

4.21(a) Separate Plane Wave and Symmetric Wave Averaging

The average of the nodes for the plane wave grids is given by

$$AV(1/R(n,m))_{PLANE\ WAVE} = [1/\ R(n,m,\theta_{NP1}) + 1/\ R(n,m,\theta_{NP2})]/2 \qquad (4.101a)$$

For the symmetric waves the average is given by

$$AV(1/R(n,m))_{SYM} = 1/(N_T-2)) \sum_{N=1}^{NT} (1/\ R(n,m,\theta_N)) \qquad (4.101b)$$

Where in the above the symmetric waves represent the full wave amplitudes discussed earlier. The final average is given by

$$AV(1/R(n,m)) = [AV(1/R(n,m))_{PLANE\ WAVE} + AV(1/R(n,m))_{SYM}]/2 \qquad (4.101c)$$

4.22(b) Grid Weight Based on Wave Energy

The second, preferred, method assigns a weight to a particular grid, taking into account the relative amounts of plane wave and symmetic energies in each cell. The first step in assigning a weight to the grid is to recognize that the weights must be normalizable, and for this we use the total energy of the TLM lines associated with the (n,m) cell. The total energy, denoted by $U_T(n,m)$, is

$$U_{TOT}(n,m) = [^{+}V_{xy}(n,m)^2 + {}^{+}V_{yx}(n,m)^2]/Z_o + [^{-}V_{xy}(n,m)^2 + {}^{-}V_{yx}(n,m)^2]/Z_o \qquad (4.102)$$

We examine the fields after completing the time iteration, and the average fields have been referenced to the original grid, allowing us to produce new plane wave correlation vectors $\mathbf{V}_{FP1}(n,m)$ and $\mathbf{V}_{FP2}(n,m)$. The energies of the planar waves $\mathbf{V}_{FP1}(n,m)$ and $\mathbf{V}_{FP2}(n,m)$ are $U_{FP1}(n,m)$ and $U_{FP2}(n,m)$, given by Eqs.(4.76a) and (4.76b). The normalized weights ,G_1 and $G_{2,}$ of these two waves are

$$G_1(n,m) = (U_{FP1}(n,m)/\ U_{TOT}(n,m) \qquad (4.103)$$
$$G_2(n,m) = (U_{FP2}(n,m)/\ U_{TOT.}(n,m) \qquad (4.104)$$

and the associated grids are N_{P1} and N_{P2}.   The total weight ascribed to the symmetric waves , denoted by $G_S(n,m)$ and comprising all the grids, is given by

$$G_S(n,m) = 1 - [(U_{FP1}(n,m) + (U_{FP2}(n,m)]/U_{TOT}(n,m) \qquad (4.105)$$

On the other hand we require the symmetric weight for each grid. Since the final symmetric field must be diluted by the number of grids, the simplest approach is to assign equal weight to each of the grids, and thus the weight for the Nth grid for the symmetric part, $G_{SN}(n,m)$, is

$$G_{SN}(n,m) = G_S(n,m)/N_T \qquad (4.106)$$

We are now prepared to obtain the node resistance, averaged over the N_T grids. The actual quantity to be averaged is $(1/R(n,m))$, i.e, the conductance, as mentioned previously. We denote the node resistance of the Nth grid by $R_N(n,m)$. The averaging then takes the form

$$AV[(1/R(n,m)] = G_1(n,m)(1/R_{N1}(n,m)) + G_2(n,m)(1/R_{N2}(n,m))$$
$$+ \sum_1^{NT} G_{SN}(n,m)(1/R_N(n,m)) \qquad (4.107)$$

The above represents the average for the node resistance, but other node properties mentioned in Chapter II may also be averaged along the same lines. For pure plane waves the last term in the above vanishes, and conversely, for pure symmetric waves the first two terms vanish.

4.22 Comparison of Standard Numerical Methods and TLM Methods Incorporating TLM Correlations/Decorrelations and Grid Orientation

Without wave correlations and grid orientation effects we should expect the TLM and numerical solutions of Maxwell's equations to yield identical simulation results. With the introduction of wave correlations and grid orientation effects, however, we will start to see departures from the standard numerical techniques. The departures will be most evident in extremely fast electromagnetic pulses and in the description of the initial field profiles. These situations cannot possibly be described by standard numerical methods without taking into account the significant effects of plane wave correlations and grid orientation. The result should be greater accuracy in the prediction of electromagnetic energy dispersal.

As mentioned before the wave correlations employed represent a classical treatment of phenomena which, strictly speaking, should be described by quantum mechanical arguments in which plane wave components in adjacent TLM lines effectively share the same quantum state. An elementary treatment of wave correlations, using quantum mechanical arguments, is given in App.4A.2

Appendices

App.4A.1 3D Scattering Corrections For Plane Waves(Wave Correlations)

The 3D correction for plane wave effects follows the same general outline as that for 2D case given previously. We provide a concrete example in which the wave propagates in the + x direction. For a perfect plane wave the wave amplitudes will be uniform be in the yz plane; under more typical conditions, the wave amplitudes will be non-uniform and plane wave effects will be exhibited only partially. The degree of plane wave effects at a particular site will depend on the degree of amplitude uniformity, or "correlation" among the amplitudes at the site.

For this situation we consider waves polarized in the y direction, Since we now are considering 3D effects, there will be four nearest neighbors instead of two, as was noted in Fig.4.2. Assume the selected site is the (n,m,q) cell and the TLM line is $Z_{xy}(n,m,q)$. The corresponding forward wave is $^+V_{xy}(n,m,q)$. The four neighboring waves are $^+V_{xy}(n,m-1,q)$, $^+V_{xy}(n,m+1,q)$, $^+V_{xy}(n,m,q-1)$, and $^+V_{xy}(n,m,q+1)$. As with the 2D case it is important to point whether the neighboring amplitudes exceed or are less than $^+V_{xy}(n,m,q)$. As before, the degree of plane wave correlation remains at unity. when a neighboring amplitude exceeds $^+V_{xy}(n,m,q)$,and again an amplitude with opposite sign of $^+V_{xy}(n,m,q)$ is assumed to have zero plane wave correlation. Since there are now five amplitudes to compare (instead of three) the number of correlation Categories is 120 , instead of 6, based on the factorial 5!. For illustrative purposes, we consider the case in which the following is satisfied

$$^+V_{xy}(n,m,q+1) \geq {}^+V_{xy}(n,m+1,q) \geq {}^+V_{xy}(n,m,q) \geq {}^+V_{xy}(n,m-1,q) \geq {}^+V_{xy}(n,m,q-1) \quad (4A.1)$$

Having defined the wave and its neighbors we can the proceed to find the degree of plane wave correlation. The first step, as before, is to break up the wave in each line, this time into *four* equal component waves, in which each component will be assumed to interact with a particular neighbor. Using the quadrature type partitioning (as before), the amplitude for each of the four components, for the $^+V_{xy}(n,m,q)$ wave, is designated by

$$^+V_{xyD}(n,m,q) = {}^+V_{xy}(n,m,q) /2 \qquad (4A.2)$$

with similar relations for the neighboring lines (Note the denominator is 2 instead of $2^{1/2}$, due to the presence of four neighbors in 3D)). We now start the correlation process by first looking at the correlation between $^+V_{xyD}(n,m,q)$ and $^+V_{xyD}(n,m-1,q)$ and between $^+V_{xyD}(n,m,q)$ and $^+V_{xyD}(n,m,q-1)$. In both cases, $^+V_{xyD}(n,m,q)$ exceeds its partner wave in the neighboring lines. We therefore must split up $^+V_{xyD}(n,m,q)$ accordingly:

$$^+V_{xyD}(n,m,q) = {}^+V_{xyD}(n,m-1,q) + {}^+\Delta_{xyD}(n,m-1,q) \qquad (4A.3)$$

$$^+V_{xyD}(n,m,q) = {}^+V_{xyD}(n,m,q-1) + {}^+\Delta_{xyD}(n,m,q-1) \qquad (4A.4)$$

$^+\Delta_{xyD}(n,m-1,q)$ and $^+\Delta_{xyD}(n,m,q-1)$ are defined by Eqs.(4A.3)-(4A.4) and represent the differences between the correlation waves . We are now in a position to obtain the plane wave and symmetric correlations among the partitioned D waves, at least in regard to lines $Z_{xy}(n,m-1,q)$ and $Z_{xy}(n,m,q-1)$. Thus

$$^+V_{xyP1}(n,m,q) = [{}^+V_{xyD}(n,m-1,q) \; {}^+V_{xyD}(n,m,q)]^{1/2} \qquad (4A.5)$$

$$^+V_{xyS1}(n,m,q) = [{}^+\Delta_{xyD}(n,m-1,q) \; {}^+V_{xyD}(n,m,q)]^{1/2} \qquad (4A.6)$$

$$^+V_{xyP2}(n,m,q) = [{}^+V_{xyD}(n,m,q-1) \; {}^+V_{xyD}(n,m,q)]^{1/2} \qquad (4A.7)$$

$$^+V_{xyS2}(n,m,q) = [{}^+\Delta_{xyD}(n,m,q-1) \; {}^+V_{xyD}(n,m,q)]^{1/2} \qquad (4A.8)$$

where ${}^{+}V_{xyP1}(n,m,q)$ and ${}^{+}V_{xyS1}(n,m,q)$ are the plane wave and symmetric correlations respectively with the $Z_{xy}(n,m-1,q)$ line. Similarly ${}^{+}V_{xyP2}(n,m,q)$ and ${}^{+}V_{xyS2}(n,m,q)$ are the correlations with $Z_{xy}(n,m,q-1)$. We now obtain the correlations with the $Z_{xy}(n,m+1,q)$ and $Z_{xy}(n,m,q+1)$ lines. These are easier to calculate since the amplitudes in these lines exceed ${}^{+}V_{xy}(n,m,q)$. The correlations with these lines are purely of the plane wave type. If we designate these correlations by ${}^{+}V_{xyP3}(n,m,q)$ and ${}^{+}V_{xyP4}(n,m,q)$, for the $Z_{xy}(n,m+1,q)$ and $Z_{xy}(n,m,q+1)$ lines respectively, then

$$
{}^{+}V_{xyP3}(n,m,q) = {}^{+}V_{xyD}(n,m,q) \tag{4A.9}
$$

$$
{}^{+}V_{xyP4}(n,m,q) = {}^{+}V_{xyD}(n,m,q) \tag{4A.10}
$$

while the symmetric components vanish. This completes the plane wave and symmetric correlations among the four D waves. The *total* plane wave component in $Z_{xy}(n,m,q)$ is obtained by adding in quadrature ${}^{+}V_{xyP1}(n,m,q)$, ${}^{+}V_{xyP2}(n,m,q)$, ${}^{+}V_{xyP3}(n,m,q)$, and ${}^{+}V_{xyP4}(n,m,q)$. Designating the total component by ${}^{+}V_{xyP}(n,m,q)$ (without a numerical index following P),

$$
\begin{aligned}
{}^{+}V_{xyP}(n,m,q) = {}&[({}^{+}V_{xyP1}(n,m,q))^2 + ({}^{+}V_{xyP2}(n,m,q))^2 + ({}^{+}V_{xyP3}(n,m,q))^2 \\
&+ ({}^{+}V_{xyP4}(n,m,q))^2]^{1/2}
\end{aligned} \tag{4A.11}
$$

Similarly the total symmetric component ${}^{+}V_{xyS}(n,m,q)$ is given by

$$
{}^{+}V_{xyS}(n,m,q) = [({}^{+}V_{xyS1}(n,m,q))^2 + ({}^{+}V_{xyS2}(n,m,q))^2]^{1/2} \tag{4A.12}
$$

As with the 2D we can also express ${}^{+}V_{xyP}(n,m,q)$ and ${}^{+}V_{xyS}(n,m,q)$ in the original TLM variables.

App.4A.2 Consistency of Plane Wave Correlations With a Simple Quantum Mechanical Model

In this Appendix we demonstrate the consistency of quantum mechanics with our treatment of wave correllations in a TLM matrix. To simplify the discussion we select two isolated horizontal lines , $Z_{xy}(n,m)$ and $Z_{xy}(n,m+1)$. Further we assume the amplitude in the m+1 line is twice that in the m line, and in fact we assume the the wave in the m+1 line may be replaced by two identical, independent waves, each of which is equal in amplitude to the wave in the mth line. We designate each of these waves as a unit wave. We now make the important assumption that each unit wave may be represented by a quantum mechanical wave function. Conceptually, we regard the wave function as representing Bose particles which are photonic in nature. Indeed in the subsequent discussion, it will be helpful to substitute "photon" for "unit wave"; thus, we regard the m+1 line as having two two photons while the mth line has a single photon.

The total wave function ψ_T will consist of the linear combinations of the three individual wave functions for the three photons(or unit waves). ψ_T will remain exactly the same when photons are interchanged, and in addition the individual wave functions will be allowed to share the same quantum numbers; in this case two photons will share any quantum numbers associated with the m+1 line.

A typical combination in ψ_T may be represented by

$$\text{Typical Combination} = \psi_{m+1}(1)\,\psi_{m+1}(2)\,\psi_m(3) \tag{4A.13}$$

where photons 1 and 2 are in the m+1 line and the third photon is in the mth line. The total wave function ψ_T is then obtained by summing the six permutations of the photon numbers. Thus

$$\psi_T = N[(\psi_{m+1}(1)\,\psi_{m+1}(2)\,\psi_m(3) + \psi_{m+1}(1)\,\psi_{m+1}(3)\,\psi_m(2) + \psi_{m+1}(2)\,\psi_{m+1}(1)\,\psi_m(3)$$
$$+ \psi_{m+1}(2)\psi_{m+1}(3)\psi_m(1) + \psi_{m+1}(3)\psi_{m+1}(1)\psi_m(2) + \psi_{m+1}(3)\psi_{m+1}(2)\psi_m(1) \tag{4A.14}$$

where N is a nomalization factor. Next the probability function $\psi\psi^*$ is calculated. This will result in a total of 36 terms each of which consists of the product of the six individual wave functions. Thus for example one of the terms is

$$\text{Typical Term} = N^2 [\psi_{m+1}(1)\ \psi_{m+1}(2)\ \psi_m(3)]\ [\psi_{m+1}(1)\ \psi_{m+1}(2)\ \psi_m(3)]^* \qquad (4A.15)$$

which is nothing more than the probability that photons 1 and 2 are in line m+1 and photon 3 is in line m. Our interest is not in these terms but in the cross-terms, i.e., the terms in which there is a photon interchange between the m and m+1 states. For example consider the combination

$$\text{Typical Cross-Term} = N^2[\psi_{m+1}(1)\psi_{m+1}(2)\psi_m(3)][\psi_{m+1}(1)\psi_{m+1}(3)\psi_m(2)]^* \qquad (4A.16)$$

Note the interchange of photons , in the in the second bracket, of photons 2 and 3 , which occurs between the m and m+1 lines. The above therefore represents an "overlap", or wave interference effect, between the m and m+1 lines. A virtually identical cross-term may also be written:

$$\text{Typical Cross-Term} = N^2[\psi_{m+1}(1)\psi_{m+1}(2)\psi_m(3)][\psi_{m+1}(3)\psi_{m+1}(2)\psi_m(1)]^* \qquad (4A.17)$$

In the above photons 1 and 3 participate in the interchange rather than photons 2 and 3. We then make the important observation that the overlap *per photon* in the m+1 line has half the weight of the total overlap between the mth and (m+1)th states. This is because only one of the two m+1 occupants changes at a time. The total overlap to the m+1 photons is therefore distributed equally among the two occupants. We should also point out that although we have considered only one type of permutation, the same comments apply of course to the other permutations.

A useful image for understanding the concept, looking for example at Eq.(4A.16), is to regard photons 2 and 3 as jumping between the m and m+1 states. The probability that one of the two photons in the m+1 state undergoes this transition is the same as that of the sole occupant in the mth state.

In the light of the above discussion, it is easier to see what we have done in treating plane wave correlations earlier in the Chapter. In this case, instead of distributing the overlap to both photons, we have arbitrarily concentrated the overlap onto one of the two photons in the m+1 line. The other photon in the m+1 line is then completely isolated and does not feel the overlap from the mth line.

We also emphasize though we have only considered the simplified case of three photons distributed in the m and m+1 lines, we may extend the discussion to arbitrary amplitudes in the two lines, in which the number of photons in each line is proportional to the amplitude. The zero order wave functions, and their associated probability functions, possess the potential for a natural source of cross-coupling between waves in adjacent regions(TLM lines). What we have not accomplished is to further model the nature of the coupling between waves in adjacent states. To do this a more detailed use of quantum mechanical methods must be employed.

V. Boundary Conditions and Dispersion

In this Chapter we incorporate the boundary conditions and dispersion into the TLM method. At first glance it may seem odd to lump together these two subject areas in the same Chapter. The two subjects, however, share a common feature in as much as both phenomena involve variations in the propagation velocity. The changes in the propagation velocity, however, have different physical origins. In the case of boundary conditions the situation often involves adjoining media with unequal dielectric constants and therefore differing propagation velocities. In the case of dispersion, of course, the differing propagation velocities arises from their frequency dependence. As we shall see, the methods used to deal with dispersion and adjoining dielectrics are entirely different. This is because with dispersion the propagation velocity is assumed to vary continuously with frequency whereas in the case of adjoining dielectrics, the velocity most often is assumed constant in each dielectric region. The amount of additional information needed to characterize dispersion is considerably greater compared to that of adjoining dielectrics. We first discuss the incorporation of various spatial boundary conditions including the aforementioned dielectric- dielectric interface.

In the following we discuss several kinds of boundary conditions one is likely to encounter in the TLM analysis. The two most common boundary conditions are the dielectric-dielectric and dielectric-conductor interfaces. It is difficult to imagine an electromagnetic configuration without the presence of one or the other of these two interfaces. However other kinds of boundaries are possible. For simulation purposes the idealized dielectric-infinite permeability interface, or "open circuit", provides a convenient means for obtaining total reflection of the wave energy from a given surface The open circuit is often useful since it may be used to approximate experimental conditions in which the radiated wave energy at a given surface is small , due to a very large mismatch in impedance levels. A conducting surface also will provide total reflection but with a resultant field inversion.

Other boundary conditions relate to the input/output of the electromagnetic signal. Here we specify the conditions under which electromagnetic energy is introduced into or leaves the region of interest, whether it be a closed device or an antenna. The difference between the input and output energies results in either the dissipation or storage of energy in the region of interest. In the following we describe ways in which to simulate the boundary conditions, using the TLM cell matrix and appropriate values of node resistors and transmission line impedances.

5.1 Dielectric-Dielectric Interface

There are several choices for positioning the TLM boundary of a dielectric-dielectric interface, two of which are shown in Fig.5.1. The side view shows the dielectric - dielectric interface with constants ε_1 and ε_2. The permeability μ is assumed to be the same in both regions. In contrast, we assume $\varepsilon_1 > \varepsilon_2$. The propagation velocities in each region satisfy

$$(v_1/ v_2) = [\varepsilon_2/\varepsilon_1]^{1/2} \tag{5.1}$$

where v_1 and v_2 are the velocities in regions 1 and 2 respectively. If we wish to retain the same time step in each region, Δt, then we are forced to adopt different cell sizes in each region. Since the linear dimension of each cell is inversely proportional to the velocity, the transmission line lengths in each region, Δl_1 and Δl_2, satisfy

$$(\Delta l_1 /\Delta l_2) = [\varepsilon_2/\varepsilon_1]^{1/2} \tag{5.2}$$

The particular choice of boundary dictates what value of characteristic impedance is selected for the transmission lines at the interface and parallel to it. In Fig.5.1(a) the horizontal lines at the interface, $Z_{xy}(n,m)$, $Z_{xy}(n+1,m)$,

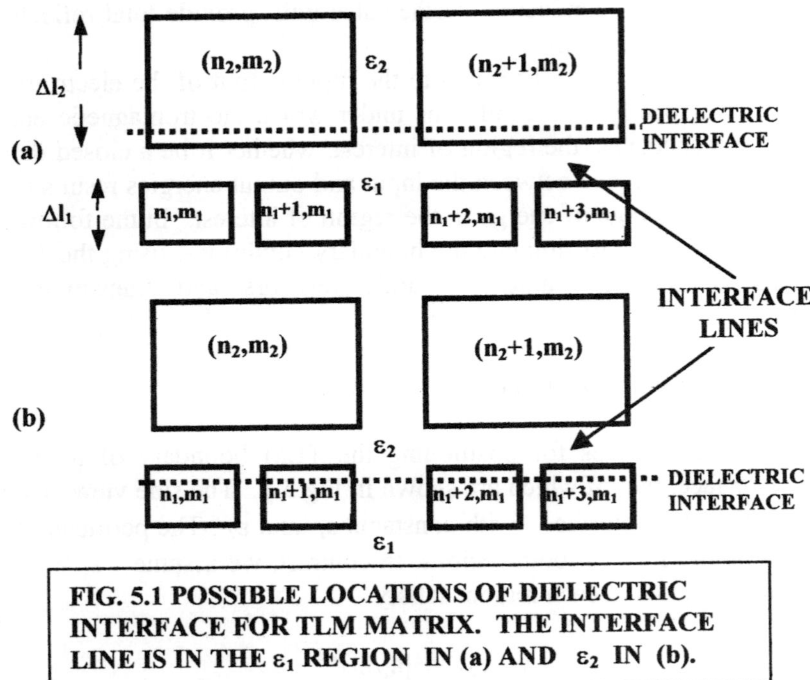

FIG. 5.1 POSSIBLE LOCATIONS OF DIELECTRIC INTERFACE FOR TLM MATRIX. THE INTERFACE LINE IS IN THE ε_1 REGION IN (a) AND ε_2 IN (b).

etc..., are situated on the ε_1 side of the boundary, and thus its interface impedance is determined by ε_1, as are all the lines beneath the interface. Similarly Fig.5.1(b) depicts a boundary wherein the horizontal interface lines are situated on the ε_2 side of the boundary and thus the interface impedance is determined by ε_2. One may also invoke a mixture of the two previous boundaries, as in Fig. 5.2. In (a) the dielectric constant of the line alternates between ε_1 and ε_2. Another approach is shown in (b), where the entire question of where to situate the boundary is avoided by choosing a suitable dielectric average for the interface region,

$$\varepsilon_{AV} = 2(\varepsilon_1\varepsilon_2)/[\varepsilon_1 + \varepsilon_2] \tag{5.3}$$

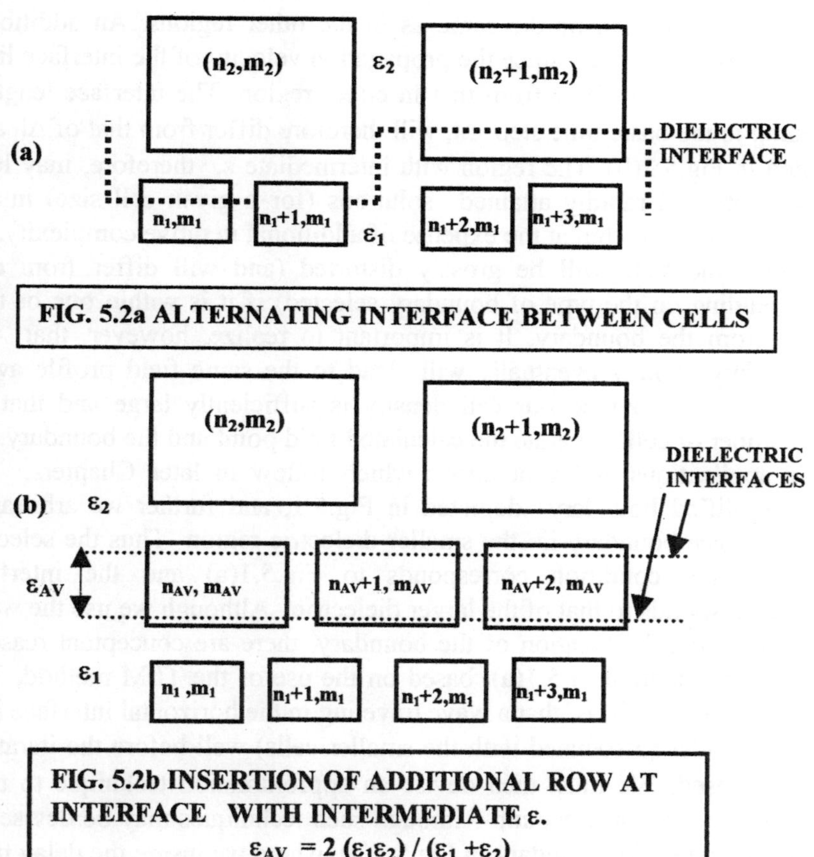

FIG. 5.2a ALTERNATING INTERFACE BETWEEN CELLS

FIG. 5.2b INSERTION OF ADDITIONAL ROW AT INTERFACE WITH INTERMEDIATE ε.

$$\varepsilon_{AV} = 2\,(\varepsilon_1\varepsilon_2)\,/\,(\varepsilon_1 + \varepsilon_2)$$

where ε_{AV} is inserted between the ε_1 and ε_2 regions, as shown in Fig.5.2(b). ε_{AV} is obtained by calculating the capacitance consisting of two layers of dielectric materials, of equal thickness, ε_1 and ε_2 between the electrodes. Note that when $\varepsilon_1 = \varepsilon_2 = \varepsilon$ then $\varepsilon_{AV} = \varepsilon$. When $\varepsilon_1 \gg \varepsilon_2$, then $\varepsilon_{AV} = 2\varepsilon_2$. In effect the capacitance in the vicinity of the interface is equal to 2X that of the smaller dielectric constant, ε_2. In addition, since the equivalent dielectric is given by Eq.(5.3), the impedance of the vertical lines in the intermediate region is

$$Z_{yx}(n,m,q) = (\mu/\varepsilon_{AV})^{1/2} \tag{5.4}$$

and μ is again assumed to be the same as in the other regions. An additional complication arises, however, since the propagation velocity of the interface line, $v = (\mu \, \varepsilon_{AV})^{-1/2}$, will differ from that in either region. The interface length , corresponding to the basic time step Δt, will therefore differ from that of Δl_1 and Δl_2, as noted in Fig.5.2(b). The region with intermediate ε, therefore, may lead to more accurate and rapidly attained solutions (for a given cell size) in the vicinity of the boundary, but at the expense of additional iterative complexity.

Obviously the field will be grossly distorted (and will differ from one another, depending on the type of boundary selected) if it is within one or two cell lengths from the boundary. It is important to realize, however, that the various boundary choices eventually will lead to the same field profile away from the boundary provided our cell density is sufficiently large and that a sufficient number of cells separate the calculated field point and the boundary.

In the applications and simulations which follow in later Chapters, we select the simplified boundary depicted in Fig.5.1, and further we arbitrarily imbed the interface boundary in the smaller dielectric region Thus the selected interface boundary condition corresponds to Fig.5.1(a) and the interface impedance corresponds to that of the larger dielectric. Although we use the word arbitrary in selecting the location of the boundary, there are conceptual reasons for selecting the situation in 5.1(a), based on the use of the TLM method, The main objection with 5.1(b) is that a wave traveling in the horizontal interface line will encounter nodes (associated with the smaller cells) well before the iteration time Δt has elapsed. We must then devise an approximation technique to take these scattering events into account. Although such techniques may be devised it is easier to simply use the boundary in Fig.5.1(a), where we insure the delay time between nodes, in the interface line , is maintained at Δt.

Regardless of the boundary location selected, the inequality in the dielectric constant at the interface presents a problem in the iterative procedure. The nodes in region 1 will in general not coincide with those in region 2 at the interface. This invites the question of how one directs the energy flow from region 1 to region 2 and vice versa. We will see that the energy flow may be described using approximations which become more and more accurate as the cell density is increased. Before delving into these approximations, we describe certain special situations which greatly simplify the energy flow. The special situation occur when Δl_1 and Δl_2 satisfy

$$(\Delta l_1 / \Delta l_2) = \Delta n \qquad (5.5)$$

where Δn is a positive integer. This automatically lines up the node of every large cell(small ε) with a node of the smaller cell, noted in Fig.5.3 for the case of $\Delta n = 2$ (For illustrative purposes this special situation was used as well in Fig. 5.1). This interface will leave some nodes of the small cells unaffiliated with large cell nodes, such as nodes A and D in Fig.5.3. In contrast , at nodes B and

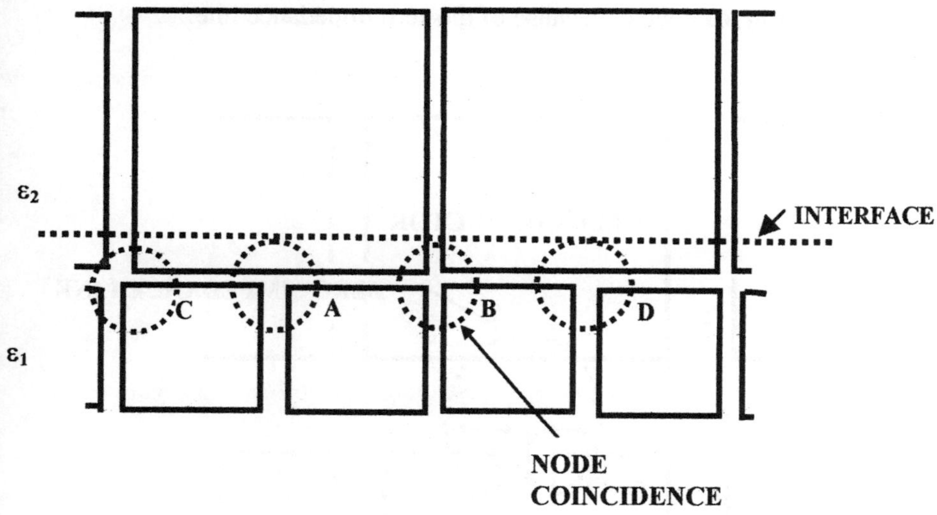

FIG. 5.3 NODE COINCIDENCE FOR $[\varepsilon_1/\varepsilon_2]^{1/2} = 2$

C the transmission lines from both dielectric regions converge exactly, and is designated as a regular node. The load impedance seen by incident waves at nodes A and D will of course differ from that at B and C. At nodes A and D the incident wave $V_{yx}(n,m)$ will transmit energy into the horizontal lines $Z_{yx}(n,m)$ and $Z_{yx}(n+1,m)$, but will not convey energy into the dielectric region ε_2 until an additional time step has elapsed, at which time energy is transferred via nodes B and C. Nodes such as A and D, located at dielectric-dielectric interfaces, in which there exists only one TLM line perpendicular to the boundary, are designated partial nodes. Another way of thinking of a partial node is to regard a zero impedance line to exist in the ε_2 region, as noted in Fig.5.4. The load impedances seen by the various waves at the interface, for partial and regular nodes, will differ, of course, because of the zero impedance line.

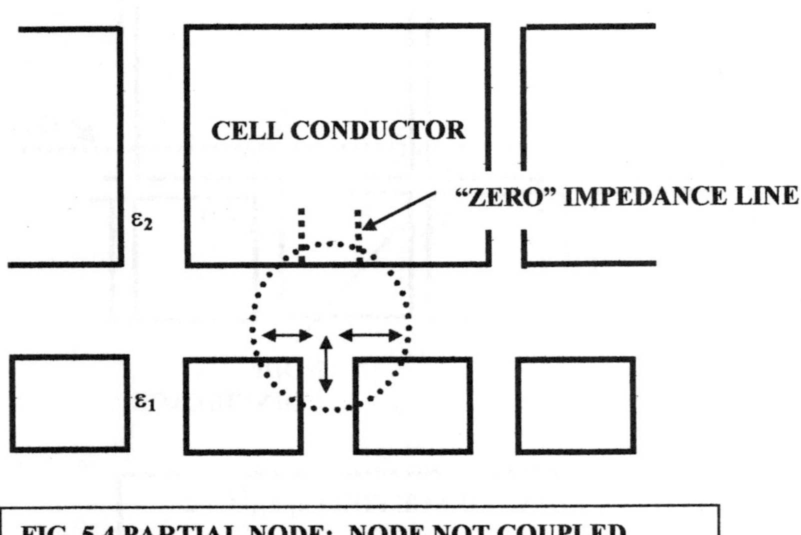

FIG. 5.4 PARTIAL NODE: NODE NOT COUPLED DIRECTLY TO NEIGHBORING DIELECTRIC ε_2.

Node Coupling: Nearest Node And Multi-Coupled Node Approximations

For the general situation we look at Fig.5.5, noting that as a rule there are only partial nodes in regions 1 and 2 and no regular nodes, i.e., the vertical lines in the two regions do not "line up" at the interface, as seen in the various partial nodes at the boundary. We now make the following approximation. We first identify the partial nodes in the *larger* cell region, i.e., region 2, and then determine what partial node in region 1 is closest to the partial node in region 2. Thus, viewing Fig.5.5 , the partial node A in region 2 is closest to partial node B in region 1. Likewise D is closest to C. We then assume that the combination of A and B, as well as C and D, etc..., form regular nodes. We neglect any

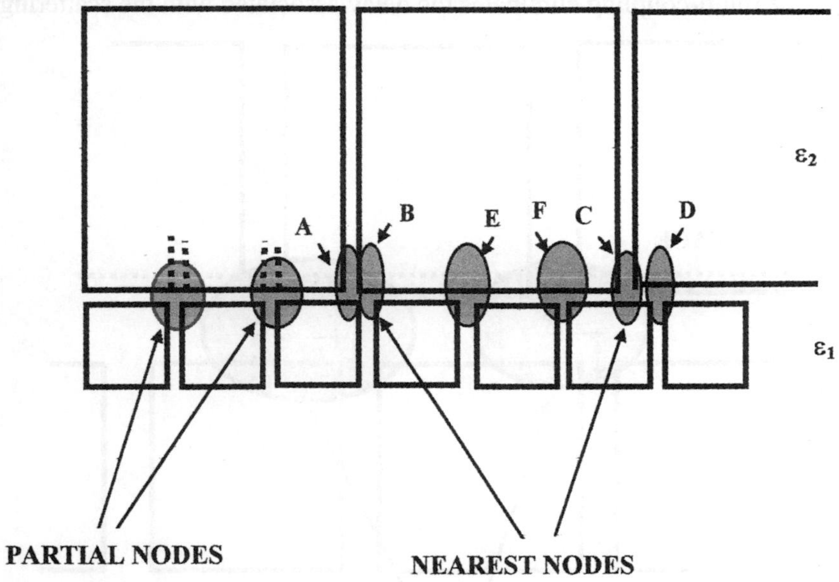

FIG. 5.5 DIELECTRIC INTERFACE SHOWING PARTIAL NODES. NODES A,B AND C,D ARE NEAREST PARTIAL NODE PAIRS. WITH THE MULTICOUPLING METHOD F JOINS C AND D WHILE E JOINS A AND B.

phase delay between A and B, as well as C between and D, etc... This is the nearest node approximation. Fig.5.6 shows a closer view of the nearest nodes, indicating the zig-zag path the wave must take in order to go from the vertical TLM line in ε_1 to the nearest vertical TLM line in ε_2.

As indicated previously in Section 4.9 and Fig.4.4(b), one may also employ multi-coupling nodes, rather the nearest nodes. This method is probably more effective when the the adjoining dielectrics differ greatly with respect to the dielectric constant($\geq 10:1$)). The location of the nodes participating in the multi-coupling relies on the same notation and methods employed previously for the nearest node approximation. With this approximation all partial nodes existing on the high dielectric side join a regular node. Thus, looking at Fig.5.5 , the partial node F joins with C and D , while E joins with A and B. However, although the multi-coupling eliminates the delay associated with the scattering

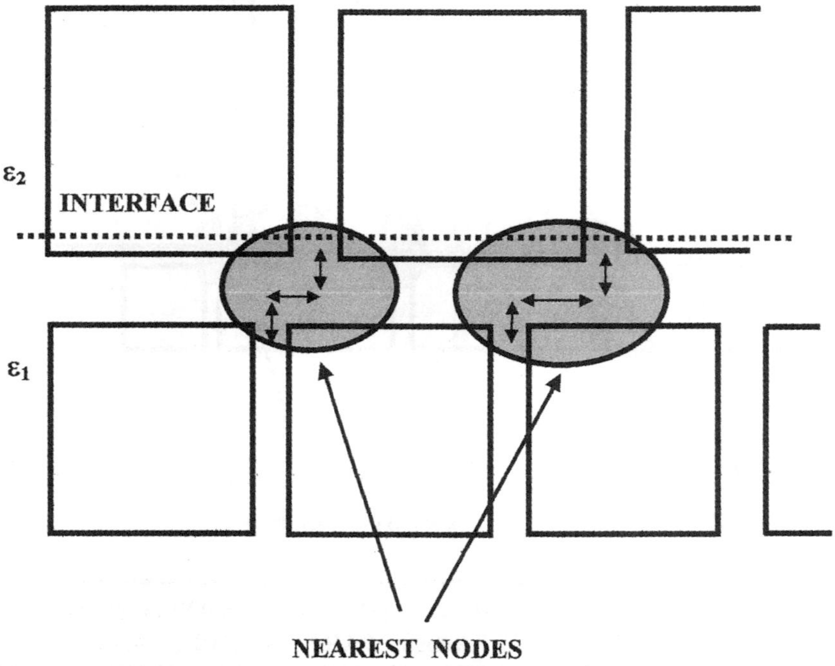

FIG. 5.6 NEAREST NODE APPROXIMATION.

to the interface line, it also introduces complications having to do with wave reconstruction, For example, if we focus on nodes F and D, this method requires us to combine the waves in the associated vertical lines(on the ε_1 side), using a quadrature approach described in the last Chapter. For this reason, we employ the more straightforward nearest node method in the remainder of this Chapter (and in Chapter VII where we present a computer iteration for a dielectric interface).

5.2 Nearest Nodes for 1D Interface

The next step is to quantify the location of the partial nodes, allowing us to locate the nearest nodes, which we may then use for iterative purposes. For illustrative purposes, we first consider a boundary with a 1D interface. The interface itself may be formed, e.g., by a pair of adjoining 2D regions. For convenience we regard the interface as formed by a pair of parallel 1D lines with nearest neighbors in the x direction, as indicated in Fig.5.7. If n_2 is the x index for the n_2th cell in region 2, then the distance to the n_2 node is $n_2 \Delta l_2$. Dividing by Δl_1, we obtain

$$(n_2 \Delta l_2 / \Delta l_1) = n'_1 + RM \tag{5.6}$$

where n'_1 is the largest integer and RM is the *fractional* remainder(i.e., the decimal portion). We assume both lines start at x=0. The nearest neighbor in region 1 to the n_2th partial node, denoted by n_1, is thus

$$n_1 = n'_1 + 1 \quad \text{if } RM \geq .5 \tag{5.7a}$$

$$n_1 = n'_1 \quad \text{if } RM < .5 \tag{5.7b}$$

In Chapter VII we will make use of these linear boundary conditions when formulating the iterative equations.

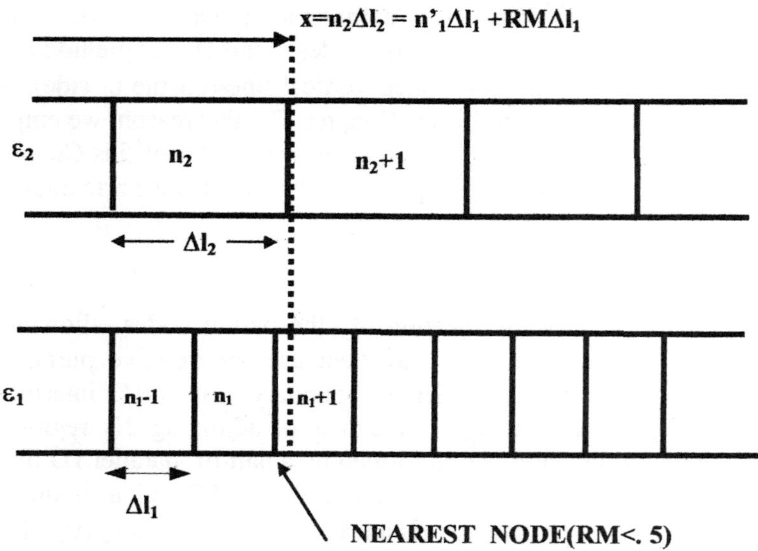

**FIG 5.7 NEAREST NODE FOR LINEAR BOUNDARY.
NEAREST NODE IS n$_1$ FOR RM < 0.5 AND n$_1$+1 FOR RM≥0.5**

5.3 Nearest Nodes at 2D Interface

We now extend Eq.(5.7) to a 2D surface boundary condition, and in particular ,
to a planar surface. Fig.5.8 shows a top view of the interface where the dotted,
smaller cells apply to region 1 and the solid line, larger cells belong to region 2.
We now determine the nearest neighbor in region 1 to the node (n_2, m_2) located
in region 2. For the xy plane the equations analogous to Eq.(5.6) are then

$$(n_2\Delta l_2 / \Delta l_1) = n'_1 + RM_x \tag{5.8a}$$

$$(m_2\Delta l_2 / \Delta l_1) = m'_1 + RM_y \tag{5.8b}$$

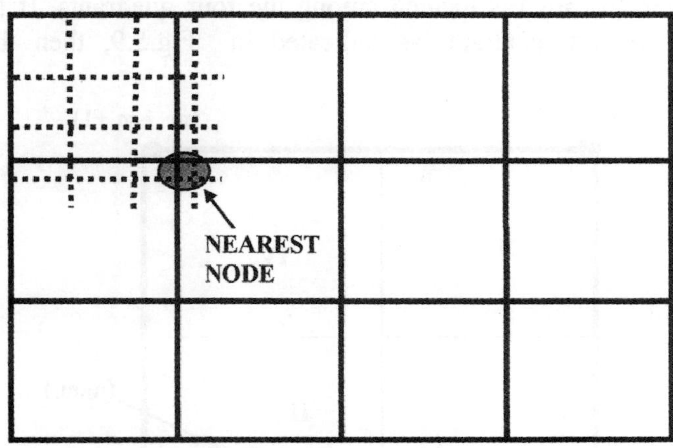

FIG. 5.8 NEAREST NODE FOR 2D DIELECTRIC INTERFACE.

where as before n_1 and m_1 are the largest integer values and RM_x and RM_y are the remaining fractional values. These two fractions will determine which of four possibilities is the nearest neighbor. The four possibilities are

$$\text{I:} \qquad n_1 = n'_1 \,, m_1 = m'_1 : \ \ \text{For } RM_x < .5, \ RM_y < .5 \qquad (5.9a)$$

$$\text{II:} \qquad n_1 = n'_1 + 1, m_1 = m'_1 : \ \text{For } RM_x \geq ..5, RM_y < .5 \qquad (5.9b)$$

$$\text{III:} \qquad n_1 = n'_1, \ m_1 = m'_1 + 1 : \text{For } RM_x < .5, RM_y \geq ..5 \qquad (5.9c)$$

$$\text{IV:} \qquad n_1 = n'_1 + 1, \ m_1 = m'_1 + 1 : \text{For } RM_x \geq .5, RM_y \geq .5 \qquad (5.9d)$$

The four possibilities may be summarized with the aid of Fig.5.9. The four nodes in region 1 form the corners of a square of length Δl_1. The square consists of four quadrants, each with its own node. The nearest neighbor in region 1 will depend on where $(n_2, m_2,)$ is located among the four quadrants. If the node is located in the second quadrant, as indicated in   Fig.5.9, then the node is (n_1+1,m_1), etc...

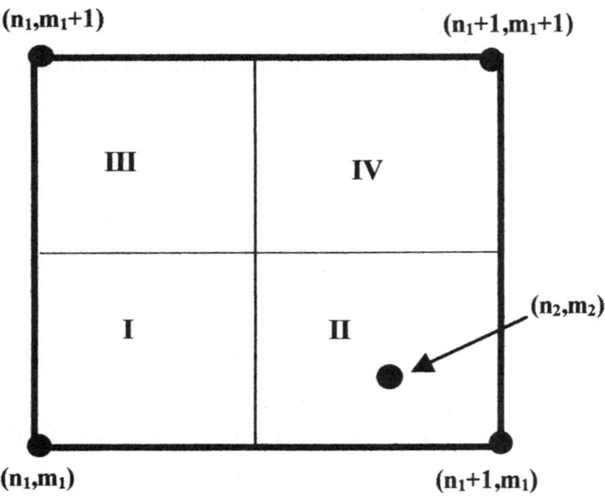

FIG. 5.9 2D INTERFACE: NODE (n_2,m_2) , SITUATED IN THE ε_2 REGION, IS CLOSEST TO (n_1+1,m_1) IN THE ε_1 REGION.

5.4 Truncated Cells and Oblique Interface

Thus far we have only considered complete cells bordering the interface. In general this will not always be possible and truncated cells may be required at the interface, as shown in Fig.5.10(a). The easiest solution to this problem is to either modify the cell size or the boundary conditions slightly so as to remove the truncated portions. A simple approach is to have the truncated portion deleted or made whole in the iteration, depending on whether the truncated portion is less than or more than $\Delta l/2$. For example, in Fig.5.10(a), the truncated

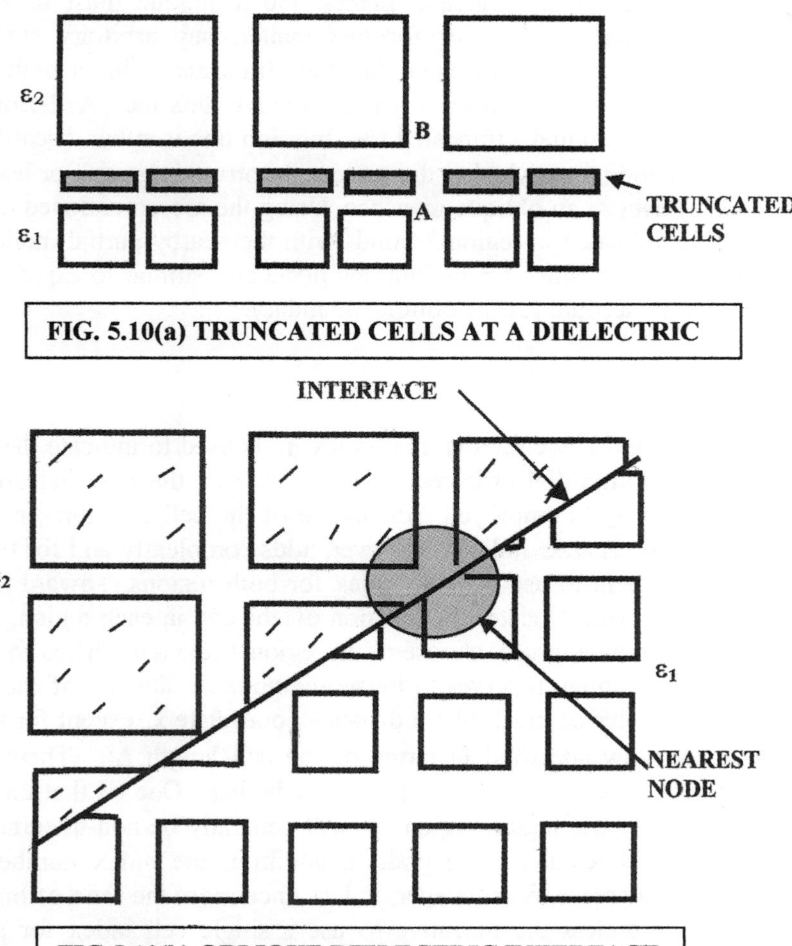

FIG. 5.10(a) TRUNCATED CELLS AT A DIELECTRIC

FIG 5.10(b) OBLIQUE DIELECTRIC INTERFACE

portion of the cell in region 1 is less than $\Delta l_1/2$ and thus node B bonds directly with A, essentially eliminating the truncated cells of ε_1.

Although nearest neighbors have been calculated for a simple planar surfaces, parallel to the principal planes, the technique must be modified to account for oblique planes, or for that matter, any arbitrary surface. If we consider oblique planes, for example, then truncated cells in both regions are inevitable and slight adjustments will not remedy this fact. As before we again employ the criterion that a truncated transmission line is either discarded or made complete depending on whether the truncated portion is greater or less than $\Delta l/2$. Fig. 5.10(b) depicts an oblique interface. Using the aforementioned criterion, the partial node indicated in region 2 bonds with the nearby partial nodes in region 1. The node relationships for the oblique plane are similar to Eq.(5.9), but must be modified to account for the oblique boundary.

5.5 Single Index Cell Notation

Referring to Fig.5.7, we see that the index n_1 is used to indicate the position of any cell in the high dielectric region. Likewise for the low dielectric region a different index n_2 is employed, indicative of the cell position in that region. The use of two separate indices, however, adds complexity and for this reason it is often convenient to use a single index for both regions. Toward this goal we select an index which locates the position of the cell in each region, and for this purpose we continue to use choose n_1 in region 1 and a modified form of n_1 in region 2. In so doing n_1 serves to locate the node in the cell of each region. In this regard n_1 behaves much like a distance coordinate x, except for the fact that distances are now specified in terms of the cell length Δl_1 There are certain disadvantages, however, to this type of re-labeling. One is that the cell index, for the low dielectric region(region 2) will generally be non-integral, unless the ratio of the cell lengths is integral. In addition, the index numbers will not consecutively increase by an integer, unless once more the ratio of line lengths is integral. Nevertheless the incentive to use a single cell index for purposes of locating each and every cell is very strong and we adopt this labeling when describing the computer iteration in Chapter VII.

The re-labeling for parallel chains is shown in Fig.5.11(a). Note that the high dielectric region, represented by the bottom chain, has exactly the identical cell labeling as that in Fig.5.7. The cell indices for the low dielectric chain, on the other hand, will differ. The label for the first cell, since it overlaps the n_1 cell, is $(n_1 + RM)$. Similarly the second cell is (n_1+2+RM), since the large cell overlaps that of (n_1+2), and so forth. As discussed previously, RM is the remainder term satisfying Eq.(5.6). We may also approximate the cell position in the upper chain , to the nearest integer, using the nearest node approximation. As shown in Fig.5.11(a), the fractional remainder overlapping n_1 is less than 0.5 and therefore the index is simply n_1. For the next cell, n_1+2, the remainder is >0.5 and thus the cell index for the ε_2 line is n_1+3, and so forth.

The notation simplifies considerably if $(\Delta l_2/\Delta l_1)$ is integral. For example if $(\Delta l_2/\Delta l_1) = 3$ the indices of the larger cells are 3, 6, 9 etc... Note that by adapting this notation, the cell index numbers no longer increase one, but by the factor $(\Delta l_2/\Delta l_1)=3$. The single index notation for the larger cells, $(\Delta l_2/\Delta l_1) =3$, is employed in the iterative program used to analyze a photoconductive switch, given in Chapter VII.

The single index cell notation is also useful when the dissimilar cells are in series rather than in parallel , as shown in Fig.5.11(b), where a 1D example of a non-uniform line is provided. In this case we assume the line consists of two separate line segments , with differing cell lengths, joined together without cell truncation at the boundary. Again note that the n_1 index, may be used to identify cells in both dielectric regions. In the large cell region the indices are $n_1 +(\Delta l_2/\Delta l_1)$, $n_1+2(\Delta l_2/\Delta l_1)$, etc... Again we note that the cell designations will in general be non-integral and the jumps in the index from cell to cell will likewise be non-integral. Finally, although the single index labeling has been applied to only one dimension(in the x direction)it is easily extended to boundaries with 2D and 3D cells.

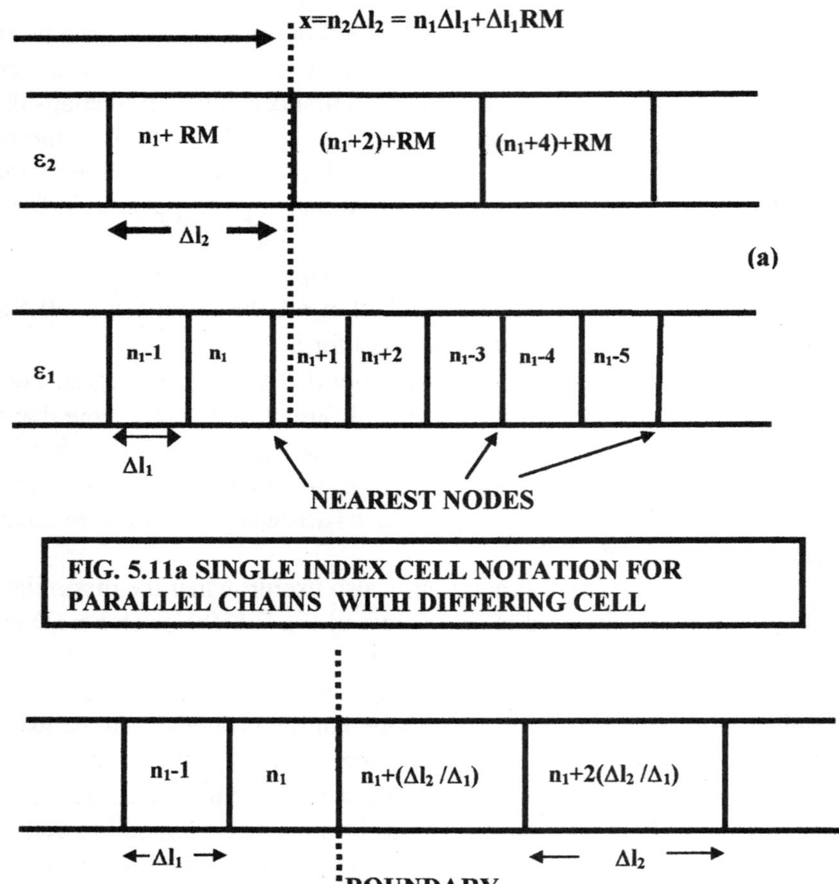

FIG. 5.11a SINGLE INDEX CELL NOTATION FOR
PARALLEL CHAINS WITH DIFFERING CELL

FIG. 5.11b SINGLE INDEX CELL NOTATION FOR SERIES
CHAIN WITH DIFFERING CELL LENGTHS

5.6 Simplified Iteration Neglecting the Nearest Node Approximation

We digress for a moment to point out an elementary version of the iteration method which requires little or no additional effort in the modification of the basic iterative program, unlike the modifications needed to implement, e.g., the nearest node method. Suppose we use the single index cell labeling of the previous Section to characterize two bordering dielectric regions, and set-up the iterative equations for each region. We then purposely neglect incorporating the nearest node approximation into the program and allow the program to proceed. At the border waves will go from one region to the other indirectly, first traveling along the interface line. The TLM lines will automatically adopt a neighbor in the adjacent region. Exactly which neighbor is selected depends on the "round-off" convention utilized by the computer language in treating non-integral array elements. As long as the cell density is sufficiently large, however, the results probably will not differ a great deal.

In view of the above comments one may question why it is at all necessary to invoke a nearest node(or multicoupling node) approximation method. The reason for implementing this approach has to do with the calculation of the scattering coefficients for the boundary nodes. When we neglect the nearest nodes, no attempt is made to correctly calculate the scattering coefficient at the boundary; and accuracy suffers compared to the nearest node method. We should point out that in the limit of small cell size, away from the boundary, both methods should lead to essentially the same results. Near the boundary, however, the nearest node method will provide more accurate results for a given cell size, thereby easing the computational burden.

In Chapter VII we briefly compare the results with and without the nearest node approximations for the case of a light activated semiconductor switch. The resultant field profiles, located two cells away from the border, demonstrate the differences between the two methods for the given cell size.

5.7 Non-Uniform Dielectric. Use of Cluster Cells

In the event that the dielectric is non-uniform throughout the medium, then there is no clear cut boundary condition, but instead a matrix of cells will appear as, for example, in Fig.5.12 where the dielectric varies in the y direction. This means of course that the cell size will vary throughout the medium as well. Just as in the case of an abrupt boundary, however, we can use a nearest node approach to obtain the scattering.

In Fig.5.12 the change in the dielectric constant is along one of the principle axes, in this case the y direction, which makes the "tilework" relatively easy to perform. In general, however, the *change* in dielectric constant will be along an arbitrary direction and, in addition , the change will be non-uniform . The technique then becomes more complicated and iterative procedures must be employed to lay out the tilework in as compact fashion as possible. Iterative procedures must then be formulated, while applying the nearest node or multi-node approximations to the non- uniform dielectrics.

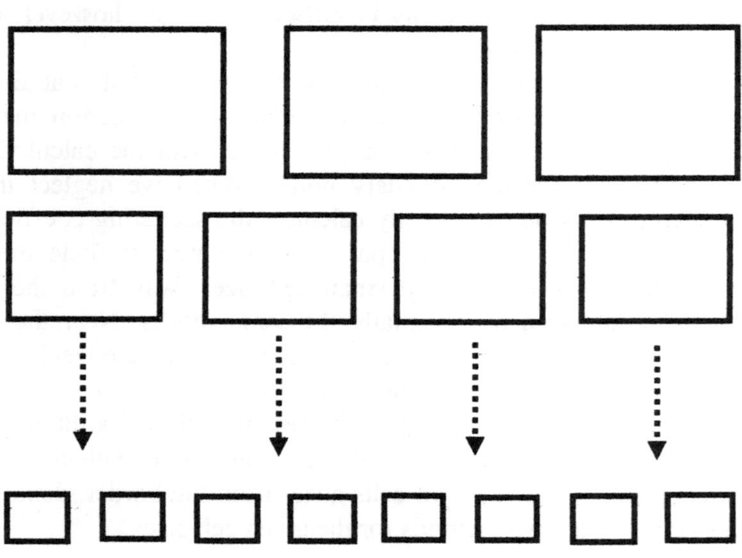

FIG. 5.12 VARIATION OF DIELECTRIC CONSTANT IN Y DIRECTION, DEPICTED AS CHANGE IN CELL SIZE.

Needless to say, the tilework procedure becomes more amenable as we resort to smaller and smaller cells. If the spatial, change in dielectric constant changes only a small amount over the selected cell length , then we may use a cluster method. As noted in Fig.5.13 , a cluster cell is an enlarged cell whose dimension is ΔL_o The dielectric constant is considered the same throughout the cluster cell; the value of ε employed corresponds to that at the center of the cluster cell, so that in a sense the value represents an average value throughout the cell. Each cluster cell will contain a number of TLM cells, all of same dimension. The size of the TLM cells, will of course depend on the dielectric constant in the cluster cell, so that with increasing dielectric constant, the cluster cell will contain more and more TLM cells. Inevitably, the cluster cell will contain truncated cells at the cluster cell border, since there is no reason to assume the cluster cell is made up of an integral number of TLM cells.

In principle one may use cluster cells to treat non-uniform dielectrics via the TLM method. Assuming a knowledge of the spatial variation of the dielectric constant, the dielectric is the divided into small cluster cells; within each cluster cell the TLM cells are identical. Fig.5.13 shows four cluster cells, each having a different ε(and therefore different sized cells) bordering one another. Scattering then proceeds as usual among the TLM cells As discussed previously, we may employ the nearest node/multi-node approximations along the borders of the cluster cells. Since truncated cells are likely to exist along the borders, we either delete or make whole these cells, depending on the degree of truncation. Both the truncation and nearest node/multi-node approximations gain in accuracy as both the TLM and cluster cells are made smaller.

5.7(a) Use of Existing Software to Generate Cell Matrix

In all of the prior situations discussed, whether it was differing dielectrics bounding one another, or nonuniform dielectric regions, we saw that the first task was the generation of a cell matrix in the given dielectric region. The cell size was determined by the value of the dielectric constant. Each cell, regardless of its location, was given a particular number label in order to identify it in the iterative process. The establishment of such a cell matrix, of course, is not a unique procedure, and there is no reason why existing software (for other

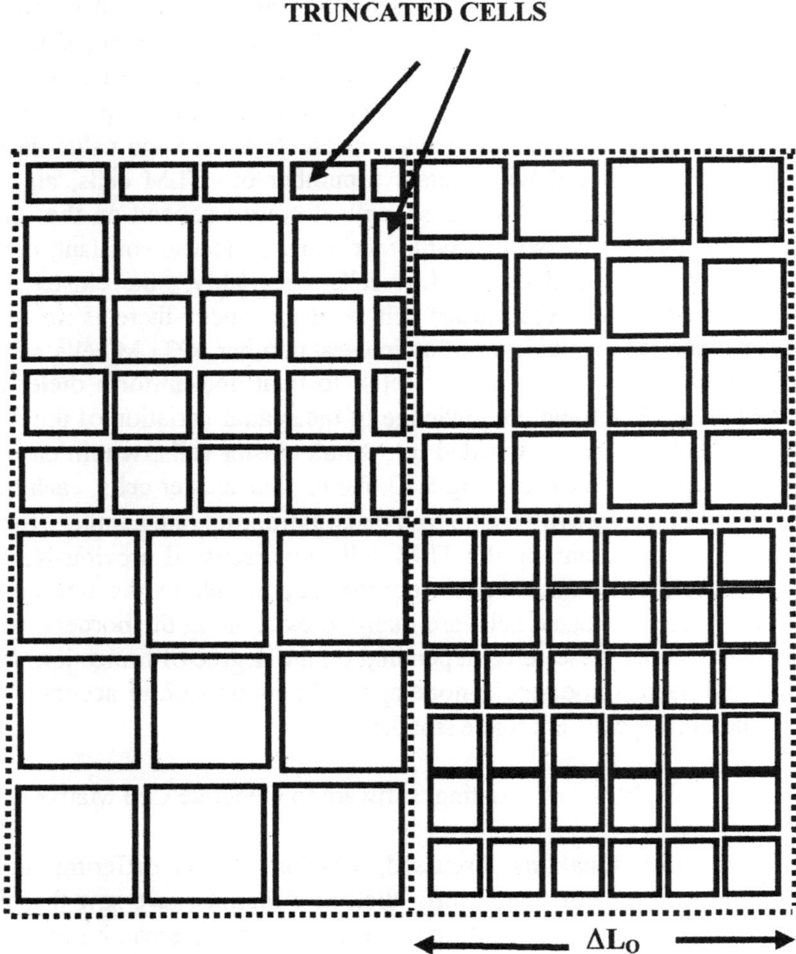

FIG. 5.13 DEPICTION OF FOUR NEIGHBORING CLUSTER CELLS, OF LENGTH ΔL_O , WITH DIFFERENT DIELECTRIC CONSTANTS. NEAREST NODE AND TRUNCATED CELL APPROXIMATIONS ARE ALONG CLUSTER CELL BORDERS.

applications) cannot be used toward the existing problem. In these other applications both uniform and non-uniform grids are established throughout the space. This software must then be adopted by attaching a node, cell, and surrounding TLM lines to each of the grid points. In the final analysis, the decision must then be made as to whether it is more expedient to adopt existing software to establish the TLM matrix, or whether it is easier to develop the necessary TLM matrix from the outset.

Other Boundary Conditions

5.8 Dielectric - Open Circuit Interface

On occasion, we may wish to constrain or direct electromagnetic energy within a particular region, by means of an extremely high permeability constant material at a given interface. Such an interface may correspond to an actual representation of the experimental facts, where an extremely large positive mismatch exists; or else the interface may correspond to a mathematical simplification of a radiation problem in which we impose the reflection of electromagnetic energy by means of an "open circuit". This is in contrast to the field reversal experienced by arbitrarily placing a conducting region at the interface. Since the relatively high impedance of the transmission line often arises from a large permeability, μ, we must include this in the determination of the cell sizes. Thus, Eq.(5.2) becomes

$$(\Delta l_1 / \Delta l_2) = [\ \varepsilon_2\mu_2/\varepsilon_1\mu_1]^{1/2} \tag{5.10}$$

In the limit of open circuit impedances, i.e., very large μ , the number of cells in the slow region will grow very large, thus adding an unnecessary degree of complexity. A simpler way to simulate an open circuit is to maintain the same cell sizes on both sides of the interface, but to assign extremely large characteristic values to $Z(n,m,q)$ in the open circuit region , as indicated in Fig. 5.14. The implication of having the same cell size is that the dielectric constant

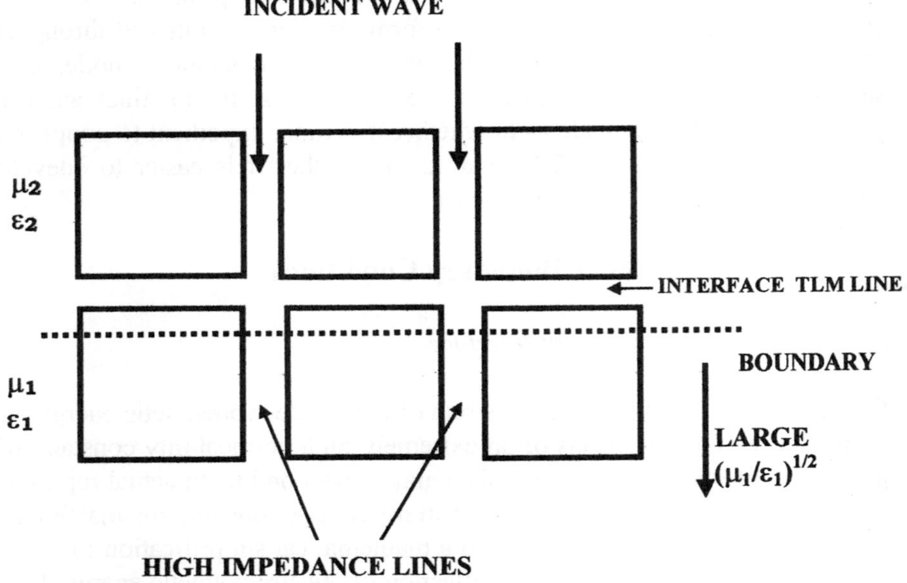

FIG. 5.14 DIELECTRIC-OPEN CIRCUIT INTERFACE. THE INCIDENT WAVE ENERGY IS PREVENTED FROM PENETRATING THE HIGH μ_1 REGION.

decreases in the proportion that the permeability increases, thereby further increasing the TLM line impedance. Thus a wave emanating from the ε_2 , μ_2 region, and arriving at the interface node will see a very large impedance, i.e., an open circuit. If the wave attempts to enter the high impedance lines, The wave will be totally reflected (with the same polarity) with no energy entering the high impedance region..

5.9 Dielectric - Conductor Interface

A representation of the dielectric - conductor interface is shown in Fig.5.15. We have portrayed the conductor, in much the same manner as a dielectric. This is somewhat of an artifact, of course, since   normally we do not associate a cell with a conducting region. The use of such cells, however , allows us to simulate the conductor in a very convenient manner, since we may simply extend the same TLM matrix from the dielectric region into the conducting region with greater flexibility.

The conducting region is simulated by a combination of two effects. First the resistor nodes in the conducting region may be assigned very small values. Secondly, we can reinforce, or accelerate the conduction properties by assigning small values to the transmission lines as well.   For convenience, we have arbitrarily assumed a cell length in the conductor equal to that of the dielectric. There are two choices for the location of the interface, as indicated in Fig.5.16 by the dashed line. The first is located in the dielectric region, Fig.5.15(a), slightly above the interface nodes, and the second is located in the conducting region, Fig.5.15(b), slightly below the nodes. We compare both boundaries.

We first look at 5.15(a), in which the interface is assumed to be embedded in the dielectric. Contact with the dielectric is made via the vertical TLM lines in the dielectric. $R(n,m)$ is surrounded by the four cells (n,m), $(n+1.m)$, $(n,m+1)$, and $(n+1,m+1)$. For both boundaries the conductivity about the node is averaged over these four cells(we consider the 2D problem for simplicity). For large conductivity, the average node resistor will be dominated by the cells in the conducting region and the conductivity will be approximated by $(1/2)[\sigma(n,m)+\sigma(n+1,m)]$, which will be small. This would appear to favor the interpretation of the upper boundary (a) as the more consistent boundary. Looking at Fig.5.15(b),on the other hand, there is no reason for the impedance of the horizontal interface line to change (compared to that in the interior of the dielectric. Waves will be prevented from entering the conductor because of the small values of impedance assigned to the conductor. These low impedance vertical lines(rather than the node) carry the burden of initially creating the short circuit conditions at the interface. The advantage of this type boundary, 5.15(b), is a geometrically neat condition in which the conducting boundary exactly coincides with lower member of the horizontal interface line, as opposed to the dangling vertical lines in Fig.5.15(a). Whichever boundary is selected, (a) or

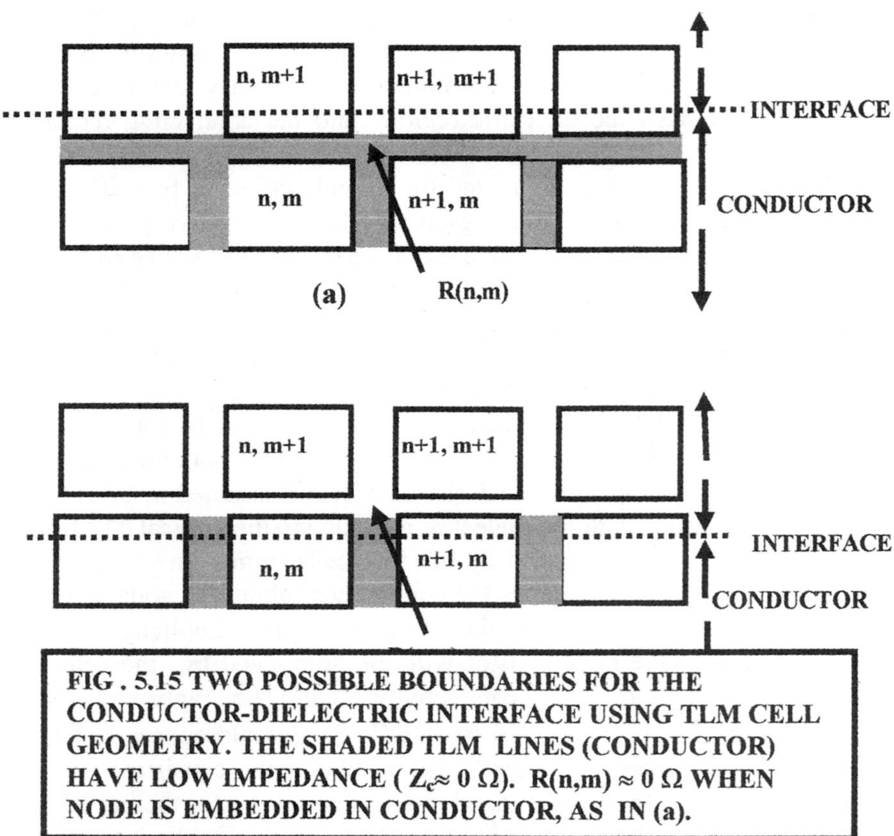

FIG . 5.15 TWO POSSIBLE BOUNDARIES FOR THE CONDUCTOR-DIELECTRIC INTERFACE USING TLM CELL GEOMETRY. THE SHADED TLM LINES (CONDUCTOR) HAVE LOW IMPEDANCE ($Z_c \approx 0\ \Omega$). $R(n,m) \approx 0\ \Omega$ WHEN NODE IS EMBEDDED IN CONDUCTOR, AS IN (a).

(b), the iterative results should be the same as the cell density is increased. One should also note that if the conductor- dielectric transition is gradual, or for moderate or low conductivity, one should not use low impedance TLM lines,but instead ascribe the conductivity to the nodes as usual. The low impedance lines should be reserved solely for the case of "hard" metal conductors.

5.10 Input/Output Conditions

Two types of input/output boundary conditions prove to be useful. In the case of the input to some electromagnetic interaction region(or "device"), Fig.5.16 , the input signal is often delivered by means of a transmission line. The input line will in general suffer reflections , however, which will then slosh back and forth

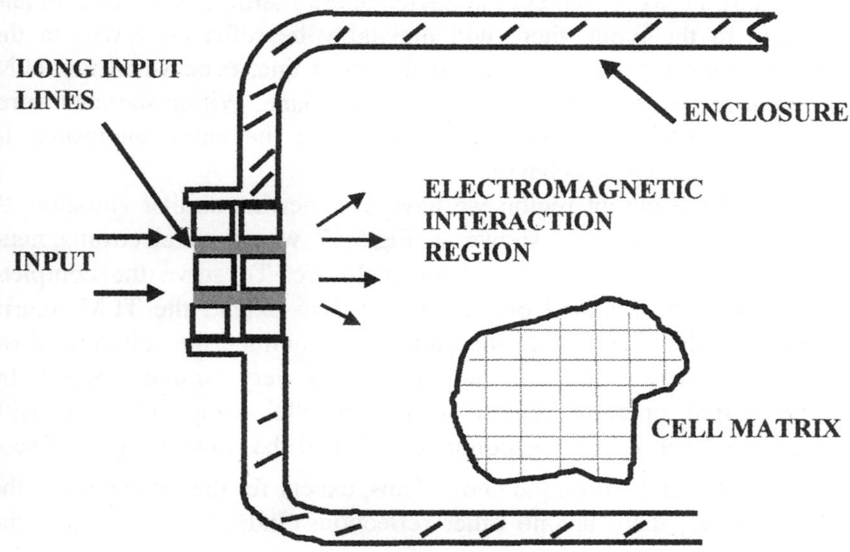

FIG. 5.16 . USE OF CELL MATRIX TO SIMULATE COUPLING TO ELECTOMAGNETIC INTERACTION REGION. LONG INPUT LINES TO THE REGION MAY BE SIMULATED WITH NODE RESISTORS WHICH TERMINATE REFLECTED WAVES.

between the input plane and the source. The multiple reflections will often introduce an unwanted disturbance into the system. One way to avoid this problem is to have an input line which is extremely long, i.e., longer than any times of interest. The reflected waves then will become "lost" and they no longer have an opportunity to disturb the system. This simplifies the input boundary condition a great deal. Of course it is completely unnecessary, for simulation purposes, to insert a long line. Instead, what we do simply is to insert node resistors which arbitrarily terminate any reflected waves at the input plane. If we wish we can also arrange (both by simulation and experiment) for the input lines to exactly match the impedance of the TLM lines in the input device region(same dielectric constant). However, unless the device interior structure is uniform and exactly matched to the input, there will inevitably be reflected waves in the device which will work their way back to the input line, even when the TLM lines at the input are matched at the input aperture plane. Within the enclosure, we use the usual TLM cells and lines, occupying the entire enclosure, to determine the electromagnetic behavior.

In the case of the output region we have a somewhat similar situation. A very simple output boundary is shown in Fig.5.17 where an electromagnetic signal is radiated from the aperture of an enclosure. To solve the complete problem, including the radiated output, we need to extend the TLM matrix (already existing in the enclosure)to the output and continue the solution. If we wish, however, we may deal with the output in a very simple manner, by replacing the output aperture region with a parallel array of lines, with characteristic impedance Z_0 , corresponding to that of the output region. These lines are then matched at their output ends. Thus, except for the reflections at the input to the Z_0 lines, there are no other reflections. Indeed, if we match the impedance of the enclosure lines to the output Z_0 lines, the electromagnetic signal will be completely absorbed, and the array of Z_0 lines will simulate an output transmission line. At this point, we have not actually specified the node resistors needed to prevent unwanted multiple reflections at the input/output. This is done in App.5A.1.

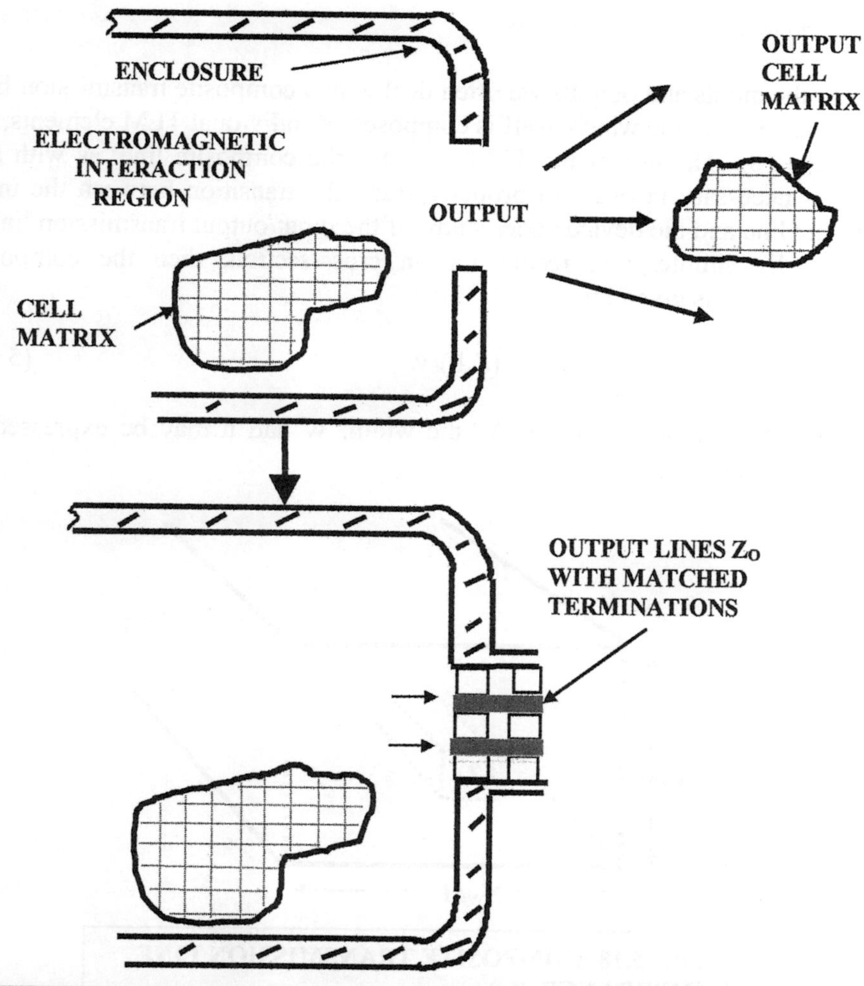

FIG. 5.17. COUPLING OF ELECTROMAGNETIC REGION TO ARBITRARY OUTPUT REGION. OUTPUT REGION IS REPLACED WITH MATCHED TERMINATION LINES.

5.11 Composite Transmission Line

For both the inputs and outputs we often deal with a composite transmission line, i.e., a transmission line which itself is composed of individual TLM elements, Z_o, as in Fig.5.18. The use of the TLM lines for the composite line, as with any region, is necessary in order to properly study the transition between the input composite line and the device under study. If the input/output transmission line is geometrically simple, i.e., rectangular in cross-section, then the composite impedance , Z_C , is given by

$$Z_C = (Z_o h)/W \qquad\qquad (5.11)$$

where is h the region height and W the width. W and h may be expressed in terms of the cell length Δl ,

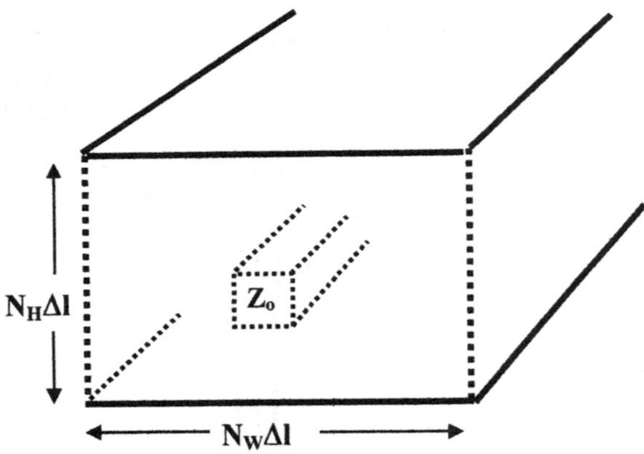

FIG. 5.18 COMPOSITE TRANSMISSION LINE
IMPEDANCE: Z_C
$$Z_C = (N_H/N_W)Z_o$$
N_H = NUMBER OF CELL LENGTHS EQUAL
    TO HEIGHT OF COMPOSITE LINE
N_W = NUMBER OF CELL LENGTHS EQUAL
    TO WIDTH OF COMPOSITE LINE.

$$h = N_h \Delta l \qquad (5.12)$$

$$W = N_w \Delta l \qquad (5.13)$$

where N_w and N_h are the number of cell lengths in W and h. The relationship between Z_c and Z_o is thus

$$Z_C = N_h Z_o / N_w \qquad (5.14)$$

The individual line impedances, Z_o, bridge the gap between the region under study and the input/output. We should point out that for purposes of achieving a "matched" condition, i.e., avoiding reflections and losses at the input/output planes of the device, the important quantity is not Z_o but the composite impedance Z_C. This impedance must be the same in the device as well as at the input/output. On the other hand, although Z_o may be the same everywhere (input, output, device) imbalances in the geometry will in general cause the device to be mis-matched.

5.12 Determination of Initial Static Field by TLM Method

As mentioned repeatedly, the TLM variables may be used to track any transient phenomena following the disturbance of the static fields. We start with a completely static case with no electromagnetic inputs, in which fixed, charged conductors give rise to a potential distribution(whose solution is assumed to be known), which then remains constant until the conductivity in the medium is altered in some manner. Before we can calculate the effects of the conductivity, we must first cast the static potential distribution in terms of the transmission line modes. As we have outlined on several occasions, this is accomplished by dividing up the medium into the usual cells and transmission lines and identifying the corresponding standing voltage waves in each line. The field description is then transferred to that of the standing waves in the lines. When the conductivity is "turned on", the standing waves become traveling waves, which may be tracked in time and so one may determine how the fields evolve throughout the region.

In the previous discussion, we assumed the initial static solution was already known, having been obtained from Laplaces' Equation. For certain simple geometries, however, it is possible to obtain the static solution, using precisely the same iteration developed for the transient case. Fig.5.19 illustrates the concept, for a pair of conductors. For the initial conditions, the wave amplitudes are taken to be zero everywhere at t = 0, except for a narrow region connecting the two conductors, labeled by A. In this region we *assume* a knowledge of the static fields, which are of course constant in time. The region may be regarded as a source of voltage which remains constant for t > 0, emitting constant amplitude waves at its boundary. After a sufficient number of time steps have elapsed the source voltage eventually gives rise to a static field throughout the space. This field , however, is not necessarily Laplaces solution since we have assumed the static solution in region A . If the assumed solution is incorrect then the solution throughout the space will be as well. The way out of this dilemma is again relies on an iterative process. We select a different region between the conductors, and assume the static solution values based on the first iteration. All other field values throughout the space, except for this region, are then set equal to zero and the emission process is repeated, but now from the new region, designated as B in Fig.5.19. The emission from B then serves as a correction to the assumed fields at A. This process may then repeated , alternating between regions A and B, until a stable solution is achieved.

A key assumption, the initial voltage distribution at A, is important in finding the fastest path toward a solution. For some simple geometries the initial selection of static field values for region A is sufficiently accurate so that the alternate iterations may be kept to a minimum or eliminated entirely. Fig.5.20 shows examples of geometries in which one may estimate the initial static field. In Fig.5.20(a) , for example, we select a region in which the two conductors are relatively close together and uniformly spaced apart. By selecting our source voltage region here, we may make the reasonable estimate that the source field is more or less constant across the gap spacing. Another obvious choice is in Fig.5.20(b), where the source region is located near the output of an infinitely

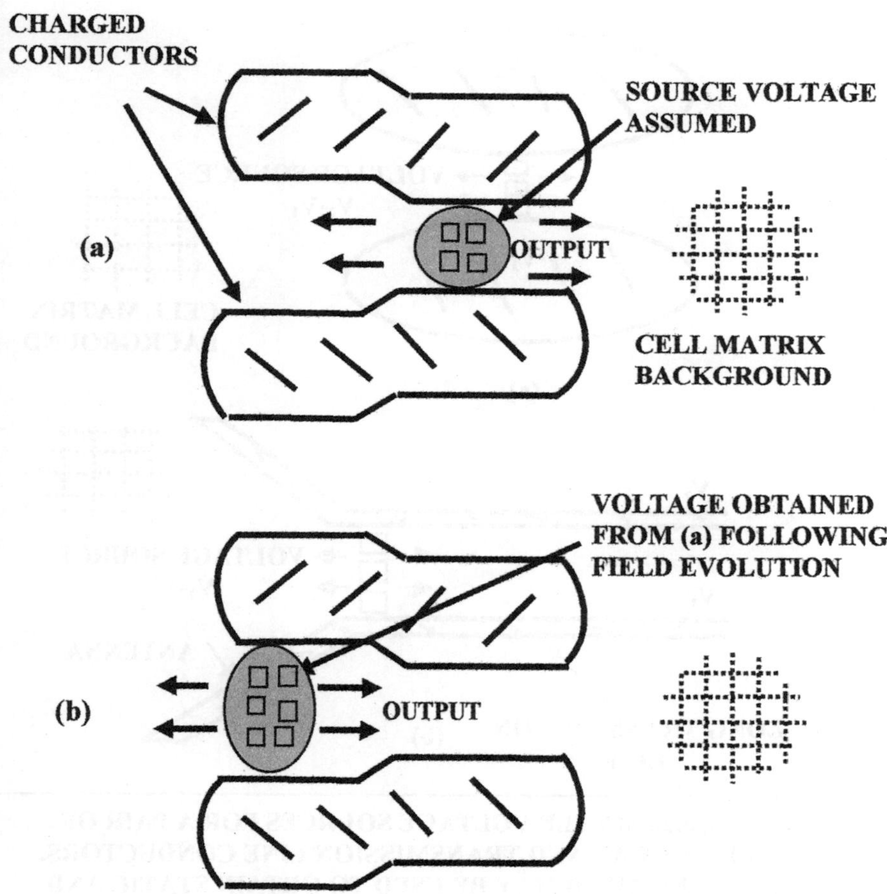

FIG. 5.19 STATIC SOLUTION BASED ON ASSUMED
VOLTAGE SOURCE AT REGION A. RESULTANT FIELDS
IN REGION B THEN CAN SERVE AS NEW SOURCE TO
CHECK INITIAL FIELDS AT A.

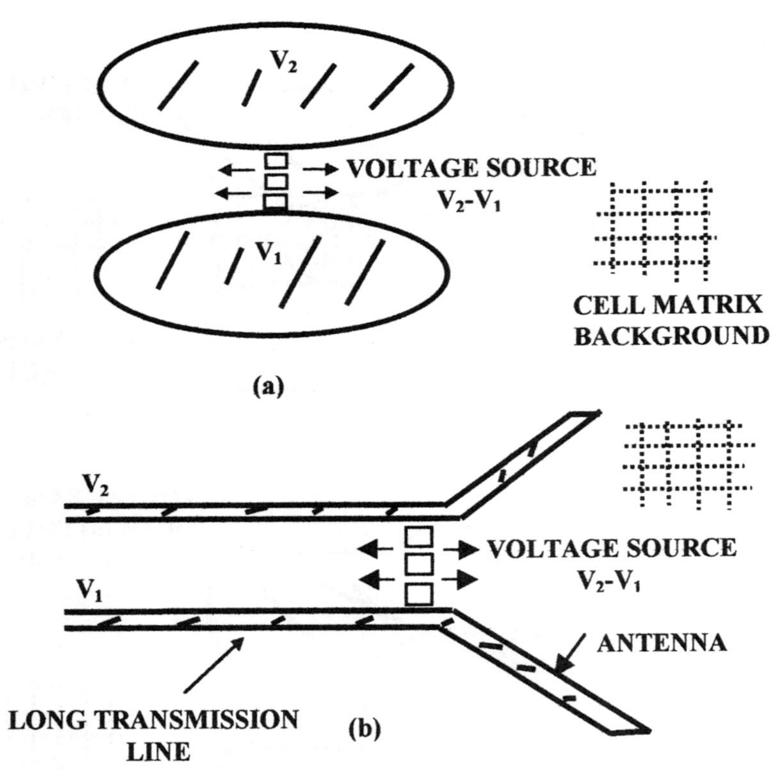

FIG. 5.20 SIMPLE VOLTAGE SOURCES FOR A PAIR OF ELLIPTICAL AND TRANSMISSION LINE CONDUCTORS. TLM METHOD MAY BE USED TO OBTAIN STATIC AND TRANSIENT SOLUTIONS.

long transmission line. Here again the field is assumed to spatially uniform, provided we ignore fringing This effect may be minimized if we locate the source voltage region further inside the transmission line. After the charge up process, which establishes the static field, the voltage source may remain in place if desired and if this conforms to the actual experimental condition. This is typical in the case of a combined antenna/switch charged up by an input transmission line. Upon initiation of the switch, the initial static fields will be disturbed, but we may use the same cell matrix as before to track the transient behavior. Reflections back into the infinite line may absorbed by placing terminating resistor loads further upstream in the line, before they reach the source region.

5.13 Time Varying Source Voltage and Antenna Simulation

In the previous Section we employed voltage sources constant in time , which made sense since we were interested in finding the static solution. In the event that the voltage source varies in time, however, then a radiative field pattern (rather than the static field) will emerge. We therefore deal with an antenna, and at large distances from the source, the radiated power away will be significant. In order to keep the simulation to manageable dimensions, matched terminations are placed at the outer boundaries, once the major trends with distance have been established. An iterative TLM example discussed in Chapter VII, for example, is easily modified so that the input represents a time varying function. A powerful advantage in using the TLM method is that the same iteration handles simultaneously the combined problem of an antenna(i.e., a time varying voltage source) while taking into account the introduction of conductivity into the medium.

Dispersion

Up until this point one might have gained the impression that the TLM method is a conceptually simple yet powerful tool , providing us with the means to solve a multitude of electromagnetic problems while at the same time providing physically intuitive framework for understanding a multitude of phenomena. For the most part, this impression is correct, but with the incorporation of dispersion into TLM method, the technique becomes a good deal more complicated. The additional complications, of course, are not native to TLM, but to any technique for dealing with dispersion. The problem is the immense amount of additional information needed to characterize dispersion, placing great demands on both computer speed and memory. Besides the needs for the additional vast amount of information, the TLM model becomes more cumbersome and its interpretation loses much of its simplicity. Other complications arise which are subsequently discussed.

There are two methods for incorporating dispersion into the TLM formulation. With the first we select a single matrix with sufficient resolution such that any variations in the field are small when observed over the selected TLM length. Thus the single matrix can accommodate all the variations , whether slow or fast, throughout the propagation region. To obtain the Fourier components, a numerical method is used which lends itself to computer methods. For convenience, the sampling length for the Transform is selected to be the same as the TLM element. Following the Fourier transformation, we deploy the same TLM matrix throughout the propagation region. There are then two interpretations for the field inside the TLM element. One is the usual constant field approximation in the TLM(representing a sort of average value) which we have employed throughout the book until this Chapter. The second interpretation allows for the variation of the field within the TLM element. The field variation is assembled from the calculated Fourier components. Which interpretation is used will be immediately evident from the context of the discussion; in the following discussion usually we will regard the fields as nonuniform within the TLM element.

During the ensuing time step(or time steps) we allow the Fourier components to move from one cell to the next, with each component having its own propagation velocity, and its own scattering properties, as dictated by dispersion. The new field , resulting from both dispersive and non-dispersive effects, is then reassembled using inverse Fourier Transform techniques. This new field is then used to produce a new set of Fourier components, and the process is thus repeated. When the final time step is attained, the non-uniform field within each cell may be replaced if desired by a constant value, representing some suitable average.

The second method for calculating dispersive effects is very similar to the first method just described except that we use two or more overlapping matrices, with each matrix representing a well defined frequency or spatial region with distinct or simplifying dispersive properties. The method is best illustrated with the interacting EM and light signals discussed in Chapter II(see Figs.2.18 and 2.19 which show a pair of 1D and 2D matrices respectively). The EM matrix, with the larger cells, will often display very little dispersion, while the light matrix , with its small matrix due to its smaller wavelength, will often have significant dispersive properties. The main advantage here is that it is unnecessary to use the same high resolution matrix for the EM signal; a lower resolution matrix may be all that is required. A separate matrix for each signal may be used, while allowing the matrices to interact at the nodes using either a nearest node or multi-node approximation. Separate Fourier transformations would be carried out for each matrix.

In the ensuing discussion we will employ the use of a single, high resolution matrix(Method I) since this simplifies the discussion. One should bear in mind, however, that the mult-matrix approach may be attractive when considering interacting signals with distinct dispersive properties.

Whatever method is used, it will be strongly dependent on the type dispersion present. Before proceeding, we will discuss dispersion sources found in electromagnetic media. One dimensional arguments will be used exclusively; the extension to 2D and 3D do not involve any additional concepts.

5.14 Dispersion Sources

Two sources of dispersion are considered in the TLM method. As one might guess the sources arise from the two elements which comprise the TLM matrix: the transmission line and the node. In the case of the transmission line, the propagation velocity v will be dispersive and this will be denoted by either $v(\omega)$ or $v(k)$ where ω is the frequency, and k is the propagation constant, and are related by $v(\omega)= v(k) = \omega/k.$ $k =2\pi/\lambda$ where λ is the wavelength in the medium. We ascribe the velocity dispersion to its dependence on the dielectric,

$\varepsilon(\omega)$, which is likewise dispersive and satisfies $v(\omega) = \{\mu\varepsilon(\omega)\}^{-1/2}$. The second dispersion resides in the node resistor and this is denoted by either $R(\omega)$ or $R(k)$. A more comprehensive labeling is given by $R(\omega,n,m,q)$ where the first index denotes the dispersion and the next three denote the location of the node.

Although we will ascribe the dispersive propagation velocity as a property of the line, we should point out that we may achieve the same effect by formally assigning a dispersive phase delay (or advance)to the property of the node. In so doing we may interpret the node as the origin of all important physical changes, while the transmission line is still responsible for the (non-dispersive)energy spread.

5.15 Dispersion Example

Examples of the dispersive properties of the dielectric constant $\varepsilon(\omega)$, as well as the node resistance, $R(\omega)$, may be demonstrated from the   wave equation, Eq.(1.2), assuming an $EXPj\{Kx-\omega t\}$ dependence, where ω is the radian frequency and $\mathbf{K}$ is a complex propagation constant. Substituting into the wave equation gives

$$\{\mu\varepsilon\omega^2 + j\omega\mu\sigma\}\mathbf{E} \; = \; \mathbf{K}^2\mathbf{E} \tag{5.15}$$

and therefore

$$\mathbf{K} \; = \; \alpha + j\beta = (\omega/c)\kappa^{1/2}[1+j\sigma/\varepsilon\omega]^{1/2} = (\omega/c)[\kappa+j\sigma/\varepsilon_o\omega]^{1/2} \tag{5.16}$$

and $\kappa = \varepsilon/\varepsilon_0$ is the specific inductive capacity.

As noted from Eq.(5.16), $\mathbf{K}$ has both real and imaginary parts and both are dispersive. A serious deficiency exists with Eq.(5.16) , however. The dispersion is based on the macroscopic, average fields, which are incomplete when describing dispersion phenomena(see , for example, References [1],[2]). The dispersive properties of σ and κ are lacking, and these can only be obtained from a knowledge of the localized electric field. With these localized fields, one can calculate the displacement of charge(including loss) within atomic systems, from which the dispersive polarization follows. From this we obtain the dispersion of the complex inductive capacity. Once we obtain the proper dispersion relation, the complex inductive capacity may be used to obtain $\mathbf{K}$, which may be compared with Eq.(5.16) to obtain the dispersive relationships for κ and σ.

Using he localized field approach[1], we can obtain the polarization P, which is now dispersive, as a function of the macroscopic field E. Thus

$$P(\omega)=a^2\varepsilon_o\mathbf{E}/(\omega_R^2-\omega^2-j\omega g) \tag{5.17}$$

adopting the notation in [1]. a^2 is proportional to the number of oscillators per unit volume, each with resonant frequency ω_o, g is the dissipative term, and ω_R is related to ω_o by $\omega^2_R = \omega^2_o -(1/3)a^2$. The phenomenological constants ω_R ,g , and a are determined from experiment. If κ' is considered to be a complex specific inductive capacity, then

$$P = [\,\kappa'-1]\varepsilon_o\mathbf{E} \tag{5.18}$$

Comparison of Eqs.(5.17) and (5.18) then provides

$$\kappa' = 1+\{a^2 / [\omega^2_R -\omega^2-j\omega g]\} \tag{5.19}$$

In terms of the complex propagation constant

$$\mathbf{K} = \alpha+j\beta = (\omega/c)\kappa'^{1/2} = (\omega/c)[1+a^2/(\omega^2_R - \omega^2 -j\omega g)]^{1/2} \tag{5.20}$$

We then separate the terms inside the radical into real and imaginary parts, and assume that the dissipative term g is much smaller than ω and $(\omega_R -\omega)$. With these assumptions,

$$\mathbf{K} = \alpha+j\beta = (\omega/c)[1+\{a^2/(\omega^2_R-\omega^2)\}+ja^2\omega g/(\omega^2_R-\omega^2)^2]^{1/2} \tag{5.21}$$

and the comparison with the macroscopic relation, Eq.(5.16), gives, for *small* losses,

$$\kappa(\omega) \;=\; 1+a^2/(\omega^2{}_R-\omega^2) \tag{5.22}$$

$$\sigma(\omega) \;=\; \varepsilon_o a^2\omega^2 g/(\omega^2{}_R-\omega^2)^2 \tag{5.23}$$

Eqs.(5.22)-(5.23) give the inductive capacity and conductivity dispersions in terms of the phenomenological constants. Note that when $\omega\to 0$, the zero frequency expression for $\kappa(\omega)$ becomes

$$\kappa(\omega{=}0) \;=\; \kappa(0) \;=\; 1+a^2/\omega^2{}_R \tag{5.24}$$

In addition, $\sigma(\omega{=}0){=}0$ (this is true even when losses are not assumed small). The fact that $\sigma(\omega){=}0$ at $\omega =0$ is an outcome of the phenomenological assumptions ; a low frequency model to augment Eq.(5.23), therefore, may be desirable.

We now return to Eq.(5.21) and decompose the real and imaginary parts, α and β, thus giving

$$\alpha(\omega) \;=\; (\omega/c)[1+ a^2/(\omega^2{}_R-\omega^2) \,]^{1/2}[[(1+A^2(\omega))^{1/2}+1]/2]^{1/2} \tag{5.25}$$

$$\beta(\omega) \;=\; (\omega/c)[1+ a^2/(\omega^2{}_R-\omega^2) \,]^{1/2}[[(1+A^2(\omega))^{1/2}-1]/2]^{1/2} \tag{5.26}$$

where

$$A(\omega) = a^2\omega g/[(\omega^2{}_R-\omega^2)^2+a^2(\omega^2{}_R-\omega^2)] \tag{5.27}$$

We can again make use of the approximation, $g \ll \omega$, and $g\ll (\omega_R -\omega)$, so that

$$\alpha(\omega) =(\omega/c)[1+a^2/(\omega^2{}_R-\omega^2)]^{1/2}[1+A^2(\omega)/8] \tag{5.28}$$

$$\beta(\omega) = (\omega/c)[1+a^2/(\omega^2{}_R-\omega^2)]^{1/2}[(A(\omega)/2)] \tag{5.29}$$

For completeness we express $\alpha(\omega)$ and $\beta(\omega)$ in terms of the conductivity and specific capacity, using Eqs.(5.22)-(5.23), or

$$\alpha(\omega)=(\omega/c)\kappa^{1/2}(\omega)[1+(1/8)\sigma^2(\omega)(\omega^2_R-\omega^2)^4/[\varepsilon_o^2\omega^2\{(\omega^2_R-\omega^2)^2+a^2(\omega^2_R-\omega^2)\}]^2]$$

(5.30a)

which simplifies , using Eq.(5.22), to

$$\alpha(\omega)=(\omega/c)\kappa^{1/2}(\omega)[1+(1/8)(\sigma^2(\omega)/\varepsilon_o^2\omega^2\kappa^2(\omega))]$$

(5.30b)

Similarly,

$$\beta(\omega) \quad = \quad (1/c)\kappa^{1/2}(\omega)[\{\sigma(\omega)(\omega^2_R-\omega^2)^2/2\varepsilon_o\{(\omega^2_R-\omega^2)^2+a^2(\omega^2_R-\omega^2)\}]$$

(5.31a)

or

$$\beta(\omega) = \sigma(\omega)[\, 2\varepsilon_o c\kappa^{1/2}(\omega)]^{-1}$$

(5.31b)

If Eqs.(5.30)-(5.31) are used one must make sure that the actual dispersion for $\sigma(\omega)$ and $\kappa(\omega)$, from Eqs.(5.22)-(5.23), are kept in mind. Two obvious points concerning $\alpha(\omega)$ and $\beta(\,\omega\,)$ should be cited: as the dissipative term goes to zero, $g\to 0$, then $\alpha \to (\omega/c)\,\kappa\,(\omega)^{1/2}$ and $\beta \to 0$, as expected. Secondly, Eqs.(5.30b) and (5.31b) reduce to Eq.(5.16) (after decomposition and allowing for small losses), as is required.

At this point we have sufficient information to obtain the propagation velocity $v(\omega)$ and the node resistance $R(\omega)$, both of which will be dispersive. First we consider $v(\omega)$, again assuming small losses. By definition, $v(\omega)$ is

$$v(\omega) = \omega/\alpha(\omega)$$

(5.32)

or more explicitly

$$v(\omega) = c\kappa^{-1/2}(\omega)[1+(1/8)(\sigma^2(\omega)/\varepsilon_o^2\omega^2\kappa^2(\omega))]^{-1}$$

(5.33)

For zero loss, the velocity is simply , from Eq.(5.22),

$$v(\,\omega\,) = c/\,\kappa^{1/2}(\omega) = c\,/[1+a^2/(\omega^2_R-\omega^2)]^{1/2}$$

(5.34a)

We note that as ω increases , the velocity decreases. As $\omega \to 0$, the velocity $v(\omega)$ approaches its zero frequency value of $c/\,\kappa^{1/2}(0)$,

$$v(0) = c/[1+a^2/\omega_R^2]^{1/2} \tag{5.34b}$$

5.16 Propagation Velocity in Terms of Wave Number

As we shall see in a moment, it is often useful to express the propagation velocity, Eq.(5.33), in terms of the propagation constant. For small losses and low frequency, substitution of $\omega \sim kv(k) \sim kv(0)$ into Eq.(5.33) approximates $v(k)$ in terms of k. For *zero* loss and low frequency $v(k)$, from Eq.(5.34a), converts to

$$v(k) = c[(1+a^2/\omega^2{}_R)+v^2(0)k^2a^2/\omega^4{}_R]^{-1/2} \tag{5.35}$$

which with further approximation gives

$$v(k) = v(0) [1- (1/2)v^2 (0) k^2a^2/\omega^4{}_R\kappa(0)] \tag{5.36}$$

Note that with the given phenomenological model and approximations, $v(k)$ decreases with k from its initial value of $v(0)$.

5.17 Dispersive Properties of Node Resistance

To obtain the dispersive $R(\omega)$, we must first examine $\beta(\omega)$, which indicates the degree of attenuation of the wave, with a dependence of $EXP[- \beta(\omega)x]$. If we choose a sufficiently small cell size, then we approximate, to first order, the exponential decay of the wave by $1- \beta(\omega)\Delta l$. From Chapter 1, however, we know from the TLM theory that the wave transmission, to first order, is governed by

$${}^{+}V(n+1) = {}^{+}V(n)T \tag{5.37}$$

where

$$T = 1- Z_0/2R(\omega) \tag{5.38}$$

We now are able to identify $Z_0/2R(\omega)$ with $\beta(\omega)\Delta l$. Equating the two terms, we obtain for $R(\omega)$

$$R(\omega) = Z_o/\, 2\beta(\omega)\Delta l \qquad (5.39)$$

with $\beta(\omega)$ given by Eq.(5.29). For small loss, $R(\omega)$ of course becomes very large. With increasing ω, as $\beta(\omega)$ increases, there is a corresponding decrease in $R(\omega)$.

5.18 Node Resistance in Terms of Wave Number

As with $v(\omega)$, we can likewise express $\beta(\omega)$, and therefore $R(\omega)$, in terms of the propagation constant k rather than the frequency. For small losses and low frequency, the changeover is simple ; we set the propagation constant equal to $k=\omega/v(0)$ in Eq.(5.31). $R(\omega)$ then becomes

$$R(k) = Z_o/\, (2\beta(k)\Delta l) \qquad (5.40)$$

where

$$\beta(k) = \kappa^{1/2}(k)\, \sigma(k)(\omega^2_R - v^2(0)k^2)^2\, /P(\omega) \qquad (5.41a)$$

$$P(\omega) = 2\varepsilon_o c[(\omega^2_R - v^2(0)k^2)^2 + a^2(\omega^2_R - v^2(0)k^2)] \qquad (5.41b)$$

and $\kappa(k)$, $\sigma(k)$ are obtained from Eqs.(5.22)-(5.23)

$$\kappa(k) = 1 + a^2/(\omega^2_R - v^2(0)k^2) \qquad (5.42a)$$
$$\sigma(k) = \varepsilon_o a^2 v^2(0)k^2 g/(\omega^2_R - v^2(0)k^2)^2 \qquad (5.42b)$$

Once more , from Eqs.(5.40)-(5.42), $\beta(k)$ may be written as

$$\beta(k) = \sigma(k)/[\, 2\varepsilon_o c\, \kappa^{1/2}(k)] \qquad (5.43)$$

Since

$$Z_o = (\mu/\varepsilon(k))^{1/2} = (\mu/\varepsilon_o\kappa(k))^{1/2} \qquad (5.44)$$

Eqs.(5.40) and (5.43) - (5.44) combine to give

$$R(k) = 1/\sigma(k)\Delta l \qquad (5.45a)$$

In terms of the modelling parameters for $\sigma(k)$, $R(k)$ becomes

$$R(k)= (1/\Delta l)[(\omega^2_R - v^2(0)k^2)^2/ \varepsilon_o a^2 v^2(0)k^2 g] \qquad (5.45b)$$

The result , $R(k) = 1/\sigma(k)\Delta l$, is exactly the same as that obtained for the non-dispersive case, indicating that we could just as well have inserted the dispersive $\sigma(k)$, Eq.(5.42b), into $R(k) = 1/\sigma(k)\Delta l$.

   In summary, we have provided examples of dispersion , based on a specific polarization model, for the propagation velocity and node resistance. The results, expressed in both wavenumber and frequency, may then be incorporated into the TLM formulation.

5.19 Anomalous Dispersion

In the previous discussion we have for the most part made the assumptions that $g << (\omega_R - \omega)$ and $g << \omega$. In this region the phase velocity decreases with increasing ω. Are there regions in which the velocity increases as ω grows? The answer is affirmative for regions where ω begins its approach toward the resonant frequency The region here is "anomalous", and the velocity, after first increasing, will begin to decline, even falling below its original value at $\omega =0$. It is quite possible, therefore, that for certain dispersion problems, frequency regions will exist in which the velocity is both smaller and larger than the original velocity at $\omega =0$. Several references discuss the more general dispersion equations for $v(\omega)$, when ω is located in the vicinity of a resonance frequency [1] , [2].

   We hasten to add that as ω becomes enveloped in the resonance region, with ω very close to the resonant frequency, the phase velocity(and even the group velocity) lose their usual meaning. In this discussion, we assume ω is still far enough from resonance, such that the phase velocity is still meaningful.

Incorporation of Dispersion into TLM Formulation

5.20 Dispersion Approximations

From the previous discussion we may infer that dispersion introduces another complication, related to the non-uniformity of the cell field, into the iteration process. Without dispersion the wave energy travels exactly one cell length for a given time step. By adding dispersion, however, certain Fourier components may travel only a fraction of a cell length while others may travel several cell lengths. Indeed, in the previous discussion, we saw that for "normal" dispersion, where $v(\omega) \leq v(0)$ at the lower frequencies, the wave components will travel a cell length or less. For the " anomalous" case, on the other hand, $v(\omega) > v(0)$ as ω approaches ω_R, and thus some components will travel more than one cell length. In the ensuing discussion we adopt a matrix utilizing the cell length determined by the "zero frequency " velocity , $v(0)$, so that $\Delta l = v(0)\Delta t$. We then continue the approximation by confining ourselves to two Categories of dispersion. In both categories we will tacitly assume the maximum velocity excursion , Δv_o is always much less than $v(0)$. Δv_o is assumed to be a dispersionless constant. To simplify matters we also assume in the following that the neighboring cells are identical, so that $v(k)$ is the same in each cell. Differing cells will of course modify the calculation, but not alter the basic technique. In Category I the velocity is always less than or equal to $v(0)$, so the individual wave components never travel more than one cell length during the time step Δt. We express the velocity limits in terms of both ω and the propagation constant k:

CATEGORY 1

$$v(0) - \Delta v_o \leq v(\omega) \leq v(0) \tag{5.46}$$

In Category II we allow for excursions above and below $v(0)$. The limits of the velocity likewise may be expressed as

CATEGORY II

$$v(0)- \Delta v_o \leq v(\omega) \leq v(0)+ \Delta v_o \qquad (5.47)$$

In Category II we will have to consider waves originating not only from the immediate cell neighbors but also from neighbors two cell lengths away. For convenience we have selected the magnitude of the positive excursion to be the same as the negative one, i.e., Δv_o. In general these excursions will of course differ.

It is natural to inquire why we should not forego the cell length of $v(0)$ Δt and simply select a length of $[v(0)+ \Delta v_o]$ Δt, thereby requiring only a Category I treatment. It is certainly allowable to introduce a matrix based on a an extreme value of the dispersion. By considering the dispersion as a perturbative effect, however, the natural starting point is the non-dispersive matrix with velocity equal to $v(0)$. This starting point is particularly convenient if the propagation constant components for k $\sim$ 0 dominate, in which case the results will be considerably easier to interpret.

5.21 Outline of Dispersion Calculation Using the TLM Method

The iterative calculation starts with the field profile in the propagation region at a given time step. At the beginning of each iteration, the total field in each cell will consist of forward and backward waves each of which will, in general, be nonuniform. Both waves must be treated separately and the results combined at the completion of the step. At the start of a particular time step, the Fourier Transform is obtained for the wave profile in the propagation region, using a numerical procedure. The procedure means segmenting the propagation region into N_F sampling elements ,where N_F is a positive integer, with each element making a contribution to the Transform. In this case we simply select the element to be identical with the TLM length; of course there is no requirement for setting the sampling element equal to the TLM length and indeed smaller elements may be used to enhance accuracy. Once the complete Transform is obtained, then, over the next time step, we allow each Fourier component to have a differing propagation velocity and for the node resistor to dispersively modify each of the

Fourier components as well. Once the iterative step is completed we then perform an inverse transform, returning to the original coordinate space, thus showing how the original profile evolves when dispersion is included. For the inverse Transform, we again use a numerical procedure, but impose maximum and minimum limits on k, so as to facilitate the calculation. Because of dispersion, the original function is allowed to "spread out" a maximum of two cells in either direction, since we are allowing both Categories I and II dispersions.

5.22 One Dimensional Dispersion Iteration

The dispersion calculation is illustrated in one dimension. The extension to two and three dimensions does not involve any additional, new concepts. As mentioned before, we limit the dispersion to two types: Category I , $(v(0)- \Delta v_o)$ $\leq v(k) \leq v(0)$, and Category II, $(v(0)- \Delta v_o) \leq v(k) \leq (v(0)+ \Delta v_o)$. Although we allow $v(k)$ to vary only slightly from $v(0)$, i.e., $v(k)$ is closely centered about $v(0)$. In order to proceed further we need to trace the development of the waves for both Categories. Fig.5.21 shows the evolution for the first Category, where $^+V(n)$ and $^-V(n)$ are the forward and backward waves in the nth cell during the kth time step. Note that there are cells on either side, (n-1) and (n+1), with similar waves, since we still allow contributions from waves in these cells. The accompanying Table gives the origin of the forward wave in the nth cell during the (k+1)th time step. Just as in the non-dispersive case, we will have contributions from the forward wave, transmitted from the (n-1) cell, as well as a reflected wave emanating from the backward wave in the nth cell. In addition, however, and this is entirely due to dispersion, we will have a contribution from the forward wave in the nth cell, resulting from the fact that for some wave numbers $v(k)<v(0)$ A similar Table showing the origin of the backward wave in the nth cell for the (k+1) th time step.

A similar description of the wave development for the Category II dispersion is shown in Fig.5.22. Note that we include two cells on either side

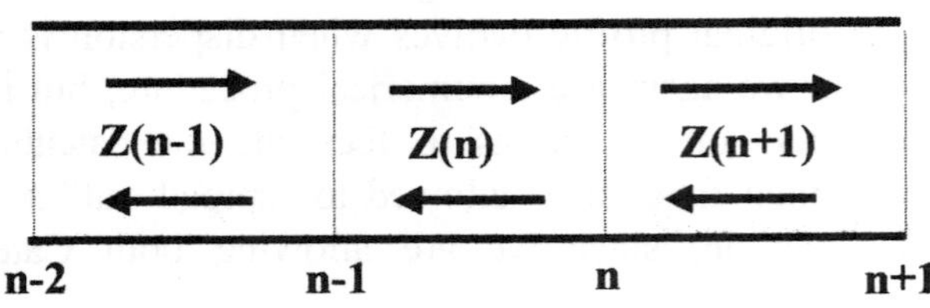

FIG. 5.21 CONTRIBUTIONS TO FORWARD WAVE IN nth CELL DUE TO DISPERSION: CATEGORY I, $\Delta l = v(0)\Delta t$, $v(0) - \Delta v_o \leq v(\omega) \leq v(0)$.

FORWARD WAVE $^+V(n)$

ORIGIN OF WAVE	SCATTERING PROCESSES
$^-V(n)$	B(n-1,2)
$^+V(n-1)$	T(n-1,2)
$^+V(n)$	NONE

BACKWARD WAVE $^-V(n)$

ORIGIN OF WAVE	SCATTERING PROCESSES
$^-V(n+1)$	T(n,2)
$^+V(n)$	B(n,1)
$^-V(n)$	NONE

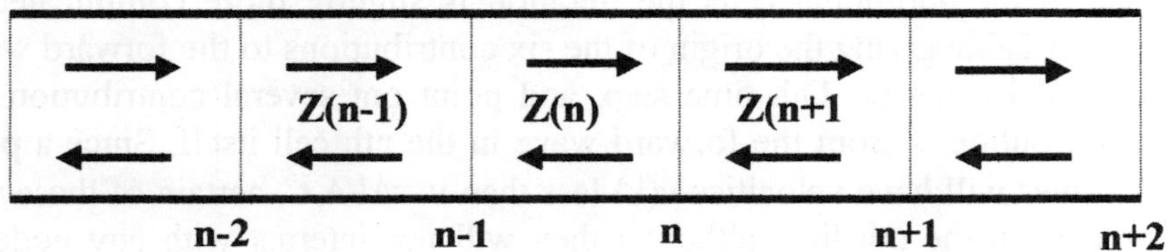

FIG. 5.22 CONTRIBUTIONS TO WAVE IN nth CELL DUE TO DISPERSION: CATEGORY II, $\Delta l = v(0)\Delta t$, $v(0)-\Delta v_o \leq v(\omega) \leq v(0)+\Delta v_o$.

FORWARD WAVE $^+V(n)$

ORIGIN OF WAVE	SCATTERING PROCESSES
$^+V(n)$	NONE
$^-V(n)$	$B(n-1,2)$
$^+V(n-1)$	$T(n-1,1)$
$^+V(n-2)$	$T(n-2,1)$, $T(n-1,1)$
$^-V(n-1)$	$B(n-2,2)$, $T(n-1,1)$
$^-V(n+1)$	$T(n,2)$, $B(n-1,2)$

BACKWARD WAVE $^-V(n)$

ORIGIN OF WAVE	SCATTERING PROCESSES
$^-V(n)$	NONE
$^+V(n)$	$B(n,1)$
$^-V(n+1)$	$T(n,2)$
$^+V(n+2)$	$T(n+1,2)$, $T(n,2)$
$^+V(n+1)$	$B(n+1,1)$, $T(n,2)$
$^+V(n-1)$	$T(n-1,2)$, $B(n,1)$

of the nth cell since we allow for contributions from the (n-2) and (n+2)th cells. Because of the additional cells the iteration is slightly more complicated. We construct a Table giving the origin of the six contributions to the forward wave in the nth cell for the (k+1)th time step, and point out several contributions. The first contribution is from the forward wave in the nth cell itself. Since a portion of the waves will have velocities v(k) less than v =Δl/Δ t , certain of these waves will remain in the nth line, although they will not interact with any node. The second contribution is from the backward wave in the nth cell wherein a portion of the wave is reflected from the R(k,n-1) node, thus contributing to the forward wave. As a final example we note the fifth contribution, where the original wave is a backward one in the (n-2)th cell, undergoing a partial reflection at the (n-2) node and then a partial transmission of the forward wave at the (n-1)th node. A similar Table is given for the backward wave contributions.

In the following we illustrate the calculations involving contributions to the forward and backward waves in the nth cell , during the k+1 step, based on the waves in the neighboring cells existing during the kth time step. For our elementary coordinate space we use five cells centered about the n th cell, i.e., cells ranging from n-2 to n+1. The five cells will allow us to consider both categories of dispersion, I and II. At this point we are prepared to write down the dispersive iteration for the nth cell. Because there are both forward and backward waves, the amount of computation is increased. For example, in Category I there are three contributions each for the forward and backward waves, for a total of six separate but similar calculations which must be performed. For Category II a total of 12 calculations must be performed. Rather than calculate all the possibilities, we outline the calculation for a single type of contribution, and afterwards point out some of the differences which occur when other contributions are present.

To illustrate the calculation we select the second entry in Fig.5.21 which represents contributions to the forward wave in the nth cell wherein the forward wave in the n-1 cell undergoes a partial transmission via the n-1 node into the nth cell. The assumed forward wave in the (n-1) cell propagates during the kth time step. If we wish we may shift the coordinate x axis so that x=0 coincides with either the start of the n-2 or n-1 cells, or for that matter the start of any of the cells.

To proceed with the calculation the element must be described completely, during the prior time step, in either the coordinate space or the corresponding Fourier transformed space (k space). If, somehow, we know the functional dependence of the wave, then we may obtain the transform using numerical techniques, and thereafter allow the Fourier components to interact with the other elements. We can remain in k space as long as we like(for as many time steps as we wish), transforming back to coordinate space when necessary. An important simplification may be invoked when static initial conditions prevail. Under these typical conditions the spatial profile is assumed constant in each TLM element. Initial, static conditions and the impact of dispersion are discussed in a later Section.

We proceed to obtain the Fourier Transform of the initial field in the propagation region. In particular, we first consider the forward wave, designated by $^{+}V(x)$, where x is an arbitrary coordinate in the propagation region, and we have omitted the time step superscript(note the difference from $^{+}V(n)$, which is the field in the nth cell). The Fourier transform for $^{+}V(x)$, designated by F(k) is[3]

$$F(k) = \int_{-\infty}^{\infty}{}^{+}V(x)\ \ EXP[\text{-}jkx]dx\ =\ \int_{xo}^{Lo}{}^{+}V(x)\ EXP[\text{-}jkx]dx \qquad (5.48)$$

where the second integral arises since x_o is the assumed start of the propagation region, and the end is L_o. In general it will not be possible to calculate the Transform in closed form and it is necessary to use numerical techniques. Accordingly we briefly describe a polygonal approximation [9] of $^{+}V(x)$ obtained by dividing the region into N_F points, equally spaced, and connecting them with straight line segments. Outside the region we assume $^{+}V(x)$ vanishes. We select the spacing between points to be Δl, the TLM length. The nth point coordinate x_n is thus

$$x_n = n\Delta l + x_o\ ;\ \ n=0,1,2,3,.....N_F \qquad (5.49)$$

The length of the propagation region $L_o\text{-}x_o$ satisfies

$$L_o\text{-}x_o = N_F\Delta l \qquad (5.50)$$

As mentioned before, $^+V(x) = 0$ when $x < x_o$ and $x > L_o$. Having defined the notation, we next employ the polygonal technique to obtain the Fourier Transform $F(k)$. In each element Δl we first calculate the two slopes about x_n, $[^+V(x_{n+1}) - {}^+V(x_n)]/\Delta l$ and $[^+V(x_n) - {}^+V(x_{n-1})]/\Delta l$ ($V(x_n)$, e.g, is the value of $V(x)$ at the edge of the nth cell). We then calculate essentially the the second derivative (without the additional Δl), given by the difference term $[^+V(x_{n+1}) - {}^+V(x_{n-1})]/\Delta l$. This difference is then utilized in the expression for $F(k)$ [3],

$$F(k) = (n=1 \text{ to } N_F) \ -\Sigma\ [(^+V(x_{n+1}) - {}^+V(x_{n-1}))/\Delta l][(1/k^2)EXP(-jkx_n)] \qquad (5.51)$$

It is convenient to cast Eq.(5.51) into polar form,

$$F(k) = {}^+A(k)\ EXP)\ [j\varphi(k)] \qquad (5.52)$$

where $A(k)$ is the amplitude and $\varphi(k)$ the phase angle. These are related to the real and imaginary parts of $F(k)$, Re $F(k) \equiv \zeta(k)$, Im $F(k) \equiv X(k)$, using Eq.(5.52). The connections are

$$^+A(k) = [\zeta^2(k) + X^2(k)]^{1/2} \qquad (5.53)$$

$$\tan\varphi(k) = X(k)/\zeta(k) \qquad (5.54)$$

We now return to the *cell description* of the fields and Fourier components. The nth cell is of course enclosed by the points x_n and $x_{n-1,}$ used in calculating the Transform. We now determine how the profile changes if we allow each of the Fourier components to propagate(in the forward direction) for one time step Δt. We proceed using a constant phase approximation. We assume that when a cell node is encountered, and the wave undergoes reflection and transmittal, that only the amplitude is affected and the phase portion is approximately constant. This assumption is reasonable since the node has no reactive components and by its very definition is infinitesimal in length. Thus the amplitude function $^+A(k)$ will undergo a change, but only for those forward waves which find themselves in the nth cell after the time step has elapsed. If $^+A(n-1,k)$ and $^+A(n,k)$ are the amplitudes in the cell lines n-1 and n respectively, and $\varphi(n-1,k)$, $\varphi(n,k)$ are the respective phase angles then the two are related by

$$^{+}A(n,k) \;\; = T(n-1) \,^{+}A\,(n-1,k) \tag{5.55}$$

$$\tan\varphi(n,k) \sim \tan\varphi(n-1,k) \tag{5.56}$$

Note that the phase factor is assumed to be unchanged as it traverses the n-1 node. $T(n-1)$ is the usual transmission coefficient at node n-1, given by(Chapter III).

$$T(n-1) = \; 2\,R(n-1)\,/\,(Z(n-1)\,+R(n-1)) \tag{5.57}$$

In order to discuss the wave delay and dispersion in the TLM line, we require the inverse Fourier Transform [9}for the wave in the (n-1)th cell, $V(n-1,x)$,

$$^{+}V(n-1,x) \;\; = (1/\pi) \int_{0}^{\infty} {}^{+}A(n-1,k)\cos(kx+\varphi(n-1,k))dk \tag{5.58}$$

The inverse transform has a somewhat simplified form because of the assumed reality of $^{+}V\,(n-1,x)$. As a consequence the integration range need only extend from 0 to ∞ .

Next we determine how to incorporate the phase changes due to the wave motion in the TLM line. The wave velocity is given by $v(k)$ with the argument indicating that we allow the velocity to have dispersion. In order to account for the phase change , we replace $\cos(kx+\varphi(n-1,k))$ with $\cos[kx\,-kv(k)t\,+\varphi(n-1,k)]$, which represents wave motion toward increasing x. Since the wave velocity is dispersive, we should expect the various Fourier components will fall out of phase with one another and the pulse will smear out as it propagates. Eq.(5.58) then becomes

$$^{+}V(n-1,x) \;\; = (1/\pi\,)\int_{0}^{\infty} {}^{+}A(n-1,k)\cos[kx\,-kv(k)t\,+\varphi(n-1,k)]dk \tag{5.59}$$

We should note that within a given cell line the kth wave moves with velocity $v(k)$ and the distance traversed δx in time δt are related by $\delta x =v(k)\delta t$. Also, when there is no dispersion, $v(k)=v(0)$ and after Δt has elapsed, $v(0)\,\Delta t =\Delta l$.

Following the time step, Δt , the wave in cell (n-1) will, for the most part move into cell n, except for effects due to dispersion. Because of dispersion a small portion of the wave may remain in cell (n-1) or may skip over into cell (n+1) if Category II dispersion exists. For the present we focus on the portion of the wave carried into cell n.

Once we perform the spatial decomposition in each cell, we then focus on the behavior of each of the waves with propagation constant k. We focus on the wave amplitude at time t in cell (n-1), denoted by $^{+}A(n-1,t, x_i, k_i)$. The arguments of ^{+}A are explained: n-1 denotes the cell (n-1) and t is the time. x_i is the position in the (n-1) cell. i is a positive integer which specifies the ith sub-element in the cell. We assume each cell is subdivided into N_C (a positive integer)sub-elements with width $\Delta l/N_C$. For the n-1 cell , x_i then equals $x_{n-2} + i(\Delta l/N_C)$. The subscripted wavenumber k_i requires further explanation as outlined in the following.

In order to perform the inverse transformation we again utilize numerical techniques, first dividing up the k space , which describes $^{+}A(n-1,k)$, into N_k equally spaced values. We then define the allowed values of k_i .

$$k_i = (i-1)\Delta k, \quad i=1,2,3,...N_k , \quad \Delta k = k_c/(N_k-1) \tag{5.60}$$

k_c is the cutoff for the wavenumber k. k_c may be estimated from analysis of $^{+}A(n-1,k)$, or from the following assumptions. We assume the high frequency components will be determined by any variations in the single cell. Since the waves are approximately uniform in the cell, we can approximate the upper limit of the wave as that used to describe a rectangular pulse within the cell, and so k_c can be put in the form $k_c \sim n_c\pi/\Delta l$, where n_c is a positive integer. A value of $n_c \sim$ 10 should be sufficient to specify the cutoff k_c. For more non-uniform pulses, higher wavenumbers may be present, in which case a larger cutoff n_c will be required. To obtain the spacing Δk, which is equivalent to stating N_k , we assume the low frequency content will be determined by the entire propagation region $L_o - x_o$; again approximating with a rectangular pulse, the wavenumber limit is $\pi/(L_o-x_o)$. We then set $\Delta k = \pi/n_L(L_o-x_o)$, where n_L is a constant used to provide resolution ; $n_L = 10$ should be sufficient . The value of N_k is then obtained from Eq.(5.60), which then allows one to completely specify k_i

Having specified k_i , we see that $^+A(n-1,t, x_i, k_i)$ may be interpreted as the k_ith component of the forward wave in cell n-1 located at x_i . Suppose v(k) may be written in the form $v(k) = v(0) + \Delta v(k)$. At time Δt later the component wave will move into cell n and x_i will transform to

$$x(n) = x_i(n-1) + [v(0) + \Delta v(k)]\Delta t \qquad (5.61)$$

where $\Delta v(k)$ is negative for this case since we have selected a Category I dispersion. Note that in accordance with Eqs.(5.46)-(5.47), $|\Delta v(k)| \leq \Delta v_o$. Note that no subscript has been used for x(n) since, because of dispersion, there is no guarantee that the new position in the nth cell coincides with one of the x_i. If dispersion is absent , $\Delta v(k) = 0$ and the wave will have moved from cell n-1 into exactly the same location in cell n a distance Δl away , where $\Delta l = v(0) \Delta t$. The location of the wave in the nth cell will then be $x_i(n)$. With the presence of $\Delta v(k)$, the new location for the wave is denoted by

$$x_i(n-1) \rightarrow x(n) \rightarrow x_i'(n) \qquad (5.62)$$

where the prime indicates that $x_i'(n)$ represents the closest member of x_i in the nth cell, satisfying Eq.(5.62). This procedure is repeated for all values of k_i corresponding to x_i in cell n-1. We then go to the next value x_i in the cell and run through the same procedure. This will produce a *distribution of points* in cell n , whose total is approximately N_f*N_k . Without dispersion the distribution will be uniform with N_k points at each position value in cell n. Dispersion will cause nonuniformity in this distribution. The task remaining is then to add up the Fourier components at each sub-element position value in cell n. This will then produce the field in cell n , at least due to the forward wave in cell n-1. To achieve this summation task, we replace dk in Eq.(5.59) by $\Delta k = k_c/(N_k-1)$, and transition to a summation (rather than an integral). We then allow the waves to proceed for an additional time step. The double summation, as alluded to previously, includes x_i and k_i in the (n-1)th cell. Thus

$$^+V(n,t+\Delta t,x_i{}'(n))=(1/\pi)(k_c/(N_k-1)_{x'_i,ki}\sum{}'T(n-1,k_i)^+A(n-1,x_i,t,k_i)\cos[k_ix_i(n-1)$$
$$+\varphi(k_i,n-1)] \tag{5.63}$$

We should note that we have reconstructed $^+V(n, t+\Delta t,x_i)$ by summing over those amplitudes closest to each $x_i(n)$; the prime in the summation symbolically indicates that the closest $x_i(n)$ in cell n is employed, using Eq.(5.61) and Eq.(5.62). Thus we insert the value of k_i in Eq.(5.61) to determine the location of the wave component in the nth cell. Note that the dispersion itself is implemented by means of Eqs.(5.61) and (5.62) for each wave in the summation. This procedure is *equivalent to making use of the velocity term* in the argument for the wave(represented by the cosine). For this reason the velocity term is missing from the argument in Eq.(5.63) and only the position and phase terms are present in the (n-1) cell. As mentioned before a total of N_F*N_k wave amplitudes are evaluated in cell n-1 at various positions x_i and various wavenumbers k_i . We also note that the wave amplitude $^+A(n-1,x_i , k_i , t)$ in Eq.(5.63) is modified by the transfer coefficient $T(n-1, k_i)$ at the n-1 node, as indicated by Eqs.(5.55) and (5.57).

After time Δt has elapsed most of these waves find themselves in cell n. After reconstructing the field we produce a new Fourier spectrum just as we did in the beginning. We then follow the same process as before starting with the decomposition of the field into a new set of Fourier components (we must also include the two other contributions to the field, to be discussed).

It is tempting to assume that it is unnecessary to reconstruct the field in cell n. Why not simply allow the Fourier components to continue to scatter in the TLM matrix, for as long as we wish, until such time that we wish to terminate the analysis.? This would be perfectly acceptable if there were no dispersion. Because of dispersion, however, the field shape will inexorably change as it propagates from cell to cell. The Fourier spectrum will therefore change; there is a continuous interplay between the Fourier spectrum and dispersion. The only question is whether it is necessary to recalculate the Fourier spectrum after each time step; if the dispersive changes are small it may be possible to wait several time steps before recalculating the spectrum. These are mathematical issues which go beyond the present scope of this discussion. The ideal approach is a conservative one, which is to recalculate the spectrum after each time step, provided such a procedure is practical in terms of computer capacity.

In the previous we confined the discussion to the Category I forward wave and in particular to the transmission of the forward wave in the n-1 cell to the nth cell. There are two other contributions of course. One is the forward wave in the nth cell during the kth time step. The bulk of this wave will have left the cell once the time step is completed. Since the velocity is slightly less than $v(0)$, however, a small portion of the wave will remain in the nth cell. The calculation follows the same course as before except that now we first obtain the Fourier Transform of the forward wave in the nth cell for the kth step. We then obtain the contributions which remain in the nth cell and add these to the previous contributions. The final contribution originates from the reflection of the backward wave at the (n-1) node. The reflected wave is related to the backward one by

$$^-A(n, k) = \ B(n\text{-}1, k) \ ^+A(n,k) \tag{5.64}$$

With the use of numerical techniques, k is of course replaced by k_i. The initial calculation differs here in that we first obtain the Fourier Transform of the *backward* wave. We then imagine the backward wave, with amplitude modified by Eq.(5.64), flowing, in the forward direction, into the nth cell at the n-1 node. For this wave the bulk of the contributions will remain in the nth cell. These contributions are then added to those of the previous waves discussed.

We may reduce the amount of computer computation, at least for the reflected wave. We do this by examining the (n-1) node and combining the transmitted and reflected waves with identical wavenumbers. We can employ this technique provided we use the same k_c cutoff in each cell , and therefore the same set of k_i By this technique, the transmitted and reflected waves are combined into a single forward wave in the nth cell. This enables us to employ analytic expressions while reducing the amount of numerical computation. This procedure works, of course , only when the adjacent cells are uniform and thus each wave in the two adjoining cells has the same velocity dispersion.

Let us denote the kth transmitted wave in the nth cell by $^+A_{TR}(n,k)$ and is related to the forward wave in the (n-1)th cell by Eq.(5.55). The forward wave resulting from the reflected wave is denoted by $^+A_{REF}(n,k)$ and is given by Eq.(5.64). The sum of these two waves is determined by sinusoidal addition , as opposed to the usual arithmetic addition when the waves are uniform throughout the cell. Denoting the sum by $^+A_S(n,k)$, the result is

$$(^+A_S(n,k))^2=(^+A_{TR}(n,k))^2+(^+A_{REF}(n,k))^2+2^+A_{TR}(n,k)^+A_{REF}(n,k)\cos(\Delta\varphi(k)) \quad (5.65)$$

where $\Delta\varphi(k)$ is the phase angle difference between the two waves, i.e., $\Delta\varphi(k)=\varphi(k,n)-\varphi(k.n-1)$.

Unfortunately we are unable to perform a similar addition for the forward wave originally in the nth cell during the kth time step. This is because we can never match the waves with the same k number. Neither the transmitted or reflected waves catch up with the original forward wave, all of which have the same velocity for the same k(or k_i) value. Waves with differing values of k may of course catch up with one another, leading to significant changes in the field profile.

5.23 Initial Conditions With Dispersion Present

The most common initial condition is that in which the field in a particular cell is both static and uniform so that both the forward and backward waves are likewise uniform ,i.e., rectangular in shape. As an example, therefore, we seek the initial Fourier Transform in each of the cells belonging to a cell chain as in Fig.5.21 or 5.22. It is relatively easy to obtain each of these transforms if we first obtain the transform of the auxiliary cell, centered about x=0, with total cell length Δl. Assume for the moment, therefore, that this cell is isolated from its neighbors and that the field profile is rectangular shaped, i.e., the field is Vo for $-\Delta l/2 < x < \Delta l/2$ and zero elsewhere. The Fourier transform for this profile is treated in all elementary texts on Fourier analysis . If the field in the auxiliary cell is V_o then the transform $F(0,k)$ is

$$F(0,k) = 2V_o\sin(k\Delta l/2)/k \tag{5.66}$$

where we understand the zero argument in $F(0,k)$ represents the transform of the *auxiliary* cell.

To obtain the corresponding transforms for the $n=1,2,3,..$ cells, we make use of a well known theorem in Fourier analysis [3]. Suppose $F(k)$ is the Fourier transform of $V(x)$. If $V(x)$ is then shifted in coordinate space by an amount Δx_0, then the transform of the shifted function is $F(k)EXP(-jk\Delta x_0)$. We are then able to state the transforms for the cells $n=1,2,3$, etc... , denoted by $F(k,n)$:

$$F(n,k) = [2V_o\sin(k\Delta l/2)/k]EXP\{-jk(n-1/2)\Delta l\} \; ; \; n=1,2,3..... \tag{5.67}$$

Note that for $n=1$, the function has been shifted by $\Delta l/2$, for $n=2$ the shift is $(3/2)\Delta l$, and so forth. From the inverse transform, we can write down the field $V(x,n)$ in terms of the transform. Thus, for the auxiliary cell as well as the nth cell, the initial fields $V(0,x)$ and $V(n,x)$ may be expressed with the following inverse integrals

$$V(0,x) = (2V_o/\pi)\int_0^\infty [\sin(k\Delta l/2)/k] \cos(kx)dk = V_o \tag{5.68}$$

$$V(n,x) = (2V_o/\pi)\int_0^\infty [\sin(k\Delta l/2)/k] \cos[(kx)+\varphi(k,n)dk = V_o \tag{5.69}$$

where

$$\varphi(k,n) = -k(n-1/2)\Delta l \; ; \; n=1,2,3... \tag{5.70}$$

Eq.(5.69) represents the total field in the nth cell. For the forward and backward waves a factor of 1/2 must be inserted in both the Fourier transform and the field for each wave.

5.24 *Stability of Initial Profiles With Dispersion Present*

In the past we have always regarded a static TLM line element as a superposition of forward and backward waves, continuously traveling back and forth in the line, and destined to remain as such in the line until the "dynamic" equilibrium is upset by changes among nodes somewhere in the region, thereby "unleashing" these modes. The use of such equilibrium modes would appear to be reasonable so long as there is no dispersion in the line. With dispersion present, however, one must question whether such modes are even appropriate, since dispersion will undoubtedly affect the nature of the assumed modes as they travel back and forth in the line element. To reiterate, can we logically continue to describe the static TLM element in terms of forward and backward waves, when dispersion is present.

In order to explore this question, we examine a centered auxiliary cell with the Fourier transform given by Eq.(5.66) and the field given by Eq.(5.68), all at t=0. V(0,x) may be split up into the usual forward and backward waves, with equal magnitude, $^{+}V(0,x) + {}^{-}V(0,x)$. In order to account for the traveling nature of these waves we make the following replacements in Eq.(5.68):

$$\text{Forward Wave} : \cos kx \rightarrow \cos[k(x-v(k)t)] \qquad (5.71)$$

$$\text{Backward Wave} : \cos kx \rightarrow \cos[k(x+v(k)t)] \qquad (5.72)$$

We initially assume the inverse transforms for the two waves are

$$^{+}V(0,x) = (V_o/\pi)\int_0^\infty [\sin(k\Delta l/2)/k][\cos\{k(x-v(k)t)]dk \qquad (5.73)$$

$$^{-}V(0,x). = (V_o/\pi)\int_0^\infty [\sin(k\Delta l/2)/k][\cos\{k(x+v(k)t)]dk \qquad (5.74)$$

The above assumes the integrand is an even function; for odd functions the integrals vanish. We may regard the above two equations as depicting a situation in which there is no dispersion to begin with, and we arbitrarily "switch on" on the dispersion immediately afterwards. Note that the numerical coefficient in front of each integral is reduced by a factor of two since each wave contributes to half the total field. Also the phase factor has been set equal to zero in each of the traveling waves of Eqs.(5.73) and (5.74). This is tantamount to saying that at t=0 the wave crest of each wave is situated at x=0, the same as the static solution. Since we exclude waves from venturing outside the cell region, we assume the forward and backward waves are confined to

$$-\Delta l/2 \; < x - v(k)t \; < \Delta l/2 \quad : \quad \text{(FORWARD WAVE)} \qquad (5.75a)$$

$$-\Delta l/2 \; < x + v(k)t \; < \Delta l/2 \quad : \quad \text{(BACKWARD WAVE)} \qquad (5.75b)$$

Eqs.(73)-(75) are incomplete, however, since they portray the wave motion only during the "first pass", and they do not account for subsequent reflections from the end points, $x = \Delta l/2$ and $\Delta l/2$, wherein the forward wave is reflected at $x = \Delta l/2$, and thereby converted to a backward wave, and vice versa for the initial backward wave reflected at $x = -\Delta l/2$. In order to analyze this situation we employ a very simple approach using "circular" boundary conditions. Because of the inherent symmetry of the forward and backward waves, it is easy to relate the reflected waves , at $x = -\Delta l/2$ and $-\Delta l/2$, to the waves at the opposite end of the TLM line. Thus , for example, after the reflection of the forward wave at $x = \Delta l/2$ we can think of the same wave (with the same amplitude) entering the TLM line as a forward wave at $x = -\Delta l/2$, but with important difference: the wave has an added phase factor $k\Delta l$, so that the wave appears as $\cos\{ kx + k\Delta l - v(k)t\}$. This is necessitated by the requirement that the reflected amplitude at $x = \Delta l/2$ maintain the same value. Continuing, the new wave will be reflected once more at $x = \Delta l/2$, leading to still a new wave entering at $x = -\Delta l/2$ with phase factor $2k\Delta l$,. Thus with each successive reflection the forward wave phase factor is $k\Delta l$, $2k\Delta l$, $3k\Delta l$, etc...   A similar treatment for the backward wave leads to a phase factor of $-k\Delta l$, , $-2k\Delta l$, $-3k\Delta l$, etc...

The change in phase factor after each reflection slightly complicates the wave motion in the cell. For a given wavenumber k and time the cell region must now be subdivided into two sub-regions, each with a different phase factor. To determine the two sub-regions we first form the ratio $v(k)t /\Delta l$, given by

$$v(k)t /\Delta l \; = \; q(k.t) + REM(k,t) \qquad (5.76)$$

where $q(k,t)$ is the highest integer and $REM(k,t)$ is the *fractional* remainder. In the case of the forward waves , the two sub-regions will therefore be occupied with waves appearing as $\cos\{kx + kq(k,t)\, \Delta l - v(k)t\}$ and $\cos\{kx + k(q(k,t) +1)\Delta l - v(k)t\}$. The transitional point separating the two sub-regions, ${}^{+}X_t(k,t)$, is

$$^+X_t(k,t) = -\Delta l/2 + REM(k,t) \text{ (FORWARD WAVE)} \qquad (5.77a)$$

We should note that the region adjacent bounded by $x=-\Delta l/2$ has the higher index, $q(k,t)+1$, while the sub-region bounded by $x=\Delta l/2$ has the $q(k,t)$ index.

The same treatment may be applied to the backward wave . The two sub regions are occupied by $\cos\{kx-kq(k,t)\ \Delta l+v(k)t\}$ and $\cos\{kx-k(q(k,t)+1)\Delta l+v(k)t\}$. The transitional point is

$$^-X_t(k,t) = +\Delta l/2 - REM(k,t) \text{ (BACKWARD WAVE)} \qquad (5.77b)$$

where the sub-region bounded by $x=\Delta l/2$ has the higher index, $q(k,t)+1$ and the other sub-region the $q(k,t)$ index.

At this point we can write down the general expressions for the forward and backward waves for arbitrary wavenumber and time. These expressions replace the "first pass" equations of Eqs.(5.73)-(5.74).   Thus

$$^+V(0,x)=(V_o/\pi)\int_0^\infty[\sin(k\Delta l/2)/k][\cos\{kx+k(q(k,t)+1)\Delta l-kv(k)t\}]dk$$
$$-\Delta l/2\leq x\leq {}^+X_t(k,t) \qquad (5.78a)$$

$$^+V(0,x)=(V_o/\pi)\int_0^\infty[\sin(k\Delta l/2)/k][\cos\{kx+k(q(k,t)\Delta l-kv(k)t\}]dk$$
$$: \qquad \Delta l/2\geq x\geq {}^+X_t(k,t) \qquad (5.78b)$$

Similarly for the backward wave

$$^-V(0,x). =(V_o/\pi)\int_0^\infty[\sin(k\Delta l/2)/k][\cos\{kx-k(q(k,t)+1)\Delta l+kv(k)t\}]dk$$
$$\Delta l/2\geq x\geq {}^+X_t(k,t) \qquad (5.79a)$$

$$^-V(0,x). = (V_o/\pi)\int_0^\infty [\sin(k\Delta l/2)/k][\cos\{kx - k(q(k,t)\,\Delta l + kv(k)t\}]dk$$
$$-\Delta l/2 \le x \le {}^-X(k,t) \qquad\qquad (5.79b)$$

The total field is of course given by $V(0,x) = {}^+V(0,x) + {}^-V(0,x)$. If the wave velocity is dispersionless then $v(k)=v_o$ and $^+V(0,x) = {}^-V(0,x) = V_o/2$ as expected.

To simplify matters for the moment assume the dispersive $v(k)$ has the form
$$v(k)=v_o + \Delta v(k) \qquad\qquad (5.80)$$

where v_o is dispersionless and the dispersive part is $\Delta v(k)$. How does this modify the forward and backward waves, which without dispersion are simply equal to Vo/2? Assume for the moment that $\Delta v << v_o$. Inserting Eq.(5.80) into Eq.(5.78), and keeping only first order quantities, we obtain for the forward wave

$$^+V(0,x)=V_o/2+(V_o/2)\int_0^\infty [\sin(k\Delta l/2)/k][\sin\{kx+k(q(k,t)+1)\Delta l-kv_o t\}]*$$
$$\sin(k\Delta v(k)t)dk: \qquad (5.81a)$$
$$-\Delta l/2 \le x \le {}^+X_t(k,t)$$

$$^+V(0,x)=V_o/2+(V_o/2)\int_0^\infty [\sin(k\Delta l/2)/k][\sin\{kx+k(q(k,t)\,\Delta l-kv_o t]\,\sin(k\Delta v(k)t)dk:$$
$$(5.81b)$$
$$\Delta l/2 \ge x \ge {}^+X_t(k,t)$$

where $V_o/2$ is the non-dispersive part and the integrals are the perturbative, dispersive contributions. Corresponding equations for the backward wave are

$$^-V(0,x). = V_o/2-(V_o/2)\int_0^\infty [\sin(k\Delta l/2)/k][\sin\{kx-k(q(k,t)+1)\Delta l+kv_o k\}]*$$
$$\sin(k\Delta v(k)t)dk: \qquad (5.81c)$$
$$\Delta l/2 \ge x \ge {}^-X_t(k,t)$$

$$^-V(0,x). = V_o/2-(V_o/2)\int_0^\infty [\sin(k\Delta l/2)/k][\sin\{kx - k(q(k,t)\,\Delta l+kv_o t]\,\sin(k\Delta v(k)t)dk$$
$$(5.81d)$$
$$-\Delta l/2 \le x \le {}^-X_t(k,t)$$

We present an argument to indicate that the dispersive contributions in Eq.(5.81) vanish. First we verify that for sufficiently small $\Delta v(k)$, the dispersive contributions vanish as expected, due to the $\sin(\,k\Delta v(k)t)$ terms in the integrand. Indeed we can show that the dispersive contributions vanish without actually setting $\Delta v(k)$ equal to zero. Instead, throughout the effective range of k, $[k\Delta v(k)]^{-1}$ may be regarded as a long period during which time the integrals in Eq.(5.81) average out to zero. Aside from $\sin(k\Delta v(k)t)$ the terms in each integrand, for example, $[\sin(k\Delta l/2)/k][\sin\{kx\ -k(q(k,t)\ \Delta l+kv_o t]$, are odd functions and therefore their contribution washes out, given the essential constancy of $\sin(\,k\Delta v(k)t)$.

An equivalent but perhaps more convincing procedure is to consider the *time average* of the dispersive integrals in Eq.(5.81). If we first perform this time averaging, then one can argue that the averaging of the dispersive contributions vanishes. For example, if we look at the integrand in Eq.(5.81a) then the term $[\sin\{kx+k(q(k,t)+1)\Delta l-kv_o t\}]$ oscillates very quickly because of the relatively large frequency associated with kv_o. On the other hand, the remaining portion of the integrand, $[\sin(k\Delta l/2)/k]\ (\sin(\,k\Delta v(k)t)$, may be considered a modulation factor since it varies more slowly because of $\Delta v(k)$. The time averaged dispersive integral therefore vanishes.

In the previous discussion we made the assumption that the velocity could split into two parts, a nondispersive part v_0 and a dispersive part $\Delta v(k)$ which remains small compared to $v(k)$ throughout the effective range of k. We then argued that the voltage throughout the TLM line remains constant while the time averaged contributions of the dispersive integrals vanish. Now suppose the general case applies, with no limitations in $v(k)$. We then ask whether after time averaging there exist dispersion relations which do not change(or at least minimally change) the initial uniform field, and if such dispersions exist, what conditions are imposed; for example a stabilization period might be required. Such dispersion relations, if they exist, offer a powerful incentive for their selection, provided of course they are verified experimentally. Further pursuit of this subject , however, requires the detailed numerical solution of Eqs.(5.78)-(5.79) , without any approximations. This task is not undertaken here.

5.25 Replacement of Non-Uniform Field in Cell with Effective Uniform Field

Once the desired time step in the calculation is achieved, then it might be useful to replace the nonuniform field in each cell with an effective uniform field, based on the energy content in each cell. This field , denoted by $V_{eff}(n)$, is

$$V_{eff}(n) \;=\; [(1/\Delta l) \int_0^{\Delta l} V^2(n) dx]^{1/2} \tag{5.82}$$

In the above we assume the cell is small enough so that there is no sign change throughout the cell and of course the sign of $V_{eff}(n)$ is the same as $V(n)$. One need not wait until the end of the final calculation to obtain $V_{eff}(n)$ for the field. The calculation may be done periodically in the iteration to provide a simple way of assessing how the field energy is being distributed.

Appendices

App.5A.1 Specification of Input/Output Node Resistance to Eliminate Multiple Reflections

5A.1(a) Input Simulation

Fig.5A.1 shows the input circuit, which simulates a long input line. A matrix of parallel TLM lines , with impedance Z_o, is employed. The specified signal is then inserted in Z_O Any reflections are then terminated in $R(n,m)$. Note that the three other lines converging on the node are assumed to have extremely high impedances. The load seen by the reflected wave is that shown in Fig.5A.1. If we designate the load as $R_L(n,m)$, then the reflected signal is perfectly matched when

$$R_L(n,m) = (3/4)R(n,m)=Z_O \tag{5A.1}$$

or $\qquad\qquad R(n,m)=(4/3)Z_o$ (Interior Line) $\qquad\qquad$ (5A.2)

We use the term interior line to indicate that the line is not adjacent a conductor. What happens when the input lines are next to a conducting boundary? This situation must occur at some point along the input. The equivalent circuit is modified, given in Fig.5A.2. Note that only two high impedance lines now converge at the node, and in fact the third line is shorted out. The load resistance, matched to Z_O, is then found from

$$R_L(n,m) = (2/3)R(n,m)=Z_O \tag{5A.3}$$

which then gives

$$R(n,m) = (3/2)Z_O \text{ (Adjacent Conducting Boundary)} \tag{5A.4}$$

5A.1(b) Output Simulation

The output simulation is very similar to that of the input. The simulation circuit is precisely the same as the input circuit, if we merely "flip" the input circuit The condition for terminating any reflections in Z_O is then the same as for the input, i.e.,

$$R(n,m) = (4/3)Z_o \ (\text{Interior Line}) \qquad (5A.5)$$

$$R(n,m) = (3/2) \ Zo \quad (\text{Adjacent Conducting Boundary}) \qquad (5A.6)$$

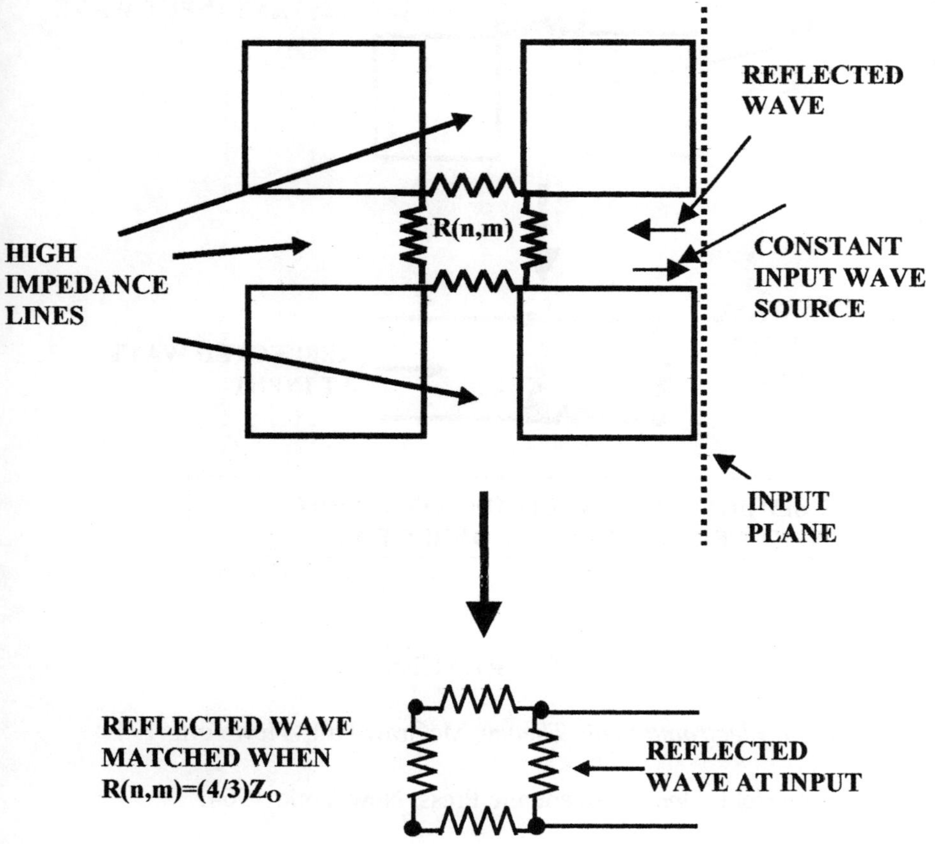

FIG. 5A.1 SIMULATION OF LONG INPUT LINE.

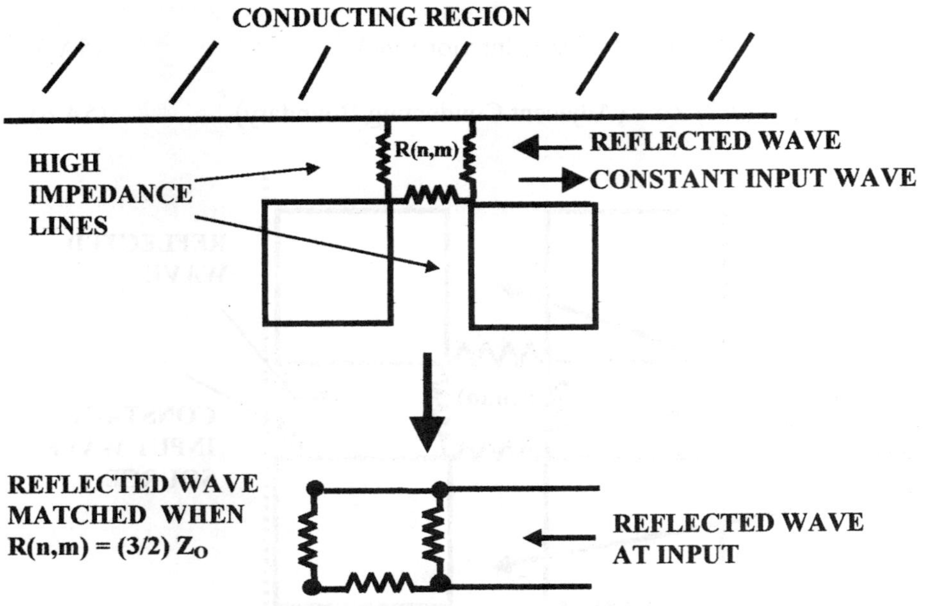

FIG. 5A.2 SIMULATION CIRCUIT OF LONG INPUT WHEN INPUT CELL IS NEXT TO CONDUCTOR.

REFERENCES

1. J.Stratton, *Electromagnetic Theory*, McGraw Hill, NewYork, 1941.

2. A. Sommerfeld, *Optics* , Academic Press, New York, 1964.

3. A.Papoulis, *The Fourier Integral and its Applications*, McGraw Hill, 1962

VI.   Cell Discharge Properties And Integration Of Transport Phenomena Into The Transmission Line Matrix

The transmission line model, which we have described thus far, is very well suited for characterizing fast electromagnetic and conductivity changes. Slower processes, however, typified by phenomena such as recombination, drift, diffusion, and space charge, may also be incorporated into the model. These transport phenomena are described in terms of charge carrier models. In almost all cases, for illustrative purposes, we will describe a semiconductor medium with hole and electron carriers. However, the same concepts apply to other situations, such as ionized gases. In this Chapter we seek to integrate the carrier model into the TLM formulation. Simple iterations, illustrating the time step changes in recombination, drift, diffusion, carrier generation, and cell charge will be described using the TLM matrix. The integration of the transport properties also will help us to understand the TLM theory. For example, the discharge of neighboring cells , via the nodal resistors between cells, may be cast in terms of the carrier drift between cells.

The integration of carrier transport into the computer iteration will improve our ability to interpret the results in a more meaningful and physical manner. If we desire such an all encompassing model, however, we will have to simultaneously track fast and slow phenomena. To implement such a model, a large number of time steps will be required; an important impact, therefore, will be the necessity for additional computer memory and speed in order to obtain solutions in a timely manner. Often, however, certain approximations may be invoked to help speed up the iteration and reduce the amount of computer time.

Incorporating the slower phenomena into the TLM matrix has the advantage that we may use the cell matrix already available for the electromagnetic analysis. In fact, for many situations, the only portion of the matrix required is the cell itself and not the TLM lines(or "tracks")separating the cells. The disadvantage in using the same matrix is that the time step being used is usually so small that very little happens, when analyzing slower phenomena during that step. In the case of drift, for example, the only change that occurs is near the cell boundary. One way of dealing with this disparity is to evaluate the slower phenomena less frequently, instead of after each step. Indeed we will see later that if we use the same electromagnetic cell size to examine transport phenomena, then it is appropriate to use a slower "sampling " speed.

Before bringing the carrier description under the umbrella of the transmission line matrix model, we will first discuss the discharge of neighboring cells, resulting from the potential difference between them. The discharge, i.e., the transfer of charge, occurs via the node resistors connecting the cells. We will see later that the description of this discharge is consistent with the introduction of carrier drift into the TLM cell matrix model.

6.1 Charge Transfer Between Cells

Within the TLM model there exists iso-potential cells which are separated by transmission lines, which represent differences in potential between cells and which also account for the conveyance of electromagnetic energy. The nodal resistors simulate the conductivity of the medium and provides the means for the cells to discharge into one another. Our scattering equations automatically take into account any changes in potential difference between cells, resulting from either a change in the wave status, or from charge transfer(i.e., current) . Equivalently this allows us to track the evolution of net charge on each cell, as we shall show in a moment.

A natural question which arises is what happens if the resistivity returns to its formerly large value, characteristic of equilibrium. Suppose, for example, the light activation process in a semiconductor, which produces conductivity, ceases at time $t = t_1$. We further assume an exponential "recovery" of the node resistance $R(n,m,q)$. Thus,

$$R(n,m,q) = [R(n,m,q)]_{t=t1} EXP((t-t_1)/\tau) \tag{6.1}$$

where $\{R(n,m,q)\}_{t=t1}$ is the node resistance value at $t = t_1$, and τ is the "recovery constant". For times $(t-t_1)$ large compared to τ one might expect Eq.(6.1) to reinstate the same situation that existed at equilibrium , before any conductivity was introduced. This turns out not to be the case, however. Although Eq.(6.1) will restore the resistivity to its equilibrium value, it will not get rid of the net charge that has evolved on the cells, as a result of the charge transfer between cells. The resultant fields, therefore, will differ from the equilibrium fields, reflecting the existence of charged up cells. The charge-up is especially evident when the semiconductor (or any other medium) is partially activated. A charge layer then develops at the interface between the activated and inactivated regions. The charge remains stationary unless we allow for carrier drift or diffusion to the electrodes.

The simple circuit shown in Fig.6.1   illustrates the concept. The circuit consists of a charged capacitor C in series with a time varying resistor R(t) and an inductance L. Crudely, C and L represent the elements of the transmission lines separating a particular cell from its neighboring cells, while R(t) represents the resistance connecting the cells. Initially, C is fully charged and R(t) is very large. When the node resistor is activated R(t) first declines in value. Once the activation ceases, R(t) then increases to its former value, as a result of recombination. During this process some of the original charge will have been lost, dissipated in R(t). The remaining charge, however, does not return to its original state. The relative charge distribution, and therefore the voltage distribution will change. Each cell will be left with a different, "trapped" net charge, which changes the original voltage difference between cells. The charging of each cell is controlled by carrier transport properties, to be addressed shortly.

The fields produced by the charged cells do not begin to impact the iteration so long as the times involved are smaller than the characteristic times associated with the transport quantities, such as carrier drift, diffusion, and recombination. Under these conditions, the solution will be a purely electromagnetic one in which the convective phenomena are considered frozen in space. For the longer times, however, the carrier transport must be incorporated into the iteration. Before addressing the issue of carrier transport and its inclusion into the model, we relate the charge accumulation on the cell to the electric field divergence for

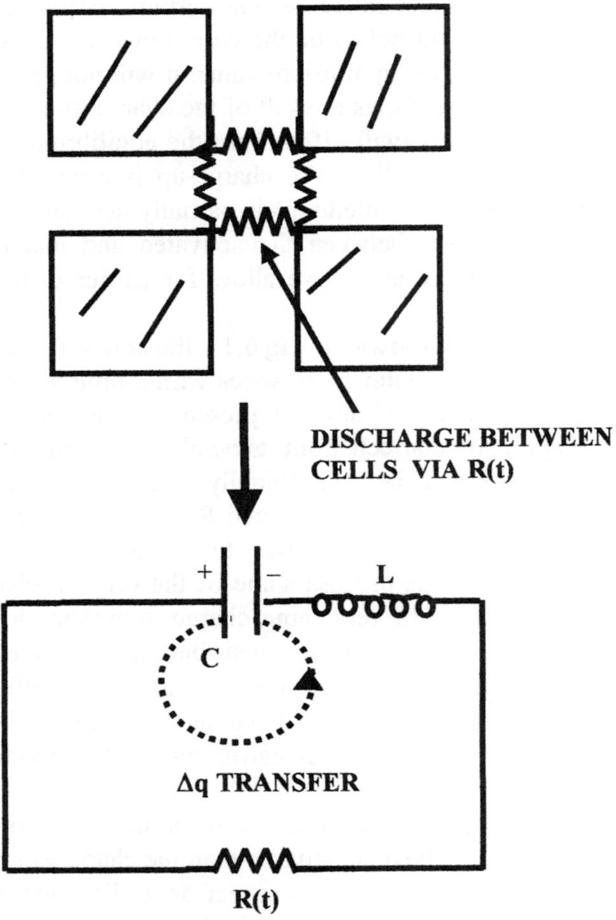

FIG. 6.1. LUMPED CIRCUIT ANALOGUE TO DESCRIBE THE
DISCHARGE BETWEEN ADJOINING TLM CELLS,
VIA THE NODE RESISTANCE. L AND C IN THE
ABOVE CORRESPOND TO THE TLM LINE.

each cell. From an iterative point of view, this information will provide us with a convenient means for determining the amount of charge in the cell for an arbitrary time step. This information will be useful when we incorporate the carrier transport into the iteration.

6.2 *Relationship between Field and Cell Charge*

In order to proceed further we must quantify the exact amount of excess charge present in each cell during each time step. One possibility involves calculating any net charge delivered to the cell, via the node resistor, and adding (or subtracting) this amount to the charge of the previous time step. This assumes of course that we know the net cell charge at a particular moment in time. Indeed we will use this method later in the Chapter when we calculate the time change in the net cell charge(i.e., the current). For the present situation, however, it is more convenient to use a different approach, namely, Gauss' Law. We simply apply Gauss' Law, i.e., the surface integral form of Poisson's Equation, to the cubical (n,m,q) cell, as shown in Fig.6.2. The averaged net outflow of the electric field, ΔE, is then related to the total net charge contained in the cell q(n,m,q) by

$$\Delta l^2 \ [\Delta E_x(n,m,q) + \Delta E_y(n,m,q) + \Delta E_z(n,m,q)] \ = \ q(n,m,q)/\varepsilon \qquad (6.2)$$

where ΔE_x, etc... are the differences of the component fields across opposite surfaces of the (n,m,q) cell

The remainder of the discussion in this Section is mostly devoted to seeing how we may re-cast Eq,(6.2), utilizing the usual TLM voltage amplitudes in the lines surrounding the cell, (n,m,q). We first replace the electric variable with the averaged transmission line variable, ΔV which are related by $\Delta V_x = -\Delta E_x \Delta l$, etc.... Thus, the charge within the cell may be expressed in terms of the net transmission line voltage, or,

$$\Delta l[\ \Delta V_x(n,m,q) + \Delta V_y(n,m,q) + \Delta V_z(n,m,q) \ ] = -q(n,m,q)/\varepsilon \qquad (6.3)$$

It is worthwhile reiterate that $\Delta V_x(n,m,q)$, etc.. is not the voltage amplitude in a particular line, but rather the *difference* in voltage between two opposite faces on either side of the cell. The next task is to calculate ΔV and this is done with the aid of Fig.6.2. We first look at ΔV_z for the two xy faces of the cell. For the face at $z=q\Delta l$, for example, we need to calculate the outwardly directed, average field perpendicular to this face. This field is simply the average of the fields contained in the four transmission lines which border the face. If we denote this field by V_{z+} then

$$V_{z+} = (1/4)\,[V_{xz}(n,m,q) + V_{xz}(n,m-1,q) + V_{yz}(n,m,q) + V_{yz}(n-1,m,q)] \qquad (6.4)$$

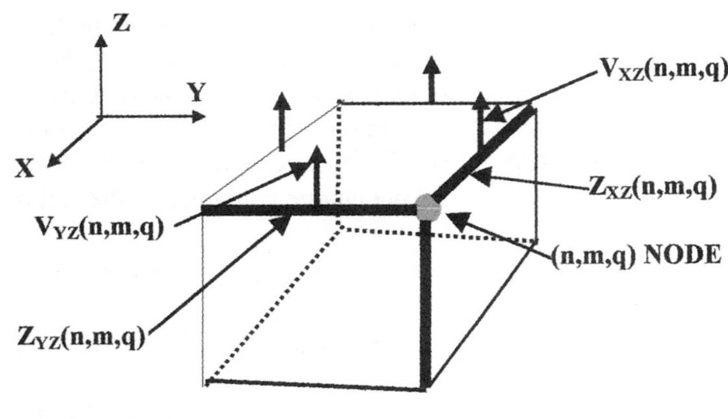

FIG. 6.2 TLM FIELDS $V_{XZ}(n,m,q)$, $V_{XZ}(n,m-1,q)$, $V_{YZ}(n,m,q)$, AND $V_{YZ}(n-1,m,q)$ EMERGING FROM THE POSITIVE XY FACE OF THE TLM CELL. FIELDS ARE AVERAGED OVER THE CELL FACE AREA. THE DIFFERENCE IN FIELDS BETWEEN OPPOSITE FACES LEADS TO THE CELL CHARGE.

and we remind ourselves of the fact that each term on the right side is the sum of a forward and backward wave, e.g., $V_{xy}(n,m,q) = V_{xy}(n,m,q) + V_{xy}(n,m,q)$, etc... Note that the second subscript in each term is z, indicating the proper polarization. Next we write down the field for the xy face located at $z = (q-1)\Delta l$. Denoting the average field for this face by V_{z-}, we have

$$V_{z-} = (1/4)\,[V_{xz}(n,m,q-1)+V_{xz}(n,m-1,q-1) +V_{yz}(n,m,q-1)+V_{yz}(n-1,m,q-1)] \quad (6.5)$$

Note that Eq.(6.5) is identical to Eq.(6.4) except for change in the q index, where q is now replaced by q-1, since the face is located at $z = (q-1)\,\Delta l$. We can now calculate the difference

$$\Delta V_z(n,m,q) = V_{z+}(n,m,q) - V_{z-}(n,m,q) \quad (6.6)$$

where V_{z+} and V_{z-} are given by Eqs.(6.4) and (6.5). We can now perform similar calculations for the x and y directions. The average voltage expressions for the yz and zx faces are easily obtained, giving

$$V_{yz+} = (1/4)\,[V_{zx}(n,m,q)+V_{yx}(n,m,q)+V_{zx}(n,m-1,q)+V_{yx}(n,m,q-1)] \quad (6.7)$$

$$V_{yz-} = (1/4)\,[V_{zx}(n-1,m,q)+V_{yx}(n-1,m,q)+V_{zx}(n-1,m-1,q)+V_{yx}(n-1,m,q-1)] \quad (6.8)$$

$$V_{zx+} = (1/4)[V_{zy}(n,m,q)+V_{xy}(n,m,q)+V_{xy}(n-1,m,q)+V_{zy}(n,m,q-1)] \quad (6.9)$$

$$V_{zx-} = (1/4)[V_{zy}(n,m-1,q)+V_{xy}(n,m-1,q)+V_{xy}(n-1,m-1,q)+V_{zy}(n,m-1,q-1)] \quad (6.10)$$

As with the z component, the net differences in voltage in he x and y directions are

$$\Delta V_x(n,m,q) = V_{x+}(n,m,q) - V_{x-}(n,m,q) \quad (6.11)$$

$$\Delta V_y(n,m,q) = V_{y+}(n,m,q) - V_{y-}(n,m,q) \quad (6.12)$$

The total net field outwardly directed from the (n,m,q) cell, denoted by $\Delta V(n,m,q)$, is then given by

$$\Delta V(n,m,q) = \Delta V_x(n,m,q) + \Delta V_y(n,m,q) + \Delta V_z(n,m,q) \qquad (6.13)$$

The total net charge contained in the (n,m,q) cell, during the kth time step, is denoted by $q^k(n,m,q)$. Thus,

$$q^k(n,m,q) = -\varepsilon\Delta l\Delta V^k(n,m,q) \qquad (6.14)$$

Eq.(6.14) allows us to calculate the net charge during any time step of the transient process . We also wish to find the time change in the net charge, i.e., the iterative change. In order to do this , we state the net charge for the k+1 step:

$$q^{k+1}(n,m,q) = -\varepsilon\Delta l\Delta V^{k+1}(n,m,q) \qquad (6.15)$$

We then take the difference between the expressions,

$$q^{k+1}(n,m,q) = q^k(n,m,q) -\varepsilon\Delta l\Delta V^{k+1}(n,m,q)+ \varepsilon\Delta l\Delta V^k(n,m,q) \qquad (6.16)$$

where it is understand that $\Delta V^{k+1}(n,m,q)$ is expressed in k state variables through the usual iterative equations relating (k+1) voltage waves to k voltage waves.

Very often, when a semiconductor is ionized, the following charge carrier conditions may prevail. We assume there exists a large background plasma of equal numbers of holes and electrons, with each charge equal to $q_o(n,m,q)$, and a small excess of either positive or negative charge , which we now denote by $\Delta q_+(n,m,q)$ and $\Delta q_-(n,m,q)$ with $\Delta q_+(n,m,q) \ll q_o(n,m,q)$ and $\Delta q_-(n,m,q) \ll q_o(n,m,q)$. The positive and negative charge in each cell is therefore

$$q_+ (n,m,q) = q_o(n,m,q) + \Delta q_+(n,m,q) \qquad (6.17a)$$
$$q_-(n,m,q) = q_o(n,m,q) \qquad (6.17b)$$

which applies if the net cell charge is positive and

$$q_+(n,m,q) = q_o(n,m,q) \qquad (6.18a)$$

$$q_-(n,m,q) = q_o(n,m,q) + \Delta q_-(n,m,q) \tag{6.18b}$$

if the net charge is negative. The total number of carriers, of each type, control the resistivity. The carrier numbers, in turn, are controlled by the transport properties. Later in the Chapter we restate the iteration in terms of the node parameters and the transport properties.

6.3 Dependence of Conductivity on Carrier Properties

Understanding the electromagnetic wave behavior in a semiconductor requires us to first understand the properties of the semiconductor. Accordingly, we briefly review the dependence of the semiconductor conductivity (from which we may obtain the node resistance) on the carrier properties[1]. In particular we employ the carrier cell occupancy, and the mobility of the carriers, expressed in cell notation. For illustrative purposes, we again choose a semiconductor with electron and hole carriers, each having different transport properties, such as differing drift velocities, diffusion constants, recombination times, etc... The following parallels the discussion in Chapter II except that now the electron and hole number occupancies are allowed to differ. First we relate the conductivity to the to the carrier properties, in terms of the TLM notation. We use the conductivity relationship corresponding to Eq(2.54), allowing for different hole and electron densities, as well as velocities. Within each cell the conductivity $\sigma(n,m,q)$ is given by

$$\sigma(n,m,q) = [e\mu_n(n,m,q)n(n,m,q) + e\mu_p(n,m,q)p(n,m,q)] / \Delta l^3 \tag{6.19}$$

where
$n(n,m,q)$ = number electrons in (n,m,q) cell
$p(n,m,q)$ = number holes in (n,m,q) cell
$\mu_n(n,m,q)$ = average electron mobility in (n,m,q) cell
$\mu_p(n,m,q)$ = average hole mobility in (n,m,q) cell
e = electron charge

and the electron and hole mobilities satisfy the relations $u_e(n,m,q) = \mu_e(n,m,q)E_{AV}(n,m,q)$ and $u_h(n,m,q) = \mu_h(n,m,q)E_{AV}(n,m,q)$, where $u_e(n,m,q)$

and $u_h(n,m,q)$ are the average electron and hole drift velocities of the (n,m,q) cell and $E_{AV}(n,m,q)$ is the average field for that cell , already determined in Chapter 2 (see Eqs.(2.35)-(2.37)). We recall that $E_{AV}(n,m,q)$ represents the average of the various fields in the transmission lines surrounding the cell. Also note that we use the italicized symbols $n(n,m,q)$ and $p(n,m,q)$, to differentiate the electron density from the cell index n.   As   shown in Chapter 2, once we know the cell conductivity , the node resistance of the (n,m,q) cell is determined by taking the conductivity average of the eight cells surrounding the node, given by Eq.(2.33).

Integration Of Carrier Transport Using TLM Notation. Changes In Cell Occupancy And Its Effect On TLM Iteration

6.4 General Continuity Equations

We see from Eq.(6.19) that the conductivity depends directly on the cell occupancy of holes and electrons. The transport phenomena controls the cell occupancy, by virtue of the carrier motion , generation, and recombination. It is appropriate therefore to track the carrier occupancy in each cell and at each time step. Toward this goal we employ the continuity equations for the cell occupancy of electrons and holes, or(omitting the cell index notation)

$$(\partial n/\partial t) = (\partial n/\partial t)_{GEN} + (\partial n/\partial t)_{RECOMB} + (\partial n/\partial t)_{DRIFT} + (\partial n/\partial t)_{DIFF} \quad (6.20)$$

$$(\partial p/\partial t) = (\partial p/\partial t)_{GEN} + (\partial p/\partial t)_{RECOMB} + (\partial p/\partial t)_{DRIFT} + (\partial p/\partial t)_{DIFF} \quad (6.21)$$

where the cell occupancy changes are due to generation, recombination, drift, and diffusion, respectively. Our task is now to recast the above equations in terms of iterative rate equations for the electron and hole occupancy numbers in each cell, i.e., the iterative equations in cell notation. We begin with the generation term, caused by light activation.

6.5 *Carrier Generation Due to Light Activation*

If $(\partial n/\partial t)_{GEN} \equiv G^k(n,m,q)$ is the generation rate of electrons in the (n,m,q) cell then the number of electrons produced during the kth time step is $G^k(n,m,q)\,\Delta t$. Similarly the holes are generated at a rate $G_p(n,m,q)$ The number of electrons and holes during the kth and kth steps are therefore related by

$$n^{k+1}(n,m,q) = n^k(n,m,q) + G^k(n,m,q)\Delta t \tag{6.22}$$

$$p^{k+1}(n,m,q) = p^k(n,m,q) + G^k(n,m,q)\Delta t \tag{6.23}$$

A simple illustration is the generation of carriers from a constant light pulse impinging on a semiconductor as discussed in Chapter II. Based on that discussion, and with the same assumptions, the generation rate for both electrons and holes is given by

$$G^k(n,m,q) = \xi\, \gamma^k(n,m,q)/\, U\Delta t \tag{6.24}$$

where
$$\gamma^k(n,m,q) = P_o EXP\{-a[(n\Delta l/l_o)-1/2]^2\}D(m)\Delta t \tag{6.25}$$

and
$$D(m) = [EXP-(m\Delta l/h_0)][EXP\,(\Delta l/h_0)-1] \quad \text{for } k \geq m \tag{6.26}$$

with the notation as given in Chapter II. We recall that $\gamma^k(n,m,q)$ is the light energy absorbed in the (n,m,q) cell while ξ is the conversion efficiency , U the photon energy, h_o the attenuation constant, l_o the semiconductor length, and a the spatial spread factor of the incident light pulse. Eqs.(6.25)-(6.26) are discussed in Chapter II . The spatial variation in Eq.(6.25) ,as well as the attenuation of the light pulse in the semiconductor, will of course change the generation rate from cell to cell. Another source of carrier generation is that due to avalanching, discussed in the following.

6.6 Carrier Generation Due to Avalanching : Identical Hole and Electron Drift Velocities

There are other means for producing electron-hole pairs, besides using light activation. If the electric field in the semiconductor is sufficiently strong, the hole or electron carrier will acquire enough kinetic energy to cause impact ionization in the lattice, i.e., to produce electron -hole pairs via collision of the lattice and the primary carriers[1]. The secondary carriers thus produced can then go on to produce additional carriers by the same process, thus causing an avalanche multiplication of carrier current . Usually the avalanche process is described in terms of the ionization coefficients of holes and electrons, α_p (x) and α_n(x), respectively, where x is the distance traversed by the carrier, and α_p (x) and α_n(x) represent the number of hole -electron pairs generated per unit distance. We can then characterize the growth of the individual hole and electron currents, $I_p(x)$ and $I_n(x)$, assuming a knowledge of α_p (x) and α_n(x). For holes the relationship describing the growth is

$$I_p(x+\Delta x)\ -I_p(x)\ =\ \alpha_p(x)I_p(x)\Delta x+\ \alpha_n(x+2\Delta x)I_n(x+2\Delta x)\Delta x \tag{6.27}$$

The right side of the above provides the increment in the hole current due to avalanching at $x+\Delta x$ after a time step Δt has elapsed. Δx is the incremental distance traversed by the hole and electron avalanche currents in time Δt. Generally we will regard Δx as much smaller than the cell length Δl and that $\Delta x/\Delta t$ may be approximated by the drift velocity. The electric field is assumed to be in the $+$ x direction. Note that the increment in the hole current is made up of two contributions; the first stems from the ionization impact of the hole current, while the second stems from that of the electron current. The $x+2\Delta x$ argument in the electron current stems from the opposite motion of the electrons In the interest of simplicity we assume, for the moment , that the hole and electron drift velocities are the same. Indeed, in order to achieve the large fields needed for avalanching, both velocities will approach their saturated values, which have comparable values.

We can also express the currents $I_p(x)$, $I_n(x)$ in terms of the hole and electron cell occupations given by

$$I_p(x) = \Delta l^2 p(x) e u(x) \qquad (6.28a)$$

$$I_n(x) = \Delta l^2 n(x) e u(x) \qquad (6.28b)$$

where $u(x)$ is the same carrier velocity for electrons and holes, $p(x)$, $n(x)$ are the carrier cell occupations, e the charge, and Δl^2 is the current cross-section. Similar equations apply of course to $I_p(x\pm\Delta x)$ and $I_n(x\pm\Delta x)$ and to $I_p(x\pm2\Delta x)$ and $I_n(x\pm2\Delta x)$. Substitution of the current equations into Eq.(6.27) gives

$$p(x+\Delta x)=\{p(x)u(x)+\Delta x[\alpha_p(x)p(x)u(x)+$$
$$\alpha_n(x+2\Delta x)n(x+2\Delta x)u(x+2\Delta x)]\}/u(x+\Delta x) \qquad (6.29)$$

where again $x+2\Delta x$ appears in the electron contribution since the electrons travel in the -x direction. We then make a simplification which allows the results to be interpreted more easily, namely, we assume the drift velocities are identical for carriers within the same cell . Thus $u(x)= u(x+\Delta x) = u(x+2\Delta x)$, etc..., provided x remains within a cell This is not a burdensome assumption since Δx is a sub-element of the cell Δl and , in any event, we usually assume all quantities, such as the drift velocity are constant within the cell. We also shift the reference so that in the previous equations $x+\Delta x \rightarrow x$, $x \rightarrow x-\Delta x$, and $x+2\Delta x \rightarrow x+\Delta x$. and apply the appropriate time step superscripts to the various quantities . Eq.(6.29) becomes

$$p^{k+1}(x) = p^k(x-\Delta x) + \Delta x\,[\alpha^k_p(x-\Delta x)p^k(x-\Delta x) + \alpha^k_n(x+\Delta x)n^k(x+\Delta x)] \qquad (6.30)$$

It is worthwhile to point the significance of each of the terms in the above. The first term represents the motion of the holes from $x-\Delta x$ to x in the absence of avalanching. The first term in the bracket is the contribution of the avalanching holes going from x-Δx to x. The second term is the contribution of avalanching electrons emanating from x+Δx .

We then proceed as before for the electron carriers, starting with the basic current equation , corresponding to Eq.(6.27), or

$$I_n(x-\Delta x) - I_n(x) = \Delta x[\alpha_n(x)I_n(x) + \alpha_p(x-2\Delta x)I_p(x-2\Delta x)] \qquad (6.31)$$

where x-Δx appears in $I_n(x-\Delta x)$ since we are dealing with electron carriers, whose motion is opposite to that of the holes. Similarly the avalanching due to the holes emanates from x-2Δx.   Substituting for the currents, and shifting position as before, the relationship for the electron carrier density is

$$n^{k+1}(x) = n^k(x+\Delta x) + \Delta x\,[\alpha^k_n(x+\Delta x)n^k(x+\Delta x) + \alpha^k_p(x-\Delta x)p^k(x-\Delta x)] \quad (6.32)$$

From Eq.(6.32) and (6.30) we can then proceed to determine the carrier time dependence in each cell(inside of which Δx is a sub-element). We postpone this step, however, until the hole and electron velocities are allowed to differ, which is the more general case, described in the next Section.

6.7 Avalanching with Differing Hole and Electron Drift Velocities

Eqs(6.30) and (6.32) represent sub-cell iterations for the hole and electron densities when the respective velocities are equal. This simplifies the iterative equations; in particular the drift velocity cancels out in the final expressions. Under general conditions, however, the velocities will differ, and accordingly we must modify Eqs.(6.30) ,(6.32). Besides the drift velocities the increment Δx must be replaced by Δx_p or Δx_n , depending on the particular carrier avalanche. These differentials may be estimated by $\Delta x_p \sim u_p(x)\,\Delta t$ and $\Delta x_n \sim u_n(x)\,\Delta t$ assuming the avalanche velocities are equal to the hole and electron drift velocities. In general this is an approximation , but the ensuing equations are still useful as long as we employ avalanche velocities which are much smaller than the electromagnetic velocity. In this regard, one should not confuse carrier creation stemming from a front of avalanching holes or electrons (whose velocities are approximated by the drift velocities) , with avalanching caused by the sudden arrival of a high intensity electromagnetic signal. Although both give rise to avalanching, the electromagnetic signal is capable of creating an avalanche region on a much faster time scale. By separating the two phenomena(and not lumping them together) we gain greater insight into avalanche effects.

Looking at the hole current equation first, we again start with Eq.(6.27). As before, we then relate the currents to the densities and drift velocities, given by

$$I_p(x) = \Delta l^2 p(x) e u_p \tag{6.33a}$$

$$I_n(x) = \Delta l^2 n(x) e u_n \tag{6.33b}$$

where u_p, u_n are the hole and electron velocities. Similar equations for $I_p(x \pm \Delta x_p)$ and $I_n(x \pm \Delta x)$ also apply, as well as $(x \pm 2\Delta x_p)$ and $I_n(x \pm 2\Delta x)$. The iteration for the hole density then becomes , using Eq.(6.30) as a guide,

$$p^{k+1}(x) = p^k(x - \Delta x_p) + \Delta x_p \left[\alpha^k_p(x - \Delta x_p) p^k(x - \Delta x_p) \right] + \Delta x_n \left[(u_n/u_p)\, \alpha^k_n(x + \Delta x_n) n^k(x + \Delta x_n) \right] \tag{6.34}$$

where we note now that the factor (u_n/u_p) now appears in the second term.(we still assume the carrier velocities do not change throughout the cell). The similar relationship for the electron iteration is

$$n^{k+1}(x) = n^k(x + \Delta x_n) + \Delta x_n \left[\alpha^k_n(x + \Delta x_n) n^k(x + \Delta x_n) \right] + \Delta x_p \left[(u_p/u_n)\alpha^k_p(x - \Delta x_p) p^k(x - \Delta x_n) \right] \tag{6.35}$$

Eqs.(6.34)-(6.35) are now modified so as to make them applicable to the TLM matrix. To simplify matters we assume the sub elements Δx_n and Δx_p are much smaller than the electromagnetic cell length, Δl. This is certainly true if the ionization is initiated solely by high energy carriers drifting into a high field region, since the drift velocity is much smaller than the propagation velocity. Thus $\Delta x_p \ll \Delta l \Delta t$ and $\Delta x_n \ll \Delta l \Delta t$.

With the aforementioned assumptions, we can   then revise Eqs.(6.34)-(6.35) to obtain the increments in the hole and electron cell occupancy numbers, thus re-casting the equations in TLM notation Assuming uniformity of the densities, drift velocities , and the ionization coefficients, throughout each TLM cell, a simple integration of Eqs.(6.34)-(6.35) then yields the sought after iterative equations for $p^k(n)$ and $n^k(n)$, caused by avalanching.

$$p^{k+1}(n) = p^k(n) + \Delta l \,[\alpha_p^k(n)p^k(n)+ (u_n^k(n)/u_p^k(n))\alpha_n^k(n)n^k(n)] \qquad (6.36)$$

$$n^{k+1}(n)= n^k(n) + \Delta l \,[\alpha_n^k(n)n^k(n)+ (u_p^k(n)/u_n^k(n))\alpha_p^k(n)p^k(n)] \qquad (6.37)$$

We reiterate that in the above iterations, $p(n)$ and $n(n)$ are the actual numbers of holes and electrons in the TLM (n) cell. and are related to the densities by dividing $p(n)$ and $n(n)$ by Δl^3. Since the differentials Δx_p and Δx_n are much smaller than Δl , any end effects at the cell boundaries are ignored. Also note that we have also assigned cell indices to the drift velocities since these may change from cell to cell as well as from one time step to another. Note that the second terms on the right side in Eqs.(6.36) and (6.37) correspond to $(\partial p/\partial t)_{GEN}$ and $(\partial n/\partial t)_{GEN}$, caused by the avalanching.

In the above iterations we have assumed a variation in only the x direction, in both field and drift velocity, and have therefore omitted the m, q indices. In general, however, the field, and hence the drift velocity will have both x , y , and z components, so that, e.g.,

$$u_p^k = u_{px}^k i + u_{py}^k j + u_{pz}^k q . \qquad (6.38)$$

Since there is no variation over the cell, however, the same type iteration applies in 3D. The method of calculation of the 3D avalanche iteration is almost identical to that given in the previous discussion, while remembering that the ratio $(u_n^k(n,m,q)/u_p^k(n,m,q))$, as in the case of Eq.(6.36), represents the ratio of the magnitudes of the two drift velocities. The simplest way to view the 3D case is to rotate the cell so that the i vector, for example, is aligned with the field and velocity, thus regaining the 1D result. Except for the (n,m,q) argument the result is the same as Eq.(6.36), or

$$p^{k+1}(n,m,q)=p^k(n,m,q) +\Delta l\ [\alpha_p{}^k(n,m,q))p^{\ k}(n,m,q)+r\alpha^k{}_n(n,m,q)n^{\ k}(n,m,q)] \quad (6.39)$$

where

$$r=\ u_n{}^k(n,m,q)/u_p{}^k(n,m,q) \quad (6.40)$$

and

$$u_n{}^k(n,m,q)=[(u_{nx}{}^k(n,m,q)^2 +(u_{ny}{}^k(n,m,q)^2+ (u_{nz}{}^k(n,m,q)^2]^{1/2} \quad (6.41)$$

$$u_p{}^k(n,m,q)=[(u_{px}{}^k(n,m,q)^2 +(u_{py}{}^k(n,m,q)^2+ (u_{pz}{}^k(n,m,q)^2]^{1/2} \quad (6.42)$$

The corresponding iteration for the electron carriers is

$$n^{k+1}(n,m,q) = n^k(n,m,q)+\Delta l\ [\alpha_n{}^k(n,m,q)n^{\ k}(n,m,q)+ (1/r)\alpha^k{}_p(n,m,q)p^{\ k}(n,mq)]$$
$$(6.43)$$

Additional corrections may be introduced to take into account the electric field dependence.   Both the drift velocity and the ionization coefficient rely on the magnitude of the electric field in the (n,m,q) cell. During each step of the iteration, the field is calculated from $E^{\ k}{}_{AV}(n,m,q)$, i.e. , the average field obtained from the transmission lines surrounding the cell (see Eqs.(2.35)-(2.37)). From $E^{\ k}{}_{AV}(n,m,q)$, one may then obtain the corrected drift velocities and ionization coefficients using semiconductor models available in the literature.

One issue that has not been addressed thus far, in regard to avalanching, has been the observed delay in the onset of the ionization process, once the avalanche field is in place about a region(in our case, a cell). In other words, the ionization coefficients may not become effective immediately at the start of a time step, but instead may experience a delay, ranging from a fraction of a time step to several time steps. Naturally, if there is a delay in the ionization, the number of carriers produced for that given time step (in a given cell) will be reduced; indeed, where the ionization delay is greater than the time step, little or no carriers will be produced for the given cell.   In cases where the ionization delay exceeds many time steps, an unusual phenomenon may occur; the ionization may occur after the high intensity portion of the avalanche field has *left* the particular cell. So far as the computer iteration is concerned , however, the incorporation of the ionization delay does not represent any fundamental problem.

6.8 Two Step Generation Process

In the previous Sections we described two sources of conductivity, avalanching and photo-ionization , and incorporated them into the TLM matrix formulation. It is important to mention , however, that either one of these sources can create an ionization region spatially removed from the original source region. The experimental conditions for this to occur are obvious. For example an ionization region, initially created by avalanching carriers or photo-ionization, can then emit high intensity light signals. The light signals may be   capable of further ionizing the semiconductor, by means of photo-ionization, away from the initial ionization region. Another form of ionization may occur when the electromagnetic field re-arranges itself in response to the initial ionization region (again caused by either avalanching or photo-ionization) such that the resultant electric field is enhanced in some region spatially removed from the original region. The newly created, field enhanced, region then undergoes avalanche breakdown. It is very likely that the two step processes play a crucial role in numerous breakdown phenomena. In addition the two processes, photoconductivity and avalanching, may co-exist, possibly reducing breakdown thresholds. We do not consider the two step processes any further in this Chapter, but we do emphasize that the TLM formulation is extremely well suited for describing such processes, especially since the phenomena involved may occur on a very fast time scale. In Chapter VII we describe field enhancement due to the partial (in a spatial sense) photo-ionization of a semiconductor gap. Also, in Chapter VIII we discuss a SPICE technique for describing the breakdown process in a semiconductor switch , incorporated into a transmission line, in which a progressive breakdown occurs caused by field enhancement.

6.9 Recombination

In this Section we include recombination effects in the iteration. As is well known, there are a large number of recombination mechanisms, many of which occur simultaneously in semiconductors. Typically the time scales involved in the recombination process will vary over a wide range but are usually much longer than the electromagnetic delay time. For illustrative purposes we select a single quite common mechanism.

For concreteness, we assume recombination of carriers is achieved via the existence of a single energy level in the midgap region. The midgap energy level serves as an indirect means for carrier recombination, i.e., these deep level sites achieve the recombination by a two step process: first an electron is captured followed by the capture of a hole. The capture and emission rates, involved in the recombination process, are assumed to differ for holes and electrons and holes, and to be field dependent as well. First we set forth the following definitions[1]

$N_T(n,m,q) =$ Number of recombination sites in (n,m,q) cell
$n_T(n,m,q) =$ Number of recombination sites filled with electrons in (n,m,q) cell
$p_T(n,m,q) =$ Number of empty recombination sites in (n,m,q) cell

and which satisfy

$$N_T(n,m,q) = \; n_T(n,m,q) + p_T(n,m,q) \tag{6.44}$$

With these definitions we are able to write down the rate equation for electrons

$$\partial n / \partial t \;)_{\text{RECOM}} = \; e_n n_T(n,m,q) - c_n p_T(n,m,q) \, n(n,m,q) \tag{6.45}$$

e_n is the emission coefficient representing the transition from the trap to the conduction band. c_n is the capture coefficient for an electron, representing a transition from the conduction band to the trap. As is often done, We can make use of the fact that e_n is related to c_n , using equilibrium arguments . We then assume the emission coefficient does not change under non-equilibrium conditions.

A similar rate equation for holes may be expressed:

$$(\partial p/\partial t)_{RECOM}= e_p p_T(n,m,q) - c_p n_T(n,m,q) p(n,m,q) \tag{6.46}$$

e_p and c_p are the emission and capture coefficients for the holes. We recall from the semiconductor background that the emission of a hole to the valence band is equivalent to the emission of an electron from the valence band to the trapping site, while the capture of a hole represents just the inverse process(see, for example, Ref [1]). Eqs.(6.45) and (6.46) give the change in the electron and hole numbers for the (n,m,q) cell, arising solely from indirect recombination with a single deep trap.

The iterative change in the electron number assuming for the moment that only the recombination process is active, is thus

$$n^{k+1} (n,m,q) = n^k(n,m,q)+ (\partial n/\partial t)_{RECOM} \Delta t \tag{6.47}$$

Eq.(6.47)simply expresses the number of electron carriers, during the (k+1)th interval, in terms of $n^k(n,m,q)$ and a first order correction term at time Δt later. A similar iteration for the hole number yields

$$p^{k+1}(n,m,q) = p^k(n,m,q)+(\partial p/\partial t)_{RECOM}\Delta t \tag{6.48}$$

We again stress the fact that the recombination iteration depends entirely on the particular mechanism, and we have chosen one particular example, a single trap, with rate expressions given by Eq.(6.45) and (6.46). Emission and capture coefficients for a single trap, for example, EL2 in GaAs, are discussed in the semiconductor literature. In general, of course, many traps will exist simultaneously , in which case the number of rate equations will multiply. The identification of traps, their energy levels, and their emission and capture coefficients, are the subjects of ongoing investigation among many semiconductor workers, with the aim of characterizing recombination properties.

6.10 *Limitations of Simple Exponential Recovery Model*

In Eq.(6.1) we assumed an exponential recovery of the resistivity, without relying, for example, on the solutions to Eqs.(6.45)-(6.48) to obtain the recovery. Given the convenience and simplicity of the exponential recovery, it is worthwhile to give examples under what conditions such a recovery is valid. One example is provided by a semiconductor with indirect recombination(as described in the previous Section) in which excess carriers are injected into a depletion region containing traps, i.e., a region with a deficit of carriers created by a high resistivity bulk semiconductor or a reversed biased diode. Under these condition holes and electrons recombine at a constant rate and a lifetime constant, τ , may be ascribed to the exponential growth of the resistivity.

Another example of exponential recovery is provided by the low level injection of carriers into an equilibrium plasma. The problem with low level injection is that the equilibrium background carrier densities, which produces conductivity, is assumed very high and thus the background conductivity often dominates the *electromagnetic* behavior. As a result, during any transient phase, which is our main interest, it is impossible to differentiate the effects of the injected carriers from that of the "equilibrium" carriers. This means that under transient conditions a simple exponential recovery based on low level injection is inadequate, and we must rely on numerical techniques to accurately model the recovery. The numerical methods take into account the total conductivity, as well as the differing transient properties of holes and electrons, including the recombination coefficients, the drift velocities, and the differing dependence on the electric field.

6.11 *Carrier Drift*

We next consider the contribution of drift to the carrier density. We continue to use the same grid, wherein the spacing is determined by the electromagnetic velocity. The drift velocity is about three orders of magnitude smaller than the electromagnetic velocity. Thus, during the delay time, Δt, the carriers will move only a very short distance relative to the transmission line length, Δl. We will therefore resort to certain approximations which make use of this disparity between the drift and electromagnetic velocities. We first consider the motion of holes and assume that t = kΔt, and the hole number is p^k(n,m,q). The average electric field for the (n,m,q) cell is calculated on the basis of the transmission line voltages surrounding the cell, as indicated in Chapter 2. This enables us to calculate the total average field E_{AV}(n,m,q), which has the components

$$E_{AV}(n,m,q) = E_{AV,x}(n,m,q)\ i\ + E_{AV,y}(n,m,q)j + E_{AV,z}(n,m,q)\ k \qquad (6.49)$$

where i ,j , k, are the unit vectors in the x,y, and z directions. If we focus on the hole carriers, for the moment, the velocity of the hole carriers u_p(n,m,q) is related to the field by

$$u_p(n,m,q) = \mu_p(n,m,q)\ E_{AV}(m,m,q) \qquad (6.50)$$

and the velocity may   be decomposed accordingly,

$$u_p(n,m,q) = u_{px}(n,m,q)\ i\ + u_{py}(n,m,q)\ j\ u_{pz}(n,m,q)\ k \qquad (6.51)$$

The changes in the hole number may be explained with the help of Fig.6.3, which shows two adjoining cells. At the beginning of the k time step the hole number is p_k(n,m,q). At the end of the next time one may regard the holes as having moved uniformly in the direction of E with velocity u_p (n,m,q). Thus, a portion of the holes, which were originally contained in the (n,m,q) cell volume, will have exited the volume after the next time step.   The holes that are about to exit are shown contained in the shaded volume(right side) . The number of holes exiting the cell is easy to estimate, assuming u_p(n,m,q) << v(n,m,q). The number exiting in the x direction may be estimated by $(u_{px}$(n,m,q) $\Delta t/\Delta l)p^k$(n,m,q), with these holes now residing in the (n+1,m,q) cell during the (k+1) step. Similarly, the number that have exited in the y direction is given by $(u_{py}$(n,m,q)t/Δl)p^k(n,m,q) (now in the (n,m+1,q) cell) and that for the z direction

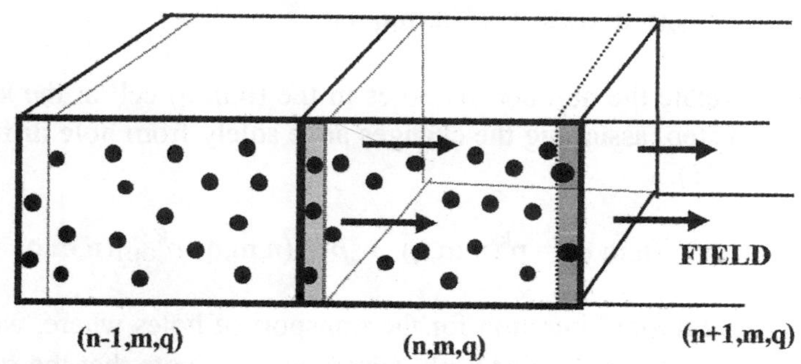

FIG. 6.3 HOLE CARRIERS LEAVING (n,m,q) CELL AND ENTERING THE (n+1,m,q) CELL DURING Δt (SHADED, RIGHT SIDE), DUE TO ELECTRIC DRIFT IN X DIRECTION.

NUMBER OF HOLE CARRIERS LEAVING (n,m) IS
$$u_{Px}(n,m,q)\,p(n,m,q)\,\Delta t/\Delta l$$
WHERE:
$u_{Px}(n,m,q)$ = HOLE DRIFT VELOCITY
$p(n,m,q)$ = NUMBER HOLE CARRIERS IN (n,m,q) CELL

SIMILARLY, HOLE CARRIERS ENTERING CELL IS
$$u_{Px}(n-1,m,q)\,p(n-1,m,q)\Delta t/\Delta l$$

is $(u_{pz}(n,m,q)t/\Delta l)p^k(n,m,q)$ (now in the $(n,m,q+1)$ cell), The total number of holes leaving the cell, $p^k{}_{OUT}$ is thus

$$p^k{}_{OUT}(n,m,q)=(u_{px}\Delta t/\Delta l)p^k(n,m,q)+(u_{py}\Delta t/\Delta l)p^k(n,m,q)+(u_{pz}\Delta t/\Delta l)p^k(n,m,q)$$

$$(6.52)$$

Having obtained the holes that have left the cell, we must now obtain the number of holes entering the cell during the k th step. The arguments are identical, with the incoming holes coming from the $(n-1,m,q)$, $(n,m-1,q)$, and $(n,m,q-1)$ cells. The number of cells entering , $p^k{}_{IN}$, is

$$p^k_{IN}(n,m,q) = (u_{px}(n,m,q)\Delta t/\Delta l)p^k(n-1,m,q) + (u_{py}(n,m,q)\Delta t/\Delta l)p^k(n,m-1,q) +$$
$$(u_{pz}(n,m,q)\Delta t/\Delta l)p^k(n,m,q-1) \tag{6.53}$$

We can now relate the number of holes in the (n,m,q) cell at the k time step to that at the kth step, assuming the changes arise solely from hole drift. Thus, from Eqs.(6.52)-(6.53),

$$p^{k+1}(n,m,q) = p^k(n,m,q) + p^k_{IN}(n,m,q) - p^k_{OUT}(n,m,q) \tag{6.54}$$

Eq.(6.54) is the desired iteration for the transport of holes where, we identify the last two terms on the right side with $(\partial p/\partial t)_{DRIFT}$. Note that the right side contains neighboring cells , as well as the principal cell, all evaluated during the kth step. The velocities are of course are field dependent, wherein $u_p(n,m,,q)$ will rely on the total field, given by Eq.(6.49) at the kth time step

A similar iteration may be developed for the electrons. The major difference is that the electrons will respond fashion in opposite because of their negative charge. The number of electrons exiting and entering the cell, n^k_{OUT} and n^k_{IN} during the kth time step are then

$$n^k_{OUT}(n,m,q) = (u_{nx}(n,m,q)\Delta t/\Delta l)n^k(n,m,q) + (u_{ny}(n,m,q)\Delta t/\Delta l)n^k(n,m,q) +$$
$$(u_{nz}(n,m,q)\Delta t/\Delta l)n^k(n,m,q) \tag{6.55}$$

$$n^k_{IN}(n,m,q) = (u_{nx}(n,m,q)\Delta t/\Delta l)n^k(n+1,m,q) + (u_{ny}(n,m,q)\Delta t/\Delta l)n^k(n,m+1,q) +$$
$$(u_{nz}(n,m,q)\Delta t/\Delta l)n^k(n,m,q+1) \tag{6.56}$$

Note that for n^k_{IN} the incoming electrons emanate from the n+1, m+1, q+1 cells because of their sign. Making use of Eqs.(6.55) and (6.56) then provides the iteration for the number of electrons in the (n,m,q) cell, or

$$n^{k+1}(n,m,q) = n^k(n,m,q) + n^k_{IN}(n,m,q) - n^k_{OUT}(n,m,q) \tag{6.57}$$

and the last two terms on the right may be identified with $(\partial n/\partial t)_{DRIFT}$

The iterative equations, Eqs.(6.54) and (6.57), rely on a nearest neighbor approximation, which becomes more and more accurate as the ratio of the drift to electromagnetic velocities decreases. The nearest neighbors are the six cells surrounding the (n,m) cell which share cell faces (in Fig. 6.3 only one of the nearest cell neighbors is shown). If one wishes to gain greater accuracy, then one must consider additional neighbors. For example one may consider all 26 neighbors surrounding the (n,m,q) cell, which come into contact with the (n,m,q), either at a face , edge, or corner.

6.12 Cell Charge Iteration. Equivalence of Drift and Inter-Cell Currents

As was stated earlier in the Chapter the carrier occupancy change due to drift is equivalent to that from the current flowing between cells via the node resistors. This is discussed in more explicit fashion, in which first we calculate the current in terms of the carrier properties and then calculate the change in the cell charge from one step to the next. We begin by calculating the current entering and leaving the (n,m,q) cell, in much the same fashion that we obtained the particle current. To simplify matters, we assume the field is applied in the z direction. The current at the z+ face denoted by $I_{z+}(n,m,q)$ is then

$$I_{z+}(n,m,q) = I_{n,z-}(n,m,q) + I_{p,z+}(n,m,q) \tag{6.58}$$

and , similarly for the z- face the current is

$$I_{z-}(n,m,q) = I_{n,z-}(n,m,q) + I_{p,z-}(n,m,q) \tag{6.59}$$

The change in the cell charge $\Delta q(n,m,q)$ in the (n,m,q) cell is

$$\Delta q(n,m,q)\,/\Delta t = I_{z-}(n,m,q) \, - \, I_{z+}(n,m,q) \tag{6.60}$$

Note the minus sign at the z+ face, which indicates positive charge leaving the (n,m,q) cell (or, equivalently, negative charge entering the cell at z+). To relate the current to the carrier properties we use the relationships, $j_p = \sigma_p E$, $j_n = \sigma_n E$ to obtain

$$I_{z-} = \Delta l^2 \{ \sigma_{p,z-}(n,m,q) + \sigma_{n,z-}(n,m,q) \} E_{z-}(n,m,q) \} \qquad (6.61)$$

$$I_{z+} = \Delta l^2 \{ \sigma_{p,z+}(n,m,q) + \sigma_{n,z+}(n,m,q) \} E_{z+}(n,m,q) \} \qquad (6.62)$$

where the $z+$, $z-$ subscripts indicate the conductivities centered about the $z+,z-$ cell faces. Since the conductivities are located at the $z-$ and $z+$ faces, we may form the averages

$$\sigma_{p,z-}(n,m,q) = [\sigma_p(n,m,q) + \sigma_p(n,m,q-1)]/2 \qquad (6.63)$$

$$\sigma_{p,z+}(n,m,q) = [\sigma_p(n,m,q) + \sigma_p(n,m,q+1)]/2 \qquad (6.64)$$

$$\sigma_{n,z-}(n,m,q) = [\sigma_n(n,m,q) + \sigma_n(n,m,q-1)]/2 \qquad (6.65)$$

$$\sigma_{n,z+}(n,m,q) = [\sigma_n(n,m,q) + \sigma_n(n,m,q+1)]/2 \qquad (6.66)$$

As discussed in Chapter II the conductivities at the faces represent auxiliary cells , which are related to the usual cell conductivities via Eqs.(6.63)-(6.66). We still have not expressed the conductivities in terms of the carrier properties, and these are obtained from the standard transport relationships

$$\sigma_p(n,m,q) = (e/\Delta l^3) \, \mu_p(n,m,q) p(n,m,q) \qquad (6.67)$$

$$\sigma_n(n,m,q) = (e/\Delta l^3) \, \mu_n(n,m,q) n(n,m,q) \qquad (6.68)$$

We then express the fields at the cell faces, $z+$ and $z-$, in terms of the TLM voltage waves associated with the cell,

$$E_{z+}(n,m,q) = -\Delta l \, V_{z+}(n,m,q) \tag{6.69a}$$

$$E_{z-}(n,m,q) = -\Delta l \, V_{z-}(n,m,q) \tag{6.69b}$$

where $V_{z+}(n,m,q)$, $V_{z-}(n,m,q)$ are the z directed voltage waves , which with the help of Fig.6.2 are given by

$$Vz+(n,m,q)=(1/4)[V_{xz}(n,m,q)+V_{xz}(n,m-1,q) +V_{yz}(n,m,q)+V_{yz}(n-1,m,q) \tag{6.70}$$

$$Vz-(n,m,q)=(1/4)[V_{xz}(n,m,q-1)+V_{xz}(n,m-1,q-1)+V_{yz}(n,m,q-1)+V_{yz}(n-1,m,q-1)$$
$$\tag{6.71}$$

Finally we return to the cell charge iteration based on Eq.(6.60) , or

$$q^{k+1}(n,m,q) = q^{k}(n,m,q) + \Delta t[I^{k}_{z-}(n,m,q) - I^{k}_{z+}(n,m,q)] \tag{6.72}$$

The current terms $I^{k}_{z-}(n,m,q)$, $I^{k}_{z+}(n,m,q)$ in Eq.(6.72) may then be expressed in terms of either the TLM formulation (voltage waves and node resistors), or in terms of the equivalent semiconductor properties. Continuing with Eq.(6.72), we substitute Eqs.(6.61)-(6.62) into (6.72) to obtain

$$q^{k+1}(n,m,q)-q^{k}(n,m,q) =\Delta l^{2}\Delta t[\sigma_{p,z-}(n,m,q) + \sigma_{n,z-}(n,m,q)] \, E^{k}_{z-}(n,m,q)$$
$$- \Delta l^{2}\Delta t[_{p,z+}(n,m,q)+ \sigma_{n,z+}(n,m,q)] \, E^{k}_{z+}(n,m,q) \tag{6.73a}$$

where $\sigma_{p,z-}(n,m,q)$, etc..., are given by Eqs.(6.63)-(6.66).

To recast the above in terms of TLM parameters, we relate E^{k}_{z-} and E^{k}_{z+} to the voltage waves using Eqs.(6.69)-(6.71). and then relate the face centered conductivities to the node resistors $R(n,m,q)$ and $R(n,m)$, using the 2D expressions in Chapter II. For small changes , the net charge iteration is then found to be equivalent to the TLM iteration given in Eq.(6.16). This result is found by first calculating $\Delta V^{k+1}(n,m,q)$ (in terms of the k state values) in Eq.(6.16) and then expanding under the condition of small loss. Use is made of the TLM relation, $Z_{o}=1/v\varepsilon$. If the drifting carriers traverse only the z+ nodes, then the change in cell charge is quickly evident. The subject is further explored in Chapter VII.

We can also cast Eq.(6.73a) in terms of the semiconductor properties, if we substitute Eqs.(6.63)-(6,66),(6.67)-(6.68) and expand $\mu_p(n,m,q-1)$, $\mu_n(n,m,q+1)$, $E_{z+}(n,m,q)$, and $E_{z-}(n,m,q)$ about the central cell(n,m,q). The approximate change in cell charge is then

$$q^{k+1}(n,m,q) - q^k(n,m,q) = (eu_{pz}(n,m,q)\Delta t/\Delta l\,)[\,p^k(n,m,q-1)-p^k(n,m,q)]$$
$$- (eu_{nz}(n,m,q)\Delta t/\Delta l\,)[\,n^k(n,m,q+1)-n^k(n,m,q)] \qquad (6.73b)$$

We remind ourselves that the right side of the above is a small quantity based on the fact that $u_{pz}(n,m,q)$, $u_{nz}(n,m,q)$ are each much smaller than the electromagnetic velocity $\Delta l/\Delta t$.

We now reconcile the cell charge iteration in Eq.(6.72) with the results of the drift model given in Section 6.11. We should not expect any differences since both results are based on the same semiconductor drift properties. Consider the carrier drift model used in Section 4.11. By definition the change in the cell charge is

$$q^{k+1}(n,m,q)-q^k(n,m,q)=e[p^{k+1}(n,m,q)-p^k(n,m,q)]-e[n^{k+1}(n,m,q)-n^k(n,m,q)] \quad (6.73c)$$

We then insert the equations (6.52)-(6.57) from the drift discussion in the above, where for purposes of comparison we consider only the z component of the drift velocity. This gives precisely the same result as Eq.(6.73b), as expected, since both results derive from the semiconductor model.

Finally we point out that although there is a large disparity in the carrier drift and wave propagation velocities, and although charge transport occurs only at the edges of the cells, this does not prevent the TLM waves from re-adjusting the electric fields at the speed of the propagation velocity. In the following Chapter, for example, we will show in a simulation that the fields at the boundary of a conducting region(produced by a nanosecond laser pulse) will grow at a rate commensurate with the propagation velocity.

6.13 Carrier Diffusion

As is well known, random thermal processes in semiconductors give rise to carrier flow from regions of high carrier concentration to regions of lower concentration. The hole and electron diffusion equations are

$$(J_p)_{diff} = -(e/\Delta l^3)D_p\nabla p(n,m,q) \tag{6.74}$$
$$(J_n)_{diff} = (e/\Delta l^3)D_n\nabla n(n,m,q) \tag{6.75}$$

where $(J_p)_{diff}$, $(J_n)_{diff}$, are the hole and electron current densities, caused by diffusion, $p(n,m,q)$ and $n(n,m,q)$ are the number of holes and electrons occupying the cell of volume Δl^3. D_p, D_n are the diffusion constants for holes and electrons , and e is the electron charge(absolute value). Note the negative sign for the hole current, which arises since we assume a positive density gradient, and therefore the motion of the holes, as well as the hole current, are in the negative direction. With the electrons, of course, the current is opposite to the electron flow and no negative sign is necessary. The discussion of the diffusion constants is given in the standard semiconductor references [1].

The rate equations corresponding to Eqs.(6.74) and (6.75) are transcribed into cellular notation. For this purpose , we may visualize the diffusion taking place between adjoining cells, (n-1,m,q), (n,m,q), and (n+1,m,q), and that both the hole and electron cell occupancy increases with x(or index number n) . We now wish to determine the time change in the number of holes contained in the (n,m,q) cell, resulting from diffusion. We first calculate the diffusion of holes between (n+1,m,q) to (n,m,q). Since the number of holes in (n+1,m,q) is greater than that of (n,m,q) the holes will diffuse from (n+1,m,q) to (n,m,q). Just as with drift phenomena, we may obtain the number of holes entering (n,m,q) from (n+1,m,q) , during the kth time step. This number, for the x direction, is designated as $p^k_{x,IN}$,

$$p^k_{x\,IN}(n,m,q) = (D_p(n,m,q)/\Delta l^2)\{\, p^k_x(n+1,m,q) - p^k_x(n,m,q)\}\, \Delta t \tag{6.76}$$

Eq.(6.76) is obtained from Eq.(6.74) by remembering that the cross-section for the current density is Δl^2 and that the gradient of the hole number is the difference in hole numbers between the two cells divided by Δl. We then set $(J_p)_{diff} \Delta l^2 \Delta t$ equal to $e\, p^k_{x\,IN}(n,m,q)$ to obtain Eq.(6.76). In like manner we can then obtain the number of holes leaving the cell which occurs when the holes go from (n,m,q) to $(n-1,m,q)$ during the k step. For the x direction we call this $p^k_{x,OUT}(n,m,q)$.

$$p^k_{x,OUT}(n,m,q) = (D_p(n,m,q)/\Delta l^2)\, \{p^k(n,m,q) - p^k(n-1,m,q)\}\Delta t \qquad (6.77)$$

Note that we have provided cell indices to $D_p(n,m,q)$ to indicate that the diffusion constant should be evaluated at each cell during the time step. We are now prepared to write down the iterative equation relating the number of holes in the (n,m,q) cell, during the k+1th time step, to that during the kth step, caused by diffusion in the x direction. Combining Eqs.(6.76) and (6.77).

$$p^{k+1}(n,m,q) = p^k(n,m,q) + [\ p^k_{xIN} - p^k_{xOUT}] \qquad (6.78)$$

Since Eq.(6.78) only applies to diffusion in the x direction., we must also account the y and z directions as well, thus obtaining the complete hole diffusion iteration, or,

$$p^{k+1}(n,m,q) = p^k(n,m,q) + (P^k_{x,IN} - P^k_{x,OUT}) + (P^k_{y,IN} - P^k_{y,OUT}) + (P^k_{z,IN} - P^k_{z,OUT})$$
$$(6.79)$$

where the definitions for $P^k_{y,IN}$, $P^k_{z,IN}$, etc... are similar to Eqs.(6.76) and (6.72). An analogous iteration also exists for the electron carriers:

$$n^{k+1}(n,m,q) = n^k(n,m,q) + (N^k_{x,IN} - N^k_{x,OUT}) + (N^k_{y,IN} - N^k_{y,OUT}) + (N^k_{z,IN} - N^k_{z,OUT})$$
$$(6.80)$$

The diffusion of carriers into the (n,m,q) cell of course contributes to the conductivity and changes the recombination properties, as well.

6.14 Frequency of Transport Iteration

Since the drift and diffusion transport phenomena are generally much slower than that of the electromagnetic type, we should question whether it is essential to perform the transport iteration during each time step. The cell size, and the corresponding time step, are determined from the outset, by the electromagnetic behavior. Therefore, during successive time steps, the changes in the cell occupancy will not be noticeable , except at the border regions of the cell. This is due to the much slower carrier drift and diffusion velocities, $u(n,m,q)$ and $u_{DIFF}(n,m,q)$, compared to the electromagnetic velocity, $v(n,m,q)$. As shown in Fig.6.3, if we are dealing with a duration of Δt, then only a small region , near the cell border, is occupied by the incoming carriers from the adjacent cell. The carriers in this narrow region, however, are averaged over the *entire* cell. This means that over the next time step some carriers(though small in number) will be treated as though their drift (or diffusion) velocity is comparable to the propagation velocity. This of course is not a real result but an artifact of the type of iteration selected. If the cell size(or time step) is made sufficiently small enough, the accompanying artifact also shrinks in size.

An alternate approach is to postpone the transport iteration until such time that the carrier front traverses the cell length. If Δk is the number of time steps needed for the front to traverse the cell

$$\Delta k \sim v(n,m,q)/\, u(n,m,q) \text{ for carrier drift} \qquad (6.81a)$$

$$\Delta k \sim v(n,m,q)/\, u_{DIFF}(n,m,q) \text{ for carrier diffusion} \qquad (6.81b)$$

where $u_{DIFF}(n,m,q)$ is of course profile dependent. A useful definition of $u_{DIFF}(n,m,q)$ given in cellular notation for the hole carriers is [1]

$$u_{DIFF}(n,m,q) = -[D(n,m,q)/p(n,m,q)]\,[(p(n+1,m,q)-p(n-1,m,q))2\Delta l] \qquad (6.82)$$

and a similar expression for electrons . The second bracketed term in Eq.(6.82) is recognized as the gradient in the x direction.

The iteration for drift or diffusion, therefore , is not implemented , until a number of time steps have elapsed, specified by Eq.(6.81a) or (6.81b). As a result of this iteration, the cell will suddenly acquire carriers when Eq.(6.81a) or Eq.(6.81b) is satisfied. At the same time, the electromagnetic iteration may proceed more frequently, with the limit determined by the cell size, i.e., Δt.

6.15 Total Contribution to Changes in Carrier Cell Occupancy

For completeness we state the total iterative equation describing changes in the carrier density, taking into account the contributions of carrier generation, recombination, carrier drift, and diffusion. We confine the iteration to those results obtained in the previous Sections, bearing in mind that the carrier generation was restricted to a simplified form of light activation and that for recombination we considered only a single deep level. In any event, the important point is to convey the techniques involved in integrating arbitrary phenomena into the transmission line matrix approach.

Assuming the rate equations are valid, we may sum the rates the for holes and electrons to obtain a total iterative equation in which the contributions from the various mechanisms which add or subtract carriers from a cell are all included. Thus for each carrier the total iteration is

$$p^{k+1}(n,m,q) - p^{k}(n,m,q) = (\partial p/\partial t)^{k}_{LIGHT} + (\partial p/\partial t)^{k}_{AVALANCHE} + (\partial p/\partial t)^{k}_{RECOMB}$$
$$+ (\partial p/\partial t)^{k}_{DRIFT} + (\partial p/\partial t)^{k}_{DIFF} \tag{6.83}$$

$$n^{k+1}(n,m,q) - n^{k}(n,m,q) = (\partial n/\partial t)^{k}_{LIGHT} + (\partial m/\partial t)^{k}_{AVALANCHE} + (\partial n/\partial t)^{k}_{RECOMB}$$
$$+ (\partial n/\partial t)^{k}_{DRIFT} + (\partial n/\partial t)^{k}_{DIFF} \tag{6.84}$$

The various rates for the holes and electrons, for each process, are scattered throughout the Chapter and are summarized in the following Table.

ITERATIVE RATE EQUATIONS FOR VARIOUS TRANSPORT PHENOMENA

PROCESS	HOLES	ELECTRONS	NET CHARGE
Light Activation	Eq.(6.23)	Eq.(6.22)	Difference
Avalanche	Eq.(6.36)	Eq.(6.37)	Difference
Recombination	Eq.(6.48)	Eq.(6.47)	Difference
Drift	Eq.(6.54)	Eq.(6.57)	Difference
Diffusion	Eq.(6.79)	Eq.(6.80)	Difference
Cell Charge			Eq.(6.16)

REFERENCES

1. S. Wang, *Semiconductor Electronics*, McGraw Hill, New York, 1966.

VII.   Description of TLM Iteration

In the previous Chapters we have laid the groundwork for constructing the iterative relationships of the electromagnetic fields in a medium undergoing conductivity changes, based on the TLM formulation. The actual transcription from the physical principles to a workable computer program , however, is not entirely straightforward. In this Chapter we examine the iterative steps involved, and focus on a particular example, that of a semiconductor switch activated by a light pulse, to further illustrate the description. Fig.7.1 gives an outline of the main steps in the iteration. Each of the steps is explained in Section 7.2 and the Appendices, using the same example of the light activated semiconductor. Results of the computer iteration are given for several cases of interest. Before delving into the details of the iteration, we first select the configuration to which the iteration will correspond.

7.1 Specification of Geometry

The example geometry, Fig.7.2, consists of a parallel plate transmission line, incorporating a photoconductive switch. We consider the geometry to be two dimensional, with rectangular type boundary conditions. The height between conductors is taken to be h = 1.5 mm and the width is 5 mm. For this switch, a portion of the top conductor, of length l_o = 2.5 mm , is removed and substituted with a semiconductor of the same length and cross-section as the conductor. The gap region, containing the semiconductor, therefore represents a high impedance to any incoming wave. The impedance remains high, unless conductivity is induced in the semiconductor, either by a light signal or by avalanching. For arbitrary inductive capacity of the dielectric region , and with

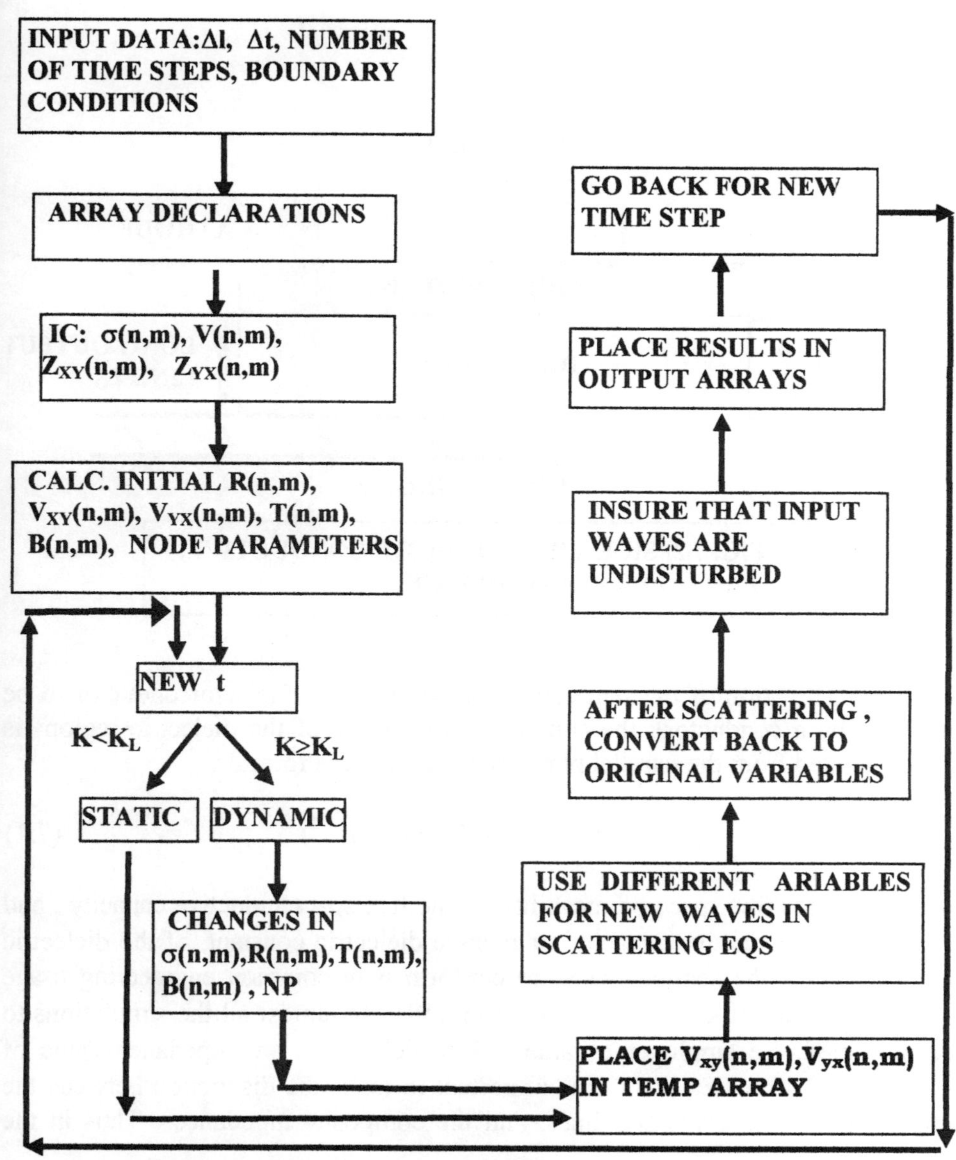

FIG. 7.1 TLM ITERATION OUTLINE(2D).

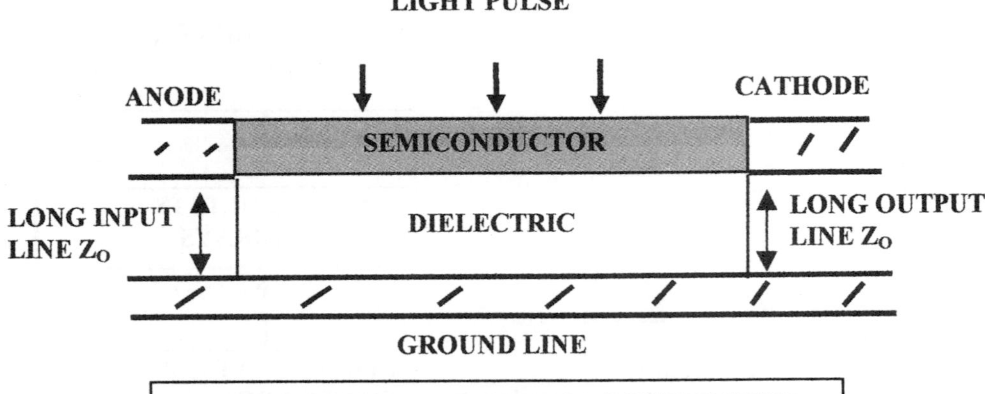

FIG. 7.2 SIDE VIEW OF PHOTOCONDUCTIVE, PARALLEL PLATE SWITCH.

complete activation of the semiconductor (we imagine the semiconductor to be replaced by a conductor), the composite impedance of the dielectric region is approximated from the parallel plate, transmission line formula,

$$Z = (\mu_o/\varepsilon_o)^{1/2}(h/w\varepsilon^{1/2}) \approx 376.7 \, (\, h/w\varepsilon^{1/2}) \tag{7.1}$$

where μ_o is the free space permeability, ε_o the free space inductive capacity , and ε the specific inductive capacity, or relative dielectric constant, of the dielectric region beneath the semiconductor(to conform with common engineering usage we use ε instead of κ in Chapters VII and VIII). In almost all the simulations to be presented, we employ the value $\varepsilon=1$, which yields an impedance value of about 113 Ω from Eq.(7.1). It is important to make the distinction between the impedance of the *individual* lines, and the composite inpedance. Thus in the case of $\varepsilon = 1$, the individual line impedance for the dielectric region is 376.7Ω while the composite impedance is 113Ω.

The input and output composite impedances, on the other hand, are assumed to be 50 Ω, a standard value associated with microwave transmission lines. The input/output cross-sections are taken to be the same as the device cross-section, meaning the same as the dielectric region beneath the activated semiconductor. This arrangement implies a dielectric constant of about 5.11 for the input/output lines. Just as we did with the dielectric region, we should then point out that, for the input/output, the impedance of the *individual* TLM lines will then be 166.7Ω, as opposed to the 50Ω composite value. In any event, the device will represent same positive mismatch to the input when $\varepsilon = 1$ in the dielectric region. One should mention that the width to height ratio of ≈ 3.3, together with the presence of the semiconductor, implies that fringing is very significant and that radiation from the sides is important as well. To deal with such limitations, we must resort to a 3D cell matrix for greater accuracy. In this example we forego the 3D treatment in the interest of minimizing the amount of detail, emphasizing instead the essential points in the iteration process.

An important preliminary in the iterative formulation involves breaking up the various regions into appropriate cells using the labeling technique described in Chapter II and Chapter V. Note the various regions in Fig.7.3, which include the semiconductor region (S), the dielectric region(D), and the input(IN) and output(O) regions. In addition we have the three conducting regions: the anode, cathode , and groundline regions, labeled (C1), C2), and C3) respectively. Finally we have the three high impedance regions: (I1),I2), and (I3). The high impedance region located immediately above the semiconductor, (I1), has the effect of turning back any wave attempting to leave the top portion of the semiconductor Regions (I2)and (I3) are adjacent the input and output regions . These regions serve as a buffer to the input and output , helping to simulate long input and output lines, which serves to prevent unwanted multiple reflections. The simulated long lines are implemented with the aid of matching resistors discussed in Chapter V and later in this Chapter.

As noted in the Figure, we first treat a simple boundary type in which the cell size in region (D) is a multiple of that in the semiconductor; in this case the multiple is three. implying that the propagation velocity in the dielectric region is 3X that in the semiconductor. This also implies a dielectric constant of 9 in the semiconductor, obtained from

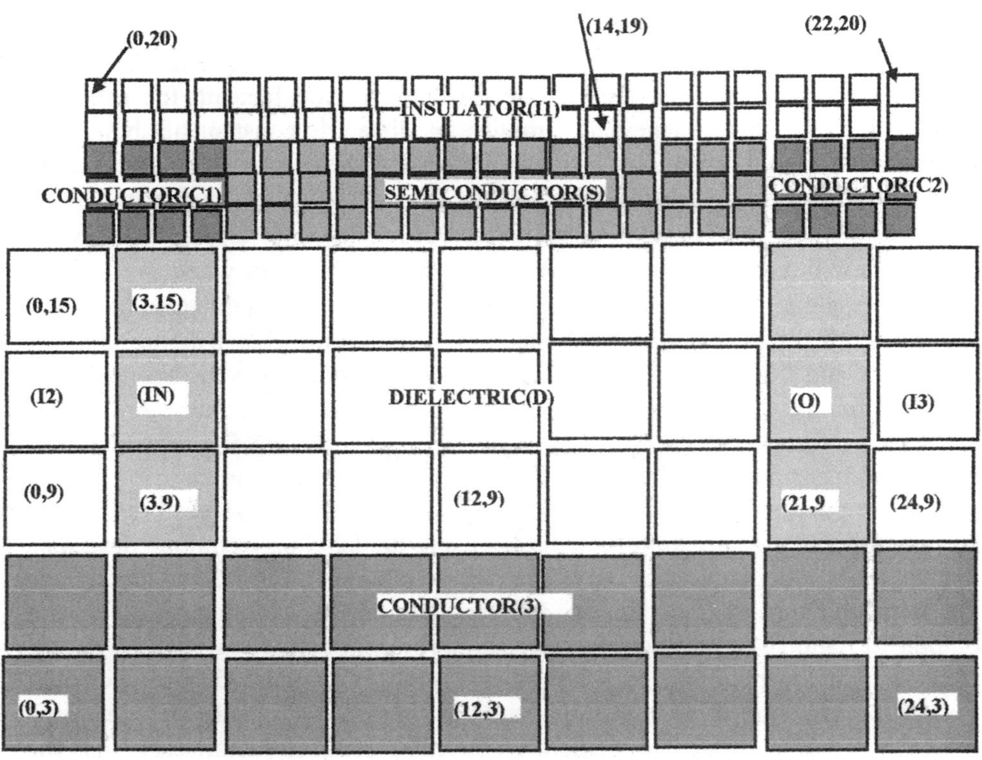

FIG. 7.3 VARIOUS REGIONS OF SEMICONDUCTOR SWITCH CELL MATRIX.

$$\varepsilon_S = (v_D/v_S)^2 \varepsilon_D \tag{7.2}$$

where the subscript denotes the particular region. The delay time in each cell TLM line (whether the cell is large or small) is equal to 1.6678ps.

The selection of propagation velocities having a multiple relationship is done as a matter of convenience since devices with arbitrary dielectric constants will in general contain non-aligned quasi nodes at the dielectric/semiconductor interface. To treat this situation, we must use the nearest node approach in the iteration, as discussed in Chapter V. Section 7.14 illustrates the results of an iteration when we are required to use the nearest node approximation. Also, as pointed out previously, the device structure is mismatched to the input/output of 50Ω, when the semiconductor is entirely activated to a state of high conductivity. The modifications to the program, needed to produce matched structures, however, are relatively minor Simulation results for the case of a matched device are discussed in Section 7.15

It may seem odd at first that we show cells in the conduction regions, as well as in the high impedance region. There is no wave propagation in these regions and therefore one would not expect the need for such cells. There are two reasons for continuing the cell structure into these regions. One is that the numbering system is easy to extend since the same cell structure is continued throughout all the regions (one possible disadvantage here is the added memory one must allot ; if the conducting regions account for a significant portion of the volume, and sufficient memory is a concern, then only cells near the boundary should be added). The second reason is that for very small cells, the wave attenuation, or reflection, may be gradual and several cells may be needed to accurately describe the waves as they transition these regions. We also note that the same cell size in the (D) and (S) regions are carried over into the adjoining conductors and high impedance regions. Thus the same size cell as that of the semiconductor is carried over to the conductors (C1),C(2), and the high impedance region (I1). Likewise, the same cell size as in the dielectric(D) region is carried over to the conductor (C3) and the high impedance regions, (I2) and (I3), and the Input/Output regions, (IN) and (O). In particular, it may seem inappropriate to choose the same length for the I/O regions, since we know the dielectric constant here is ~5.11 and one would expect an appropriately smaller length, according to the factor $(1/5.11)^{1/2}$. But in fact there is no reason to select a different sized cell, since the only function of the input/output regions is either to inject a signal or terminate a signal and there is no scattering within these regions.

Before proceeding we must decide on the number of cells to be employed in the iteration. This choice represents a tradeoff, of course, between the simulation accuracy and the computer capabilities. For the problem at hand, we choose a cell density which can provide reasonable quantitative results, without creating a burden for the computer. For regions (C1), (C2), (S), and (I1) we choose a 23X5 array , with the cell length equal to 1.667 mm. A total of 15X3 such cells are required to cover the semiconductor itself. In the regions beneath the semiconductor , (IN),(O), (D), (C3),(I2) ,and (I3), we construct 9X5 cells. Here, of course, the cells are 3X larger, with each equal to 0.5 mm in length. The cell nomenclature, following the guidelines in Chapter II and V, is shown in Fig.7.3 for several selected cells. Note that for the larger cells the indices of adjacent cells differ by three. Thus we follow the *single index* notation for the larger cells as outlined in Chapter V, Section 5.4. This insures that the indices properly locate the cell, as do the indices for the smaller cells. Looking at Fig.7.3, if n,m are the indices, this means that for the large cells, an (n,m) cell is undefined if m<15, if either m or n is *not* a multiple of three. As a result, the omnipresent iterative loops, which are used in the program, must be modified to exclude such cells. The iterative details are discussed in the Appendices.

7.2 Description of Inputs and TLM Iteration Outline

We now discuss the broad outline of the iteration, Fig.7.1, leaving the details to App.7A.1 and 7A.2. The actual program statements are given in App.7A.2. In order to start the program, various kinds of input information such as the number of time steps, the activation time, I/O impedances, the type of output data requested, etc.. , are required. The Input statements are listed in App.7A.1, Table 7A.1. The next step is to declare the various arrays needed during the course of the iteration, and these are listed in Table 7A.2. These arrays will include the cell voltages, the backward and forward waves in the horizontal and transverse lines, the node resistance, etc...

Following the declaration of arrays, the initial conditions are specified starting with the values of the cell voltages array. Very often we wish to determine the static field profile under the conditions of a constant input voltage at the anode. Under these conditions the cell voltage is everywhere zero except for the input region and possibly (C1), the anode. The word possibly is used since we can just as well set the anode cells initially equal to zero and allow the waves in the input region raise the voltage of the anode. We assume the voltage is evenly distributed among the input cells, m=9,12 and 15. In fact what is specified are *input waves* in lines $Z_{xy}(3,6)$, $Z_{xy}(3,9)$, $Z_{xy}(3,12)$, and $Z_{xy}(1,15)$, $Z_{xy}(2,15)$, $Z_{xy}(3,15)$. These lines are of course part of the 50 Ω input composite line. In each of these lines, the input wave is V_0 /8. The choice of Vo/8 is of course not accidental ; the total composite input voltage adds up to V_o/2 The individual input wave amplitudes,V_0/8, are re- introduced at every time step during the charge-up process until equilibrium is achieved, at which time the voltage on the anode(actually the anode cells bordering the dielectric) attains V_0, which is the sum of the composite forward and composite backward waves , each equal to Vo/2, at equilibrium. The backward waves are assumed to be terminated, as we will discuss in shortly.

Following the specification of the initial cell voltages and input waves , we specify the initial values of the other arrays. Certain arrays, in particular, the transmission line impedance values, normally are constant with time throughout the iteration. These include the horizontal line impedances, denoted by $Z_{xy}(n,m)$, and the transverse lines, $Z_{yx}(n,m)$, surrounding each cell. We recall that only two of the four lines surrounding each cell are associated with the (n,m) cell, these two being in the direction of increasing n and m. As noted previously the impedance in the dielectric and semiconductor is given simply by $(\mu/\varepsilon)^{1/2}$. As mentioned before, the horizontal lines in the I/O regions possess an impedance such that the composite value is 50Ω. The high impedance regions, (I1), (I2), and (I3) are assigned very high impedance values ($\sim 10^{12}\Omega$) while the conducting regions are given very low values ($\sim .01\Omega$).   Next we look at the forward and backward waves in both the transverse and horizontal lines(for a total of four arrays), denoted by $^+V_{yx}(n,m)$,$^-V_{yx}(n,m)$ and $^+V_{xy}(n,m)$, $^-V_{xy}(n,m)$ respectively. The initial wave values are obtained from the voltage cell array, i.e., by taking the voltage differences of adjacent cells, as outlined in Chapter I and elsewhere. Thus , if we examine two adjacent voltage cells V(n,m), V(n+1,m) , then the sum of the waves in the $Z_{yx}(n,m)$ line is equal to the difference in the cell voltage, or

$$V_{yx}(n,m) = {}^{+}V_{yx}(n,m) + {}^{-}V_{yx}(n,m) = V(n+1,m) - V(n,m) \tag{7.3}$$

Under equilibrium conditions, we assume the waves are equal and thus,

$${}^{+}V_{yx}(n,m) = {}^{-}V_{yx}(n,m) = [V(n+1,m) - V(n,m)]/2 \tag{7.4}$$

Similar equation for the horizontal line provide

$$V_{xy}(n,m) = {}^{+}V_{xy}(n,m) + {}^{-}V_{xy}(n,m) = V(n,m+1) - V(n,m) \tag{7.5}$$

And under equilibrium conditions

$${}^{+}V_{xy}(n,m) = {}^{-}V_{xy}(n,m) = [V(n,m+1) - V(n,m)]/2 \tag{7.6}$$

The previous assumes a knowledge of the initial cell voltages. If we assume the initial cell voltages are everywhere zero, the only initial array elements required are the wave amplitudes in the horizontal lines of the (IN) region. As previously mentioned, the wave amplitudes in fact are

$${}^{+}V_{xy}(n,m) = V_o/8 \; : \quad m=6,9,12,15 \;, \quad n=3 \tag{7.7}$$

where V_o is the assumed anode voltage.

Next, we check the node resistor array R(n,m). Again we use the same indices as that for the basic cells, and we also remind ourselves that the node for each cell is located , by definition, at the upper right hand corner of the cell. As anticipated the initial node resistance will be very large in the semiconductor and dielectric regions (10^7 - $10^{12}\Omega$) and very small in the conducting regions(.01Ω). The node resistance is also very large in region (I1), as one might expect, as well as in regions (I2) and (I3), which buffer the input and output regions. However, finite resistance values must be inserted at the nodes along the interface between the (I2) and (IN) regions and between the (O) and (I3) regions. These resistors serve as effective matching resistors to the reflected waves of the (IN) region and to the forward waves in the output (O) region. Thus the node resistors $R(0,m)$, $m=6,9,12,15$ terminate the reflected waves in the horizontal lines, $Z_{xy}(3,m)$, $m=6,9,12,15$. By inserting terminating resistors, thereby removing the effects of any reflected waves, in effect we simulate the input as an infinitely long transmission line as discussed in Chapter V. We again make the point that we are not disallowing reflections from the input plane, but that we are simply getting rid of any possible reflections from the source, which would then work their way back into the configuration under study. As indicated before, each of the four lines comprising the input has an input wave amplitude of $V_0/8$. The value of $R(n,m)$ selected is obtained is obtained from Chapter V, using either Eq.(5A.2) or Eq.(5A.4), as noted in Figs.5A.1 and 5A.2   Fig.5.A1 applies to nodes $(0,9)$ and $(0,12)$ while Fig.5A.2 applies to nodes $(0,6)$ and $(0,15)$. It should be noted that three of the TLM lines have very high impedance levels and therefore these line impedances do not contribute to the terminating resistance. The parallel combination of the resistors in line $Z_{xy}(3,9)$ is $(3/4)R(0,9)$ and for line $Z_{xy}(3,12)$ it is $(3/4)R(0,12)$. From Eq.(5A.2) the matching requirement leads to, for a uniform input dielectric,

$$R(0,9) = R(0,12) = (4/3)\, Z_{xy}(3,9) = (4/3)Z_{xy}(3,12) \qquad (7.8)$$

where the input line impedance is $\{\mu/\varepsilon\}^{1/2}$ with a dielectric constant of 5.1085, thus yielding a line impedance of 166.67Ω. The matching resistor, from Eq.(7.8) is 222.2Ω. Calculating the matching resistance for the remaining input nodes, $(0,15)$ and $(0,6)$, is similar except for the adjacent conductor. The total resistance is thus $(2/3)R(0,15)$ or $(2/3)R(0,6)$ and hence the matching requirement gives, from Eq.(5A.4),

$$R(0,6) = R(0,15) = (3/2)Z_{xy}(3,6) = (3/2)Z_{xy}(3,15) \qquad (7.9)$$

with a matching resistance of 250Ω. Similar arguments may be used to determine the matching resistors for the forward waves in the output region, providing

$$R(21,12) = R(21,9) = (4/3) Z_{xy}(21,12) = (4/3)Z_{xy}(21,9) \qquad (7.10)$$

$$R(21,15) = R(21,6) = (3/2)Z_{xy}(21,15) = (3/2)Z_{xy}(21,6) \qquad (7.11)$$

and the impedance of $Z_{xy}(21,m)$ is the same as that for the input. The output matching resistors are identical to those of the input.

Having defined the node resistor array, and its initial conditions, we call attention to several other arrays, listed in App.7A.3, which expedite the calculation and are defined in terms of the node resistor elements. Referring to Chapter III we make sure first to define the node parameter array, as well as the load impedance array. Finally, the 2 dimensional scattering coefficient arrays, $T(n,m,s)$ and $B(n,m,s)$, must be enumerated (we recall that the index s identifies the scattering route). With these auxiliary arrays defined, we can begin the iteration in both time and space,, as outlined in Fig.7.1. The iteration in time is taken as the outer loop. The first few hundred time steps are devoted to solving the static problem, i.e., we allow the input voltage to charge up the device and during this time there are no changes in any of the node resistors. Eventually the a stable solution is attained At $k= KL$, conductivity is introduced in the semiconductor and the stability ends, with waves now being both reflected and dissipated in regions where the conductivity has been added. In either case , the same basic iteration for both the "charge up" phase as well as the transient conductivity phase, is employed, the only difference being the assumed changes in the node resistor elements. As noted in Fig.7.1, there are several separate calculations for each time step. In the first, the presupposed changes in the node resistors are introduced; as noted before the changes in the node resistance reflect changes in the conductivity of a particular region, which in turn is induced in the semiconductor by either photons or avalanching. Once the node resistor array is specified, we then calculate the node parameters, which consists of the parallel resistance and the load impedance. This is followed by the calculation of the scattering coefficient arrays. This completes the first stage of calculations. In the next step we run through all the wave arrays, $^{+}V_{yx}(n,m)$, $^{-}V_{yx}(n,m)$ $^{+}V_{xy}(n,m)$, $^{-}V_{xy}(n,m)$, and place them into temporary arrays. Although this may appear to be a superfluous step, it is actually quite necessary. This is because in the next step, when we actually calculate the new wave amplitudes for the next time step, the calculation involves various wave contributions surrounding a particular transmission line and node. These contributions should all belong to the same, previous time step. If we do not resort to a temporary array, we may inadvertently use an updated wave rather than one belonging to the previous time step. Omitting the temporary array then will have the erroneous result of distorting propagation effects throughout the medium.

Following the introduction of the temporary wave array, we proceed directly to the scattering equations outlined in Chapter III. On the right side of the scattering equations we use the temporary waves previously defined. On the left side of the equations, we use *different* temporary wave variables, for the same reasons given before, i.e., we wish to avoid premature use of waves belonging to the next time step. Finally, upon completing the scattering equations, we convert the temporary variables contained in the equations to the original wave amplitudes, $^{+}V_{xy}(n,m)$, etc... We insure, during the iteration, that we have not disturbed the constant amplitude input waves , and this is achieved by adding, the requirement ,after the scattering equations, that $^{+}V_{xy}(3,m) = V_{o}/8$ for m= 6,9,12, 15, as well as for $^{+}V_{xy}(2,15)$ and $^{+}V_{xy}(1,15)$. One should note that special care is needed when m=15 in the iteration. When such occurs, then $^{+}V_{yx}(n,15)$ and $^{-}V_{yx}(n,15)$ both vanish when n is a non-integral of three. This is intuitively obvious since such lines are embedded in the conductor and no fields should be expected. On the other hand when n is an integral of three then the node (n,m) is a regular node and $V_{yx}(n, 15)$ certainly exists. The scattering at the (n,m) node then follows the normal procedure.

7.3 Output Format

Output data is saved in the form of various arrays. One important data array , as a function of k, is the pulse delivered to the output(O) region. The output array variable $V_{out}(k)$ is given by

$$V_{out}(k) = {}^{+}V^{k}_{xy}(21,6) + {}^{+}V^{k}_{xy}(21,9) + {}^{+}V^{k}_{xy}(21,12) + {}^{+}V^{k}_{xy}(21,15) \qquad (7.12)$$

Note that Eq.(7.12) is the sum of the individual lines in the output region. The right side of Eq (7.12) is recognized as the sum of the forward waves of the horizontal TLM lines in the output region. These waves see a matched impedance. Once these waves gain entry into the lines $Z_{xy}(21, 6)$, etc... , there are no reflections back into the device.

In order to save array space, while retaining sufficient accuracy, one may wish to use multiples of k. For example, if we have a 1000 time steps for k, we may decide that sufficient accuracy is attained by calculating Eq.(7.12) every 10th step, in which case the array size reduced by a factor of ten.

Another useful data array, in cell space, is the average horizontal field in the semiconductor at a specified distance from the anode, $AV(V^k_{yx}(n,m))$ defined by

$$AV\,(V^k_{yx}(n,m)) = [\,^+V^k_{yx}(n,16) + \,^-V^k_{yx}(n,16) + \,^+V^k_{yx}(n,17) + \,^-V^k_{yx}(n,17)$$
$$+\,^+V^k_{yx}(n,18) + \,^-V^k_{yx}(n,18)]/3 \qquad (7.13)$$

where we have selected the above six elements for field averaging. The right side represents the field, averaged over the three vertical TLM lines for m= 16,17, and 18 at a given instant in time, i,e, at a given time step k. Again we may choose to use the reduced time step array, or we may choose the to determine the field at each time step for greater temporal resolution. A simpler version of Eq.(7.13) for the semiconductor field is given by the centrally located single cell at m=17,

$$V^k_{yx}(n,m) = \,^+V^k_{yx}(n,17) + \,^-V^k_{yx}(n,17) \qquad (7.14)$$

For the bulk of the computer runs which follow, we will use the single cell field estimate given by the m=17 cell in Eq.(7.14).

Another useful array makes up the 16 elements of the average horizontal field throughout the entire semiconductor gap, from anode to cathode. This array provides with a profile of the semiconductor field, between electrodes. We provide for several such arrays, each array having a different time element and therefore representing a "snapshot" of the semiconductor field at any given instant. We can use Eq.(7.14) to calculate each of the thirty two field elements, ranging from n=3 to n=18 , or

$$V^k_{yx}(n,m) = {}^+V^k_{yx}(n,17) + {}^-V^k_{yx}(n,17) \quad n=3,4,\ldots 18 \qquad (7.15)$$

Simulation examples of these profiles , as well as those for the cell charge and the time evolution of the load voltage and the semiconductor field, are given in the ensuing Sections .

In order to accommodate the array data, an output data file is opened up and arrays previously mentioned, plus any others, are placed in a column format. We then use any one of several graphical software packages available, such as EXCEL, to graph the data arrays.

Output Simulation Data

7.4 Conditions During Simulation

The input parameters for the problem under consideration is shown in Table 7A.2. The time step is 1.66787ps, which corresponds to a Δl of 0.5 mm in the dielectric and 0.1667 mm in the semiconductor. The total number of time steps KM, was varied between 500 and 5000. Normally the bias voltage was specified by assigning the appropriate amplitudes to the incoming waves in the input region. The values of the initial node resistance and line impedance depended of course on the particular region. In the case of the conducting region, a value of 0.01Ω was typically selected for R(n,m) as well as for $Z_{xy}(n,m)$ and $Z_{yx}(n,m)$. The use of any non-zero number less than 0.01 resulted in very little change. Inserting an outright zero for R(n,m) or $Z_{xy}(n,m)$, $Z_{yx}(n,m)$ is not recommended , however, since it may lead to occasional spurious results. This can easily be circumvented , of course , by inserting a lower limit value , say 10^{-8} Ω , into the program code. Conversely a value of 10^{12} Ω for R(n,m) was employed in the high impedance region. Similarly an R(n,m) value of 10^{12} Ω was used at all times in the dielectric region. Except for one particular example, Section 7.7, the same initial node resistance also was used for the semiconductor during charge-up. Smaller values of R(n,m) for the semiconductor naturally may also be used. Once conductivity is introduced the value of R(n,m) in the semiconductor is lowered. In almost all the examples which follow, the values of R(n,m) were lowered suddenly at a particular time step. Of course,   a gradual risetime may be built in over several time steps, and an example of this is provided. As indicated, activated $R(n,m)$ values of 15Ω and 150Ω were employed in the examples cited. The spread factor ,G0, which is an indication of the uniformity of the node activation, was assumed to be very small. An important input is the activation time, which is the time step (k=350, in these examples) when the conductivity is introduced. The activation was initiated only after sufficient has elapsed to insure that the static solution has been obtained. Once activation began, then the recovery process went into effect. The presence of recovery mechanisms , such as recombination, dictates whether $R(n,m)$ returns to its former large value, and at what speed. The recovery times used in the examples were either k=200, or essentially no recovery($k=10^8$). Only those nodes belonging to the semiconductor were activated. For total activation, the nodes n=3-18, and m=16-18 were all illuminated. In other examples, only a partial region of the semiconductor was activated.

7.5 Behavior During Charge-up and Establishment of Static Field Profile

Initially the device, except for the input, is void of any electromagnetic energy. The input lines in region (I) then begin to inject signals into the dielectric, semiconductor, and output regions When they do so, the equilibrium fields are not immediately attained. There are two reasons for this. In the first place, the signals have a finite propagation velocity and a time delay occurs before the signal can reach every nook and cranny of the device. Secondly, each cell has a capacitance and time is required for the signal to charge up the cell. It should not come as a surprise, therefore, that transient fields, resulting from the charge-up, will be observed in the device. One way to suppress such fields are to allow the input signal to come up to full amplitude very slowly, instead of suddenly injecting the input signals at full amplitude. We have avoided this procedure for the ensuring simulations, however, since much insight may be gained by examining the charge-up transients.

We first look at the signal developed at the output lines. The output transient, seen in Fig (7.4), has several noteworthy characteristics. One notable

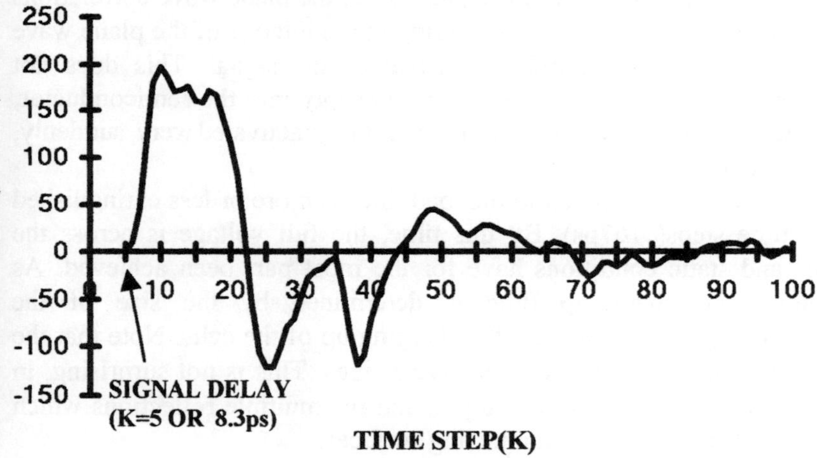

point to observe is the signal delay . The arrival of the earliest signal occurs at about 6 time steps(each time step is 1.6678 ps). This time of arrival is expected, since it represents the approximate propagation delay from the input to the output. The output transient amplitude, however, does not reach full amplitude until several picoseconds later, at approximately 12 time steps. One would expect the signal development to occur much faster since the input wave is planar (initially), and therefore one would suspect that a significant portion of the input would remain intact as it propagates to the output. As we have pointed out previously, however, the signal build-up is delayed because of the signal diversion into the transverse lines, which effectively slows down the propagation velocity of the transient signal.

We also note that the maximum amplitude of the output transient is 200V, or about 40% of the input wave. This is not unexpected, of course, since the input plane wave will begin to diverge as soon as the wave leaves the input region and is injected into the structure. Energy begins to "peel off" from the wave, with the process beginning at the wave boundary, where energy is diverted into the semiconductor region. The peel-off process continues with each time step, gradually working its way to the interior of the wave. The proper treatment of the delay actually requires us to use modified scattering coefficients, corrected for plane wave effects as discussed in Chapter IV. If the plane wave correlations are included in the analysis, transverse scattering in the interior of the plane wave is prevented, thus reducing the time of arrival of the signal. This does not prevent, however, the peeling- off of plane wave energy into the semiconductor, near the interface, unless the semiconductor is entirely activated very suddenly, as discussed in Section 7.7.

The bulk of the transient signal to the load lines is more or less extinguished by about 100 time steps(167ps). By this time, the full voltage is across the semiconductor and static conditions have for the most part been achieved. As indicated before, the charge-up time is determined by the size of the device(causing propagation delay) and the charging up of the cells. Note that the transient signal has both positive and negative swings. This is not surprising, in view of the dielectric mismatch at the output, and the multiple reflections which occur as signals are reflected from conducting surfaces.

Next we look at the field in the semiconductor as the system approaches equilibrium(the charge-up interval). Depending on the accuracy desired, the evolution of the system toward complete static conditions may require many time steps.   This is illustrated when we examine the horizontal field at the approximate center of the semiconductor, i.e., the total field equal to $V_{xy}(11,17)$. Fig.7.5 shows the time development for the first 100 time steps. Again we see that there exists a signal delay, during which time there is no field at the (11,17) cell.   This is followed by a transient signal, which is actually stronger in amplitude than the final static value, After additional time has elapsed , the field stabilizes to about 50.5 V after 100 time steps. At 100 time steps, the evolution to static conditions is far from complete, as noted in Fig.7.6, which has a longer time scale for the field. A slow oscillatory decay toward equilibrium may be noted, with an oscillation period approximately equal to the initial transient response.

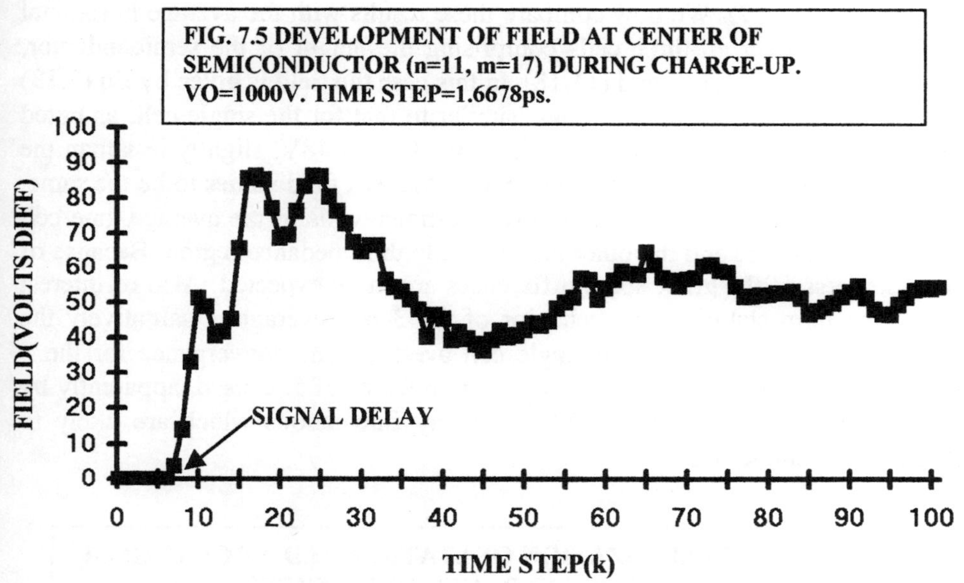

FIG. 7.5 DEVELOPMENT OF FIELD AT CENTER OF SEMICONDUCTOR (n=11, m=17) DURING CHARGE-UP. VO=1000V. TIME STEP=1.6678ps.

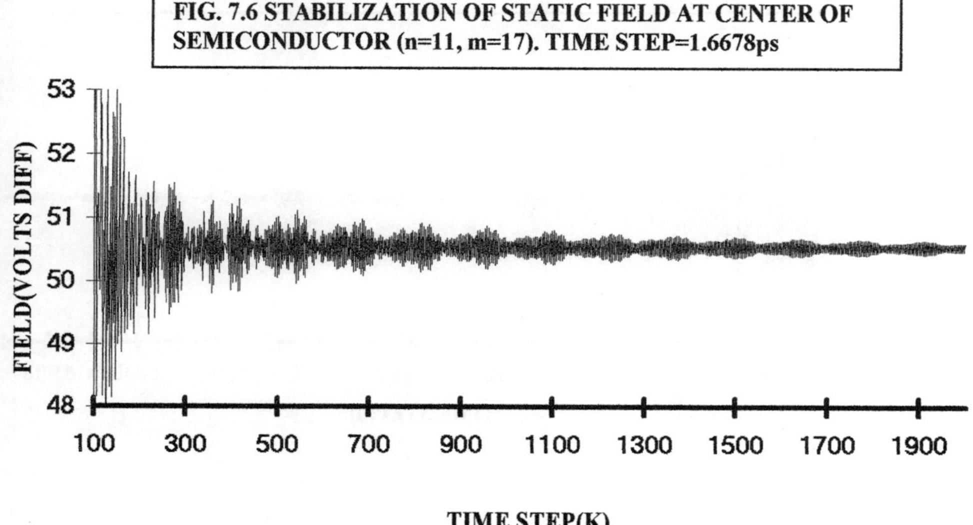

FIG. 7.6 STABILIZATION OF STATIC FIELD AT CENTER OF SEMICONDUCTOR (n=11, m=17). TIME STEP=1.6678ps

In the previous discussion, we utilized a the field associated with a single cell, namely, (11,17). We now compare these results with the average horizontal field averaged over all three cells comprising the height of the semiconductor, i.e., cells (11,16), (11,17), and (11,18). In this case the field is given by Eq.(7.13) with n=11. The field evolution looks similar to that for the single cell, as noted in Fig.7.7. The static field, however is now close to 48V, slightly less than the single cell field, There is no reason to expect the two field values to be the same, however, since we have now added two additional cells to the average, one cell next to the dielectric and the other next to the high impedance region. Because of the courseness of the grid , some differences are to be expected. Also of interest is the long term stabilization behavior of the 3 cell average. Qualitatively the behavior looks the same as the single cell average. The convergence for the 3 cells, however, appears to be better by about a factor of 5, caused, apparently by fact that the 3 cell average smoothes out any fluctuations which are likely to appear with a single cell.

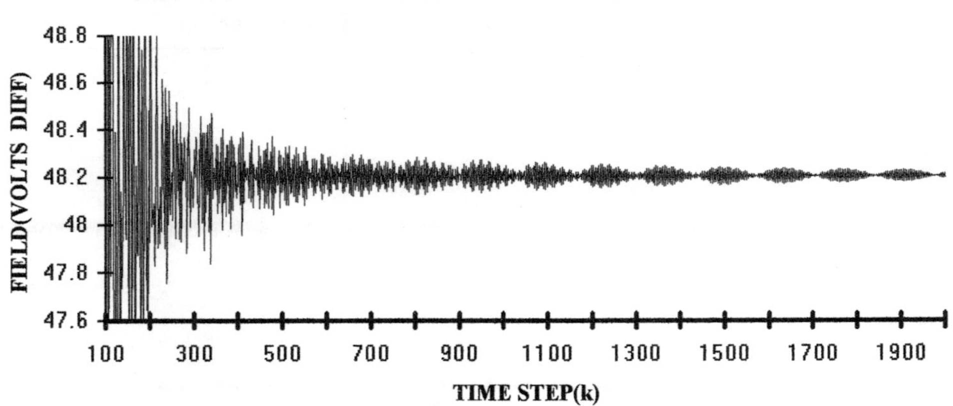

FIG. 7.7 STABILIZATION OF STATIC FIELD AT CENTER OF SEMICONDUCTOR AVERAGED OVER THREE CELLS (n=11, m=16,17,18). TIME STEP= 1.6678ps

Fig.7.8 shows the field profile over the semiconductor during the charge-up process. Initially, of course, the field is zero throughout the semiconductor. At just 41.7ps(k=25) the field has grown rapidly so that the field actually exceeds the equilibrium values. At 83.4ps , the field has settled down, and the profile is very close to the equilibrium one. Finally for times greater than 584 ps (k=350) there is very little change in the profile, and this we designate as the equilibrium profile. We note of course that the field is stronger at the anode than at the cathode, caused by the strong fringing fields at the anode to the ground line. An important internal check shows that the sum of the voltages (the "integral" of the field) is equal to the bias field of 1000V, i.e., twice the sum of the total input line voltages of 500 V.

A slight oscillatory variation of the profiles , caused by the disparity in cell size at the dielectric interface, may be observed. As expected, the superimposed oscillation peaks at nodes corresponding to full nodes at the interface (e.g., m=6, 9 , etc) while the fields are lower at nodes corresponding to partial nodes

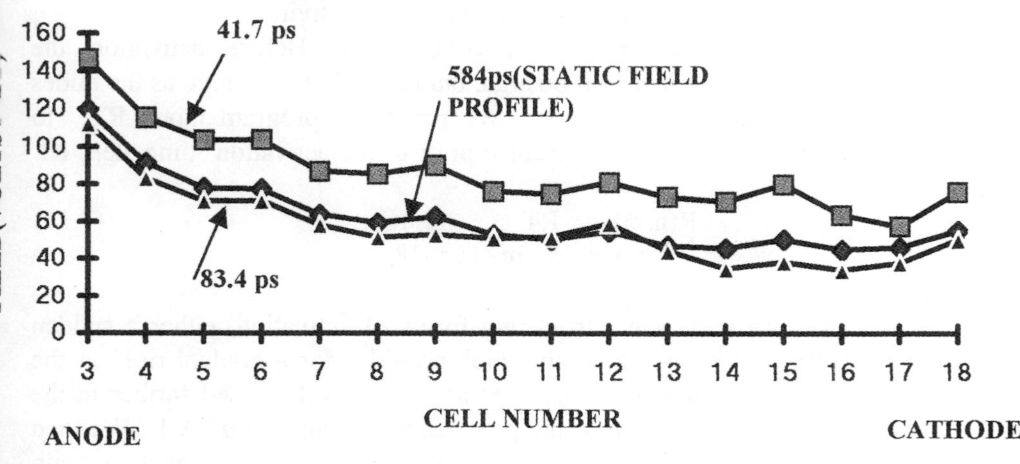

at the interface. Such variations are of course an artifact and may be eliminated by resorting to larger cell densities , and may be expedited by spatial averaging over the oscillation length.

It is worthwhile to emphasize that the semiconductor field values shown in Fig.7.8, and throughout   Chapters VII and VIII , are the total voltage amplitudes in the transmission line. The actual *electric field* is obtained by normalizing with respect to the cell dimension Δl ;the field can then be interpreted as the average field of an auxiliary cell centered about the line, as discussed in Section 2.3.

7.6 Node Resistance R(n,m) During Activation

Before describing the actual simulations which take place when R(n,m) is activated we first describe several of the mechanisms which control R(n,m), and which are specifically included in the program statements. Besides the conductivity caused by light, we also include the background semiconductor conductivity, which is simultaneous with any other source of conductivity present. In addition the program includes node recovery and spatial changes in the conductivity generation caused by the spatial dependence of the incident light pulse. Further we include a temporal dependence in the light pulse, allowing for either an instantaneous or gradual risetime in the conductivity.

We begin with the background conductivity. Before activation the semiconductor node resistance will be high, but certainly not as high as the nodes for the other non- conducting regions. We provide a program input, RT, to specify the semiconductor node resistance prior to the activation time step, KL ,or

$$R(n,m) \;=\; RT \;, \qquad K{<}KL; \qquad\qquad (7.16)$$
$$n{=}3 \text{ to } 18 \;, \; m{=}15 \text{ to} 18$$

For $K{\geq}KL$, we may select from two forms of activation, either a sudden activation of R(n,m) or an activation which provides for a gradual rise in the conductivity. The selection of either type activation is discussed further in the following and in the description of the program statements, App.7A.1. We then describe the   factors controlling R(n,m). The following discusses two main classifications: instantaneous and gradual changes in the conductivity.

7.6(a) Instantaneous Change in R(n,m)

We first consider the case in which the activation is very abrupt and the node resistance drops suddenly. Once the added conductivity reaches a specified value, which is virtually instantaneous, any additional conductivity ceases. The added conductivity may be stated in terms of a resistance, RO, and is proportional to (1/RO). RO is specified as an input. We can think of this type activation as a special case of light activation discussed in Chapter 2, wherein a high intensity light signal P_o has a duration of one time step. If the time step is very short , with P_o very large, then the light energy is transferred to the semiconductor in a single time step, and the resistance is reduced to RO. We postpone until later the effect of the background conductivity, proportional to (1/RT) , where RT is the background node resistance.

Once the node resistance is lowered , we arbitrarily assume that the added conductivity decays exponentially, so that the decay is proportional to (1/RO)*EXP(-(K-KL)/KREC), where KL is the activation time and KREC is the assumed recovery time. KL and KREC are expressed in terms of the number of elementary time steps Δt. If we again neglect the background conductivity then the node resistance becomes

$$R'(n,m) \rightarrow RO*EXP((k-KL)/KREC), \qquad (7.17)$$

where the primed R'(n,m) designates the resistance stemming solely from the light activation and recombination effects, but excludes background conductivity. RO is the initial resistance immediately following the activation. Eq.(7.17) does not take into account any spatial variation of the activation across the length of the semiconductor. If we use the results of Chapter 2, R'(n,m) must be modified by replacing Eq.(7.17) with

$$R'(n,m) = RO* [EXP(G0(n-11)^2)]* EXP((k-KL)/KREC); k>(KL-1) \quad (7.18)$$

where we again ignore RT for the moment. . The bracketed term is the spread factor ,i.e., the spread in light energy along the length of the semiconductor . The spread term has the same form as in Eq.(2.58) in Chapter 2 with the spread factor G0 = a$(\Delta l/l_o)^2$. In order to reduce the amount of detail, we do not include the modeling of the activation with depth, as expressed by the function D(m) in Eqs.(2.51)-(2.52), which is time dependent. Nor do we take into account reflections from the dielectric surface , which will also contribute to the carrier generation. R'(n,m) is therefore assumed uniform as a function of m in the semiconductor(i.e., uniform among the cells m=16,17,18) We emphasize, however , that the TLM method is very well suited for taking all these factors into account, and the variation of R(n,m) in the semiconductor with the m index may be incorporated at any time. Eq.(2.50) , for example, expresses the activation of the cells as a function of m(depth) as well as n(thickness) and k (time).We also make the important point concerning the recovery of the node resistance. We have assumed an exponential recovery throughout the entire range of k , both for small amounts of added conductivity as well as large. This is obviously a crude simplification We remark that the exponential recovery is simply being used as a handy example to explore the TLM method; In actual practice, the *physics* will dictate what recovery term will apply, which then must be cast into the TLM format, using techniques outlined in Chapter VI. Finally we take into account the background conductivity. When we take into account the background resistance RT, the new resistance after activation becomes

$$R(n,m) = (RT)(R'(n,m))/[RT+R'(n,m)] \qquad (7.19)$$

Eq.(7.19) follows from the fact that the conductivities add while the inverses combine as parallel resistors. In the examples which follow, the background resistance RT usually will be assumed to be much larger than RO, so that the initial R(n,m) after activation may be approximated by RO.

7.6(b) Gradual Change in R(n,m)

In this sub-section we assume the added conductivity is done gradually rather than all at once. In Eq.(2.58) we saw that for constant light power P_o, falling on the center cell, the conductivity change (neglecting RT) is proportional to Po(k-KL)Δt, the light energy at the completion of the kth step. (In Eq.(2.58) the light activation occurs at KL=0) Since the light power is in general a function of time, we replace Po (k-KL)Δt with the integral

$$\text{Light Energy} = \int^k_{KL} P(t)dt \tag{7.20}$$

where P(t) is the time dependent power. To illustrate the TLM method we assume P(t) has an exponential dependence, or

$$P(t) = P_o\, EXP[(t-KL\Delta t)/KRS\Delta t] \; ; k \geq KL \tag{7.21a}$$
$$P(t) =0 \; ; k<KL \tag{7.21b}$$

where k, KL,and KRS are expressed in the number of time steps and t=kΔt The light signal therefore starts from P_o at k=KL and increases with risetime KRS. The integrated light energy is then

$$\text{Light Energy} = P_o\, KRS\, \Delta t\{EXP\{(k-KL)/KRS) -1) \tag{7.22}$$

Note that if we expand the exponential term, we retrieve the constant power case, discussed in Chapter II,   so that Eq.(7.22) becomes P_o(k-KL)Δt. The node resistance associated with the light energy may in fact be obtained from Eq.(2.58) if we replace k with KRS{EXP{(k-KL)/KRS) -1). The same previous remarks regarding the depth apply here as well. For our purposes it is more useful to cast the resistance in the form

$$RO(n,m)= R_{ST} * [EXP(G0*(n-11)^2)]* [KRS*\{EXP\{(k-KL)/KRS\} -1\}]^{-1} \tag{7.23a}$$

where we use the designation RO(n,m) to indicate that we have excluded the contributions of recombination and the background resistance RT. The input constant R_{ST} has the dimensions of resistance and contains factors such as the power constant Po , the carrier mobilities, the TLM length, plus other factors discussed in Chapter II and embedded in Eq.(2.58). If k-KL $<<$ KRS then Eq.(7.23a) then becomes

$$RO(n,m) = R_{ST} * [EXP(G0*(n-11)^2)]*(k-KL)^{-1} \qquad (7.23b)$$

This is nothing more than the result for the case of the constant (in time) light power source discussed in Chapter II. When k=KL, $RO(n,m)$ is still infinite since in the absence of any accumulated light and any background conductivity, there are no contributors to, the conductivity. At the end of the first time step k-KL =1, $RO(n,m) = R_{ST} * [EXP(G0*(n-11)^2)]$, at which point $RO(n,m)$ is at its initial (finite) value. When k-KL is equal to KRS, $RO(n,m) \approx R_{ST} * [EXP(G0*(n-11)^2)]/KRS$, indicating the anticipated decline from R_{ST}.

Next we have to append the recovery term to Eq.(7.23), in much the same fashion as before, again neglecting the background conductivity. Thus, we assume the recovery attaches only to $RO(n,m)$. The result is

$$R'(n,m) = [EXP((k-kl)/KREC)]* RO(n,m) \qquad (7.24)$$

$RO(n,m)$ is given by Eq.(7.23) and $R'(n,m)$ is the node resistance which accounts for both light activation and recombination effects, but excludes background conductivity. As before, we have temporarily added a prime to indicate the neglect of the background conductivity.

The background conductivity, proportional to (1/RT), adds to the conductivity induced by the light signal. In the absence of any background conductivity, the light induced contribution, including recombination effects, is proportional to $(1/R'(n,))$, with $R'(n,m)$ given by Eq.(7.24). When background conductivity exists , the total resistance will consist of the parallel combination of RT and the resistance stemming from Eq.(7.24). Essentially repeating Eq.(7.19), the final node resistance is

$$R(n,m) = RT*R'(n,m)/[RT+R'(n,m)] \qquad (7.25)$$

with $R'(n,m)$ given by Eq.(7.24).

We should specify any bounds which may exist on $R(n,m)$, which are not explicitly addressed by the above expressions. One such problem is the risetime

function in which   there is no lower limit to R(n,m). To remedy this defect we must pre-select a value $R_M((n,m)$ below which R(n,m) is not allowed to fall, based on the fact that the lower values of R(n,m) are physically unrealizable. In the case of the semiconductor, there is the finite number of carriers available. Another limit may occur earlier because of optical screening; in this case the light signal is rejected at the semiconductor surface because of large carrier densities. Whatever the origin, when R(n,m) falls below R_M, we "turn off" the light induced contribution as indicated in the program description in App.7.A4.
Because of the presence of RT, it is unnecessary to introduce an upper limit to R(n,m). RT serves as this limit as noted from Eq.(7.25 if we allow EXP((k-kl)/krec) to increase without limit.

7.7 Output Pulse When Semiconductor is Activated

Figs.7.9   and 7.10 show the output pulse when the entire semiconductor is activated(nodes m=15 to 18 and n=3 to 18) with a constant light source.   The activation occurs at k=350, well after the transient term has died out. (In regard to the transient term, some fine structure has been lost in Figs, 7.9 and 7.10, compared to that in Fig.7.4, due to the evaluation at every 10th time step, rather than every step). No recovery of the node resistance is assumed.   The figures differ with respect to the activation resistance; one is at 15 Ω and the other at 150 Ω. We note that at 15Ω the output is close to 500 V, which is what one would expect when the output is matched to the input(both 50Ω) and the voltage drop in the semiconductor is sufficiently low so as not to interfere with the matching condition. AT RO =150 Ω we begin to notice a drop in the output, to about 400V, due to the increased voltage drop in the semiconductor.

During activation we can view the device in a simple manner, in which the input and output are separated by a section of transmission line , 2.5 mm long, with ε =1. The effect of this intervening transmission line section is twofold: there is the usual propagation delay, but in addition there is a rounding of the

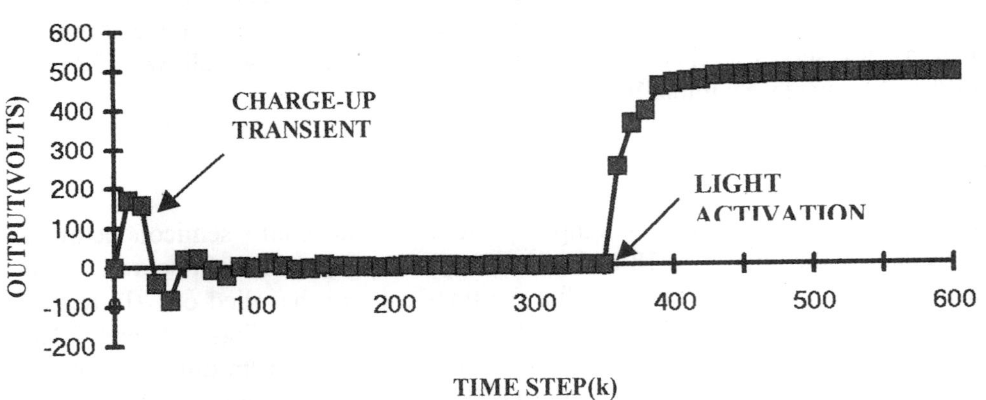

FIG. 7.9 VOLTAGE DELIVERED TO OUTPUT LINE WHEN SEMICONDUCTOR IS ENTIRELY ACTIVATED. RO=15Ω. TIME STEP= 1.6678ps. VO=1000V

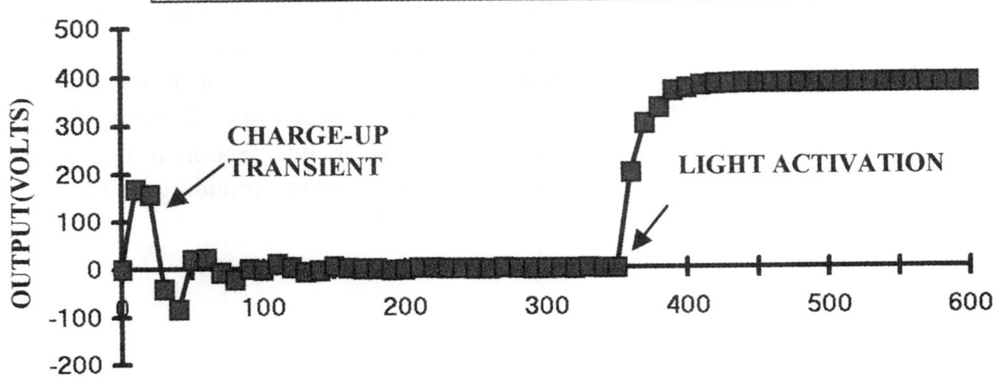

FIG. 7.10 VOLTAGE DELIVERED TO OUTPUT LINE WHEN SEMICONDUCTOR IS ENTIRELY ACTIVATED. RO=150Ω. TIME STEP= 1.6678ps. VO=1000V

leading edge of the pulse, due to the mismatch between the intervening section and the input/output. Ultimately, however , the full amplitude of 500V is delivered to the output line, since the input and output are matched. The situation changes, however, when the activation intensity is reduced, RO =150 Ω . Because of the significant series voltage drop incurred in the semiconductor the output is no longer matched to the input. This is noted by the reduced signal amplitude , about 400V, delivered to the output.

From the previous discussion one might be tempted into thinking that if the input, dielectric, and output sections are all matched, then the instantaneous activation of the entire semiconductor will result in the delivery of a pulse to the load with an infinitely fast risetime. Such is not the case however, except for certain limiting situations. The problem is that fringing exists at the device , prior to activation. Following the semiconductor activation, the fringing field profile turns into a traveling wave and the intact profile is the first to arrive at the load., which effectively constitutes a risetime. The only way to deal with this fringing field effect is to resort to small heights, h_1 , relative to the semiconductor gap spacing, l_o.

Fig.7.11 shows the effect of the conductivity risetime on the output pulse, using the expression Eq.(7.23a) with G0=0. As before the semiconductor activation begins at k=350, but with a risetime of KRS=50 (again we assume there is no node recovery).The initial light activated resistance, RST, is arbitrarily set equal to the background resistance of RT=EXP07 Ω. The lower limit of R(n,m) is taken to be 15Ω. There are two main effects to be observed concerning the output, compared to the previous case where the conductivity change is instantaneous. First, there is substantial delay in the output pulse. This is due to the fact that a several E fold decrease(i.e., several times KRS) in the resistance is required before the resistance becomes comparable to the semiconductor impedance . The second thing to observe is that the risetime is somewhat longer than KRS=50, in the output, due to the fact that the initial resistance is relatively small. In the remaining simulations to be discussed we only consider constant light sources.

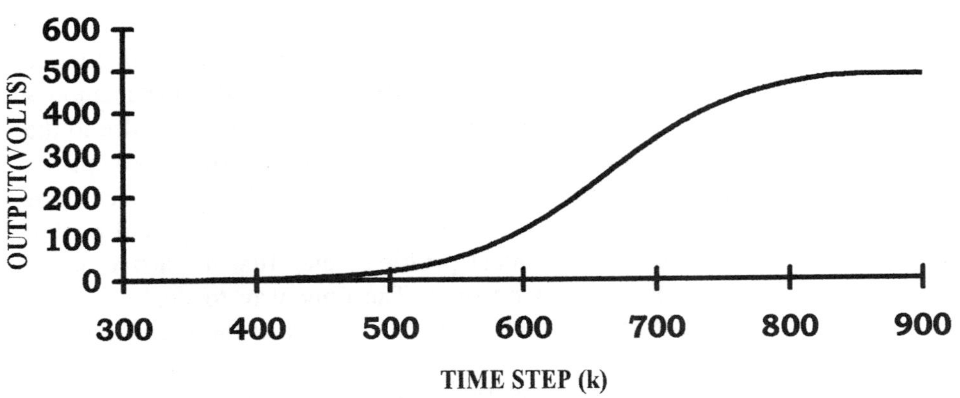

7.8 *Node Recovery and its Effect on Output Pulse*

In Figs.7.12-7.13 we see the effects of the node resistor recovery. In order to simplify the recovery process, we assumed the node resistor recovery has the form given by Eq.(7.17), i.e., we assume $R(n,m)$ is the node resistance immediately   after activation and is thereafter controlled by the assumed exponential recovery(neglecting RT, $R'(n,m) = R(n,m)$). Thus,

$$R(n,m) = RO \cdot [EXP\{(k-KL)/KREC\}] \; ; \; k. > (KL-1) \tag{7.26}$$

In Figs.7.12-7.13 we assume KREC= 200 for RO =15Ω and 150Ω respectively. Looking at the recoveries in the load voltage, particularly the 15 Ω activation in Fig.7.12, we can see that the effective recovery is much longer than k=200. This is to be expected since the effects of the recovery will not be felt until $R(n,m)$

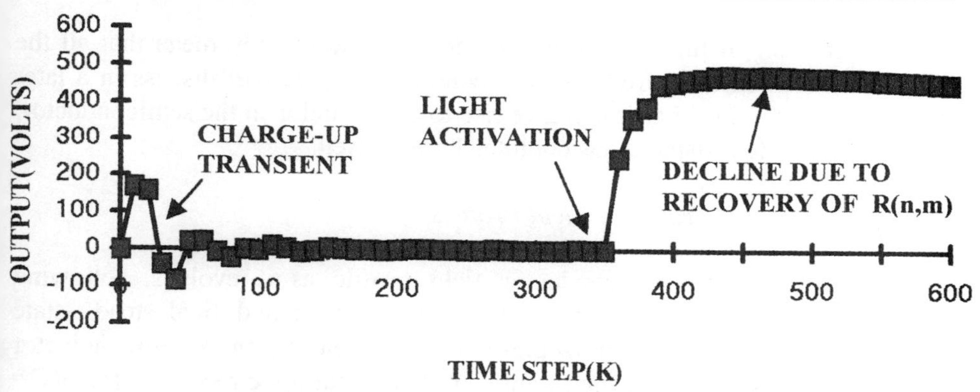

FIG. 7.12 VOLTAGE OUTPUT DELIVERED TO OUTPUT LINE WHEN SEMICONDUCTOR IS ENTIRELY ACTIVATED. RECOVERY IS INCLUDED. KREC=200, VO=1000V, TIME STEP=1.6678ps, RO=15Ω.

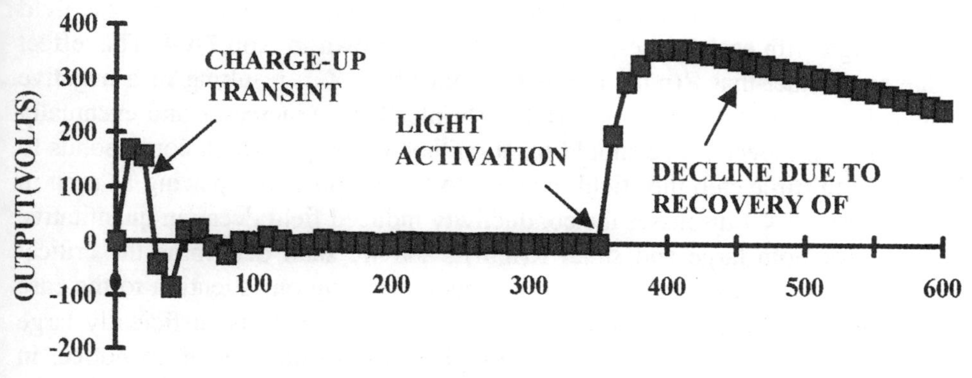

FIG. 7.13 VOLTAGE OUTPUT DELIVERED TO OUTPUT LINE WHEN SEMICONDUCTOR IS ENTIRELY ACTIVATED. RECOVERY IS INCLUDED. KREC=200, VO=1000V, TIME STEP =1.6678ps, RO=150Ω.

reaches the value of the line impedance of the semiconductor, which in this case is Z_o = 126Ω . A few E-folds are therefore required before attaining the line resistance. Compared to Fig.7.13, the recovery is longer in the case of Fig.712, since it is starting off with a smaller resistance, 15Ω as opposed to 150 Ω in Fig.7.13.

The recovery of the node resistance does not necessarily mean that all the charge has been removed from the semiconductor, as we will discuss in a later section. The remaining charge can give rise to strong fields in the semiconductor, even after the node resistance has completely recovered.

7.9 *Steady State and Transient Field Profiles*

We next consider the semiconductor field profile as it evolves, following activation, from its initial, static state to its transient and final steady state profiles. Figs.7.14 and 7.15 show these profiles when the entire semiconductor has been activated with a constant light source, without node recovery, for RO= 15Ω and 150 Ω . The conditions correspond to the output pulses in Figs.7.9 and 7.10. At RO= 15Ω, which is the high conductivity case, the fields for both the transient (599ps) and the final, steady state profile(1ns) are quite small. For the final profile the field (voltage difference between adjacent cells) is in the 1-2 volt range, resulting in an integrated voltage drop of 25-30 Volts. With RO = 15 Ω an unusual effect is observed. During the first 10 time steps or so(~17ps) after activation, the fields in each cell decay in "ringing" fashion, .i.e., the field changes sign with each successive time step, as shown in App.7A.4. This effect is due to the fact that R(n,m) is much smaller than Z_o , resulting in a negative mismatch . As the cell size is allowed to shrink, R(n,m) increases and eventually the ringing disappears. We should mention that at 599 ps, which corresponds to the k=359 th time step the field happens to have a positive upswing as seen in Fig.7.14. App.7A.4 discusses the conductivity induced field decay in quantitative fashion , for both large and small R(n,m) , and we shall determine the critical value of R(n,m) , below which ringing occurs. Turning our attention to the case when R(n,m) is 150Ω, the ringing disappears since R(n,m) is sufficiently large compared to Z_o. Also, because of the higher R(n,m) one begins to notice, in Fig.7.15 , substantial voltage drops in the semiconductor, particularly during the final steady-state profile. Indeed the integrated voltage difference across the semiconductor is ~ 250 volts at the final field.

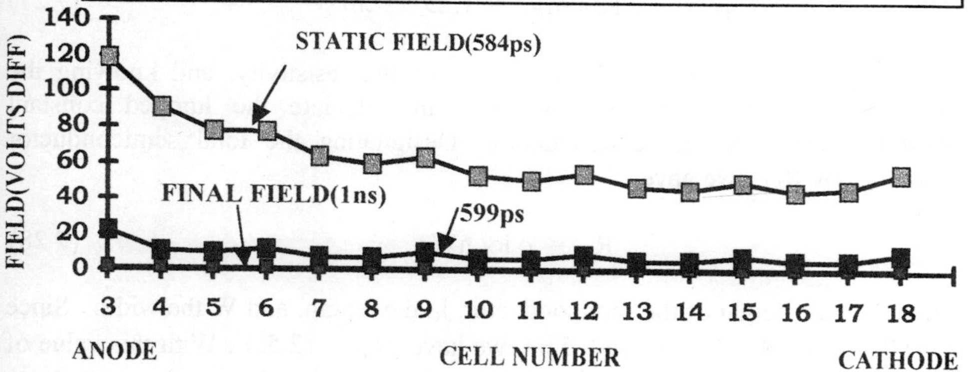

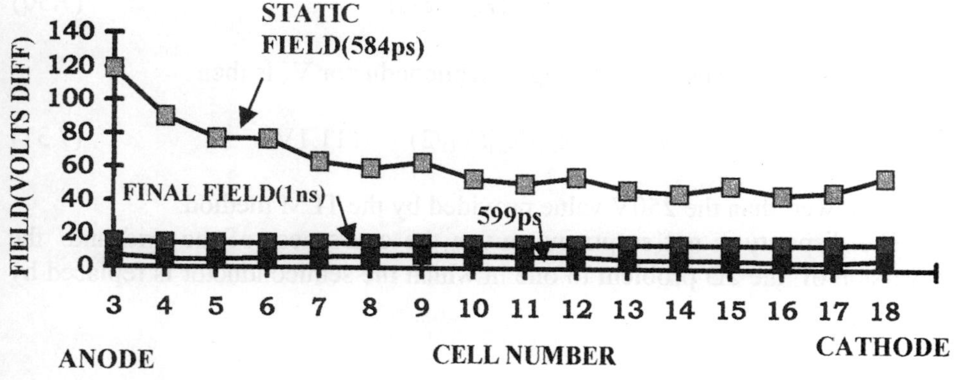

It is worthwhile to compare the voltage drop in the semiconductor at $RO=150\Omega$ with that obtained using lumped circuit variables. To do this, we convert the 2D semiconductor matrix to a lumped circuit resistance and we ignore the dielectric region. First we obtain the resistivity in the semiconductor. From Chapter II, Section 2.13, ρ(suppressing the indices) is given by

$$\rho = \Delta l\, R(n,m)/2 = 1.25\ \Omega\,cm \tag{7.27}$$

since $\Delta l = 0.0167$ cm.   Using this value of the resistivity, and knowing the dimensions of the semiconductor, we can calculate the lumped constant resistance value of the semiconductor. Designating the total semiconductor resistance by R_{ST}, we have

$$R_{ST} = \rho\, lo/\, hW \tag{7.28}$$

where h is the height of the semiconductor, l_O the length, and W the width. Since h $=0.05$cm, $l_o =2.5$cm, and w$=0.5$cm, we have $R_{ST} = 12.5\Omega$. With this value of R_{ST} we obtain the total load impedance, R_L, seen by the input wave, again regarding the semiconductor and the 50 output as a lumped impedance. Thus

$$R_L = R_{ST} + Z_{OUT} = 62.5\Omega \tag{7.29}$$

We may now calculate the transfer of voltage to the semiconductor and the output line by first obtaining the transfer coefficient T, given by

$$T = 2R_L/(R_L + Z_{IN}) \tag{7.30}$$

where Z_{IN} is 50Ω . The voltage drop in semiconductor V_S is then

$$V_S = T[\, R_{ST}/R_L](V_O/2) = 111.1V \tag{7.31}$$

which is lower than the 250V value provided by the TLM method.

   The disparity is not surprising, given the courseness of the grid and the conversion of the 2D problem to one in which the semiconductor is replaced by a lumped resistor, thus ignoring 2D fields and currents in the semiconductor. The courseness of the grid is illustrated by the overestimated height of the activated semiconductor. The m=15 nodes are part of the semiconductor, but they have not been activated(in order to clear a path for the input and output lines). In effect, this reduces the effective height of the activated region(from 05cm to perhaps 0.04cm). In addition the vertically directed currents have been ignored in the semiconductor. Assuming a sufficiently fine grid the 2D TLM description should be used(not a lumped element description) in order to obtain accurate results, especially during the transient phase.

7.10 Partial Activation of Nodes and Effect on Profiles and Output

Next we examine the *partial* activation of the semiconductor as , opposed to the complete activation. We again stress that the light source is constant in time. As an example we select nodes from n=3 to n=11 for activation, which constitutes about 1/2 of the semiconductor(the half closest to the anode). We first look at the pulse delivered to the output line, shown in Fig.7.16. We see that a transient pulse is produced when the semiconductor is partially activated. A simple interpretation of the result may be provided. Following the activation, we may regard the new situation as resembling a device of half the length, but only partially charged(to about half the voltage). This "new" device undergoes a charge -up to full voltage, but in the process a transient is delivered to the load in much the same fashion as that which occurs during the initial charge-up of the device.

Figs.7.17-7.18 show the transient and final field profiles under partial spatial activation with RO =150Ω and 15Ω. No recovery of the node resistors is assumed. The most important thing to observe the field enhancement that occurs in the region not exposed to the activation. As discussed in the previous paragraph, the device has essentially been reduced to half that of the original, with the result that the full voltage appears across 1/2 the original distance. Thus a field enhancement occurs. With this enhancement, however, an additional phenomenon enters the picture. A positive charge layer is produced on the n=12 cells , which now becomes a virtual anode. The enhancement does not continue

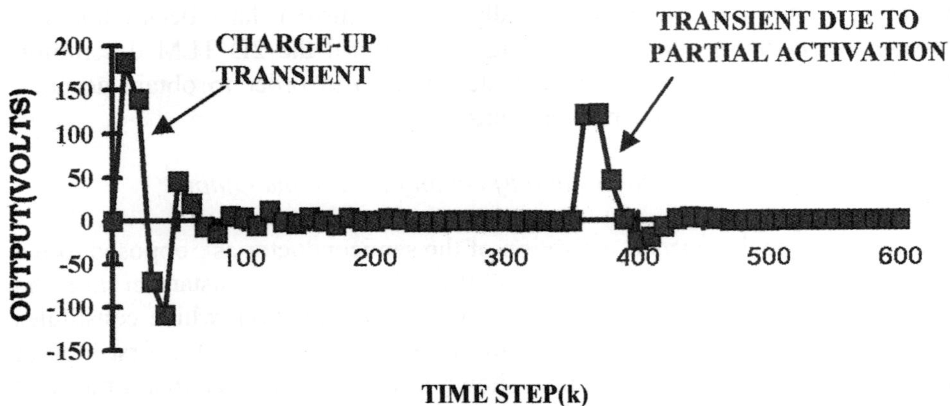

FIG. 7.16 TRANSIENT PULSE DELIVERED TO OUTPUT WHEN SEMICONDUCTOR IS PARTIALLY ACTIVATED(n=3 TO 11). RO=15 Ω, VO=1000V, TIME STEP=1.6678ps. ACTIVATION IS AT KL=350
CHARGE-UP TRANSIENT
TRANSIENT DUE TO PARTIAL ACTIVATION
OUTPUT(VOLTS)
200
150
100
50
0
-50
-100
-150
100
200
300
400
500
600
TIME STEP(k)

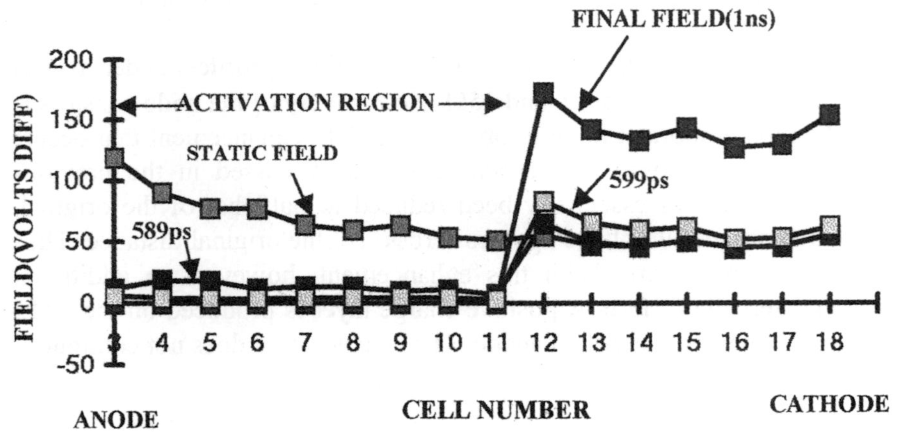

FIG. 7.17 STATIC AND TRANSIENT FIELD PROFILES OF SEMICONDUCTOR WITH PARTIAL ACTIVATION. VO=1000V. ACTIVATION AT 584ps. RO=150Ω, m=17
FINAL FIELD(1ns)
ACTIVATION REGION
STATIC FIELD
589ps
599ps
FIELD(VOLTS DIFF)
200
150
100
50
0
-50
3 4 5 6 7 8 9 10 11 12 13 14 15 16 17 18
ANODE
CELL NUMBER
CATHODE

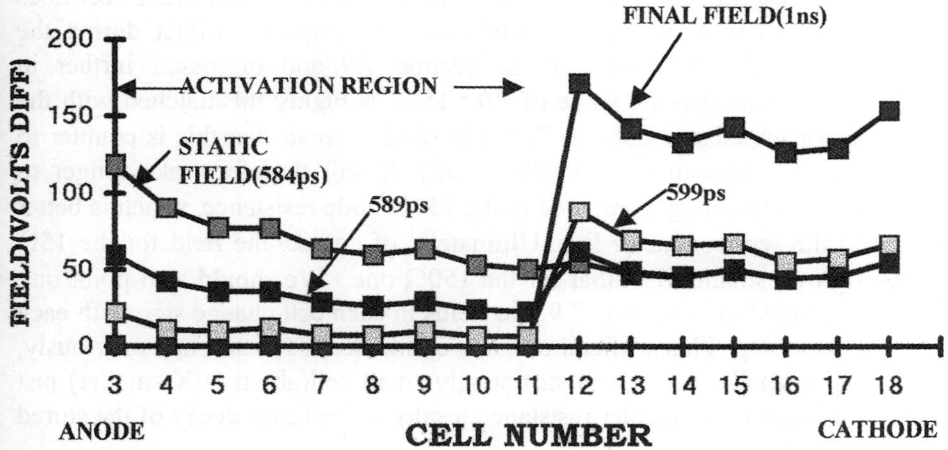

indefinitely, of course, since the charge carriers will eventually drift to the electrodes. Provided no recovery occurs, and the activation continues, the entire semiconductor then becomes filled with carriers and the field enhancement disappears.

Even with an instantaneous activation, the field enhancement does not occur immediately. We can see this if we look at the 589ps profile, which occurs three time steps after the activation. We can observe the slight field enhancement compared to the static profile occurring in the cells just ahead of m=11, i.e., the m=12 and m=13. Beyond these cells the profile coincides with the static profile. The reason for this, of course, is that the new signals produced have not yet had time to reach the cells further away and therefore no enhancement can occur.

The differences between the profiles for RO =150Ω and R0 =15Ω are relatively minor. Note that the final fields in the activated region are both very small despite the differences in the activation resistance, and unlike the case when the entire semiconductor is activated. This has to do with the lack of current flow in the device once the new equilibrium is established. We also point out the transient profiles at 589ps and note that he field is larger when $R0 = 15\Omega$ as compared to the 150Ω case. This seems contradictory since one would guess that the lower resistance would imply a lower resistance. In fact this effect does indeed occur and what we have exhibited is an computer artifact during the transient phase, already alluded to in Section 7.9 and discussed further in App.7A.4. We note that the value of $R0 = 15\Omega$ is highly mismatched with the semiconductor impedance lines of $Z_o = 125.67\Omega$ (recall that this is counter to our TLM model assumption that $R0 >> Z_o$). It will therefore take longer to dissipate the stored energy compared to the 150Ω node resistance, which is better matched to the semiconductor line. Ultimately, of course, the field for the 15Ω activation grows smaller than that for the 150Ω one. We should also point out, just as was alluded to in Section 7.9, the fields in each cell change sign with each successive time step. This artificial effect is eliminated, as mentioned previously, by choosing a smaller cell. For a sufficiently small cell(about 10X smaller) just the opposite occurs; the smaller resistance results in the faster decay of the stored field , as expected.

7.11 Cell Charge Following Recovery

If the node resistance of the partially activated semiconductor (as in Figs.7.17 and 7.18) are allowed to recover, i.e., if we introduce a recovery such as Eq.(7.26) into the iteration, an apparently anomalous situation ensues. The final fields hardly change at all from the final fields shown in Figs.7.17 and 7.18. Thus, for the first half of the semiconductor, cells n=3 to 11, the field is close to zero, while for cells n=12 to 18 the field is enhanced well above its static values. This despite the fact that the node resistors have long recovered. The reason has to do with the accumulation of real charge on the cells bounding the activated region, i.e., on the m=12 cells. During the discharge positive charge is delivered to the n=12 cell from the n=11 cell, while there is no corresponding loss of charge from n=12 cell to any other cells. Once the node resistance is recovered, the charge still remains.

Perhaps it is best to start with he static field where we know there is no charge initially present, at least in the semiconductor. We seek to find the charge in each of the m=17 cells from n=3 to n=19, i.e. , across the length of the semiconductor. To do this we use the expressions in Chapter VI for the cell charge(within a multiplicative constant)

$$Q(n,17) = [\ V_{yx}(n,\ 17)\ \text{-}Vyx((n\text{-}1),17]\ +\ [\ V_{xy}(n,17)\ \text{-}\ V_{xy}(n,16)] \quad (7.32)$$

The charge is normalized with respect to $\Delta l\ \varepsilon$, i.e., multiplying $Q(n,17)$ by $\Delta l \varepsilon$ gives the actual charge. We should also note that each term on the right side represents the sum of the forward and backward waves. Note that the terms in the first bracket provide the difference in field in the horizontal direction while the second bracket is for the vertical field The distribution of charge for the initial static field is shown in Fig.7.19. As expected the only true charge is on the m=3 and m=19 cells, which of course correspond to the anode and cathode, respectively. In contrast, any charge on the interior cells of the semiconductor

FIG.7.19 RELATIVE CHARGE OF CELLS AT ANODE AND CATHODE FOLLOWING CHARGE-UP OF SEMICONDUCTOR AT 584ps.

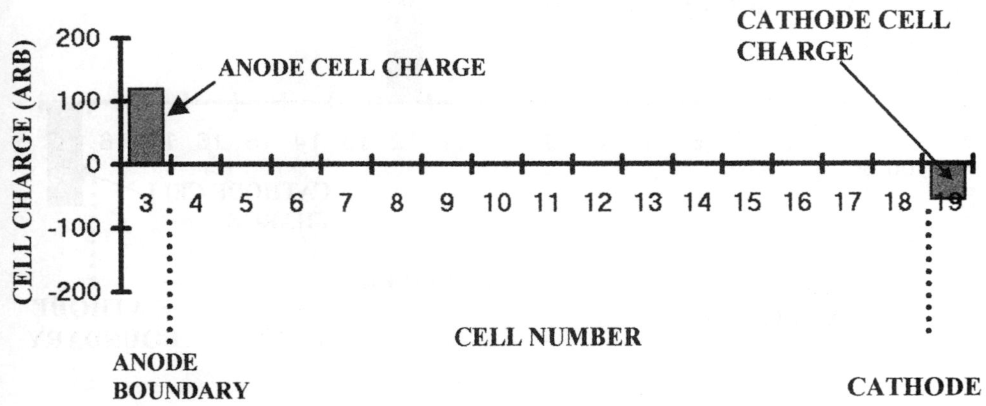

corresponds to induced charge and therefore does not contribute to Q(n,m). In cellular notation, Q(3,17) and Q(19,17), represent the charge in the conducting cells bordering the semiconductor. We should note that the charge produces a field dominated by the horizontal component as expected. At the electrodes the second bracket in Eq.(7.32), produces the vertical field component, which in this case is zero since $V_{yx}(2,17)$ and $V_{yx}(20,17)$ are embedded in the electrodes and are therefore zero.

We now partially activate the semiconductor, from cells n=3 to 11, and also include a recovery term with KREC =200, and allow the system to come to its final state. As before we look at the charge in the m=17 cells. As shown in Fig.7.20 We see that the cell charge remains on the m=12 cell even after recovery, thus becoming a virtual anode. Also note that the cell charge has increased on both the cathode and on the (virtual) anode, which corresponds to

FIG. 7.20 RELATIVE CHARGE OF m=17 CELLS IN SEMICONDUCTOR
FOLLOWING PARTIAL ACTIVATION , RO=150 Ω. VIEW IS AT 1ns
AFTER ACTIVATION(584ps). RECOVERY IS INCLUDED, KREC=200

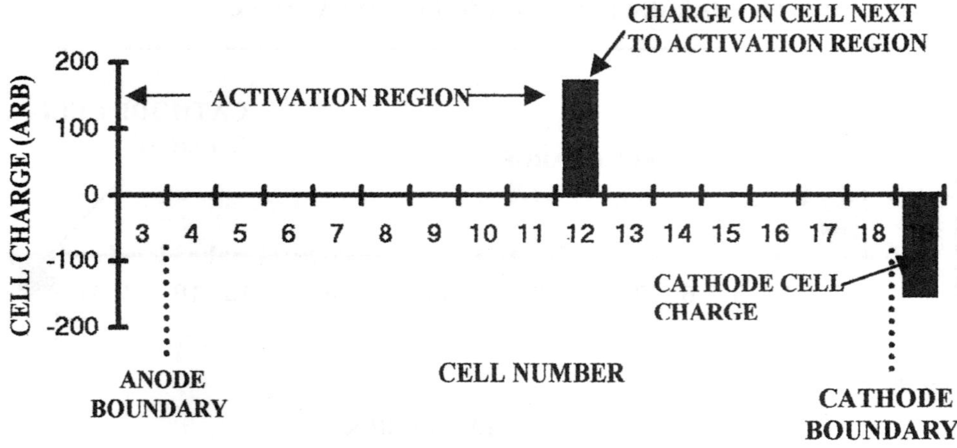

the enhanced field in the inactivated region. We also observe that the cell charge disparity between the n=12 and the cathode is much smaller owing to the fact that there is considerably less fringing from the virtual anode to the ground line. We see from Fig.7.20 that the TLM method is a powerful tool for obtaining the transient distribution of charge in the semiconductor. The charge distribution is temporary , of course, until the charge drifts to the electrodes.

Besides partial activation, it is also of interest to point out how the charge re-distributes itself when the *entire* semiconductor is activated, followed by the recovery of the nodes. When this happens another charge layer is seen to have an important role, besides that of the electrode charge. The additional charge layer is along the m=15 interface, with the cell charge diminishing as n increases from 3 to 18 . In effect, just as in the case for the partial activation, the semiconductor field in the horizontal lines has been effectively replaced with a charge layer along the m=15 boundary line. An important interpretation of this charge is that it represents, for the most part, the effect of the fringing field to the ground line. The addition of this charge layer also has an important effect on the field in the semiconductor, as seen in Fig.7.21. In contrast to the initial static field, once activated, both the transient and final fields are very uniform. The residual charge on the semiconductor/dielectric interface exactly counteracts the fringing field so that the semiconductor field is quite uniform. This phenomenon therefore provides the possibility, in a rather novel way, of obtaining temporary field uniformity in a long semiconductor where fringing is usually difficult to control.

7.12 Role of TLM Waves at Charged Boundary

It is natural to ask how the waves conspire to enforce the new boundary conditions brought about by the presence of charge at the boundary. As an example we look at the situation where the field has been totally excluded from the semiconductor, due to activation, and where the field exclusion remains after recovery , as in Fig.7.21. The situation at the boundary is depicted in Fig.7.22). Since $V_A \sim 0$, this imposes the condition,

FIG. 7.21 FIELD PROFILES IN SEMICONDUCTOR BEFORE AND
AFTER RECOVERY. ALL NODES IN SEMICONDUCTOR
ACTIVATED, VO=1000V, RO=150Ω, KREC=200

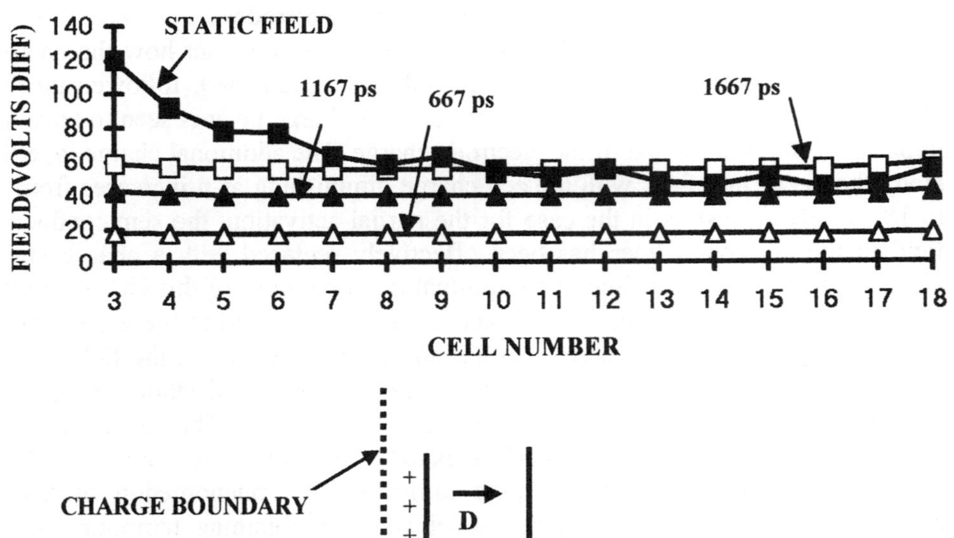

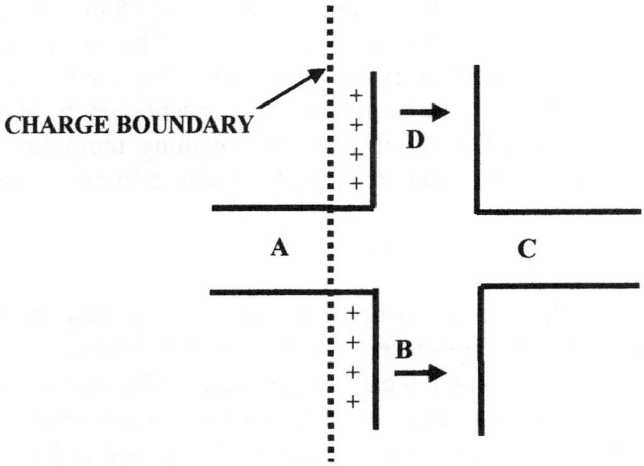

FIG. 7.22 ESTABLISHMENT OF CHARGED CELLS IN
SEMICONDUCTOR. THE PREDOMINANT FIELDS, IN D AND B,
CANCEL OUT ONE ANOTHER IF THEY ATTEMPT TO ENTER A.

$$V_B + V_C - V_D = 0 \qquad (7.33)$$

It is easiest to see how the waves enforce the boundary condition when $V_C \approx 0$, which is a good approximation since the fields normal to the charge layer predominate. This assumption also implies that $V_B \approx V_D$. In terms of the forward and backward waves, under the new equilibrium conditions, ^+V_B and ^-V_D are equal in magnitude and thus the coupling to the backward wave in A is

$$V_A = (1/2) \, {}^+V_B - (1/2) \, {}^-V_D = 0 \qquad (7.34)$$

which is to say that the waves in lines in B and D exactly cancel one another in line A. Thus the field energy does not penetrate the region to the left of the charge layer. The field configuration is of course only stable temporarily. Eventually the carrier drift becomes important, as the charged carriers move toward the cathode.

7.13 Comparison of Possible Boundary Conditions at the Semiconductor/Dielectric Interface

In the particular problem addressed , one issue, which we had to resolve at the outset, was where to imbed the TLM line corresponding to the interface between the semiconductor and the dielectric. As mentioned in Section 5.1 we arbitrarily decided to place the boundary in the low dielectric region Following this guideline, we can then assume the horizontal $Z_{xy}(n,15)$ line, which is the interface line, belongs to the high dielectric region, i.e., the semiconductor region. Given the coarseness of the grid we should try to determine the degree to which the solutions are dependent on the location of the interface TLM line, i.e., what are the differences in the solutions if the TLM line is in the semiconductor region versus the dielectric region. An estimate of this effect may be obtained if we assume the iterative equations are maintained , while we simply assign higher impedance values to the interface line (377Ω). For the simulation result, we select the field profile of the semiconductor, using the center cell approximation, $(n,17)$. In Fig.7.23, we compare the two profiles,

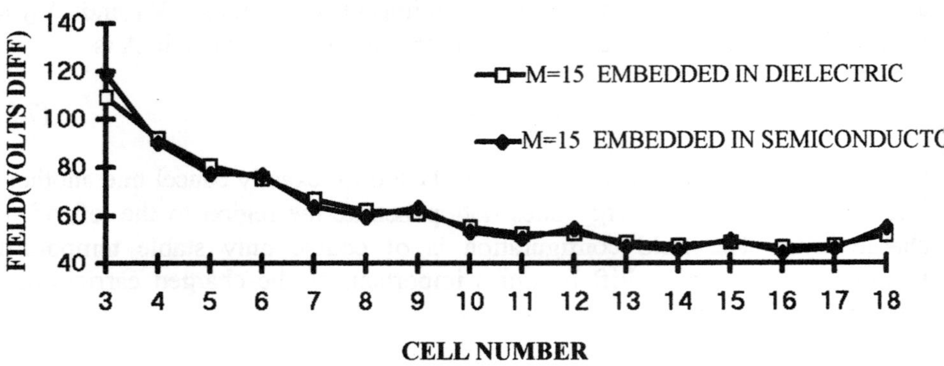

one where we use a 126 Ω interface impedance (semiconductor) and the other 377Ω (dielectric). We note that the two profiles are very close in value. The differences for other field profiles, as well as the output current, are also minor.

7.14 Simulation Results for Boundary with Non-Integral Nearest Nodes

The results of the previous examples dealt with a semiconductor/dielectric boundary in which the he ratio of propagation velocities was integral(in this case , a ratio of three) so that the vertical lines in the dielectric are perfectly aligned with the semiconductor lines (regular nodes). This simplifies the iteration considerably, but of course it does not correspond to the general situation in which the velocity ratio is non-integral. For completeness we consider the dielectric region to have a dielectric constant of ε =4 (instead of unity). This corresponds to the boundary shown in Fig.7.24. Note that every other vertical line in the dielectric region fails to coincide with a node in the

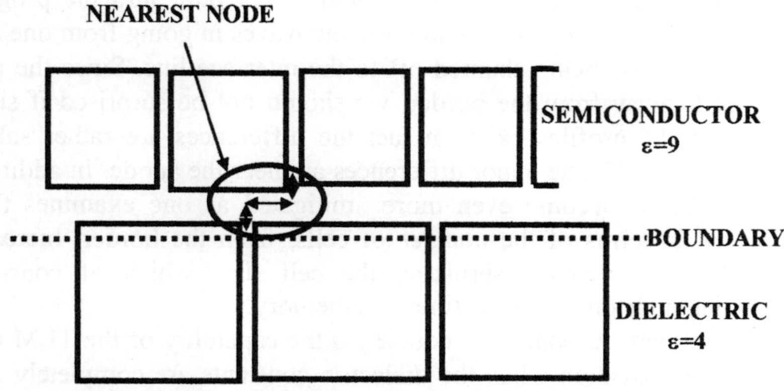

FIG. 7.24 SEMICONDUCTOR-DIELECTRIC INTERFACE WITH NEAREST NODE AS INDICATED. NEAREST NODE APPROXIMATION IS INCORPORATED INTO PROGRAM.

semiconductor region, with the line ending abruptly in the midst of the semiconductor cell. As indicated in the Figure, we invoke the nearest node approximation and assume the vertical dielectric line is part of the node located half of a semiconductor cell away. In cell notation the node consists of the line $Z_{yx}(4.5,15)$ in the dielectric region, $Z_{yx}(5,17)$ in the semiconductor, and the horizontal lines, $Z_{xy}(5,15)$ and $Z_{xy}(6,15)$. For the next nearest node, the vertical lines are $Z_{yx}(7.5,15)$, $Z_{yx}(7,17)$, and the horizontal lines are $Z_{xy}(7,17)$ and $Z_{xy}(8,17)$. The iterative equations must then be modified, taking into account the lines surrounding the nearest node, as well as the regular nodes.

Fig.7.25 shows the static field profiles in the semiconductor using a nearest node approximation for ε=4 in the dielectric region. The static horizontal field profile is very similar to that for ε=1, displaying a field concentration in the anode region and a slight concentration near the cathode. For comparison Fig.7.26 also includes the profile resulting from the neglect of the nearest node method, discussed briefly in Chapter V. We recall that with this simplified method the nearest node method is not built into the program and instead the computer decides how to route the waves in going from one region to the other, after first being shunted off to the interface line. Since the profile is only two cells away from the border, we should not be surprised if significant differences in the profiles exist; in fact the differences are rather subdued as indicated in Fig.7.25 The major differences are near the anode. In addition, such differences should become even more mitigated as one examines the fields further away, in terms of the *number* of cells, from the border. Increasing the number of cells requires shrinking the cell size, which of course places additional demands on computer time and memory.

Of paramount relevance ,of course , is the capability of the TLM matrix to simulate the propagation when the dielectric constants are completely arbitrary. The program modifications needed to simulate this situation are indicated in App.7A.3. Fortunately the changes are fairly straightforward.

FIG. 7.25 STATIC FIELD PROFILES IN SEMICONDUCTOR OBTAINED WITH AND WITHOUT THE NEAREST NODE METHODS, VO=1000V. FIELD IS AT m=17

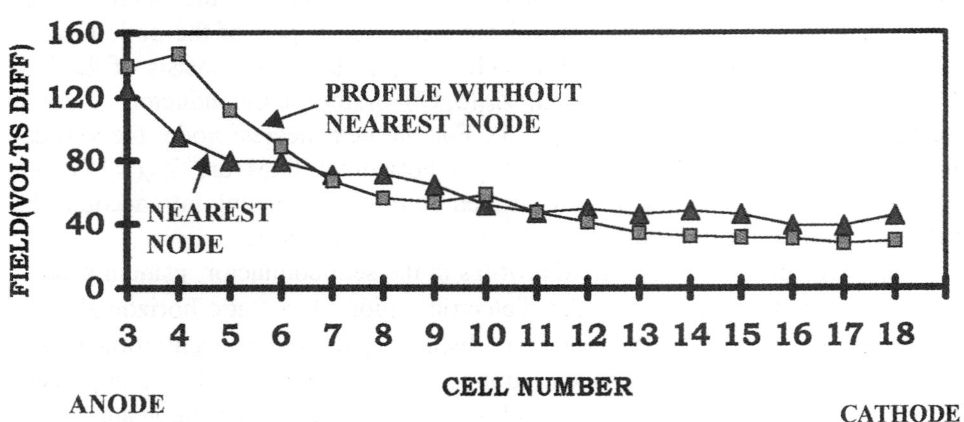

7.15 Comparison of Output With and Without Matched Input/Output Lines

In all the previous simulations, the inputs and outputs were mismatched to the device impedance. As discussed early in the Chapter, when the semiconductor is completely activated, the semiconductor portion of the device possesses an impedance of 113 Ω while the input and output lines are both 50 Ω . Assuming a device cross-section of 1.5mm X 5mm for both the device and input/output, this implies a dielectric constant of 5.11 for the input/output lines. Suppose now, we change the impedance of the input/output from 50Ω (which is a common standard for microwave transmission lines) to 113Ω , i.e., suppose we look at the transient behavior under *matched* conditions. Although we can match the device impedance to that of the input/output (50Ω), we instead choose to change the input/output impedance to that of the device(113Ω) in order to achieve the match; By utilizing the same device, we can obtain a more meaningful comparison. The program changes needed to switch to a matched device are relatively minor.

First the individual TLM lines of the input /output are changed from 167Ω to 377Ω, the same value as the device (when activated.). This involves the input horizontal lines $Z_{xy}(3,m)$, and the output lines $Z_{xy}(21,m)$ with m=6,9,12,15 in each case. Having changed the impedance levels of the input/output lines, we must then change the node resistors which terminate the lines so as to prevent unwanted multiple reflections This allows us to simulate (infinitely)long TLM lines for the input/output, by absorbing any reflections at the input and by absorbing all the signals delivered to the output lines.

One may question whether it is necessary to have matching node resistors at all at the input, since the individual line impedances are the same for both the semiconductor portion and the input, and therefore one might expect no reflections. In fact such reflections are inevitable especially since we have not taken plane wave effects into effect and thus reflections at the node will occur. The expressions for the terminating resistors are given by Eqs.(7.8),(7.10) for the interior lines and Eqs.(7.9), (7.11) for the lines adjacent the conductors. Substituting the value of 376.7 Ω into these equations we obtain a value of 502.3Ω for the interior lines and 565.1 Ω for the lines adjacent the conductors. The node resistors affected are R(0,m) and R(21,m) with m=6,9,12 ,15. With the new values of node resistance and line impedance for the input/output lines, we can now perform the simulations for the matched transmission line, semiconductor switch.

We should add that the input/output are not perfectly matched since the horizontal TLM lines corresponding to m=15 belong to the dielectric, which has a line impedance of 126Ω, instead of 377Ω. The substitution of 126Ω lines at the input/output, in any event, does not change the waveforms in any important way. In the limit of small cell size the degree of mismatch at the conducting boundary will of course diminish.

The major differences between the matched and unmatched cases are best illustrated by examining the output pulse , shown in Fig.7.26. In this Figure the device parameters remain the same but the individual input/output lines are

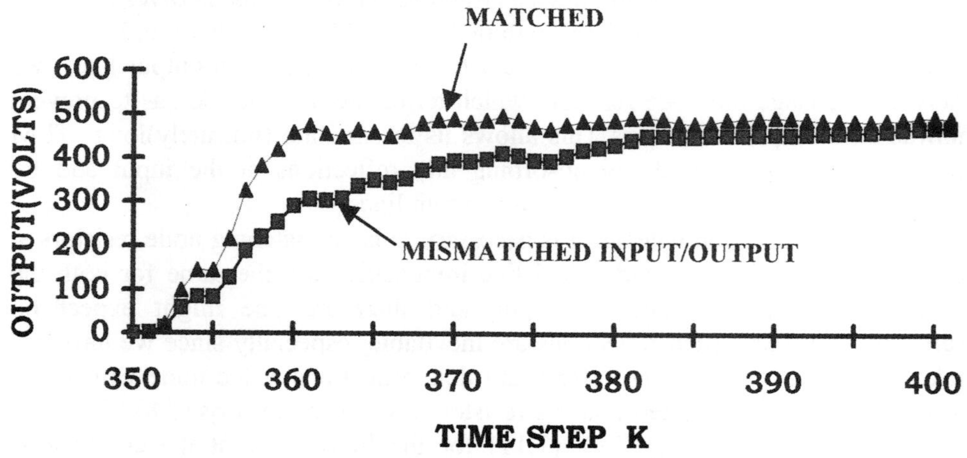

FIG. 7.26 COMPARISON OF OUTPUT PULSE WITH AND WITHOUT MATCHED INPUT/OUTPUT. VO=1000V. SEMICONDUCTOR ENTIRELY ACTIVATED. TIME STEP=1.6678ps. RO=15Ω

either matched(377Ω) or unmatched(113Ω). The waveforms shown are the output pulses following the activation of the entire semiconductor at k=350.We expand the time scale about the risetime regions. The first thing to note is the faster risetime associated with the matched device, compared to that of the mismatched device. The risetime of the matched device is approximately $\Delta k=10$, or 16.7ps. The transit time delay between anode and cathode, on the other hand, is $\Delta k=5$, or 8.3 ps. This should not come as a surprise since plane wave effects have not been included in the simulation. Thus, as discussed in the l Sections 3.14 to 3.19 of Chapter III, the wave energy will be diverted into the transverse TLM lines, as the wave progresses down the length of the switch. As pointed out previously, the full amplitude of the signal will not be attained until twice the transit delay has elapsed. The simulated risetime(*full* amplitude) of $\Delta k=10$ is therefore to be expected. The increased risetime of the mismatched device, with a risetime of $\Delta k\sim 30$, arises mainly from the mismatch between the 377Ω lines and the 167Ω lines of the input/output. As a result, some time will be required before the signal builds up to steady state amplitude in the device. The reader may notice that in both cases the there is detectable signal at the output at times *less* than $\Delta k =5$, the transit time of the device. What is the source of this wave energy? It certainly cannot come from the input at the cathode side, since the signal has not yet had time to reach the output. The origin of this energy stems from the energy stored in the region of the cathode just prior to the activation. Once the semiconductor is activated, the energy in this region is redistributed and as a result a transient signal, shortly after activation, is delivered to the output.

7.16 Simulation of Plane Wave Effects . Effect of Alternating Input

Much insight is gained by showing the effects of plane wave correlation   on the simulated output pulse, utilizing the same   semiconductor device   as before. Specifically we focus on the charge-up phase during which the system moves input. This will provide some insight as to how the plane waves affect the charge-up phase. We also include a simulation of a simplified time varying input .

Once we employ plane wave correlations, we must decide on how to treat the iteration when sign disparities appear. The full decorrelation discussed in

Chapter IV does add complexity to the program, and it may be worthwhile to employ very simple techniques, if only to gain insight into the correlation process. One way to treat this disparity, in a very approximate way, without actually performing a decorrelation, is to utilize the sign identity index γ_D discussed in Section 4.6. As noted previously γ_D varies from 0 to 1. We recall that γ_D is inverse to the magnitude of the disparity; a value close to zero has a large disparity(amplitudes are equal in magnitude but opposite in sign), while a value of γ_D close to 1 has a small sign disparity. During the iteration, we arbitrarily adopt the rule that if γ_D is less than some pre-selected value, γ_{D1}, then we create a new wave from the plane and symmetric components, wherein the new amplitude is equal to the quadrature sum of the plane and symmetric components, and the sign is equal to that of the dominant component, i.e., the sign is determined by the sign of $[V_{xyP}(n,m)+V_{xyS}(n,m)]$. If we select a value of γ_{D1} close to or equal to 1, therefore, all the waves with a sign disparity are treated in the aforementioned manner. For γ_D greater or equal to γ_{D1} the sign disparity effects are less and presumably we may invoke some other approximation(perhaps ignoring the disparity and redefining the correlation process). Given such uncertainties, we select $\gamma_{D1} = 1$ for the simulations shown.

In the following we consider several simulations. The first is that of the purely symmetric wave situation, which completely ignores plane wave effects, and was discussed previously in Section 7.5 . The second simulation is one in which we include plane wave correlations, for the usual case of constant input, but do not take into account any decorrelation effects. In this case any sign disparity is treated in the manner previously described. The final simulation has to do with an entirely different input. Instead of a constant input each of the four input lines delivers an alternating voltage, alternating between VO/8 and -VO/8 for successive time steps. This type input is examined both with and without plane wave correlation(but no decorrelation). Any sign disparities encountered in the correlated outputs are dealt with in an approximate manner, as outlined previously, using $\gamma_{D1}=1$.

The first two simulations shown in Fig.7.27 are interesting since they compare a symmetric wave, and the same wave but with plane wave correlations . The purely symmetric case is the same as that in Fig.7.5. The plane wave output, with no decorrelation, has an interesting feature which is not unexpected. We see that with an input of 500 volts, a continuous output of about 200 volts is delivered to the load. The wave correlation in the absence of any decorrelation, forces a portion of the wave energy toward the output load. The value of the sign index parameter selected is $\gamma_{D1} = 1$, so that all waves having a sign disparity were treated in the same manner, outlined previously. Also, we hasten to add that the simulation shown in Fig.7.27, given the approximate nature of the correlation should be taken purely as a qualitative description of the correlation process. When decorrelation effects are added the plane wave output presumably goes to zero(for uniform, time invarient input),

FIG. 7.27 WAVEFORM COMPARISON DURING CHARGEUP:
SYMMETRIC AND PLANE WAVE CORRELATED WAVEFORMS.
VO=1000V, TIME STEP=1.6678ps

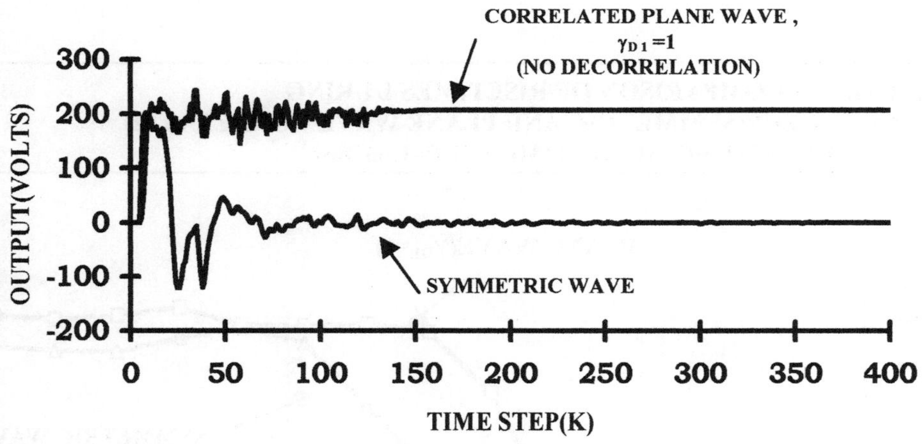

Just as the symmetric wave. In these simulations we do not include modifications to the *correlation* , due to the presence of a conductor interface , or to an interface between dissimilar dielectrics.

In Fig.7.28 we expand the time step scale in order to observe the risetimes of the two curves in Fig.7.27, the symmetric wave and the wave with plane wave correlation(with no decorrelation). As expected we see that the plane wave has a shorter risetime, compared to that of the pure symmetric wave. Roughly speaking the risetime of the symmetric wave is twice as long as the risetimes of the correlated waves($\Delta k \approx 4$ versus $\Delta k \approx 2$). The residual risetime remaining in the correlated output stems from several reasons. One reason is the "peel-off" of wave energy at the edges of the wave front, as progresses away from the input. Other reasons have to do with shortcomings in the simulation, rather than the underlying physics. For one thing, the correlation enhancement due to the conductor, at m=6, was not included(at least for the plane wave with

FIG. 7.28 COMPARISON OF RISETIMES DURING CHARGEUP:SYMMETRIC AND PLANE WAVE CORRELATED WAVEFORMS.VO=1000V. TIME STEP=1.6678ns

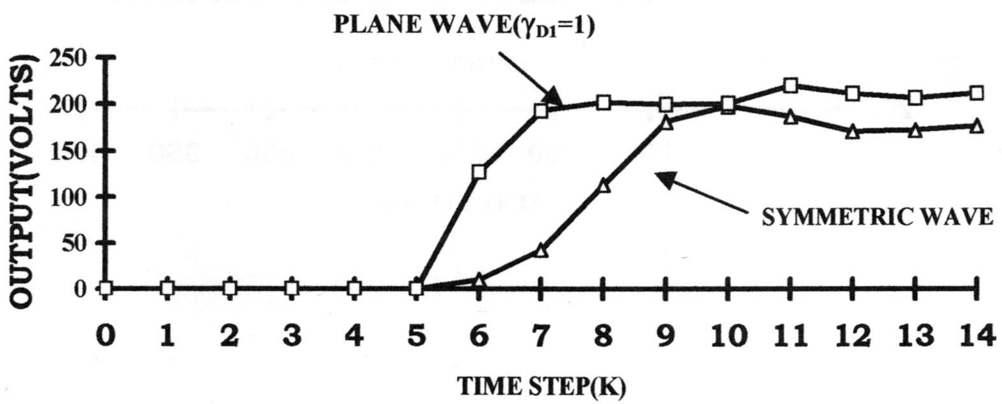

no decorrelations) Another important issue, which affects the risetime, has to do with the fact that the output receives a significant amount of its energy along the slower (high dielectric) line at m=15, which contributes to the slower risetime , and also upsets the plane wave condition. As the cell density increases this slowdown effect will be minimized.   Fig.7.29, again with $\gamma_{D1} = 1$, simulates the response when the input alternates in sign (rather than being constant in time) , both with and without plane wave correlation(but not including decorrelation effects).   In the case where there is no correlation(i.e., purely symmetric waves) the output is essentially "washed out" , with a small amount of energy distributed throughout the device, and very little energy delivered to the output. After 50 time steps or so, almost all the energy is reflected at the input. The simulation is entirely

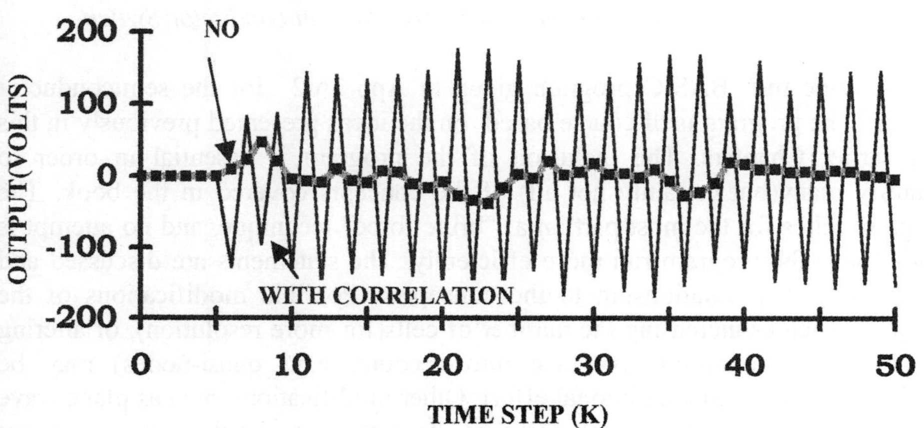

FIG. 7.29 COMPARISON OF ALTERNATING INPUTS WITH AND WIHOUT PLANE WAVE CORRELATION. INPUT WAVE IS 500V. TIME STEP IS 1.6678ps

different when , for the same input, plane wave correlations are included. Note that a strong alternating signal is delivered to the output. The output, however, is not a replica of the input because of the reasons outlined in the previous paragraph. In any event it is important to recognize that the correlated output in Fig.7.29 corresponds to a *high frequency* input. In the extreme case this would correspond to a laser signal simply traversing the device and exiting with hardly any interaction with its surroundings. Be that as it may, there is no reason why the TLM formulation cannot be applied to high frequency signals, provided sufficiently small TLM cells are selected.

We emphasize the fact that the plane wave simulations shown in Figs.7.27 - 7.29 are approximate in nature ; they do not account for any decorrelations, or for that matter corrections to the correlations.   Still, the approximations employed are valuable in at least providing a qualitative idea of the correlation processes, particularly in regard to the output current, which represents an "integrated" quantity. For obtaining the field profiles, or, for that matter quantitative estimates of both the field and current, the exact correlation;/decorrelation methods described in Chapter IV must be employed.

Appendices

App. 7A.1. Discussion Of Program Statements For Semiconductor Switch

We describe the BASIC program given in App.7A.2   for the semiconductor switch . The program is of course based on the ideas presented previously in this and earlier Chapters. The inclusion of the program is essential in order to quantify many(but certainly not all) of the concepts covered in the book. The program relies for the most part on a "brute force" technique, and no attempt is made make the program run more efficiently. The statements are discussed and we identify their relationship to the concepts. Possible modifications of the program, such as increasing the number of cells(for more resolution), or altering the boundary conditions (to take into account, e.g., quasi-nodes) may be implemented with little additional effort. Other modifications such as plane wave correlations require additional software development. Further, we have not attempted to express the Program in more advanced computer languages. Our main aim has been directed toward the understanding the main concepts and iterations. In any event, the transcription of the results to more advanced languages should not be difficult.

Lines 10-110 provide input data such as the total number of time steps, the bias voltage, the background node resistance, and so forth. Table 7A.1 explains the input parameters and the symbols used. Typical values for the inputs are listed. Note that included in the input is the region of activated nodes , specified by the range values NS1,NS2 and MS1,MS2 where the activated nodes are in the range   NS1<n< NS2 and MS1<m<MS2. Before activation, RT is the background node resistance in the semiconductor, which will be generally smaller than REQ, the background resistance in the dielectrics, (D) and (I1). Lines 40 and 90-110 specify whether the node resistance is allowed to collapse suddenly to RO(in which case INSTAN =10), or whether the resistance is allowed to decrease gradually, INSTAN < >10. In the event of the latter case, we must also specify the conductivity risetime, KRS , LIT and RST. LIT is a limiting node resistance of the semiconductor below which we do not allow the resistance to fall(such limitations may be ascribed physically, e.g., to optical screening effects). RST may be regarded as the initial light activated resistance of the semiconductor , following the first time step after KL. RST also may be modeled from the semiconductor physics, as discussed in Chapter II.

Lines 120-210 declare the array sizes needed to run the program. The various arrays are listed in Table 7A.2 Several of these are obvious such as the cell voltage VO(n,m) , the scattering coefficients, B(n,m) and T(n,m), as well as the node resistance R(n,m), and the corresponding node parameters , R1(n,m), R2(n,m) ,R3(n,m), R4(n,m), RL1(n,m),RL2(n,m), RL3(n,m), and RL4(n,m). The forward and backward waves have the following computer symbols: $^{+}V_{xy}(n,m) \rightarrow$ VPM(n,m), $^{-}V_{xy}(n,m) \rightarrow$ VNM(n,m), $^{+}V_{yx}(n,m) \rightarrow$ VPN(n,m) , and $^{-}V_{yx}(n,m) \rightarrow$ VNN(n,m). Similarly the line impedances are identified by $Z_{xy}(n,m) \rightarrow$ ZM(n,m) and $Z_{yx}(n,m) \rightarrow$ ZN(n,m). The arrays VPMS(n,m) , etc.., are temporary arrays assigned to the waves as discussed earlier. The remaining arrays, VS(n,m), VS2(n,m), VS1P(n,m), VS2P(n,m), VOUT(n,m), and VGC(n,m) are output arrays for the field and current , to be defined later in the program.

TABLE 7A.1

INPUT ITERATION PARAMETERS

<u>TYPE</u>	TYPICAL <u>VALUES</u>
TIME STEP ,Δt	1.6678ps
TOTAL N0. TIME STEPS , KM	500 -5000
BIAS VOLTAGE, VO	1000 V
R(n,m), Z(n,m) OF CONDUCTING & DIELECTRIC REGIONS	.01Ω, $10^{12}\Omega$
R(n,m) AFTER ACTIVATION, RO	15 - 150Ω
SPREAD FACTOR, G0	10^{-6}
ACTIVATION TIME, KL	k= 350
RECOVERY TIME,KREC	k=300, 10^{8}
ACTIVATED NODES NS1-NS2, MS1-MS2 RANGE	n=3-18 m= 15-18,
BACKGROUND R(n,m): RT, REQ IN SEMICONDUCTOR, DIELECTRIC	10^{7}, 10^{12} Ω
LINE IMPEDANCES, Z_D , Z_S, ZIN, ZOUT	126-377Ω
RISETIME CONSTANTS: RST, LIT, KRS	$10^{6}\Omega$, 15Ω K=50

TABLE 7A.2

VARIOUS TLM ARRAYS APPLICABLE TO LIGHT ACTIVATED SEMICONDUCTOR SWITCH

TEXT SYMBOL	PROGRAM SYMBOL	DESCRIPTION	DIMENSION
$V(n,m,)$	$VO(n,m)$	CELL VOLTAGE	25 X 21
$^+V_{xy}(n,m), ^-V_{xy}(n,m)$	$VPM(n,m), VNM(n,m)$	HORIZONTAL WAVES	25 X 21
$^+V_{yx}(n,m), ^-V_{yx}(n,m),$	$VPN(n,m), VNN(n,m)$	VERTICAL WAVES	25 X 21
N/A	$VPMS(n,m), VNMS(n,m)$	TEMP WAVES	25X21
N/A	$VPNS(n,m), VNNS(n,m)$	TEMP WAVES	25 X 21
$Z_{XY}(n,m), Z_{YX}(n,m)$	$ZM(n,m), ZN(n,m)$	TL LINES	25 X 21
$R(n,m)$	$R(n,m)$	NODE RESISTANCE	25x21
$R1(n,m), R2(n,m)$	$R1(n,m), R2(n,m)$	NODE PARAMETER	25X21
$R3(n,m), R4(n,m)$	$R3(n,m), R4(n,m)$	NODE PARAMETER	25X21
$RL1(n,m), RL2(n,m)$	$RL1(n,m), RL2(n,m)$	NODE PARAMETER	25X21
$RL3(n,m), RL4(n,m)$	$RL3(n,m), RL4(n,m)$	NODE PARAMETER	25X21
$T(n,m,s)$	$T(n,m,s)$	TRANSFER COEFF	25X21X12
$B(n,m,s)$	$B(n,m,s)$	REFLECTION COEFF	25X21X4
$VOUT(K)$	$VOUT(K)$	OUTPUT VOLTAGE	$10^2\text{-}10^5$
$V_{YX}(n,m)$	$VGC(K)$	SC FIELDS(TIME)	$10^2\text{-}10^5$
$V_{YX}(n,17)$	$VS(L), VS2(L)$	SC FIELD(PROFILE)	18
$V_{YX}(n,17)$	$X(L)$	SC CELL NUMBER	18

The initial cell voltages are given in lines 225-250. Note that in the case of the larger dielectric cells , a STEP 3 is required in the for-next loop. Once we have the cell voltages we can determine the initial forward and backward waves in the lines, since these depend on nothing more than the differences in cell voltage of the adjacent cells. The forward and backward waves are given in lines 300-350. In the simulations provided the initial cell voltages are zero, so that in fact VOI(n,m)=0, and therefore the initial waves vanish as well.

The transmission line impedances and node resistance values are stated in lines 600-1660. Because there are several regions, each with a different dielectric constant and loss property, extra care must be exercised assigning the correct line impedance and node resistance. A piecemeal approach was used in determining these values, wherein the appropriate TLM impedances and node resistors are inserted row by row, starting with m=3.

In going through the spatial loops for the   line impedances and node resistances, we could have chosen to use separate FOR - NEXT loops to account for the difference in cell size between the dielectric and the semiconductor, as was done in assigning values to the cell voltages. Instead we use a special technique for using the same index source for both large and small cell regions, as indicated in lines 680-700. The 680-700 statements are equivalent to the step 3 process in the dielectric process while in the semiconductor region   the usual step 1 process applies. Thus, lines 680-700 are a convenient way for using a single FOR-NEXT loop for the entire space. The same type index source is used repeatedly throughout the program.

Except for the I/O regions, where there exist matching resistors, the nodes initially have either very large resistance value, as in the dielectric or semiconductor, or very small values, as in the conducting regions. The node resistors in the semiconductor will be smaller that of the dielectric, due of course to the presence of   background conductivity which always exists in the semiconductor. Before beginning the main iteration we insure that the constant input waves, lines 2022-2024, are introduced in the m= 3 TLM lines. Each of the waves has an amplitude VO/8.

The main iteration begins with line 4000, with the iteration over both the time and spatial steps. Nothing of interest has happened thus far, however. This is because the initial cell voltages, as well as the waves derived from the cell voltages, are zero(assuming VOI(n,m)=0). In order to kick-start the program we introduce the aforementioned input waves in lines 2022-2024, just before the start of the iteration.

Line 4060 assigns node resistance value to the semiconductor region during the charge-up period, during which time no external conductivity has been introduced(to an extent this is redundant since we have introduced the same resistance values during the initialization). Even without activation the semiconductor node resistance RT will be small compared to that of the dielectric REQ(e.g., RT~E107 ohms versus REQ~E1012 ohms). Lines 4160-4340 are responsible for the node activation in the semiconductor region, Note that there are two choices: When INSTAN = 10 the semiconductor region undergoes a sudden conductivity change at k=kL. When INSTAN<>10 an exponential risetime is assumed starting at k=kL. Assuming negligible background conductivity, the node resistance is equal to E1*E2*E3, where E2 is the recovery term, line 4280, E3 is the spatial term, line 4295, and E1 is the induced conductivity. For the sudden conductivity change, E1=RO , and for the exponential risetime E1 is given by line 4210 where the constant RST is discussed in Section 7.6 Lines 4300 and 4310 combine the background and induced conductivities , which then yields R(n,m). Lines 4220 ,4290, limit the size of the exponential numbers and 4320 and 4340 assign bounds to the node resistance.

Once we change the node resistance, this triggers a sequence of other changes. Lines 5020-5100 calculates the various node parameters, all based of course on the new values of R(n,m). Following the calculation of the node parameters, we then obtain the transfer and reflection coefficients from lines 5130-5310. Note that the node activation, calculation of the node parameters , and the scattering coefficients all belong to the same for-next loop.for n,m. We now have all the information necessary to implement the scattering equations, at a given time step.

Before we can hand off the new values of R(n,m) , we must first create a set of temporary variables for the forward and backward waves, which we denote by appending an S to the old variable. Thus, VPM(n,m)→ VPMS(n,m), etc...The transformation is accomplished in lines 5450-5480. In lines 5500-5530, we insure that the phantom lines, located at the quasi nodes on the dielectric side equal to zero. This occurs at m=15 when n is not divisible by three, or when (n-3*INT(n/3))>0. This procedure is to some extent redundant since the scattering coefficients at the boundary presumably take into account the zero impedance of the phantom line and therefore there should be an absence of wave energy in these lines. Just prior to the scattering equations, we again insert the omnipresent lines 6010 -6025, which insures that in the medium with the lower dielectric constant, we perform the proper index skipping. We also have need of line 6030, as well. The parameter dt , equal to either 1 or 3, identifies whether the scattering is from lines in the dielectric or semiconductor.

Lines 6050-6165 represent the scattering equations. Lines 6050-6102 give the scattering for VPM(n,m) and VNM((n,m), the horizontal lines, using temporary variables. For the transverse lines we need to be more careful. If m= 15 then we set VPN(n,m) and VNN(n,m) =0 for those n yielding a quasi node, as indicated in lines 6104 and 6105. . This then completes the iteration for these particular waves with line 6106 essentially bypassing the remaining scattering equations, and returning to the start of the for-next loop. Line 6108   treats a special case in which m=15, with n divisible by three, in which case we wish to determine the new forward wave VPN(n, 15). The waves scattering into the $Z_{yx}(n,15)$ line emanate from lines with m=12, at the (n,12) node, so that the parameter dt=3. We make this explicit by routing the forward wave VPN(n,15) to line 6132. Lines 6132-6134 take into account the proper scattering at the (n,12) node. The backward wave ,VNN(n,m), is treated in lines 6140-6160, and requires no special treatment since the scattering node situated   at (n,15). Suppose m< >15? We then return to lines 6110-6130 where the scattering for VPN(n,m) and VNN(n,m) is completed . This completes the iteration. Note the iterative equations use temporary variables throughout until the iteration is ended.. Having essentially finished the iteration we then transform back to the original variables as shown in line 6165. We have not totally completed our task, however. The previous iteration does not provide for any input waves. We must not forget to re-establish the input waves, given in lines 6180-6222.

The remainder of the program deals with tabulating the output data and then putting into a form suitable for graphing. Keep in mind that we are still inside the k loop At a given time step, the total output pulse VOUT(k), delivered to the output line, is given in line 6250. Note that VOUT(k) is an array, as are the other outputs, so that the data is being stored. Also note that the output is the sum of the four waves in series. Likewise. the field at the centrally located cell, (11,17) , denoted by VGC(k). is also calculated and stored. We recall that the field, calculated from VPN(11,17) + VNN(11,17), involves only the horizontal component. A small vertically directed field is also present, which may be calculated from VPM(11,17)+VNM(11,17).   Line 6310 displays the value of k during the calculation, while lines 6320 and 6330 display either VOUT(k) or VGC(k), or both.

We next create data arrays of the field profiles in the semiconductor. Each of the profile arrays contain 16 elements (one more than the number of cells). Lines 6232 -6335 creates an array profile of the horizontal field, denoted by VS2(L) just prior to the light activation at time KL-2,   using the eighteen cells available.   By this time equilibrium has been achieved and this represents , therefore, the static field profile (since equilibrium has been achieved, any time step just prior to activation may be used, e.g., KL-1 or KL-3) We also make use of our input parameters, KP1 and KP2 , which are selected moments of time at which we take a "snapshot" of the field profile, either during the charge-up or after activation. These are described in lines 6336-6339, and are denoted by VSP1(L) and VSP2(L), where L is an integer, 3 to 18, specifying the cell number (corresponding to the n index). We can specify as many of these snapshots as we like, of course. The iteration in time then comes to an end. Immediately afterwards, we create a final array, VS(L), representing the final profile. We also create a number array for the number of profile array elements, denoted by X(L).

We next place our data files in column format so as to make them more suitable to most types of graphical software. This is implemented in lines 7050-7390 The output file bunk.xl is opened and made available for the data array elements. The three columns of data, k, VOUT(K), VGC(K) are placed in the file, as indicated in lines 7200-7272. This is followed by the columns of the profile elements consisting of X(L), VS(L), VS2(L), VSP1(L), and VSP2(L).

App. 7A.2 Program Statements

```
REM App.7A.2: 2D PHOTOCONDUCTIVE SEMICONDUCTOR SWITCH
PROGRAM



10   INPUT "BIAS VOLTAGE : VO="; VO
     INPUT "MAXIMUM TIME STEP: KM="; KM
     INPUT "SEMICONDUCTOR BACKGROUND RESISTANCE: RT=";RT
20   INPUT "CONDUCTOR LINE IMPEDANCE:ZL="; ZL
     INPUT "CONDUCTOR NODE RESISTANCE:RLL=";RLL
     INPUT "DIELECTRIC, INSULATOR NODE RESISTANCE:REQ=";REQ
30   INPUT "INSTANTANEOUS (INSTAN=10) OR GRADUAL
(INSTAN<>10) RISETIME: INSTAN=";INSTAN
40   IF INSTAN =10 THEN INPUT "INSTANTANEOUS NODE
RESISTANCE;RO="; RO
50   INPUT "SPREAD FACTOR: G0=";G0
REM IN THIS EXAMPLE THE DIELECTRIC CONSTANT OF THE
SEMICONDUCTOR AND THE
REM DIELECTRIC FILL REGION ARE 9 AND 1 RESPECTIVELY,WITH
THE CORRESPONDING
REM  TLM LINE IMPEDANCES EQUAL TO ZS=125.6 OHMS AND
ZD=376.7
OHMS.
     ZS=125.6: ZD=376.7
     INPUT "COMPOSITE INPUT LINE IMPEDANCE:ZINC=";ZINC
     INPUT "COMPOSITE OUTPUT LINE IMPEDANCE:ZOUTC=";ZOUTC
REM SINCE THE CROSS-SECTION OF THE INPUT/OUTPUT IS EQUAL TO
THAT OF THE
REM DIELECTRIC FILL(5MMX1.5MM) THE LINE IMPEDANCES, ZIN AND
ZOUT, ARE
     ZIN=5*ZINC/1.5:ZOUT=5*ZOUTC/1.5
60   REM THE INPUT/OUTPUT NODE RESISTANCES ARE
70 RIN1=(4/3)*ZIN:RIN2=(3/2)*ZIN:ROUT1=(4/3)*ZOUT:ROUT2=
(3/2)*ZOUT
```

```
80   INPUT "LIGHT ACTIVATION STEP:KL="; KL

     INPUT "STATUS AT TIME STEPS: KP1,KP2=";KP1,KP2
     INPUT "RECOVERY TIME(NUMBER OF TIME STEPS):KREC=";KREC
INPUT "CELL NODE ACTIVATION; LOWER, UPPER M
INDEX:MS1,MS2=";MS1,MS2
INPUT " CELL NODE ACTIVATION; LOWER, UPPER N
INDEX:NS1,NS2=";NS1,NS2
90   IF instan<>10 then input "LIGHT AMPLITUDE
RISETIME(NUMBER OF TIMESTEPS):krs=";krs
100 IF instan<>10 then input "LOWER NODE RESISTANCE
BOUND:LIT=";LIT
110 IF instan<>10 then input "INITIAL NODE RESISTANCE=";RST


120 DIM VO(25, 21), VPM(25,21),
VPMS(25,21),VNMS(25,21),VPNS(25,21), VNNS(25,21)
125 DIM VOI(25,21)
130 DIM VPN(25, 21), VNM(25, 21), VNN(25, 21)
140 DIM ZN(25, 21), ZM(25, 21)
150 DIM R(25, 21), R1(25, 21), R2(25, 21)
160 DIM R3(25, 21), R4(25, 21), RL1(25, 21)
170 DIM RL2(25, 21), RL3(25, 21), RL4(25, 21)
180 DIM   VS(18), X(18), VS2(18)
190 DIM B(25, 21, 4), F(25, 21, 12)
195 DIM   VS1P(18), VS2P(18)
200 DIM VOUT(KM)
210 DIM VGC(KM)


225 REM    INITIAL BIAS VOLTAGES OF EACH CELL
    REM    FOLL ASSUMES VOI(N,M) OF EACH CELL IS SPECIFIED;
ELSE VOI(N,M)=0

230  FOR M=0 TO 15 STEP 3
235  FOR N=0 TO 24 STEP 3
     VO(N,M)=VOI(N,M)
     NEXT N
```

```
     NEXT M


240 FOR M=16 TO 20
    FOR N= 0 TO 22
    VO(N,M)=VOI(N,M)
    NEXT N
    NEXT M

REM  FOR M= 15 INSURE THAT SIB-CELL VOLTAGE SAME AS LARGE
CELL

    FOR N=3 TO 24 STEP 3
    FOR D=1 TO 2
    VO(N-D,15) =VO(N,15)
    NEXT D
    NEXT N

250 REM

300 REM WAVE CALC

REM WAVES TRAVELING ALONG VERTICAL TRACKS

310  FOR M= 0 TO 15 STEP 3
    FOR N= 0 TO N=21 STEP 3
    VPN(N,M)=  (VO(N+3,M) -VO(N,M) )/2
    VNN(N,M)=VPN(N,M)
    NEXT N
    NEXT M

320  FOR M= 16 TO 20
    FOR N= 0 TO 21
    VPN(N,M)=(VO(N+1,M) -VO(N,M) )/2
    VNN(N,M)=VPN(N,M)
    NEXT N
    NEXT M


REM   INSURE FIELDS IN PHANTOM LINES ARE ZERO
```

```
      FOR N=0 TO 24
      IF (N-3*INT(N/3))>0 THEN VPN(N,15)=0
      IF (N-3*INT(N/3))>0 THEN VNN(N,15)=0
      NEXT  N


REM WAVES TRAVELING ALONG HORIZONTAL LINES

330   FOR M= 0 TO 12 STEP 3
      FOR N=0 TO 24 STEP 3
      VPM(N,M)=(VO(N,M+3)-VO(N,M))/2
      VNM(N,M)=VPM(N,M)
      NEXT N
      NEXT M

      FOR M= 15 TO 19
      FOR N= 0 TO 22
      VPM(N,M)  =(VO(N,M+1)-VO(N,M))/2
      VNM(N,M)=VPM(N,M)
      NEXT N
      NEXT M
350 REM


600 REM TRANSMISSION LINE AND RESISTANCE VALUES

HH=REQ
REM
670 FOR M=1 TO 20
675 FOR N=0 TO 24

680      IF M<15  AND (M-3*INT(M/3))>0 THEN M=M+1
685      IF M<15  AND (M-3*INT(M/3))>0 THEN GOTO 680
695      IF M<15  AND (N-3*INT(N/3))>0 THEN N=N+1
700      IF M<15  AND (N-3*INT(N/3))>0 THEN GOTO 695

IF M=3 THEN
ZM(N,M)=ZL
ZN(N,M)=ZL
R(N,M)=RLL
END IF
```

```
IF M=6 AND N=0 THEN
ZM(N,M)=HH
ZN(N,M)=ZL
R(N,M)=RIN2
END IF

IF M=6 AND N=3 THEN
ZM(N,M)=ZIN
ZN(N,M)=ZL
R(N,M)=REQ
END IF

IF M=6 AND N>3 AND N<19 THEN
ZM(N,M)=ZD
ZN(N,M)=ZL
R(N,M)=REQ
END IF

IF M=6 AND N=21 THEN
ZM(N,M)=ZOUT
ZN(N,M)=ZL
R(N,M)=ROUT2
END IF

IF M=6 AND N=24 THEN
ZM(N,M)=HH
ZN(N,M)=ZL
R(N,M)=REQ
END IF

IF M>8 AND M<13 AND N=0 THEN
ZM(N,M)=HH
ZN(N,M)=HH
R(N,M)=RIN1
END IF

IF M>8 AND M<13 AND N=3 THEN
ZM(N,M)=ZIN
```

```
ZN(N,M)=ZD
R(N,M)=REQ
END IF

IF M>8 AND M<13 AND N>3 AND N<19 THEN
ZM(N,M)=ZD
ZN(N,M)=ZD
R(N,M)=REQ
END IF

IF M>8 AND M<13 AND N=21 THEN
ZM(N,M)=ZOUT
ZN(N,M)=HH
R(N,M)=ROUT1
END IF

IF M>8 AND M<13 AND N=24 THEN
ZM(N,M)=HH
ZN(N,M)=HH
R(N,M)=REQ
END IF

IF M=15 AND N=0 THEN
ZM(N,M)=HH
ZN(N,M)=HH
R(N,M)=RIN2
END IF

IF M=  15 AND N=3 THEN
ZM(N,M)=ZIN
ZN(N,M)=ZD
R(N,M)=RT
END IF

IF M=15 AND N>0 AND N<3 THEN
ZM(N,M)=ZIN
ZN(N,M)=ZL
R(N,M)=REQ
END IF
```

```
IF M=15 AND N>3 AND N<19 THEN
ZM(N,M)=ZS
ZN(N,M)=ZD
R(N,M)=RT
END IF

IF M=15 AND N>18 AND N<21 THEN
ZM(N,M)=ZOUT
ZN(N,M)=ZL
R(N,M)=REQ
END IF

IF M=15 AND N=21   THEN
ZM(N,M)=ZOUT
ZN(N,M)=HH
R(N,M)=ROUT2
END IF

IF M=15 AND N>21 AND N<24 THEN
ZM(N,M)=HH
ZN(N,M)=ZL
R(N,M)=REQ
END IF

IF M=15 AND N=24 THEN
ZM(N,M)= HH
ZN(N,M)=HH
R(N,M)=REQ
END IF

IF M>15 AND M<18 AND N<3 THEN
ZM(N,M)=ZL
ZN(N,M)=ZL
R(N,M)=RLL
END IF

DOP=0
IF M=16 OR M=17 THEN DOP=10
```

```
IF DOP=10 AND N=3 THEN
ZN(N,M)=ZS
ZM(N,M)=ZL
R(N,M)=RT
END IF

IF M>15 AND M<18 AND N>3 AND N<19 THEN
ZM(N,M)=ZS
ZN(N,M)=ZS
R(N,M)=RT
END IF

IF M>15 AND M<18 AND N>18 THEN
ZM(N,M)=ZL
ZN(N,M)=ZL
R(N,M)=RLL
END IF

IF M=18 AND N<3 THEN
ZN(N,M)=ZL
ZM(N,M)=ZS
R(N,M)=RT
END IF

IF M=18 AND N=3 THEN
ZN(N,M)=ZS
ZM(N,M)=ZS
R(N,M)=RT
END IF

IF M=18 AND N>3 AND N<19 THEN
ZM(N,M)=ZS
ZN(N,M)=ZS
R(N,M)=RT
END IF

IF M=18 AND N>18 THEN
ZN(N,M)=ZL
ZM(N,M)=ZS
```

```
R(N,M)=RT
END IF

IF M=19 THEN
ZN(N,M)=ZS
ZM(N,M)=HH
R(N,M)=REQ
END IF

IF M=20 THEN
ZM(N,M)=HH
ZN(N,M)=HH
R(N,M)=REQ
END IF

IF M=15 AND (N-3*INT(N/3))>0 THEN
ZN(N,M)=ZL
END IF

1660 REM

NEXT N
NEXT M

REM INPUT WAVES
2022 VPM(3,6)=VO/8 : VPM(3,9)=VO/8: VPM(3,12)=VO/8
2024 VPM(1,15)=VO/8: VPM(2,15)=VO/8:VPM(3,15)=VO/8



4000 REM NODE ACTIVATION
4002 FOR K = 1 TO KM

REM ***** THE FOLLOWING 8 STATEMENTS APPLY WHEN INPUT
ALTERNATES IN SIGN
REM IF (K-2*INT(K/2))=0 THEN
REM VPM(1,15)=-VO/8 :VPM(2,15)=-VO/8:VPM(3,15)=-VO/8
REM VPM(3,12)= -VO/8:VPM(3,9)= -VO/8:VPM(3,6)=-VO/8
REM END IF
```

```
REM IF  (K-2*INT(K/2))>0 THEN
REM VPM(1,15)= VO/8:VPM(2,15)=VO/8:VPM(3,15)= VO/8
REM VPM(3,12)= VO/8:VPM(3,9)= VO/8:VPM(3,6)= VO/8
REM END IF
REM  ******

4004 FOR m = 1 TO 19
4006 FOR n = 1 TO 21
4010 IF m < 15 AND (m - 3 * INT (m / 3)) > 0 THEN m = m + 1
4020 IF m < 15 AND (m - 3 * INT(m / 3)) > 0 THEN GOTO 4010
4030 IF m < 15 AND (n - 3 * INT(n / 3)) > 0 THEN n = n + 1
4040 IF m < 15 AND (n - 3 * INT(n / 3)) > 0 THEN GOTO 4030
4050  RR2=0
4060 if n>2 and n<19 and m>14 and m<19 then R(n,m)=RT
4065  if R(n,m)>(hh+.1) then R(n,m)=hh
4080 if k>(KL-1) then goto 4160
4130  rem
4150 if k<KL then goto 4345
4160 if k>(KL-1) and n>NS1 and n<ns2 and m>MS1 and m<MS2
then RR2=10
4180 if instan=10 then goto 4240
4200  rem
4210 E1  =  RST/(KRS*(EXP((k-kl)/krs)-.99999))
4220 if ((K-KL)/KRS)>40 then E1=RST/(KRS*(EXP(40)-1))
4240 if instan =10 and rr2=10 then E1=RO
4260 rem
4280 E2 = EXP((K-KL)/KREC)
4290 IF ((K-KL)/KREC)>40 THEN E2=EXP(40)
4295 E3 = EXP(G0*(n-11)^2)
4300 if RR2=10 then RQ = E1*E2*E3
4310  IF RR2=10 then R(n,m) = RT*RQ/(RT+RQ)
4320   if RR2=10 AND R(n,m)>RT THEN R(n,m)=RT
4325  if instan=10 then goto 4345
4340   if RR2=10 and R(n,m)<LIT then R(n,m)=LIT
4345  rem

5015 REM  CALC OF NODE PARAMETERS
5020 if m<15 then dt=3 else dt=1
5030 R1(n, m) = ZM(n, m) * R(n, m) / (ZM(n, m) + R(n, m))
```

```
5040 R2(n, m) = ZN(n, m) * R(n, m) / (ZN(n, m) + R(n, m))
5050 R3(n, m) = ZM((n + dt), m) * R(n, m) / (ZM((n + dt),
m) + R(n, m))
5060 R4(n, m) = ZN(n,(m+dt)) * R(n, m) / (ZN(n,(m+dt)) +
R(n, m))
5070 1RL1(n, m)=(R2(n, m)+R3(n,m)+R4(n, m))*R(n, m)/((R2(n,
m)+R3(n,m)+R4(n, m))+R(n, m))
5080 RL2(n, m)=(R1(n, m)+R3(n, m)+R4(n, m))*R(n, m)/((R1(n,
m)+R3(n, m)+R4(n, m))+R(n,m))
5090 RL3(n, m)=(R1(n, m)+R2(n,m)+R4(n,m))*R(n, m)/((R1(n,
m)+R2(n, m)+R4(n, m))+R(n, m))
5100 RL4(n, m)=(R1(n, m)+R2(n, m)+R3(n, m))*R(n, m)/((R1(n,
m)+R2(n, m)+R3(n,m))+R(n, m))
5110 rem
5120 rem
5130 REM CALCULATION OF TRANSFER/REFLECTION COEFF
5140  rem
5150  rem
5160 B(n, m, 1) = (RL3(n, m) - ZM((n+dt), m)) / (RL3(n, m)
+ ZM((n+dt), m))
5170 B(n, m, 2) = (RL1(n, m) - ZM(n, m)) / (RL1(n, m) +
ZM(n, m))
5180 B(n, m, 3) = (RL4(n, m) - ZN(n,(m+dt))) / (RL4(n, m) +
ZN(n ,(m+dt)))
5190 B(n, m, 4) = (RL2(n, m) - ZN(n,m)) / (RL2(n, m) +
ZN(n,m))

5200 F(n, m, 1)=2*RL1(n, m)*R3(n, m)/((RL1(n, m)+ZM(n,
m))*(R2(n, m)+R3(n, m)+R4(n, m)))

5210  F(n,m,2)=2*RL2(n, m)*R3(n, m)/((RL2(n, m)+ZN(n,
m))*(R1(n, m)+R3(n, m)+R4(n, m)))

5220  F(n, m,3)=2*RL4(n,
m)*r3(n,m)/((RL4(n,m)+ZN(n,(m+dt)))*( R1(n,m)+R2(n,m)+R3(n,
m)))
```

```
5230  F(n,m,4)=2*RL3(n,
m)*R1(n,m)/((RL3(n,m)+ZM((n+dt),m))*(R1(n, m)+R2(n,
m)+R4(n, m)))

5240 F(n, m,5)=2*RL4(n,m)*R1(n,m)/((RL4(n,
m)+ZN(n,(m+dt)))*(R1(n, m)+R2(n, m)+R3(n, m)))

5250  F(n,m,6)=2 * RL2(n, m)*R1(n, m)/((RL2(n, m)+ZN(n,
m))*(R1(n,m)+R3(n, m)+R4(n, m)))

5260 F(n, m,7)=2 * RL1(n, m)*R4(n, m)/((RL1(n, m)+ZM(n,
m))*(R2(n, m)+R3(n, m)+R4(n, m)))

5270 F(n, m, 8)=2*RL2(n, m)*R4(n, m)/((RL2(n, m)+ZN(n,
m))*(R1(n, m)+R3(n, m)+R4(n, m)))

5280 F(n, m,9)=2*RL3(n,m)*R4(n,m)/((RL3(n,
m)+ZM((n+dt),m))*(R1(n, m)+R2(n,m)+R4(n, m)))

5290 F(n,m,10)=2*RL1(n, m)*R2(n, m)/((RL1(n, m)+ZM(n, m))*(
R2(n, m)+R3(n,m)+R4(n, m)))

5300 F(n,m,11)=2*RL4(n, m)*R2(n,m)/((RL4(n,
m)+ZN(n,(m+dt)))*(R1(n, m)+R2(n,m)+R3(n,m)))

5310 F(n,m,12)=2*RL3(n,m)*R2(n,
m)/((RL3(n,m)+ZM((n+dt),m))*(R1(n,m)+R2(n,m)+R4(n, m)))

5315 rem
5320  rem
5325 next n
5330 next m

REM  TEMPORARY FIELD REPLACEMENT
5430 FOR m=0 to 20
5440 FOR n=0 to 22
5450 if m<15 and (m-3*INT(m/3))>0 then m=m+1
5455 if m<15 and (m-3*INT(m/3))>0 then goto 5450
5460 if m<15 and (n-3*INT(n/3))>0 then n=n+1
```

```
5470 if m<15 and (n-3*INT(n/3))>0 then goto 5460
5480
VPNS(n,m)=VPN(n,m):VNNS(n,m)=VNN(n,m):VPMS(n,m)=VPM(n,m):VN
MS(n,m)=VNM(n,m)
5490 next n
5495 next m

REM SET TEMPORARY PHANTOM FIELDS EQUAL TO ZERO
5500    FOR N=1 TO 21
5510    IF  (N-3*INT(n/3))>0 THEN VPNS(n,15)=0
5520    IF (N-3*INT(n/3))>0 THEN  VNNS(n,15)=0
5530    NEXT N


REM MAIN ITERATION
6000 for m= 1 to 19
6005 for n=1 to 21
6010 if m<15 and (m-3*INT(m/3))>0 then m=m+1
6015 if m<15 and (m-3*INT(m/3))>0 then goto 6010
6020 if m<15 and (n-3*INT(n/3))>0 then n=n+1
6025 if m<15 and (n-3*INT(n/3))>0 then goto 6020
6030 IF m < 15 THEN dt = 3 ELSE dt = 1
6035 REM FOLLOWING ARE ITERATIVE EQS
6040 rem
6050     X = F((n - dt), m, 1) * VPMS((n - dt), m) -F((n -
dt), m, 2) * VPNS((n - dt), m)
6060     Y = F((n - dt), m, 3) * VNNS((n - dt), (m + dt)) +
B((n - dt), m, 1) * VNMS(n, m)
6070     Z1 = X + Y

6080     X = F(n, m, 4) * VNMS((n + dt), m) - F(n, m, 5) *
VNNS(n, (m + dt))
6090     Y = F(n, m, 6) * VPNS(n, m) + B(n, m, 2) * VPMS(n,
m)
6100     Z2 = X + Y

6102     VPM(n,m)=z1 : VNM(n,m)=z2
6104     if m=15 and (n-3*INT(n/3))>0 then VPN(n,15)=0
6105     if m=15 and (n-3*INT(n/3))>0 then VNN(n,15)=0
```

```
6106        if m=15 and (n-3*INT(n/3))>0 then goto 6167
6108        if m=15 and (n-3*INT(n/3))=0   then goto 6132
6110        X = -F(n, (m - dt),7) * VPMS(n,(m - dt))+F(n, (m -
dt), 8)*VPNS(n, (m - dt))
6120        Y = F(n,(m - dt),9)*VNMS((n+dt),(m-dt))+B(n, (m -
dt), 3) * VNNS(n, m)
6130        Z3 = X + Y
6131        goto 6140
6132        X=-F(n,12,7)*VPMS(n,12)+F(n,12,8)*VPNS(n,12)
6134        Y=F(n,12,9)*VNMS((n+3),12)+B(n,12,3)*VNNS(n,15)
6136        Z3=x+y

6140        X = F(n, m, 10) * VPMS(n, m) + F(n, m, 11) *
VNNS(n, (m + dt))
6150        Y = -F(n, m, 12) * VNMS((n + dt), m) + B(n, m, 4)
* VPNS(n, m)
6160        Z4 = X + Y

6165        VPM(n,m)=Z1:VNM(n,m)=Z2:VPN(n,m)=Z3:VNN(n,m)=Z4
6167        rem
6180        VPM(3,  6) =VO/8
6200        VPM(3,  9) =VO/8
6220        VPM(3, 12) =VO/8
6222        VPM(3,15)=VO/8:VPM(2,15)=VO/8:VPM(1,15)=VO/8
6224        REM

REM  ******* THE FOLLOWING 8 STATEMENTS ARE USED FOR
ALTERNATING INPUT
REM IF (K-2*INT(K/2))=0 THEN
REM VPM(1,15)= -VO/8:VPM(2,15)= -VO/8:VPM(3,15)= -VO/8
REM VPM(3,12)= -VO/8:VPM(3,9)= -VO/8:VPM(3,6)=-VO/8
REM END IF
REM IF (K-2*INT(K/2))>0 THEN
REM VPM(1,15)= VO/8:VPM(2,15)=VO/8:VPM(3,15)= VO/8
REM VPM(3,12)= VO/8:VPM(3,9)= VO/8:VPM(3,6)= VO/8
REM END IF
REM  ********

6230        NEXT n
```

```
6240        NEXT m
6241        KQ=0
6242     rem        if ((K/10)-INT(K/10))=0 then KQ=K/10
6250     VOUT(K) = VPM(21, 6) + VPM(21, 9) + VPM(21, 12)
+VPM(21,15)
6260     rem  VGC(K) =
(VPN(11,16)+VNN(11,16)+VPN(11,17)+VNN(11,17)+VPN(11,18)+VNN
(11,18))/3
6270     VGC(K) = VPN(11,17)  +VNN(11,17)
6280     rem
6290     rem
6300        if ((k/100)-INT(k/100))=0 then print
6310        if ((K/100)-INT(k/100))=0 then print k
6315     REM print
VPNS(18,15);VNNS(18,15);VPNS(17,15);VNNS(17,15)
6320     rem      print VGC(k);
6330     print VOUT(K);
6332   FOR L=3 to 18
6333   REM
6334   IF K=KL-2 THEN VS2(L)  =VPN(L,17)+VNN(L,17)
6335    next L
6336   FOR L=3 to 18
6337   if k=KP1 then VS1P(L)=  VPN(L,17)+VNN(L,17)
6338   if k=kp2 then VS2P(L)=VPN(L,17)+VNN(L,17)
6339   next L
6340   NEXT K
6345   rem
6350 for L=3 to 18
6355 REM
6360   VS(L) = VPN(L,17)+ VNN(L,17)
6365 next L
6370 for L=3 to 18
6375 x(L)=L
6380 next L
6385 rem


REM  SETUP OUTPUT FILE
7050 OPEN "bunk.xl" FOR output AS #1
```

```
7190 rem
7195 print #1, "k"; CHR$(9); "VOUT(k)";CHR$(9); "VGC(k)"
7200 FOR K = 1 TO KM
7210 rem    K=10*KQ
7220  rem
7230 PRINT #1, k;
7240 PRINT #1, CHR$(9);
7250  PRINT #1, VOUT(k);
7260  PRINT #1, CHR$(9);
7270  PRINT #1, -VGC(K)
7272 next k
7273 print #1, "   "
7274 print #1, "   "
7275 print
#1,"X(L)";CHR$(9);"VS(L)";CHR$(9);"VS2(L)";CHR$(9);"VS1P(L)
";CHR$(9);"VS2P(L)"
7280  for l=3 to 18
7290 PRINT #1, X(L);
7300 PRINT #1,CHR$(9);
7310 PRINT #1, -VS(L);
7320 PRINT #1, CHR$(9);
7330 PRINT #1,-VS2(L);
7340 PRINT #1, CHR$(9);
7350 PRINT #1, -VS1P(L);
7352 PRINT #1, CHR$(9);
7354 PRINT #1,-VS2P(L)
7360 next L
7365 rem
7370 CLOSE #1
7380 rem
7390  end
```

App.7A.3 Program Changes For Arbitrary Dielectric Constant , Cell Density, and Device Size

When the dielectric constants and cell density are arbitrarily selected, the changes to the program code are not large, at least from a conceptual viewpoint. The most significant changes occur at the dielectric interface, where we must employ a nearest node approximation.

We consider the same type design switch as before, i.e, a rectangular type switch in a parallel plate transmission line. Designating the dielectric constants of the semiconductor and dielectric regions by ε_S and ε_D, and the corresponding cell lengths by Δl_S and Δl_D , we have the usual relation

$$(\Delta l_S/\Delta l_D) = (\varepsilon_D/\varepsilon_S)^{1/2} \tag{7A5.1}$$

The cell lengths satisfy

$$\Delta l_S = (1//\varepsilon_S/\varepsilon_o\mu_o)^{1/2}\Delta t \;\; ; \Delta l_D = (1//\varepsilon_D/\varepsilon_o\mu_o)^{1/2}\Delta t \qquad (7A5.2) \, ; (7A5.3)$$

where Δt is the selected time step.

Since we are selecting arbitrary dielectric constants, we should not expect that there will be an integral number of cell lengths in each of the switch dimensions(height and length of the semiconductor and dielectric regions). One must therefore expect truncated cells at the boundaries. By increasing the cell densities we can reduce the effect of the truncated cells to an arbitrarily small amount. In the following we "round off" the truncated cells(either deleting the partial cell or replacing it with a whole cell, depending on whether the truncated cell $>= 0.5$). In a way this is equivalent to changing the dimensions very slightly.

In addition, since the ratio of dielectric constants is arbitrary, we should not expect the cell index, or for that matter the increment in the index to be an integer in the dielectric region(this assumes the dielectric region has a smaller dielectric constant with larger cells). As for the semiconductor region(again assuming the semiconductor has the larger dielectric constant and thus the smaller cells), we are free to adopt an index which *increments* by unity. Such a selection is assumed throughout.

The use of non-integral indices , unfortunately, introduces complications in the in the array variables. The arrays only consider the indices as integers, When a non-integer is encountered , the array variable usually(depending on the language) rounds off the index to the closest integer. Ordinarily this property is desired, but when using the nearest node approximation there will be occasions when this selection is incorrect; in this case program revisions must be incorporated so as to make the correct selection for the array variable. The nearest node method therefore results in a program which is more involved (compared to the previous example) but is naturally of greater utility.

In Fig.7A.1 we provide sample results of simulations which allow for arbitrary dielectric constant and cell density, using the nearest node approximation(the program revisions are based on the discussion in Chapter V)). The Figure shows three static profiles of the semiconductor, i.e., the TLM solution of LaPlaces equation. Curves I and II are the same except for the dielectric constant of the fill region; for I it is $\varepsilon_D = 3$ while for II it is $\varepsilon_D = 9$. In III , $\varepsilon_D = 3$ but the height of the fill H1 is 3 mm instead of 1.5 mm. In each of the three curves a slightly different time step was selected in order to insure that the number of cells in the semiconductor was the same; in this case each has exactly 15 cells in the semiconductor (or 16 field values). This was done to facilitate the comparison between the three curves. Also, the cell index numbers are shown as integral; in fact the indices differ slightly and in general are non-integral. The index *changes* in the semiconductor, however, are unity and therefore it is convenient to label the cells with integers.

We should note that in curves I and II, where the dielectric constant of the fill is changed, from 3 to 9, the profiles do not change greatly. When the height of the dielectric fill is doubled, however. from 1.5 mm to 3.0 mm, the fields near the anode change significantly. This is expected since, for the smaller height. the main fringing to the ground line occurs in the first couple of cells next to the anode. When the distance from the anode to the ground line is increased , the fringing is cast out further, so that the field profile in the

FIG. 7A.1 STATIC FIELDS USING NEAREST NODE APPROXIMATION. THE DIELECTRIC CONSTANTS OF THE FILL AND SEMICONDUCTOR ARE ε_D AND ε_S RESPECTIVELY. H1 IS THE FILL HEIGHT.

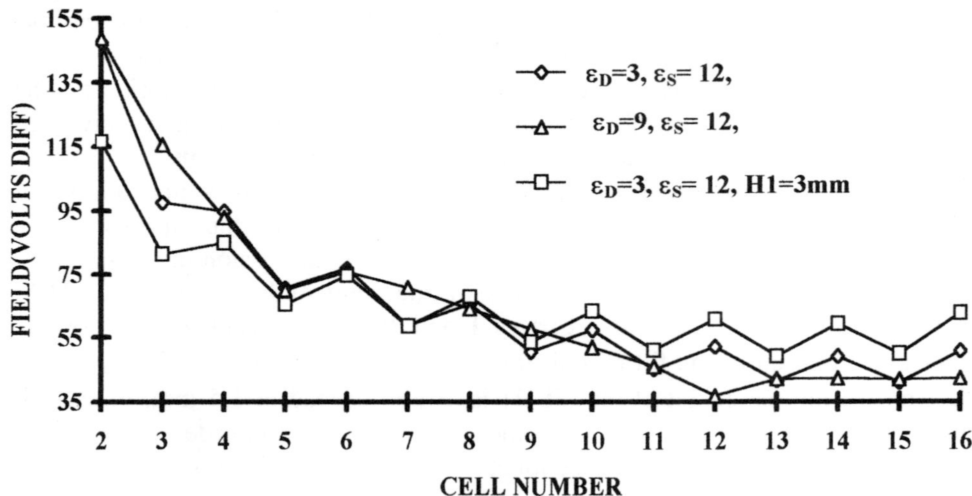

semiconductor is flattened out to some extent. Alternatively, when the semiconductor length is made large compared to the height of the dielectric fill, the field is strong near the anode but relatively weak near the cathode. As mentioned earlier, the oscillatory behavior of the profiles(especially for ε_D =3) is an artifact of the simulation; it may be removed by resorting to larger cell densities and with the assistance of spatial averaging over the oscillation length(approximately equal to the cell length in ε_D).

App.7A.4　Field Decay in Semiconductor Using the TLM Formulation

As indicated on several occasions, the enhancement in resolution of the TLM solution may be obtained by reducing the cell length.(assuming sufficient computer capacity). Although important for the TLM method, the small cell length is critical in the case of numerical methods   The numerical methods can only guarantee a solution if the cell size is sufficiently small ; in TLM language they require that the cell transit time be roughly equal to or smaller than the capacitive time constant, which is equivalent to $Z_o \leq R(n,m)$. In contrast to numerical methods, however, the TLM method allows $R(n,m) < Z_o$, although the accuracy will be affected, particularly for the cells within the conductor and cells near the conducting interface. In allowing this condition, it behooves us therefore to understand how the fields will behave in the TLM lines.

In this Appendix we explicitly indicate the field decay for arbitrary values of R and Zo, using the TLM formulation. To simplify matters we consider a one dimensional chain of cells biased to the same voltage V_o, with $R(n,m)=R$ We then activate the cell chain uniformly. As first noted in Section 1.4, under uniform conditions, and utilizing Fig.1.14, the backward and forward waves combine, together with the reflected signals so that there is no net transfer of energy outside the cell; we may regard each cell as self contained as one might expect from symmetry considerations. This is now made more explicit.

During the static conditions the forward and backward waves in each cell are identical and the same from cell to cell. Upon activation we then determine the change in the forward wave during the next time step. The same analysis will of course apply to the backward wave because of symmetry. The transfer coefficient at the node is

$$T = 2R_L/[Z_o+R_L] \qquad (7A6.1)$$

where R_L is

$$R_L = RZ_o/[Z_o+R] \qquad (7A6.2)$$

Combining the above equations,

$$T = 2R/[2R+Z_o] \qquad (7A6.3)$$

Similarly the reflection coefficient is

$$B = (R_L - Z_o) / (R_L + Z_O) \qquad (7A6.4)$$

where again

$$R_L = RZ_o/[Z_o+R] \qquad (7A6.5)$$

Combining the last two equations gives

$$B= -Z_o/[2R+Z_o] \qquad (7A6.6)$$

We can now determine the forward wave during the follow on time step Denoting the initial forward wave by $^+V^0$ and the wave during the follow on step by $^+V^1$.. From the usual scattering equations,

$$^+V^1 = {}^+V^0 [T+B] \qquad (7A6.7)$$

where the above takes into account the equality of the forward and backward waves and the wave symmetry among the neighboring cells. Combining the previous equations,

$$^+V^1 = {}^+V^0 [2R-Z_o]/[2R+Z_o] \qquad (7A6.8)$$

From symmetry considerations, however, the backward wave $^-V^1$, is exactly equal to $^+V^1$. But the above equation is exactly identical to that of a reflected wave in an *isolated* cell with end resistors 2R. The same factor is applied with each successive time step; thus, for $k=2$, $^+V^2={}^+V^0\{[2R-Z_o]/[2R+Z_o]\}^2$. The description of the cell chain may therefore be replaced with that of a single isolated cell with an effective load of 2R, thus making the interpretation far simpler. In the remainder of the analysis we employ the isolated cell for convenience.

For $2R>Z_o$, the isolated cell will see a positive mismatch and the fields will decay with no change in polarity. For $2R>>Z_o$, the decay will take many time steps, but since a correspondingly small Δt is implied, the "RC" time constant, or equivalently ε/σ, is the same. When $2R<<Z_o$, the field will decay in a ringing fashion, with the field changing polarity with each successive time step. With $2R=Z_o$, which is the matched condition, the field would appear to

instantaneously vanish. Bear in mind that the TLM method can only be applied with accuracy when R>>Zo, and the situations occurring in the conductivity region when $Z_o \geq 2R$ must be treated with caution. As stated numerous times before, the situation is corrected by selecting a smaller cell size

Next we determine the number of time steps needed to attenuate the field within each cell. As a criterion we say that the field is attenuated, if at the Kth time step, the field is at half its value. From Eq.(7A6.8) therefore we set

$$(1/2) = \{ [2R-Z_o]/[2R+Z_o]\}^K \tag{7A6.9}$$

K is the determined from

$$K = \mathfrak{I}_n(1/2)/\mathfrak{I}_n (B_{EFF}) \tag{7A6.10}$$

where

$$B_{EFF} = \{ [2R-Z_o]/[2R+Z_o]\} \tag{7A6.11}$$

As R is made large relative to Z_0, the quantity $K\Delta t$ approaches the RC time constant, i.e., $K\Delta t \rightarrow RC$, or in terms of the material constants, $K\Delta t \rightarrow \varepsilon/\sigma$.

The field behavior just discussed may be illustrated using the same semiconductor switch example and program code employed previously in the Chapter, in which the the entire semiconductor is activated. The simulations in Figs.7A.2 and 7A.3 show the field in the center cell(n=11,m=17) of the semiconductor immediately after activation , reducing the node resistance instantaneously to either 150Ω, as in 7A.2, or 15 Ω, as in 7A.3. The conditions correspond to those in Figs.7.15 and 7.16 .We can interpret the simulations in light of the above results derived in this Appendix; keep in mind , however, that we have only considered 1D situations and we have not made any corrections for 2D.

At 150Ω , 2R>Z_o (since Z_o=126Ω), and therefore we should probably not expect any ringing, and indeed the field in Fig.7A.2 appears to be well behaved in that respect. Since 2R is close to the matching conditions the field is seen to

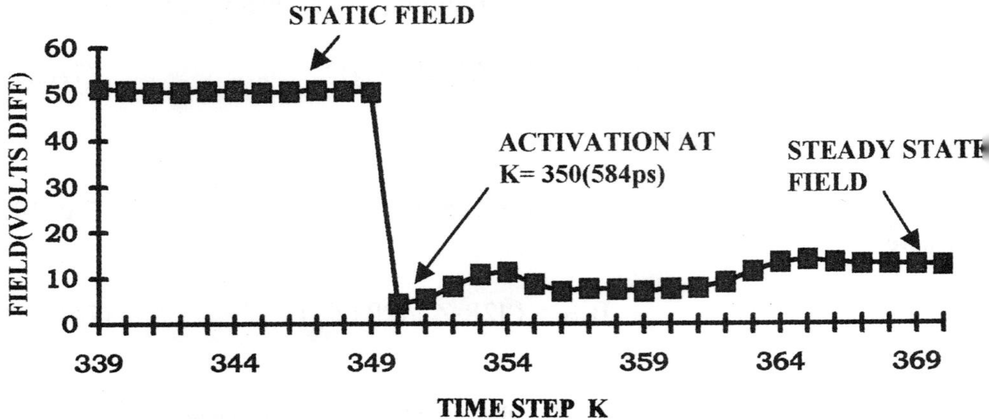

dissipate very quickly, and after one time step the field is not very far from its steady state value of ~15V.

When R = 15Ω, however , Fig.7A.3, the number of transit times needed to dissipate the wave is larger, because of the strong negative mismatch. From the simulation shown in the figure a K value of ~ 3 appears to reduce the field by about a factor of two. The 1D expression from Eq.(7A.6.1), on the other hand, yields a value of ~ 1.5. The discrepancy is due to the use of a 2D matrix instead of a linear chain.   We also note the alternate changes in the field polarity, again brought about by the negative mismatch. The steady state field value is smaller than that for the case of 150 Ω, as expected, because of the smaller semiconductor resistance. We again emphasize that the transient response in the activated region, certainly for the case when R=15Ω, is a computer artifact which we may eliminate by selecting a smaller size for the cell.

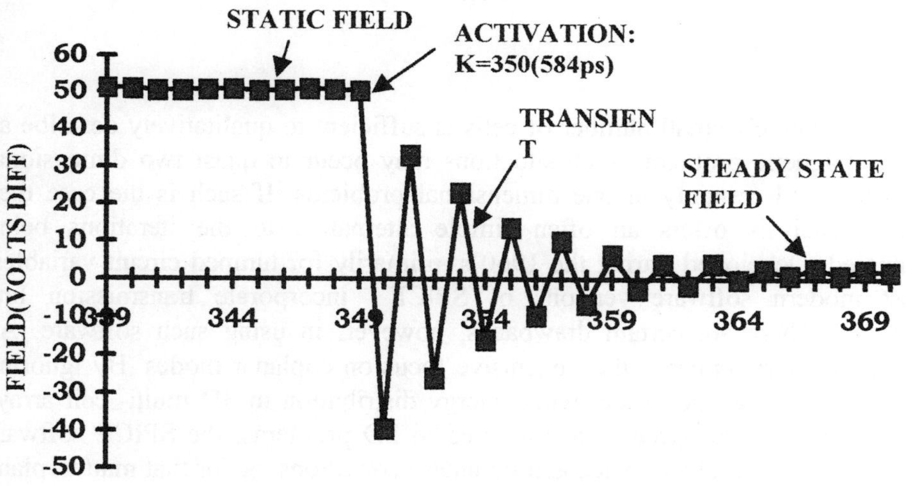

FIG. 7A.3 FIELD AT CENTER CELL n=11,m=17. SAME CONDITIONS
AS BEFORE BUT WITH LOWER NODE RESISTANCE DURING
ACTIVATION:RO=15Ω.
STATIC FIELD
ACTIVATION:
K=350(584ps)
TRANSIENT
STEADY STATE
FIELD
FIELD(VOLTS DIFF)
60
50
40
30
20
10
0
-10
-20
-30
-40
-50
339
344
349
354
359
364
369
TIME STEP K

VIII. SPICE Solutions

Often a relatively small number of cells is sufficient to qualitatively describe an electromagnetic problem. Such situations may occur in quasi two dimensional problems and certainly in one dimensional problems. If such is the case then SPICE analysis offers an often simple alternative to the iterations being discussed. Developed during the 1960's, primarily for lumped circuit variables, most modern software versions of SPICE incorporate transmission line elements. There are certain drawbacks, however, in using such software. An important shortcoming is their exclusive focus on coplanar modes. By ignoring normal scattering, the proper wave energy distribution in 3D multi- cell arrays cannot be obtained. Even when applied to 2D problems, the SPICE software does not properly take into account boundary conditions, or for that matter plane wave correlations or grid anisotropy effects. At the very least, the result is a diminution in accuracy. In addition there exists a major practical disadvantage having to do with the maximum number of cells one may employ with SPICE. Exactly how many cells can be analyzed by SPICE depends on the particular software. It appears, however, that when the number of cells exceeds 100 or so, the programs, if not specifically prohibited, run into practical difficulties(at least for the SPICE programs *currently* tested). Nevertheless, despite these crucial limitations, SPICE offers a useful and simple tool for understanding the nature of the TLM model. The SPICE method is applied to 1D devices and simple 2D devices(especially devices which are relatively" long" in one direction, so the device may be considered quasi one dimensional),   where the cell number capability is not exceeded. In this Chapter, we employ SPICE to analyze, in particular, avalanching photoconductive switches, transmission line Marx generators,   Darlington pulsers, and various types of transformers and pulse generators.

8.1 Photoconductive Switch

Fig.8.1 shows a conceptual view of a parallel plate photoconductive switch, with an input transmission line of impedance Z_o. A portion of the top conductor is removed and replaced with a semiconductor, which also makes contact with the ground line. To simplify matters we assume the composite impedance of the semiconductor portion, when entirely activated, is also equal to Z_o. The output at the cathode end is likewise matched to the input Z_o. The semiconductor gap is then charged up to voltage V_o via the input line which is assumed to be extremely long compared to the device dimensions. Once the charge up process is completed the top surface of the semiconductor region, representing a small fraction of the height, is then illuminated, creating a conductive layer in the top region of the semiconductor. The entire length of the semiconductor need not be activated , however. Indeed, in the example to be discussed only the region near the anode is optically activated while the remaining length is activated by avalanching due to the build-up of the electric field.

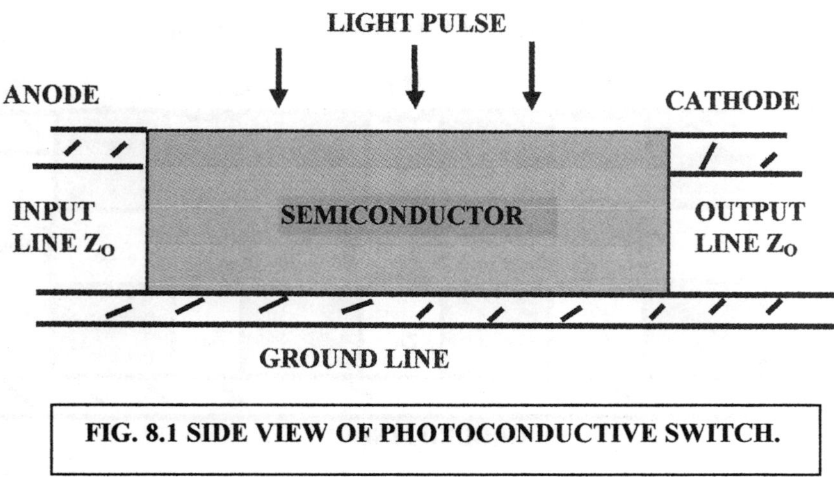

FIG. 8.1 SIDE VIEW OF PHOTOCONDUCTIVE SWITCH.

This device is analyzed using SPICE , replacing the semiconductor with an eight cell simulation shown in Fig.8.2. There are a total of ten TLM lines, all with impedance Z_O, representing the semiconductor. We note that there is only a single row (consisting of four lines)of TLM lines. Also note that at the bottom of the device the nodes are at zero voltage since they are in contact with the ground line. The top nodes (A, B, and C) represent the activated region of the semiconductor. These node resistors are not activated until photoconductivity or avalanching is introduced.

The first task is to obtain the static solution , which will give us the voltage of each cell. In cases where there are many cells(as in the iterative process) the static solution is normally obtained by initially setting all the cell voltages to zero and charging up the cells via the input transmission line, assuming the node resistors are   infinitely large, as was done in the last Chapter. In this case, however, the limited number of cells grossly distorts the static field. Were we to use this charge-up procedure (using either the TLM   or SPICE methods)   we would find that the input merely charges up the first vertical line (below A in Fig.8.2 to the full input voltage of 3 kV. This is due to the fact that there is essentially only one active row of cells , each connected by an infinite

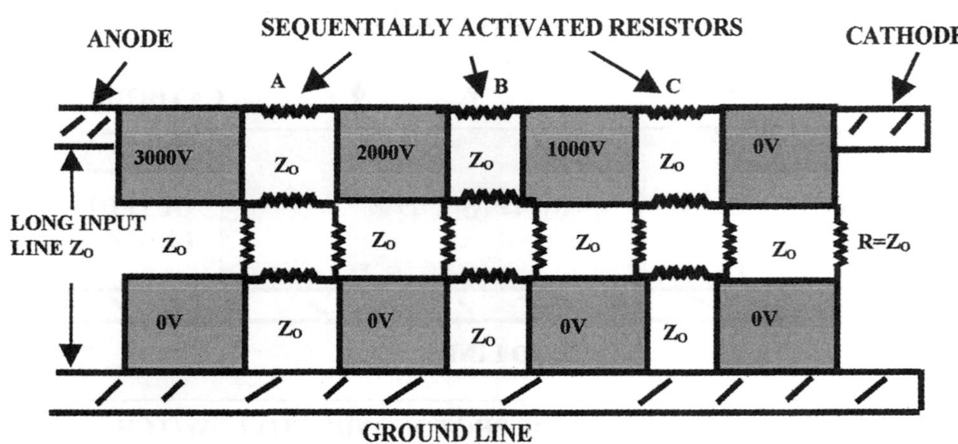

FIG. 8.2 EIGHT CELL MATRIX APPROXIMATION OF TRANSMISSION LINE SWITCH USING SPICE

resistance. In order to circumvent this limitation (other than adding additional rows, i.e., additional cells), we simply assume a linear gradient for the upper row of four cells. If the charge up voltage is V_o then the cell in contact with the anode is V_o , the adjacent cell is $(2/3)V_o$, the next is $V_o/3$, and the cell in contact with the cathode is at zero voltage. Next we assign specific values to the parameters, representative of a millimeter sized semiconductor switch. We take $V_o = 3kV$ and $Z_o=104$ Ω(corresponding to a dielectric constant of ~13.1).   The transmission line lengths are all 10 ps each , except for the input line , which is assumed to be long compared to 10ps, and which provides a reservoir of energy for the semiconductor. Another assumption is the avalanche delay time. Once the semiconductor is exposed to a large electric field, the conductivity induced by avalanching is not instantaneous and a delay in the build-up of conductivity exists. An avalanche delay time of 10 ps is assumed.

The only resistors activated are those at A, B, and C, and these are done sequentially. The resistor at A, near the anode, is first turned on optically. This is followed by B and C which are presumed to turn on via avalanching. Once the switch is turned on at A , a larger field develops at B, which causes the switch there to turn on (by avalanching). The activation at B in turn causes a higher field to develop at C, thus turning on C. Following the activation at C, the enhanced pulse to the load is delivered The simulated SPICE fields which develop across A, B C, and the load are shown in Fig. 8.3. At t =1 ps, switch A is turned on, and an enhanced voltage difference, close to 2kV, then appears across B approximately 50 ps later. After an additional 10 ps avalanche delay, switch B is activated and after another 30 ps, an enhanced field of 2.6 kV appears at switch C. Finally, after switch C is activated, the load pulse is delivered at t=120 ps. The total delay time, for delivery of the load pulse, is caused by a combination of propagation and avalanche delay. Since SPICE does not take plane wave effects into account, the propagation delay of the main signal, between anode and cathode, will be approximately twice that of the straight line delay(See Sections Chapter IV , where this delay is termed the earliest arrival time and the issue is discussed further). Since the straight line delay is 40ps(consisting of the 4 horizontal TLM lines, each 10ps long), the full strength signal should not arrive at the output until at least 80ps has elapsed. As discussed in Chapter III, the

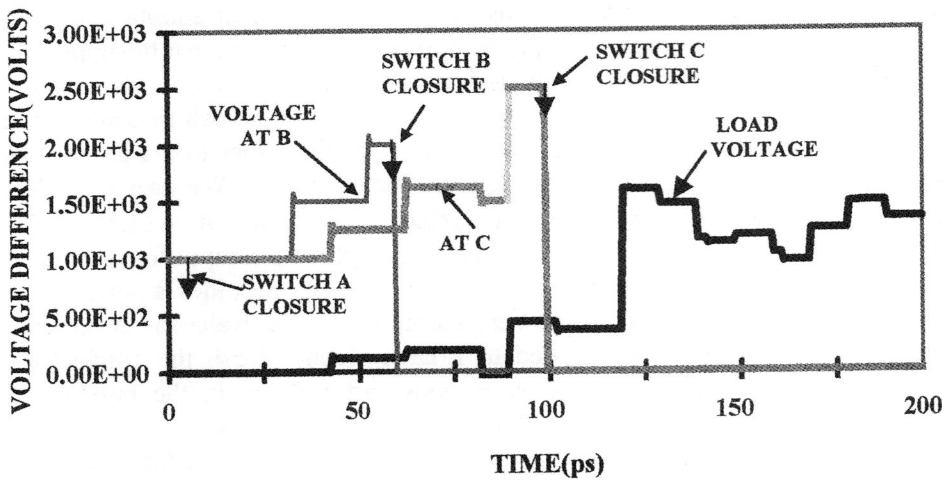

FIG. 8.3 SPICE SIULATION OF VOLTAGE SIGNAL DEVELOPED AT VARIOUS POINTS IN THE SEMICONDUCTOR, PRIOR TO AND AFTER SWITCHING.

effective slowing down of the signal is due to the filling up of wave energy in the transverse lines. When this delay is added to the avalanche delay of 30ps, the appearance of the full strength signal at 120ps is not surprising. The pulse plateau eventually settles down to half the charging voltage, as expected, in the limit of long input lines. The SPICE FORMAT for the light activated semiconductor is discussed in App.8A.1.

In the previous discussion we saw that it was possible to obtain field growth, caused by the imposition of the same voltage over progressively decreasing distances, and initiated by the introduction of conductivity, produced by either a light signal or avalanching. Efficient operation in this mode relies on the synchronization of the process so that the cell resistor is instantaneously activated once the maximum field signal arrives at the cell, i.e., avalanche delay must be minimized. Losses, of course, must also be controlled. The minimization of these two factors, avalanche delay and losses, can result in enormous field growth.

Such field growth is analogous to that in a traveling wave , Marx generator discussed in the next Section.

8.2 Traveling Wave Marx Generator

The field growth alluded to in the previous paragraph may be made clearer by analyzing the traveling wave Marx generator. First, however, we describe the conventional, lumped circuit Marx generator shown in Fig.8.4. An array of energy storing capacitors is charged in parallel as noted in (a). The charging resistors, R_1, R_2 , etc...,are selected to be much larger than the load impedance. Now, suppose the switches S_1 , S_2 , etc...in the diagonal lines of each cell are turned on simultaneously. The N capacitors, initially in parallel, will find themselves in series, as shown in (b), with a resultant N times multiplication of the charging voltage. Note that the large charging resistors essentially isolate the pulse circuit , preventing any unwanted discharge into these resistors.

An important defect, associated with the circuit in Fig.8.4, is that the pulse risetime may be unacceptably slow, a direct result of the fact that the circuit does not behave as a transmission line. Thus the electrical lengths in the switch segments, as well as in the capacitor sections, represents an effective inductance which tends to smear out the risetime. In order to overcome this defect, we employ a transmission line circuit [1], with characteristic impedance Z_o, as shown in Fig.8.5(a). The series combination of the capacitor and the switch is placed periodically in the line, with spacing Δl, and with a delay time interval equal to Δt. We assume the electrical length, i.e., inductance, of this combination is small compared to that of Δl, and also assume the switching time is very short compared to Δt. Note that the cells are all identical except for the first and last ones where input and output matching resistors terminate the line. We also point out that there is no requirement that the energy storage be restricted to a lumped capacitor. Indeed the capacitor may be replaced by a low impedance transmission line, Z_n , with $Z_n \ll Z_N$, as noted in Fig.8.5(b).

The sequence of operation for the transmission line Marx is as follows. Initially it is assumed each of the energy storage elements is charged to voltage V_o. At t=0 , switch S_1 is activated. This gives rise to forward and backward waves of amplitude $V_o/2$ (assuming no losses). The backward wave is absorbed in the matched resistor $R_{IN} = Z_o$. The forward wave, however progresses to the next cell. When the wave front reaches S_2 , this switch is quickly activated. Just as before, a forward and backward waves are launched from cell #2. The backward wave meets the same fate as before, being absorbed in the matched

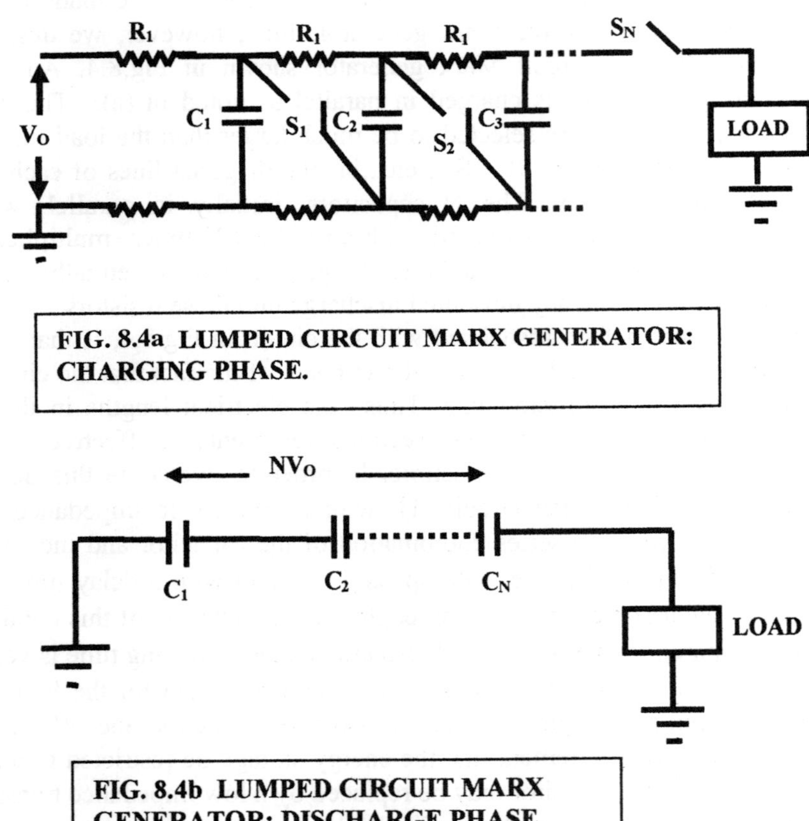

FIG. 8.4a LUMPED CIRCUIT MARX GENERATOR: CHARGING PHASE.

FIG. 8.4b LUMPED CIRCUIT MARX GENERATOR: DISCHARGE PHASE.

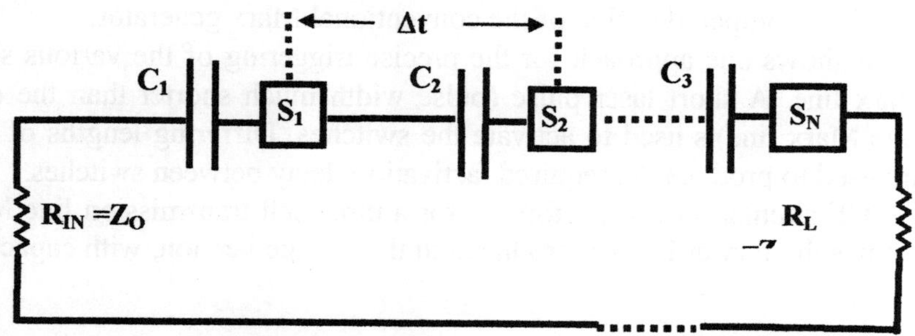

FIG. 8.5a TRAVELING WAVE MARX GENERATOR.

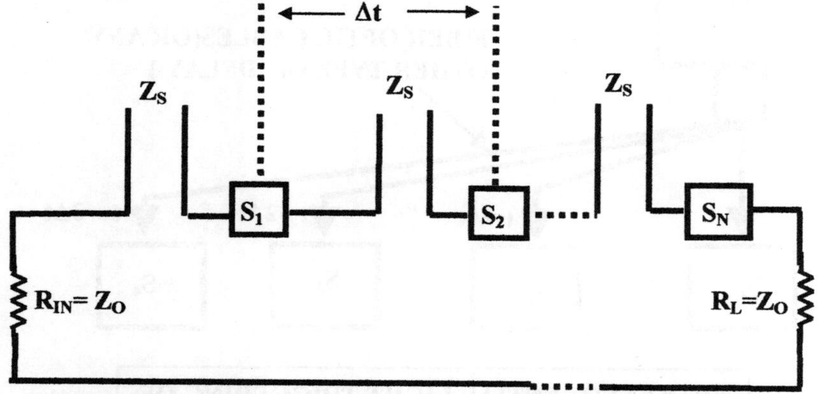

FIG.8.5b TRAVELING WAVE MARX GENERATOR WITH TRANSMISSION LINE STORAGE ELEMENTS Zₛ.

in the matched resistor $R_{IN} = Z_o$. The forward wave, however, is superimposed onto the previous wave. The total amplitude of the forward wave, therefore , is V_o, instead of $V_o/2$. The process continues until the end of the line is reached. If there are N stages then the total amplitude attained is $N(V_o/2)$. The important point is that the precise sequential triggering of the switches gives rise to a fast risetime pulse , compared to that of the conventional Marx generator.

Fig.8.6 shows one approach for the precise triggering of the various stages in the Marx line. A short laser pulse (pulse width much shorter than the delay time in the Marx line)is used to activate the switches. Differing lengths of fiber cables are used to produce the required activation delay between switches.

A SPICE simulation was performed for a three cell transmission line Marx. Fig.8.7 shows the forward wave, produced in the 3 stage version, with capacitor

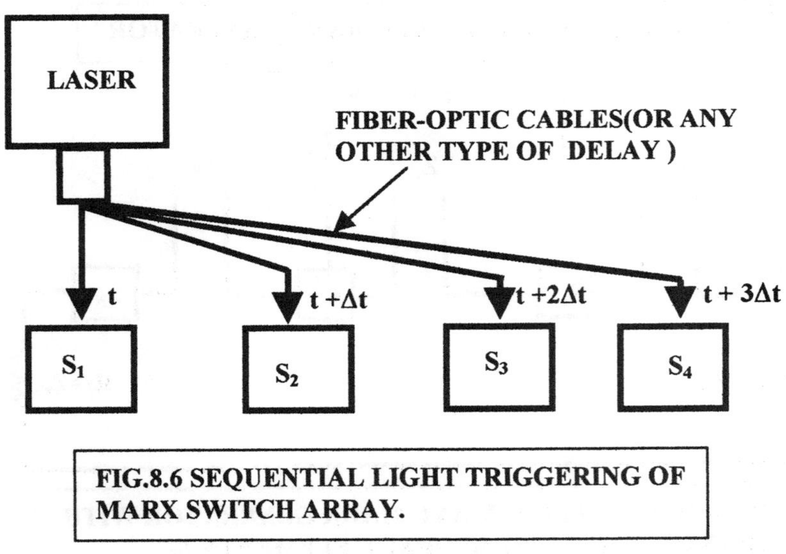

FIG.8.6 SEQUENTIAL LIGHT TRIGGERING OF MARX SWITCH ARRAY.

storage, initially charged to 200V, with a 10 ns delay between stages. The fast risetime and 3X multiplication are evident. In contrast, the backward wave (not shown) is a series individual pulses with amplitude $V_o/2$. Fig.8.8 shows how to modify the Marx circuit if we include losses, stemming from either   the conducting wires or the medium. Note that the conductor loss is represented by a series resistance while the medium loss is represented by the parallel resistance.

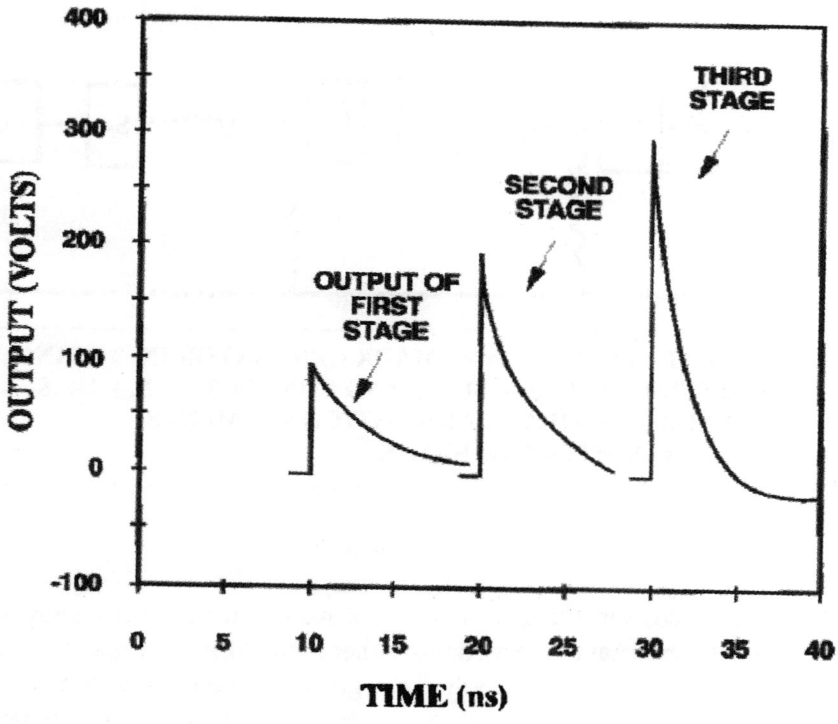

FIG. 8.7 SIMULATION SHOWING GROWTH OF FORWARD WAVE IN TRANSMISSION LINE MARX CIRCUIT, BIAS=200V AND Δt=10ns.

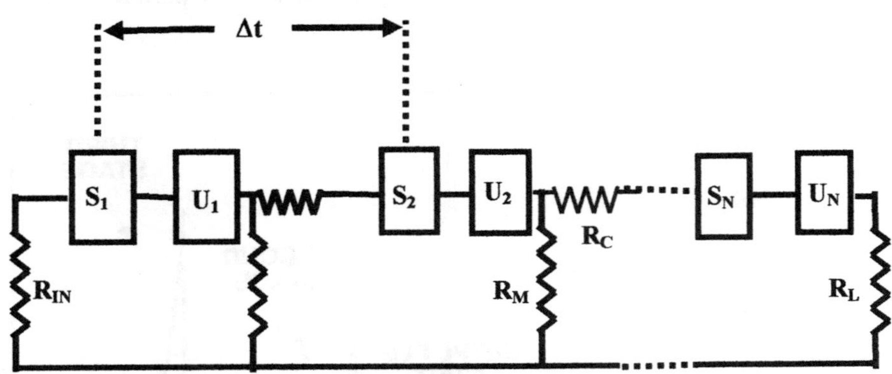

FIG. 8.8 TRAVELING WAVE MARX GENERATOR INCLUDING LOSSES DUE TO MEDIUM(R_M) AND CONDUCTOR(R_C). U_N ,S_N ARE ENERGY STORAGE AND SWITCH ELEMENTS ASSOCIATED WITH NTH STAGE.

8.3 *Traveling Marx Wave in a Layered Dielectric*

The previous discussion seems to imply that we require distinct energy storage circuit elements, external to the medium, either in the form of capacitors or input pulse forming lines, in order to achieve a growing Marx wave. Similar waves may be obtained, however, without relying on such external circuit elements. In fact, a growing wave may be derived from the electric field energy distributed throughout the medium itself, although the conditions are somewhat restrictive, as we shall see. Fig.8.9(a) shows a TLM matrix representation of the medium in which the transverse fields are assumed to be initially static. Note that the medium consists of a material with a dielectric constant ε_2, and containing a narrow channel, made up of another dielectric material, ε_1. In this configuration , the ε_2 material plays the role of the energy storage elements of the circuit model. Under static conditions, the field storage is assumed to be primarily in the transverse lines of the ε_2 matrix, at which time there is no field growth in the channel. If we focus on the node (n,m) we see that the equilibrium

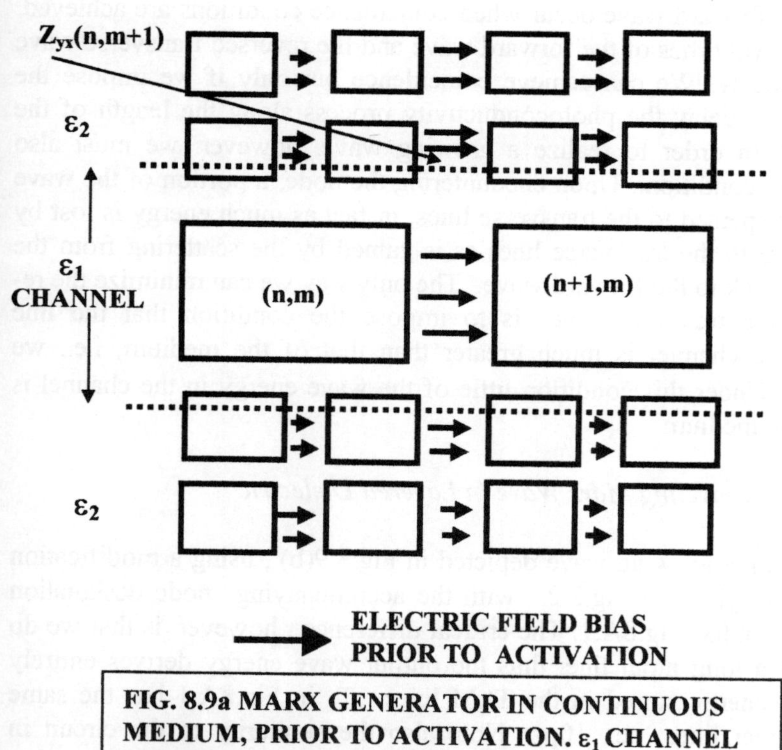

FIG. 8.9a MARX GENERATOR IN CONTINUOUS MEDIUM, PRIOR TO ACTIVATION. ε_1 CHANNEL EMBEDDED IN ε_2 MEDIUM WITH $\varepsilon_2 \gg \varepsilon_1$.

can be upset if we are able to reverse the field direction in the $Z_{yx}(n,m+1)$ line in the ε_2 region but leave the $Z_{yx}(n,m)$ and $Z_{yx}(n,m-1)$ lines alone We can imagine the field reversal occurring by means of photoconductivity which produces conductivity at one end of the transverse line, as indicated in Fig.8.9(b). The reversed field wave, in combination with the wave in $Z_{yx}(n,m,)$ now contributes to the forward horizontal wave in the channel. The fastest risetime and largest amplitude of the forward wave occur when coincidence conditions are achieved, i.e., when the arrival times of the forward wave and the reversed transverse wave occur simultaneously. We can achieve coincidence but only if we impose the condition that we delay the photoconductivity process along the length of the semiconductor. In order to realize a growing wave, however, we must also specify one more condition. Upon encountering the node, a portion of the wave energy will be dispersed to the transverse lines; in fact as much energy is lost by the forward wave to the transverse lines as is gained by the scattering from the transverse lines back to the forward wave. The only way we can minimize the re-scattering into the transverse lines is to impose the condition that the line impedance of the channel is much greater than that of the medium, i.e., we require $\varepsilon_2 >> \varepsilon_1$. Under this condition little of the wave energy in the channel is scattered into the medium.

8.4 Simulation of Traveling Marx Wave in Layered Dielectric

We can simulate the growing wave depicted in Fig.8.9(b) , using a modification of the cell matrix given in Fig.8.2, with the accompanying node designation given in the Appendix, Fig.8A.1. The critical difference , however, is that we do not provide for a long input line; thus the output wave energy derives entirely from the initial energy stored in the TLM lines. T_1 in Fig.8A.1 has the same length as the other lines, i.e., 10ps Note also the similarity to the circuit in Fig.8.5(b) with the main change being the location of the switches, which are in series with the storage elements.

Referencing the designations in Fig.8A.1, we begin the modification of the matrix by first   assigning low impedance values to the six transverse lines, T3, T6,T9,T2,T5, and T8. In the simulation to be presented, the transverse lines have an impedance of 5Ω (instead of 104Ω)while the horizontal lines, T1,T4, T7, and

T10 have an impedance of 104Ω. Next we assign the delay times for the activation of the switches, GS1, GS2, and GS3. These switches simulate the light activation discussed previously. In order to achieve synchronism, the

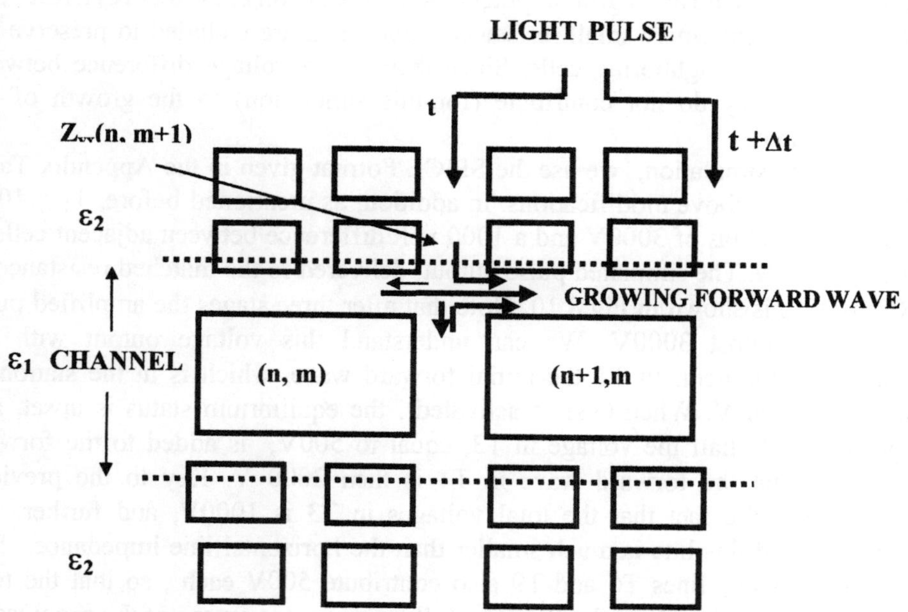

FIG. 8.9b MARX GENERATOR IN CONTINUOUS MEDIUM FOLLOWING ACTIVATION WAVE. GROWTH OCCURS IN EMBEDDED ε_1 CHANNEL, DUE TO SEQUENTIAL PHOTOCONDUCTIVE SIGNALS.

switches must be delayed from one another by one transit time. Thus if we arbitrarily cause GS1 to turn at 1ps, and we assume the delay time of each TLM line is 10 ps, then GS2 is turned on at 11ps and GS3 at 21 ps. Avalanche delay is not included in this simulation(Also note that the lengths of the low dielectric constant, horizontal lines are not drawn to scale). In using the matrix of Fig.8A.1, which corresponds to a transmission line, we should note that the lower four cells are all at ground potential; the connecting resistors RS1,RS2, and RS3 all have extremely small resistance values, and are included to preserve the identity of the neighboring cells. Since there is no voltage difference between these cells, they do not contribute (for this simulation) to the growth of the forward wave.

For the simulation, we use the SPICE Format given in the Appendix Table 8A.1 with the above modifications. In addition, as mentioned before, $T_1 = 10$ps. We assume a bias of 3000V and a 1000 volt difference between adjacent cells in the upper row. The simulated pulse output, delivered to the matched resistance of RL=104Ω , is shown in Fig.8.10. Note that after three stages the amplified pulse output is almost 3000V. We can understand this voltage output with the following argument. In T1 the initial forward wave, which is in the stationary state, is 1500 V. When GS1 is activated , the equilibrium status is upset, and approximately half the voltage in T3, equal to 500V, is added to the forward wave, so that the forward wave in T4 is then 2000 V. Key to the previous statement is the fact that the total voltages in T3 is 1000V, and further the impedance of this line is much smaller than the horizontal line impedance. In a similar manner , lines T6 and T9 also contribute 500V each , so that the total voltage delivered is about 3000V. The follow-on pulses represent the progressive decay of energy in both the transverse and input lines. The follow-on pulses eventually decay to zero, as expected, since the output is derived from the field energy stored in the TLM lines. Input resistors, for terminating backward waves, have not been included; their absence will contribute to the follow-on pulses as well.

Fig.8.2 does not quite correspond to Fig.8.9, in as much a ground line exists on one side of the ε channel in Fig.8.2 whereas a low dielectric channel is

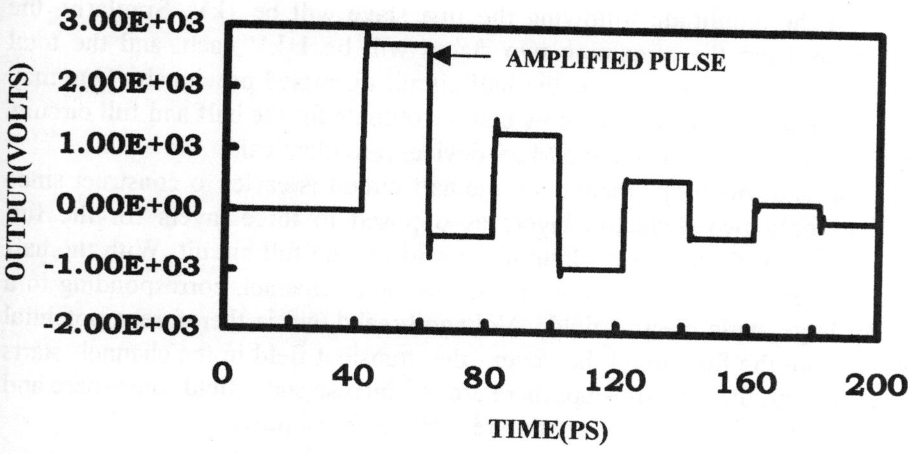

FIG.8.10 SIMULATED OUTPUT FOR A 3 STAGE TRAVELING MARX WAVE IN A LAYERED DIELECTRIC. FORWARD WAVE PULSE IS AMPLIFIED FROM INITIAL 1.5KV TO 3.0 KV. ACTIVATION DELAY IS 10 ps,

sandwiched between two high dielectric regions in Fig.8.9 . In other words, the results just described represent a half circuit version of the traveling wave Marx wave for a layered dielectric. Suppose, therefore, we were to remove the ground line in Fig.8,2 and allow the horizontal field to also exist in the lower four cells of circuit, so that the situation corresponds more closely to that of Fig. 8.9. Will the output voltage change from its previous value of 3 kV? The answer is no , but the reason is not immediately evident. Since we have removed the ground line , and the voltages in the lower row cells are the same as the upper one, the initial *transverse* field (i.e., the field in the horizontal line)in the ε_1 region is zero. , compared to 1500 V for the previous situation. Because of the double layer, however, the *growth factor* in the ε_1 region will be twice as large. Thus, when switch GS1 is triggered, waves from both T2 and T3 will contribute to the growth and the amplitude following the first stage will be 1kV. Similarly, the contributions from the remaining two stages will be 1 kV each, and the total output will be 3 kV, the same as the half circuit discussed previously. One may quickly generalize the results to show that the outputs for the half and full circuits , representing the Traveling Wave Marx devices, are identical.

So far as device implementation, the half circuit is easier to construct since it involves only two dielectric layers as opposed to three layers for the full circuit. However, there is something to be said for the full circuit. With the half circuit, the entire bias voltage exists across the ε_1 channel, corresponding to a relatively large static electric field. Voltage breakdown is therefore a potential problem. With the full circuit, however, the transient field in the channel starts out small and gradually builds up; there are no intense static fields anywhere and therefore a lesser likelihood of premature voltage breakdown .

Pulse Transformation And Generation Using Non-Uniform Transmission Lines

SPICE provides a convenient and effective tool for analyzing transformers using non-uniform transmission lines. Not surprisingly we can convert the transformer to a pulse source by adding a switch, and SPICE is very effective for analyzing these devices as well. As far as the pulse sources are concerned, we analyze two types. In the first a pulse is first produced externally and simply delivered to the input terminals of the transformer, usually the low impedance side. The pulse then undergoes an impedance transformation, emerging at the output terminals. Usually the output is at the higher impedance, in which case the pulse amplitude is transformed upward. The second type of pulse source is more self contained, wherein the non-uniform line itself stores the energy(rather than being stored in an external line)and a switch is provided either at the input or output of the line to trigger the wave(When the switch is situated at the input , the circuit must be augmented with an auxiliary, uniform section of transmission line, as we will discuss). The SPICE analysis for both these pulse sources, as well as the solitary transformer, are similar. We begin the analysis by considering the transformer by itself.

8.5 Use of Cell Chain to Simulate Pulse Transformer

The pulse transformer, Fig.8.11, usually consists of a transformation line whose impedance varies slowly (and usually monotonically) with length. Typically the impedance is varied by changing some dimension of the line, e.g., the height between conductors in a strip transmission line. Provided that the propagation time of the transformer is much longer than the pulse width, pulse deformation may be kept to a minimum. For a long, lossless transformer, the output pulse, V_{OUT} , is ideally related to the input V_{IN} , by the transformer relationship,

$$V_{OUT} = [\ Z_{OUT}/Z_{IN}]^{1/2}\ V_{IN} \qquad (8.1)$$

Eq.8.1 is derived analytically, using a TLM formulation, in App.8A.3. Inevitably, the finite length of the transformer, as well as losses and parasitics, degrade the output pulse to some extent. SPICE is able to take into account these additional effects.

The SPICE(as well as the TLM) technique involves conceptually breaking up the transformer into a number of segments of equal length , each of which is approximated by a constant impedance(see Reference [1]) The impedance of

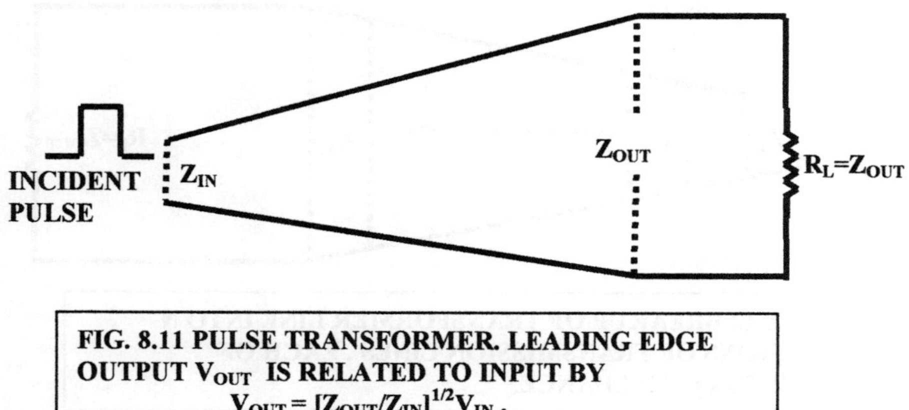

FIG. 8.11 PULSE TRANSFORMER. LEADING EDGE OUTPUT V_{OUT} IS RELATED TO INPUT BY $V_{OUT} = [Z_{OUT}/Z_{IN}]^{1/2}V_{IN}$.

each section follows the functional impedance of the transformer with length. Once this is done, then the SPICE software may be used to determine the response to an input pulse. Fig.8.12 demonstrates the technique for a transformer where the impedance is a function of length $Z(l)$. Although the transformer shown has a linear taper, the following method also applies to non-linear transformers where the conductors are of equal length and symmetric about the x axis. As a first step, the transformer is divided into N sections (N= a positive integer). This will enable us to define a set of lengths l_n given by

$$l_n = l_i + n(l_f - l_i)/N \tag{8.2}$$

where l_i is the initial value of the transformer length, l_f is the final value, and n is an integer 0,1,2, ...N. The length of each section is $\Delta l = (l_f - l_i)/N$. We then define a midpoint of each section, l'_n

$$l'_n = (l_n + l_{n-1})/2 \tag{8.3}$$

where n = 1,2,3, ...N. The average impedance of the nth section, Z_n , is then

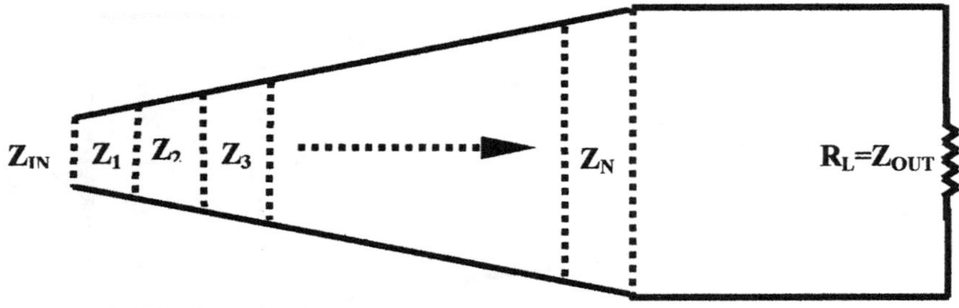

FIG. 8.12 BREAKUP OF TRANSFORMER LINE INTO N SECTIONS OF TRANSMISSION LINES , EACH OF CONSTANT IMPEDANCE.

$$Z_n = Z(l'_n) \tag{8.4}$$

where $Z(l'_n)$ is the impedance of the line evaluated at l'_n , which in turn, is proportional to the height $h(l'_n)$ between the conductors, which is a function of l'_n .

In the case of a linear taper, for example, $h(l'_n)$ is proportional to l'_n . If we use the simple strip transmission line relationship, then

$$Z(l'_n) = 376.7 \, h(l'_n)/ (\varepsilon^{1/2}W) \tag{8.5}$$

where W is the transmission line width and ε is the relative dielectric constant. Since the two conductors are symmetric about the x axis, as in Fig,8.11, then we may approximate the height for each cell as the vertical line connecting the two conductors.

The SPICE simulation just described applies to transformers in which both conductors are of equal length and symmetric about the x axis. If the two conductors are not symmetric about the x axis, extra care must be taken in calculating the impedance of each cell. Thus, in Fig.8.13, the lower conductor is a straight line, but the upper conductor is altogether different, displaying significant curvature. We consider the case in which the total line lengths of the two conductors are the same. Because of the curved conductor, the cells

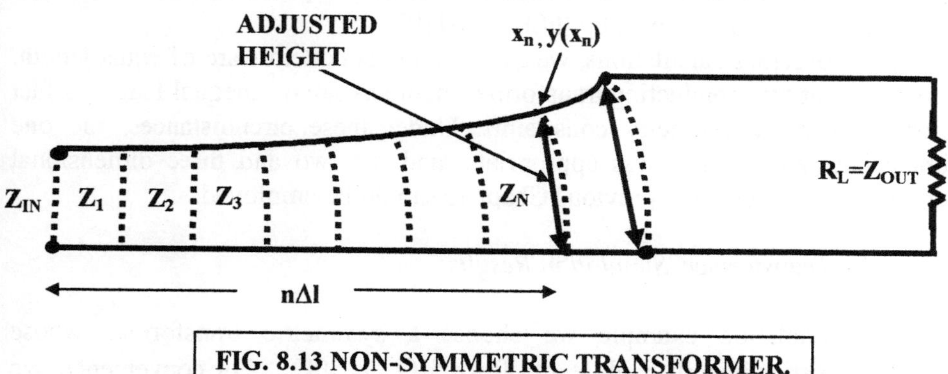

FIG. 8.13 NON-SYMMETRIC TRANSFORMER.

themselves have curvature. The extra complication then arises in the calculation of Z(x), and in particular, the height between conductors, which now bends. In order to proceed further, we assume the curved conductor has the functional dependence, $y(x)$. Then for the nth cell, we wish to obtain the two heights enclosing the cell. To do this we need the coordinates of $y(x)$ intersecting the heights, and to obtain these we calculate the line integrals of $y(x)$. For the nth cell, we have , due to the equality of the conductor lengths,

$$n \, \Delta l \; = \; l_i \text{ to } x_n \int \; [1+ \; (dy(x)/dx)^2 \,]^{1/2} dx \qquad (8.6)$$

where x_n is the x intercept of the adjusted height of the nth cell and the curved conductor, and l_i is the initial x value of the transformer. The value of x_n is calculated using Eq.(8.6). We can then obtain the adjusted height, $h(n)$, which is simply

$$h(n) \; = \; \; [(\, n \, \Delta l - x_n)^2 \; + y(x_n)^2 \,]^{1/2} \qquad (8.7)$$

A similar calculation is performed for $h(n-1)$:

$$h(n-1) \; = \; \; [(\, (n-1) \, \Delta l - x_{n-1})^2 \; + y(x_{n-1})^2 \,]^{1/2} \qquad (8.8)$$

$h(n)$ and $h(n-1)$ are then averaged to obtain the cell impedance, i.e., the average height for the nth cell is given by $[h(n)+h(n-1)]/2$

In the previous calculations, we assumed the conductors are of equal length. It may be that two conducting transformer members are of unequal length, a fact often dictated by geometry constraints. Under these circumstances, the one dimensional approach is less appropriate, and the two and three dimensional iterations described in the previous Chapters should be employed.

8.6 *Pulse Transformer Simulation Results*

For our simulation example we choose a symmetric transformer whose impedance varies linearly with length from 5 Ω to 50Ω . For convenience, we break up the transformer into ten stages, N=10. The total delay time is 10ns, representing a delay time of 1ns in each stage($\Delta t = \Delta l/v$). Fig.8.14 shows the average impedance of each cell, using Eqs.(8.2)-(8.4). Two SPICE simulations were performed for this transformer, as shown in Fig(8.15). In the first simulation , the 50 volt input pulse injected is 10 ns wide. Although the initial portion of the pulse reaches the anticipated value, ~ 150 volts(using Eq.(8.1)), much of the pulse is attenuated. The pulse degradation should not be surprising

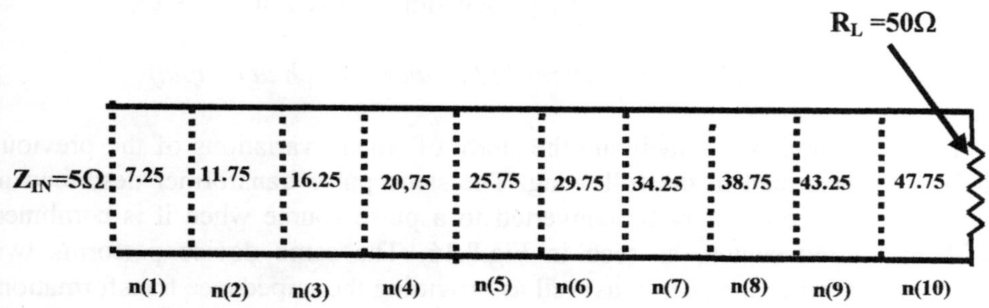

FIG. 8.14 IMPEDANCE VALUES FOR LINEAR TRANSFORMER 5→50Ω), DIVIDED INTO TEN CELLS.

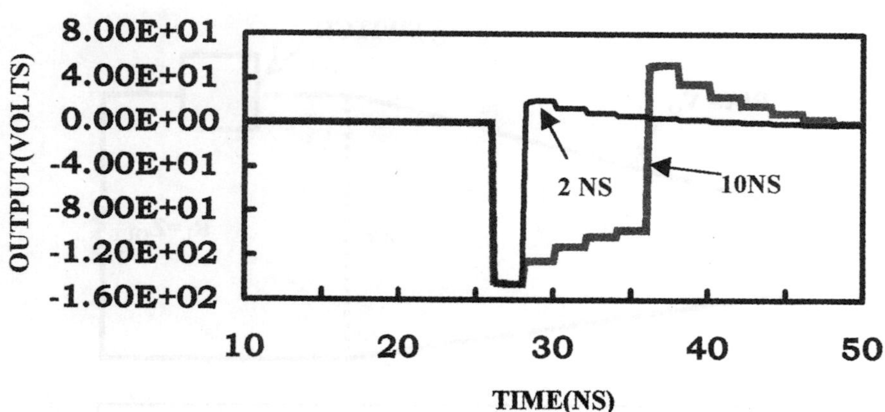

FIG. 8.15 OUTPUT FROM 10:1 TRANSFORMER WITH 10NS LONG DELAY. INCIDENT PULSE AMPLITUDE IS 50V. INCIDENT PULSEWIDTHS ARE 10NS AND 2NS.

since the input pulsewidth is comparable to the delay time of the transformer. Indeed for very long input pulsewidths, relative to the transformer, the voltage gain will drop to unity. In contrast, the narrower input pulse, 2ns, will transform upward to~150 volts, over the entire pulsewidth, as noted in Fig. 8.15.

8.7 Pulse Source Using Non-Uniform TLM Lines (Switch at Output)

The SPICE simulations facilitate the study of many variations of the previous pulser, as described in the following. The same pulse transformer described in the previous Section may be converted to a pulse source when it is combined with an *output* switch, as seen in Fig.8.16. The same device performs two functions: it stores the energy as well as providing the impedance transformation. The switch allows the non-uniform impedance line to store energy prior to triggering the switch. For a given charging voltage, the line produces the pulse shown in Fig.8.17. For this simulation we used the same transformer as in the previous discussion, with the same delay time of 10 ns, number of sections(10) , and impedance values for each section.

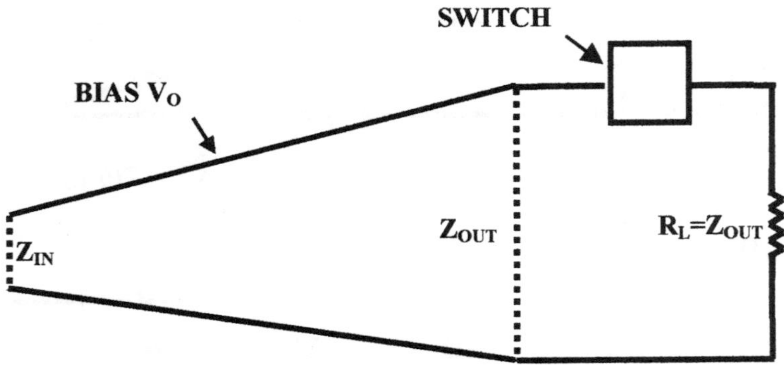

FIG. 8.16 PULSE SOURCE COMBINING FUNCTIONS OF IMPEDANCE TRANSFORMATION AND ENERGY STORAGE (BIAS VO) , WITH SWITCH S AT OUTPUT. VOLTAGE GAIN IS LIMITED.

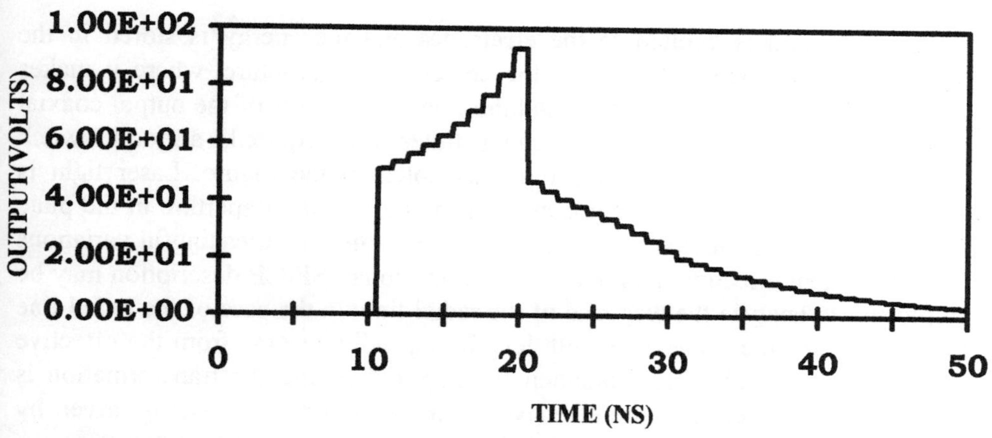

FIG.8.17 SIMULATED OUTPUT FROM ENERGY STORING TRANSFORMER WITH OUTPUT SWITCH. LINEAR IMPEDANCE CHANGES FROM 5Ω TO 50Ω. 100V BIAS.

We note in Fig.8.17 that initially the pulse rapidly climbs to half the charging voltage; this is to be expected since the line impedance is close to 50Ω, which matches the load resistance. Following this region in time, the pulse climbs to about the full charging voltage, before beginning its decline. It may seem surprising at first that higher voltages are not obtained since the impedance transformation ratio is ten. However, we have no control over the pulsewidth being produced, and as we have seen before, when the pulsewidth is comparable to the transformer transit time, the pulse transformation is severely curtailed. The gain, therefore, is a definite drawback with this type of source. In Section 8.9 we shall see how we may modify the circuit in Fig.8.16 so as to achieve significant transformer gain.

8.8 Radial Pulse Source(Switch at Output)

A particularly compact design of an energy storing device is that of the radial transmission line, for which SPICE analysis may be employed. As shown in Fig. 8.18, the source consists of a pair of circular electrodes separated by a dielectric medium. A field is applied to the electrodes and the energy is stored in the dielectric. The switch is located at the center of the structure, where it makes contact with the high voltage plate and the center conductor of the output coaxial transmission line. Since fast switching times are required, a good switch candidate is the photoconductive type as noted in the Figure. Laser light is introduced into the semiconductor medium by means of an aperture in the plate electrode. Since the device has circular symmetry, the only meaningful variations are in the radial direction, and thus a one dimensional SPICE description may be employed to analyze the response of the radial line. In the case of the radial line the variation of the impedance with length(i.e., radius) stems from the effective changes in the width. The fundamental wave undergoing the transformation is concentric in nature. Here the effective width of the line is changing, given by $W(r') = 2\,\pi\,r$. If h is the fixed height between the radial plates then the impedance formula , comparable to Eq.(8.5), is

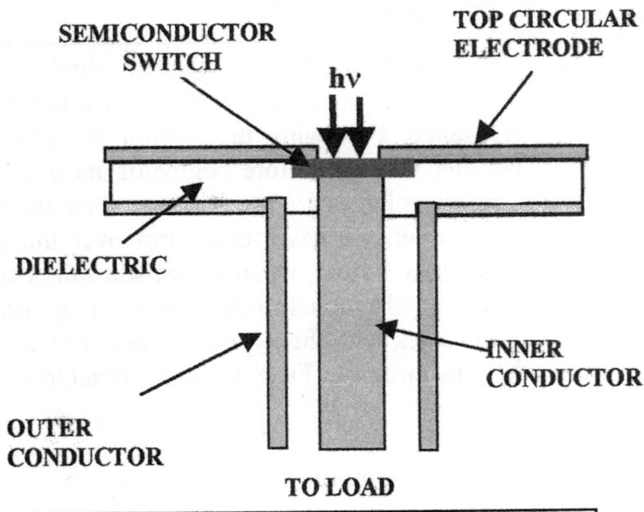

FIG. 8.18 RADIAL LINE PULSER WITH SWITCH AT OUTPUT.

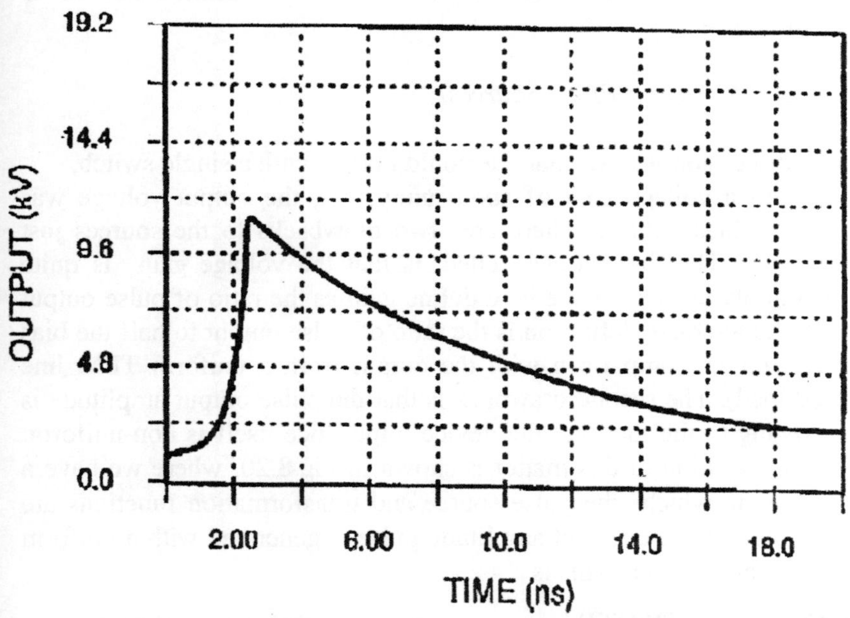

FIG. 8.19 EXPERIMENTAL OUTPUT FROM ENERGY STORING
RADIAL LINE WITH OUTPUT SWITCH AND POSITIVE
MISMATCH. OUTPUT RESEMBLES "RF CAPACITOR".

$$Z_m(r') = 376.7 \ h/[\varepsilon^{1/2}(2 \ \pi \ r')] \tag{8.9}$$

where we identify $2\pi r'$ with l'. When used as a pulse source with the switch at
the output, this type of design lacks true voltage gain; the peak output pulse is
about equal to the bias voltage, as mentioned before. An interesting special case
occurs when the there is a very large positive mismatch between the load
resistance and the final impedance of the radial line, located near the center (R_L
much larger than the line impedance at output). With this circumstance, we have
a 2D "RF" capacitor, wherein the output has a very fast initial risetime followed
by an exponential decay. Fig.8.19 shows the experimental output from

such a radial line ; the SPICE simulation is in very good agreement with the result. The following Section discusses the design modification needed to enhance the gain.

8.9 Pulse Sources With Gain(PFXL Sources)

In the previous discussion we saw that one could obtain, with a single switch, pulse sources with a voltage gain of about unity, i.e., the output voltage was about equal to the bias voltage. There are two drawbacks to the sources just described, however. The first, already cited, is that the voltage gain is quite inadequate, with only unity gain(We here define gain as the ratio of pulse output to bias voltage. An alternate definition is the ratio of pulse output to half the bias voltage, which implies comparison with the output from a uniform TLM line with a matched load). The second drawback is that the pulse output amplitude is non-uniform, owing to the fact that the device impedance itself is non-uniform. A straightforward solution to this matter is shown in Fig.8.20, where we have a composite device in which the pulse source and transformation functions are separated. The low voltage constant amplitude pulse is generated with a uniform transmission line and a switch. This is

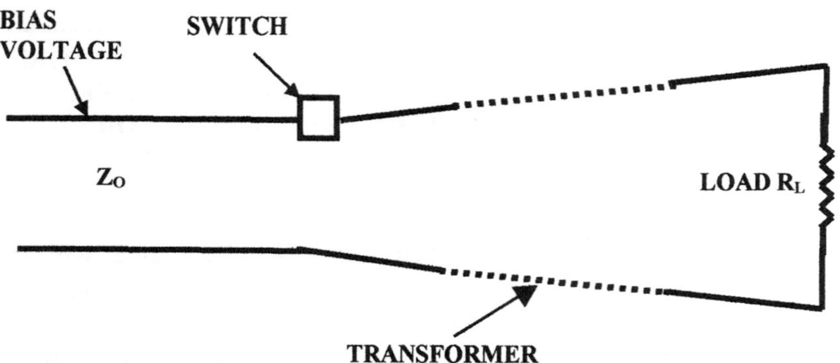

FIG. 8.20 PULSE SOURCE USING SEPARATE ENERGY STORAGE LINE (Z_O) AND TRANSFORMER SECTION. INPUT OF TRANSFORMER IS Z_O AND OUTPUT IS EQUAL TO R_L.

followed with a pulser transformer to obtain the voltage gain. In order to maintain the constant amplitude, the transformer length must be much longer than that of the original energy storage element, and as a result the size of the composite pulser will be enlarged.

The preceding invites the question as to whether it is possible to obtain a high gain, constant amplitude pulse source which combines the energy storage and transformation functions, again using a single switch. We use the designation "PFXL" to describe such a circuit element, and to differentiate it from the conventional "PFL" , or pulse forming line , which is used to produce a constant amplitude pulse, but without gain(In Reference [1] the PFXL designation also was inadvertently applied to sources of the type in Fig.8.16. Because of the aforementioned gain limitations with such sources, however, the PFXL designation is not appropriate). If we forego the goal of constant amplitude, then we can certainly obtain a PFXL pulser as shown in Fig.8.21. Here we insert the switch at the low impedance end of the transformer section.

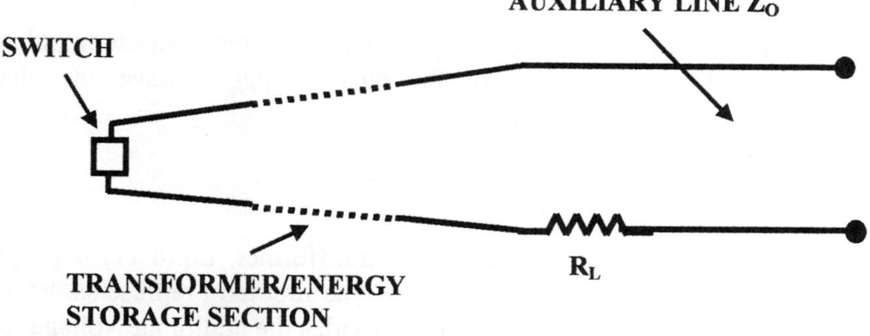

FIG. 8.21 PULSE SOURCE USING TRANSFORMER/ENERGY STORAGE ELEMENT AND AUXILIARY LINE Z_O (PFXL). SWITCH IS LOCATED AT LOW IMPEDANCE END OF TRANSFORMER. $R_L = Z_O + Z_F$ WHERE Z_F IS HIGH IMPEDANCE OUTPUT OF TRANSFORMER.

Firing the switch then launches an inverted wave toward the output end of the transformer, in which the final impedance is Z_F. We include an auxiliary element, a uniform section of transmission line Z_o at the output, in series with the load R_L This auxiliary line serves to provide voltage hold-off (since we have moved the switch to the input), but also contributes to the voltage gain. Since the switch is at the low impedance end side, the generated wave in some ways resembles a wave injected into pure transformer, at least for the leading edge of the wave. We should contrast this with the situation wherein the switch is at the output of the transformer; in this case the generated wave must make a round trip in the transformer, thus nullifying the possibility of any significant gain

   We may roughly estimate the output of the PFXL, using the following arguments. The output consists of the load R_L in series with the auxiliary section of uniform transmission line of impedance Z_o . Because of the inverted wave in the transformer, Z_o and Z_F, which are in series, provide current in the same direction, and so discharge into R_L. In order to maximize the energy transfer, we arrange to have matching conditions at the output

$$Z_o + Z_F = R_L \qquad\qquad (8.10)$$

As implied before, the PFXL does not produce a constant amplitude pulse, but we expect the the leading edge of the pulse output to have an enhanced amplitude, V_{OUT}, approximately equal to

$$V_{OUT} = (V_o/2) \ (Z_F / Z_i)^{1/2} + \ (V_o/2) \qquad\qquad (8.11)$$

where Z_i is the initial low impedance of the transformer. Eq.(8.11) is proposed, based on the following simple considerations: the first term represents the effect of the transformer, while the second term provides the additional voltage of the final storage line, Z_o. In fact we shall see that Eq.(8.11) underestimates the gain, by about 10-15%, and that an accurate estimate of the leading edge, as well as the entire pulse shape, can only be obtained by a SPICE simulation or the equivalent TLM analysis, to be discussed.

   Figs.8.22(a) - 8.22(b) show SPICE simulations for a PFXL, using the same type of linear transformer section as before, but with differing line lengths, 1ns and 250ps in the auxiliary line. As with the previous simulations , we arbitrarily divide the transformer section into 10 sections, each with a one ns delay. With a 100 volt bias, we see that in both cases the leading edge output is ~ 243 Volts, which is slightly higher than the estimate from Eq.(8.11), which yields ~ 210 volts. The reason for the higher output in the simulation has to do with the fact that Eq.(8.11) does not completely portray the actual situation since it deals solely with the leading edge in a pure transformer, augmented by the output from Z_o. In fact the PFXL transformer is subject to a bias voltage and the leading edge output contains contributions from backward waves which have been reflected toward the output direction, throughout the entire transformer. This effectively increases the voltage gain. In App.8A.4 we obtain the same result as the simulation using TLM analysis for calculating the leading edge.

Comparing Figs.8.22(a) and 8.22(b) we see that the pulsewidth of the auxiliary line controls the pulsewidth of the output pulse. For a transformer cell size of 1ns, an auxiliary line of 1ns (Fig.8.22(a)) results in an output pulse of about 2 ns. When the auxiliary line is reduced to 250ps (Fig.8.22(b)), we

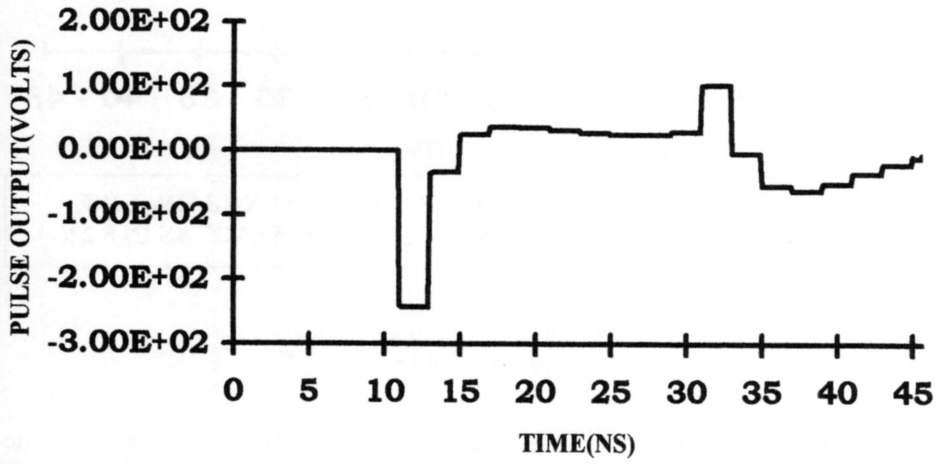

FIG. 8.22(a) OUTPUT FROM PULSER WITH COMBINED TRANSFORMER/ENERGY STORAGE SECTION(PFXL). TAPER IS LINEAR, 10 NS LONG , 5Ω TO 50 Ω. 100V BIAS. DELAY TIME OF AUXILIARY LINE =1NS.

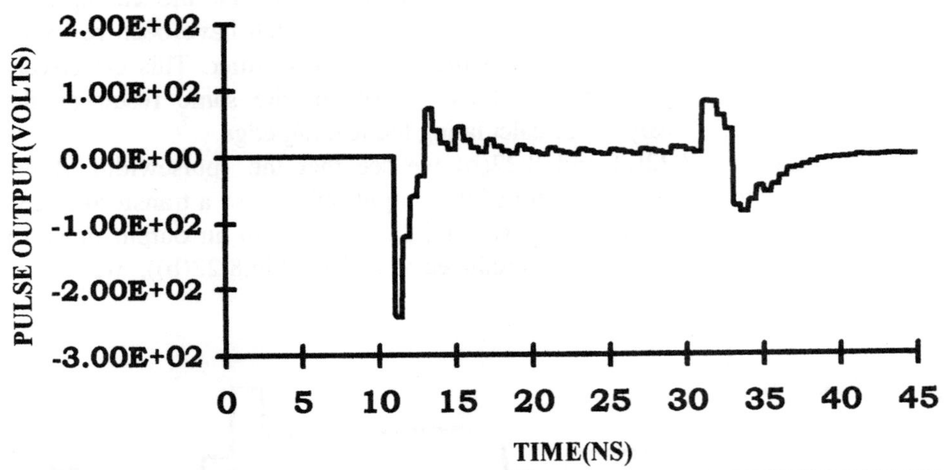

FIG. 8.22(b) OUTPUT FROM PFXL BUT WITH AUXILIARY LINE DELAY OF 0.25NS. OTHER CONDITIONS ARE SAME AS IN 8.22(a).

see that the output pulsewidth is reduced accordingly. The auxiliary line pulsewidth therefore controls the output pulsewidth, at least when Z_o and Z_F are comparable. We also note, as expected, that the leading edge amplitude of the output is the same in both cases.

In both cases, the oscillatory behavior due to multiple reflections is significant , as seen in Figs.8.22. No attempt has been to optimize the output pulse shape. It is very likely that tradeoffs in in pulse amplitude and pulse shape may be accomplished using various kinds of tapers in the transformer as well as various auxiliary impedance values and pulsewidths. These tradeoffs can be examined using pure simulation methods or a combination of simulation and analysis

A similar PFXL design modification to the radial line pulser is shown in Fig. (8.23). Instead of having the switch at the ouput high impedance region, the switch is located at the outer circumference, which constitutes the low impedance region. By its very nature, of course, a single switch is not possible and instead a series of switches, located at the outer periphery, and turned on simultaneously, is required. This launches a circular wave, directed toward the center. Upon reaching the center, both the radial line and the auxiliary line Z_o contribute to the total output, R_L. As before the load is matched to the sum of the radial impedance, Z_F and Z_o. Qualitatively, the output simulation is very similar to that of the linear PFXL. The pulse delivered to the load will consist of the transformed pulse plus the auxiliary line wave.

We see from Fig.8.22 that the resolution of the output PFXL wave is determined by the smallest element in the circuit, in this case either the cell size of the transformer or pulsewidth of the auxiliary line. To observe structure in the leading edge of the output, we need to increase the number of cells, i.e., shrink the cell size. Using only ten cells, the simulation can approximate the amplitude of the pulse, but it is likely that some of the structure in the output pulse shape is lost. Under these circumstances the only remedy is to increase the number of cells. We then return to the central question as to whether SPICE can accommodate large numbers of cells, or whether it is preferable instead to use the basic TLM method.   As seen in the previous Chapters, there are no basic obstacles to using the TLM method for solving such problems, especially 1D type.

Darlington Pulser

8.10 TLM Formulation of Darlington Pulser

In the previous pulse sources described, we saw that it was possible to achieve significant voltage gain in devices which both store the energy as well as provide the impedance transformation, using only a single switch. The output,

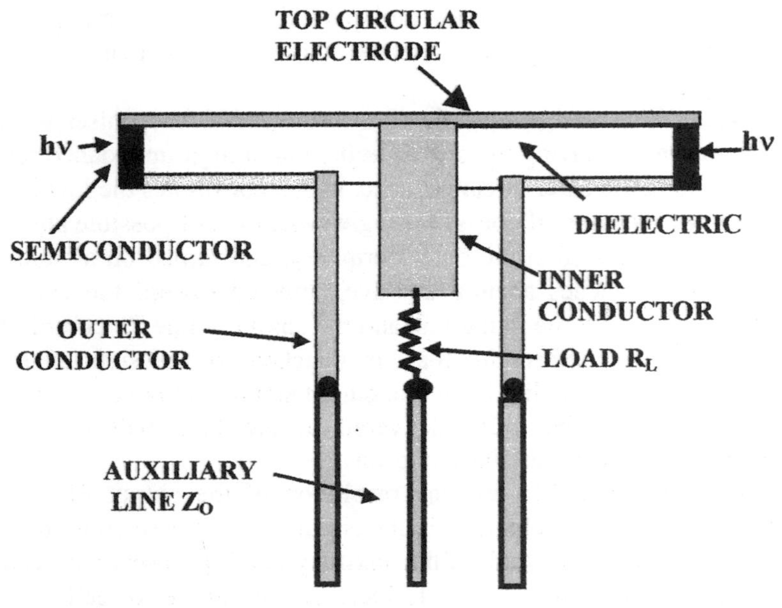

FIG. 8.23 RADIAL LINE PULSER (PFXL) WITH AUXILIARY LINE Z_O AND WITH LIGHT ACTIVATION AT LOW IMPEDANCE PERIMETER. $R_L = Z_O + Z_I$ WHERE Z_I IS THE IMPEDANCE OF RADIAL LINE AT THE INNER RADIUS.

however, did not have constant amplitude, and we inquire under what circumstances a constant amplitude may be achieved while still retaining the aforementioned properties. We shall see that it is indeed possible to obtain such a source, but only if we resort to discrete sections of pulse forming lines rather than to a continuously varying transformer line. Such a pulser is designated as a Darlington pulser[2]. Although it is certainly possible to employ a SPICE simulation to describe a Darlington pulser, it will be instructive to derive its basic properties from the TLM formulation. The results are simple design formulas, which are are very easy to apply. In the follow-on Section, we then add losses to the Darlington , for which the SPICE simulation is easier to use.

The new design is prompted by our previous considerations and thus we consider the design in Fig.8.24, in which we have a chain of N transmission line sections Z(1), Z(2),....Z(N). A single switch is located in the first section and the load R_L is between the Nth and (N-1)th sections. All the sections are biased to voltage V_o. Now suppose the impedance sections satisfy

$$Z(n) = n(n+1)R_L/ N^2 \quad : n = 1, 2, ... (N-1) \qquad (8.12)$$

$$Z(N) = R_L/N \qquad (8.13)$$

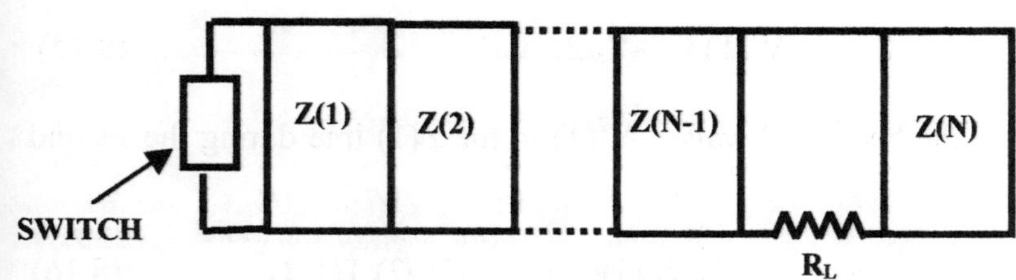

FIG. 8.24 N STAGE DARLINGTON CIRCUIT. Z(n) SATISFIES: $Z(n) = n(n+1)(R_L/N^2)$; n=1,2,....N-1
$Z(N) = R_L/N$, GAIN = $N(V_o/2)$

With the impedance sections given by Eqs.8.12-8.13, we can show that , upon initiation of the switch, a growing wave is propagated toward the load, and that with each section, the wave amplitude is increased by an amount $V_o/2$. The total gain is then $NV_o/2$, i.e., the voltage pulse delivered to the load , V_{RL}, is

$$V_{RL} = NV_o/2 \qquad (8.14)$$

We can demonstrate Eq.8.14 by the method of induction.   We briefly summarize this method. We show that at the n=1 node, the energy in Z(1) is completely transferred to Z(2), while at the same time there is no energy transfer whatever from Z(2) to Z(1). The wave growth results in an additional amplitude of $V_o/2$ in Z(2) during the second time step. We now *assume* that for the nth impedance(after 2n time steps have elapsed), the amplitude is $nV_o/2$ in the $Z_o(n)$

line, and all the wave energy in the prior lines (less than n) has been emptied. If now we can show that all the energy in $Z(n)$ is transferred to $Z(n+1)$ with a resultant $V_o/2$ enhancement in the amplitude, then by virtue of induction, the same behavior is true for all values of n. A separate verification for the final impedance section completes the proof.

The proof begins by assuming the switch at $Z(1)$ has been turned on and the growth process has begun. Because of the initial switching action the forward wave at the end of the first time step   will be inverted and the wave amplitude will be

$$^+V^1(1) = \ -V_o/2 \tag{8.15}$$

We then calculate the backward wave $^-V^2(1)$ in the $Z(1)$ line during the second step. Thus

$$^-V^2(1) = \quad -\ ^+(V_o/2)\, B(1,1)\ +\ ^-(V_o/2)\, T(1,2) \tag{8.16}$$

We remind ourselves that the two arguments in the scattering coefficients identify the node and the incident wave direction, respectively. From the expressions for the 1D scattering coefficients, and the Darlington impedance values from Eq.(8.12)

$$B(1,1) = (1/2) \tag{8.17a}$$
$$T(1,2) = \ (1/2) \tag{8.17b}$$

and thus $^-V^2(1) = 0$. Next we determine the wave transfer from the $Z(1)$ to $Z(2)$ lines during the second time step. The forward wave $^+V^2(2)$ in the $Z(2)$ line is

$$^+V^2(2) = -\ ^+(V_o/2)\, T(1,1) + \ ^-(V_o/2)\, B(1,2) \tag{8.18}$$

Since

$$T(1,1) = \ 1/2 \tag{8.19a}$$
$$B(1,2) = -1/2 \tag{8.19b}$$

the forward wave $^+V^2(2) = -V_o$ and thus the forward wave has increased by $-V_o/2$. Next we assume this process continues and that in the nth line the forward wave is $(-nV_o/2)$ during the nth time step. The forward wave in the (n+1)th line is

$$^+V^{n+1}(n+1) = -\,^+(nV_o/2)\,T(n,1) + \,^-(V_o/2)\,B(n,2) \tag{8.20}$$

Using Eq (8.12) for the line impedances, the coefficients are

$$T(n,1) = (n+2)/(n+1) \tag{8.21a}$$
$$B(n,2) = 1/(n+1) \tag{8.21b}$$

Substituting into Eq.(8.20) we obtain

$$^+V^{n+1}(n+1) = -(n+1)V_o/2 \tag{8.22}$$

A similar analysis shows that the backward wave in $Z(n)$ vanishes. This completes the proof except for the last cell which is different because of the presence of R_L, and therefore requires a separate treatment. We note that the impedances $Z(N) = R_L/N$ and $Z(N-1) = R_L N(N-1)/N^2$ add in series such that

$$Z(N) + Z(N-1) = R_L \tag{8.23}$$

Because of the inverted wave in $Z(N-1)$, the combination of waves in $Z(N)$ and $Z(N+1)$ is matched to R_L. Thus

$$V_{RL} = V(N-1) + V(N) = -(N-1)(V_o/2) - V_o/2 = -NV_o/2 \tag{8.24}$$

which completes the proof.

8.11 *SPICE Simulation of Lossy Darlington Pulser*

It should be noted that for the first few sections of the Darlington pulser, the impedance levels are small, and therefore the design is very sensitive to small amounts of inductance and loss. SPICE(or the equivalent TLM analysis) can play an important role in understanding the effects of losses and other parasitics in transformers and pulse sources. Indeed, in the case of the lossless Darlington pulser, for example, the design equations are extremely simple and SPICE solutions are unnecessary. It is only when we try to determine the effects of losses or deviations from the line impedance , when closed form solutions are impractical, that SPICE becomes essential. Fig.8.25 shows how an arbitrary 1D circuit is modified by losses, such as a transformer, PFXL, or Darlington

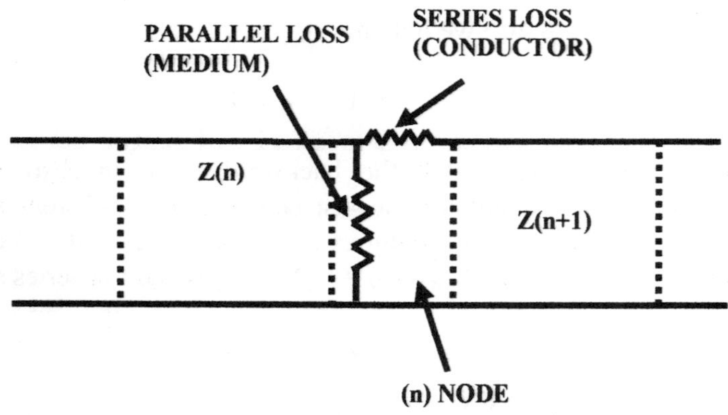

FIG. 8.25 CONDUCTOR AND MEDIUM LOSS REPRESENTATIONS IN 1D CIRCUIT (e.g., DARLINGTON, TRANSFORMER, OR PFXL CIRCUITS).

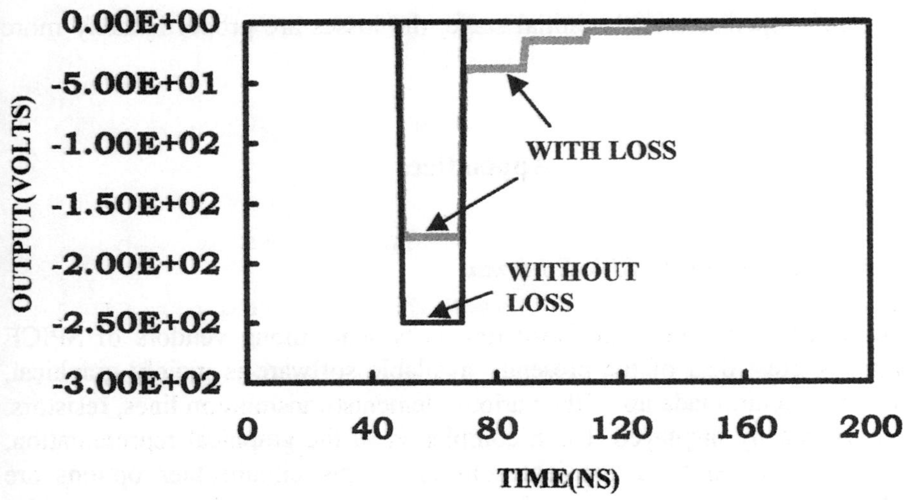

FIG. 8.26 OUTPUT FROM 5 STAGE DARLINGTON CIRCUIT WITH AND WITHOUT LOSS. LOSS IS DUE TO 1Ω SERIES DROP IN EACH STAGE. BIAS =100V AND OUTPUT IS ACROSS 50Ω.

circuit. In the Figure we have included the effects of both medium losses, represented by the usual shunt resistors, as well as losses in the "wires" represented by the series resistors. The inclusion of series losses points out the fact that the TLM formulation may be extended to account for such losses. Both parallel and series resistors, as usual, are located at the node. The width of the node in Fig.8.25 is of course exaggerated to display the resistors.

As an example, in Fig.8.26 we compare the simulations of a lossless Darlington circuit with that when loss is included. The selected circuit is a 5 stage Darlington, and a 1Ω series loss per stage is included(we neglect parallel losses). We see that in the lossless case, the output obeys Eq.8.24, and the 100 volt bias results in a 250 volt output. When loss is included, substantial decay begins to appear in the output. Much of the loss occurs in the initial stage (n=1) where the TLM impedance value is only 4Ω. Since the series resistance is comparable to the TLM impedance in the initial stage, the losses are proportionately more important.

Appendices

App. 8A.1 Introduction To SPICE Format

The following discussion and formatting is typical among vendors of SPICE software. A great deal of the presently available software is mainly graphical, wherein the circuit, made up of the various elements(transmission lines, resistors, etc), is graphically displayed. Upon completion of the graphical representation, the user then requests a transient solution of the circuit(other options are available, such as a steady state solution, Fourier analysis, etc..) and provides the numerical limits for the analysis. The computer then automatically assigns node numbers to the circuit and proceeds to grind out the solution. The tabulated results may then be displayed using the graphics package provided with the software, or else the tabulated results also can be delivered to an external graphics package for display. One should also mention that most commercial versions of SPICE present the format listing (which identifies the various components at the nodes) in a graphical manner. This greatly facilitates the use of the SPICE methods.

App.8A.2   Discussion of Format for Photoconductive Switch

In this Section we discuss the SPICE format for the photoconductive switch given in Table 8A.1. Only the most essential statements, corresponding to the eight cell representation in Fig.8.2, are provided. We add node numbers(which are arbitrary) and the labels for the circuit elements as shown in Fig.8A.1.

Line 1 of the format is simply the title. Line 2 specifies the transmission line of the input, designated as T1. The first pair of node numbers following T1, 1 2 , specifies the input port nodes of the line; the second pair, 3 4, specifies the node numbers of the output port.   This is followed by the characteristic impedance(ohms) of the line, ZO=104Ω. Finally the length of the input line, TD, is specified, which for this program was 120ps (a longer input line is probably more desirable, so long as the software can provide a solution). The other nine lines, T2, T3, etc follow the same format. The value of TD for the other lines, however, is 10ps, which is the selected matrix length for the semiconductor. Line 5 specifies one of three shorting resistors. RS1 is identified by the pair of nodes 6 11. The third entry 1E-6 provides the resistance value(ohms). The other two shorting resistors, RS2 and RS3 are assigned the same value. We next complete the description of the other resistors: the load resistor(line 21) and the resistors attached to each node(lines 22-52). The load resistor, connecting nodes 30 31, terminates the transmission line so that the value is 104 ohms, as indicated. Many of the iterations require a dc path from the node site to ground. Therefore, as a precaution, we insert a large resistance , 10^7 ohms, between each non-zero node and ground. Thus, lines 22-52 enumerate all the non zero nodes, inserting 10^7 ohms to ground. For example, the first entry in line 22 is for R1, where the 1 indicates the node number. The specification is therefore R1 1 0 1E7. Note that we have not actually specified the zero node, i.e., ground , anywhere in the Figure. This is completely arbitrary, but often the ground node selected is one of the two nodes enclosing the load resistor, either 30 or 31. Next we discuss the switches, GS1, GS2, and GS3 found on lines 6-7, 12-13, 18-19. The switches are used to simulate the added conductivity, induced by either a light signal or avalanching. The actual switching elements, however, are voltage controlled current sources, with the voltage control signals designated by VIN1, VIN2, and VIN3 for each of the three sources. Looking at GS1 (line 6), for example, the GS1 is followed by the nodes enclosing the element, 5 8. The next portion of the statement POLY(2), assumes a two dimensional polynomial expansion of the current in terms of the two variables, the source voltage, V(5,8), and the control voltage, V(7,0). Note that the control voltage is referenced to ground. These four nodes make up the first four entries following POLY(2). The remaining entries are the coefficients of the expansion for the current. If the coefficients are designated by P_0 , P_1 , P_2 etc.., then the first few terms for the current I are

$$I = P_0 + P_1* [V(5,8)] + P_2*[V(7,0)] + P_3*[V(5,8)]^2 + P_4*[V(5,8)]*[V(7,0)]$$

$$+ P_5*[V(7,0)]^2 + P_6*[V(5,8)]^3 + P_7*[V(5,8)]^2*[V(7,0)] + \text{ etc...} \qquad (8A.1)$$

We arbitrarily select $P_4 = 1$ as the only non-zero coefficient of POLY(2). Unspecified coefficients are assumed to be zero. The coefficient is a convenient choice because we may then attach a very simple interpretation to the result. Thus

$$I = V(5,8)* V(7,0) \qquad (8A.2)$$

The equivalent resistance R(5,8) across the source nodes, 5 8 , is therefore

$$R(5,8) = 1/V(7,0) \qquad (8A.3)$$

(More generally, $R(5,8) = 1/ P_4V(7,0)$ if $P_4 \neq 1$). We may therefore change the resistance via the control voltage; the higher the control voltage the smaller the resistance. The statement for the control voltage, VIN1, is shown in line 7. The first two entries are the node numbers , followed by the function PULSE. This indicates the control voltage is a flat top pulse with the capability of non-zero rise and fall times. The first two entries under PULSE indicate the initial and plateau values, which are zero and 100 volts respectively. From Eq.(8A.3), we see that initially the resistance is extremely large . When the 100V pulse is applied, the resistance then falls to 1/V100 or .01 ohms. The next entry, 1ps, is the delay time for the onset of the pulse. Note that the delay time, for each of the three switches, increases with distance from the anode, corresponding to a progressive triggering of the switches. The next two entries, each 1.0ps, are the rise and fall times respectively. The final two entries, each 200ps in length, are the pulsewidth and the period. In fact the latter two values do not play a direct role but are merely selected large enough so as to insure that the resistance remains at its low value(.01 ohms) throughout the time domain of interest. Lastly we point out the initial conditions for each of the noes, lines 53-57. Note that all the initial node voltages are zero except for those sitting on one of the three "charged cells", i.e., 3000V(anode), 2000V, or 1000V.

The SPICE format for the pulse transformer , together with the node diagram, are shown in Table 8A.2 and Fig.8A.2, respectively. The simulation of the transformer consists of elements already described for the photoconductive switch. Note that the all the transmission lines are identical in length, 1ns, except for T2 , which is very long, 16ns, by comparison. The purpose of this line is to simplify the device, essentially isolating the transformer from any reflections from the pulse source.

TABLE 8A.1. SPICE FORMAT FOR PHOTOCONDUCTIVE SWITCH

[1] **PHSW.CIR**
[2] **T1 1 2 3 4 ZO=104 TD =120PS**
[3] **T2 6 11 3 10 ZO=104 TD=10PS**
[4] **T3 4 9 5 8   ZO=104 TD=10PS**
[5] **RS1 6 11 1E-6**
[6] **GS1 5 8 POLY(2) 5 8 7 0 0 0 0 0 1**
[7] **VIN1 7 0 PULSE(0 100 1PS 1PS 1PS 200PS 200PS)**
[8] **T4 10 9 12 13 ZO=104 TD=10PS**
[9] **T5 15 20 12 17 ZO=104 TD=10PS**
[10] **T6 13 18 14 19 ZO=104 TD =10PS**
[11] **RS2 15   20 1E-6**
[12] **GS2 14 19 POLY(2) 14 19 16 0 0 0 0 0 1**
[13] **VIN2 16 0 PULSE(0 100 60PS 1PS 1PS 200PS 200PS)**
[14]   **T7 17 18 21 22 ZO=104 TD=10PS**
[15] **T8 24 29 21 26 ZO=104 TD = 10PS**
[16] **T9 22 27 23 28 ZO=104 TD=10PS**
[17] **RS3 24 29 1E-6**
[18] **GS3 23 28 POLY(2) 23 28 25 0 0 0 0 0 1**
[19] **VIN3 25 0 PULSE(0 100 100PS 1 PS 1PS 200PS 200PS)**
[20] **T10 26 27 30 31 ZO = 104 TD =10PS**
[21] **RL 30 31 104**
[22] **R1 1 0 1E7**
[23] **R2 2 0 1E7**

(CONT)
[24] R3 3 0 1E7
[25] R4 4 0 1E7
[26] R5 5 0 1E7
[27] R6 6 0 1E7
[28] R7 70 1E7
[29] R8 8 0 1E7
[30] R9 9 0 1E7
[31] R15 15 0 1E7
[32]R10 10 0 1E7
[33] R11 11 0 1E7
[34] R12 12 0 1E7
[35] R13 13 0 1E7
[36] R14 14 0 1E7
[37] R15 15 0 1E7
[38] R16 16 0 1E7
[39] R17 17 0 1E7
[40] R18 18 0 1E7
[41] R19 19 0 1E7
[42] R20 20 0 1E7
[43] R21 21 0 1E7
[44] R22 22 0 1E7
[45] R23 23 0 1E7
[46] R24 24 0 1E7
[47] R25 25 0 1E7
[48] R26 26 0 1E7
[49] R28 28 0 1E7
[50] R29 29 0 1E7
[51] R30 30 0 1E7
[52] R31 31 0 1E7
[53] .IC V(0)=0 V(1)=0 V(2)=3000 V(3)=0 V(4)=3000V(5)=3000 V(6)=0
[54] .IC V(7)=0 V(8)=2000 V(9)=2000 V(10)=0 V(11)=0 V(12)=0 V(13)=2000
[55] .IC V(14)=2000 V(15)=0 V(16)=0 V(17)=0 V(18)=1000 V(19)=1000

(CONT)

**[56] .IC V(20)=0.ICV(21)=0 V(22)=1000 V(23)=1000 V(24)=0 V(25)=0
     V(26)=0**

[57].IC V(27)=0 V(28) =0 V(29)=0 V(30)=0 V(31)=0

[58] .END

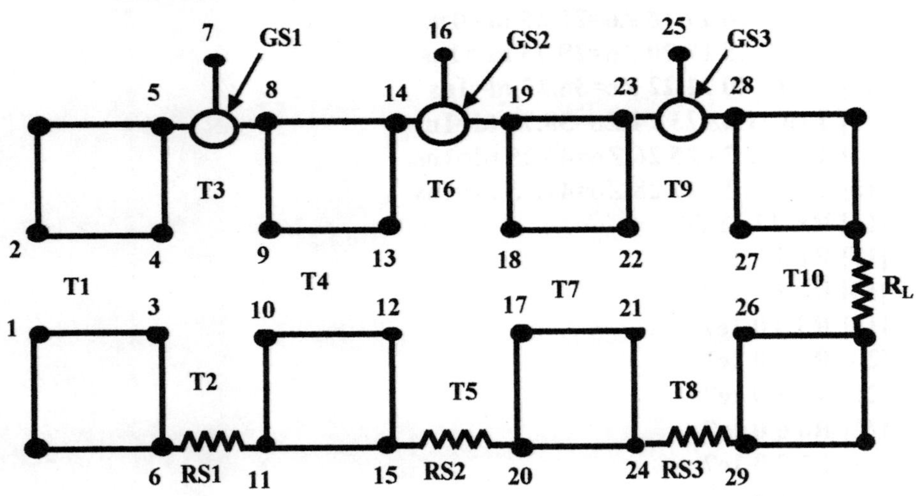

**FIG. 8A.1 SPICE NODE IDENTIFICATION FOR
PHOCONDUCTIVE SWITCH.**

TABLE 8A.2. TYPICAL SPICE FORMAT FOR PULSE TRANSFORMER

[1] XMFR.cir
[2] T1 1 2 3 4 zo=5 td=1ns
[3] Gs1 4 6 poly(2) 4 6 5 0 0 0 0 0 1
[4] Vin1 5 0 pulse(0 100 1ns 1ns 1ns 51ns 51ns)
[5] T2 3 6 7 8 zo=5 td =15ns
[6] T3 7 8 9 10 zo=7.5 td=1ns
[7] T4 9 10 11 12 zo=11.75 td=1ns
[8] T5 11 12 13 14 Zo=16.25 td=1ns
[9] T6 13 14 15 16 Zo=20.75 td=1ns
[10] T7 15 16 17 18 Zo=25.25 td =1ns
[11] T8 17 18 19 20 Zo=29.75 td =1ns
[12] T9 19 20 21 22 Zo=34.25 td=1ns
[13] T10 21 22 23 24 Zo=38.75 td=1ns
[14] T11 23 24 25 26 Zo=43.25 td=1ns
[15] T12 25 26 27 28 Zo=47.75 td=1ns
[16] RL 27 28 50
[17] R1 1 0 1e7
[18] R2 2 0 1e7
[19] R3 3 0 1e7
[20] R4 4 0 1e7
[21] R5 5 0 1e7
[22] R6 6 0 1e7
[23] R7 7 0 1e7
[24] R8 8 0 1e7
[25] R9 9 0 1e7
[26] R10 10 0 1e7
[27] R11 11 0 1e7
[28] R12 12 0 1e7
[29] R13 13 0 1e7
[30] R14 14 0 1e7

[31] R15 15 0 1e7
[32] R16 16 0 1e7
[33] R17 17 0 1e7
[34] R18 18 0 1e7
[35] R19 19 0 1e7
[36] R20 20 0 1e7
[37] R21 21 0 1e7
[38] R22 22 0 1e7
[39] R23 23 0 1e7
[40] R24 24 0 1e7
[41] R25 25 0 1e7
[42] R26 26 0 1e7
[43] R27 27 0 1e7
[44] R28 28 0 1e7
[45] .ic V(0)=0 V(1)=0 V(2)=100 V(3)=0 V(4)=100V(5)=0 V(6)=0 V(7)=0
[46] .ic V(8)=0 V(9)=0 V(10)=0 V(11)=0 V(12)=0 V(13)=0: V(14)=0
[47] .ic V(15)=0 V(16)=0 V(17)=0 V(18)=0 V(19)=0 V(20)=0 V(21)=0
[48] .ic V(22)=0 V(23)=0 V(24)=0 V(25)=0 V(26)=0 V(27)=0 V(28)=0
[49] .end

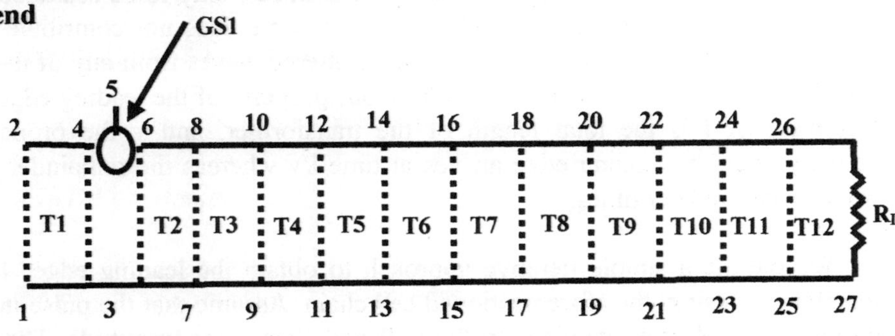

FIG. 8A.2 SPICE NODE IDENTIFICATION FOR PULSE TRANSFORMER. T1(PULSE SOURCE) AND T2(ISOLATION LINE) HAVE THE SAME IMPEDANCE. T2 IS ASSUMED TO BE MUCH LONGER COMPARED TO THE OTHER CELL LINES.

8A.3 TLM Analysis of Leading Edge Pulse in a Transformer

In this Appendix we calculate the leading edge of the electromagnetic pulse as it progresses through the transformer, The results obtained of course are not in conflict with the SPICE results and in fact the SPICE simulation is implicitly based on the discussion here and that in the next section where the general 1D TLM scattering equations are restated. What the TLM approach does offer, of course, is the insight into the operation of the transformer and other devices.

The leading edge of the propagating pulse, to which we have referred previously, is important since it provides the output amplitude representing the ideal transformer. We employ the cell designation as shown in the example provided in Fig.8.15. Suppose a pulse with an infinitely fast risetime is delivered to the input of the first cell n(1). The leading edge output(at the 50 Ω load) is the earliest arriving signal, and represents that portion of the signal uncontaminated by any wave components which have been back-scattered, only to be re-scattered later in the forward direction. As an example, we know that the forward wave will backscatter into the n(8) cell from the n=8 node , only to be scattered again in the forward direction at the n=7 node. This wave does not contribute to the leading edge, nor do any of the other back scattered waves from any of the other nodes throughout the transformer. An obvious property of the leading edge is the following: if l is the total length of the transformer, and v the propagation velocity, then the leading edge arrives at time l/v whereas the remainder of the wave arrives at later times.

We outline a simple iterative approach to obtain the leading edge for the transformer, using the aforementioned cell chain. Assume that the pulse incident on the n(1) cell (i.e., incident on the n=0 node)has unity amplitude. The initial transmission coefficient, T(0,1), (the argument 1 indicates a forward wave)determines the enhancement of the leading edge as the wave travels from the 5Ω TLM line into the 7.25 Ω cell. We know from Chapters 1 and 3 that

$$T(0,1) = 2Z(1)/(Z(0) +Z(1)) = 2(7.25)/(5+7.25)=1.224 \qquad (8A.4)$$

The leading edge amplitude in the n(1) cell is therefore 1.224. We next calculate the T(1)transmission coefficient, which will tell us how the wave is enhanced in going from the n(1) cell to the n(2) cell. Thus the wave amplitude in n(2) is

$$^+V(2)=1.224T(1,1)=(1.224)2Z(2)/(Z(1)+Z(2)) \qquad (8A.5a)$$

or

$$^+V(2)=(1.224)2(11.75)/(7.25+11.75)=1.513 \qquad (8A.5b)$$

This process continues , with repeated field enhancement at each cell interface, the wave reaches the 50 Ω load, where the final field is

$$^+V_{OUT} = T(0,1)\ T(1,1)\ T(2,1)\ ....T(10,1) \qquad (8A.6)$$

where

$$T(n,1)=2Z(n+1)/(Z(n)+Z(n+1)) \qquad (8A.7a)$$

$$T(10,1)=2(50)/(Z(10)+50) \qquad (8A.7b)$$

Using the linear transformer values in Fig.8.15, we obtain a leading edge value of
$$^+V_{OUT} = 3.01\ \text{volts} \qquad (8A.8)$$

It is reassuring that this output agrees with the SPICE simulations in Fig.8.16 where the 50 volt leading edge wave is transformed into 150 volts(3X input).

We know that with the ideal pulse transformer, the leading edge of pulse satisfies

$$(^+V_2/^+V_1) = (Z_2/Z_1)^{1/2} \qquad (8A.9)$$

where Z_1, Z_2 are transformer impedance values and $^+V_1$, $^+V_2$ are the corresponding field values which the wave assumes as it progresses along the transformer. The transformer relationship, Eq.(8A.9), is very well known, but we have not yet formerly and explicitly shown that the TLM theory leads to Eq.(8A.9)(or to some relationship closely related to Eq.(8A.9)). Of course, this has important implications for SPICE, which is based on the TLM theory. In the following we fill in the gap between the TLM analysis and simple transformer theory. We begin by shrinking the size of the transformer cells to infinitesimal dimensions of two neighboring cells, designated by z, and $z+\Delta z$ (we have changed the notation for the cells , for obvious reasons). Suppose the forward wave $^+V(z)$ is just incident on the node separating the z and $z+\Delta z$ cells. The transmission coefficient is $2(z+\Delta z)/(2z+\Delta z)$. The transmitted wave $^+V(z+\Delta z)$ is therefore

$$^+V(z+\Delta z) = 2\,^+V(z)\,(z+\Delta z)/(2z+\Delta z). \qquad (8A.10)$$

If we expand the denominator to first order

$$^+V(z+\Delta z) = \,^+V(z)\,(1+\Delta z/z)(1-\Delta z/2z) \qquad (8A.11)$$

Retaining only terms up to first order, we have

$$^+V(z+\Delta z) = \,^+V(z) +\,^+V(z)\,\Delta z/2z \qquad (8A.12)$$

or in differential form,

$$d\,^+V(z)\,/dz \; = \; ^+V(z)\,/2z \qquad (8A.13)$$

The above has the simple solution

$$^+V(z) = C\,z^{1/2} \qquad (8A.14)$$

Assuming $^+V(z) = \,^+V_1$ at at $z=z_1$, $C = \,^+V_1\,/z_1^{1/2}$, then

$$^+V(z)\,/\,^+V_1 \; = \; (z/z_1)^{1/2} \qquad (8A.15)$$

which is identical to Eq.(8A.9). If we use Eq.(8A.15) to obtain the leading edge amplitude, we will obtain a slightly larger result (by about 5%) compared to the 10 stage simulation performed earlier ; this is not surprising since Eq(8A.15) presupposes an infinite number of cells, thereby approaching the theoretical limit of the transformer, so that some differences should be expected.

Finally we address the entire pulse output from the transformer, not just the leading edge. Of course the SPICE simulations performed previously already take into account the multi-scattering processes needed to describe the entire pulse, based on 1D, TLM iterations. The question posed is whether it is worthwhile to do a TLM analysis as in the previous, allowing the cell size to shrink to infinitesimal dimensions, or outline the TLM iteration using cells which are of finite extent, and are not infinitesimally small. We choose the latter approach , since the pure analysis becomes cumbersome (especially when losses are added) and , in any event, the SPICE are based on the TLM iterations. As noted before, Table 8A.2 gives the SPICE format for the transformer and Fig.8A.2 is the accompanying node designation. The modification for the PFXL is straightforward, merely requiring an input switch and an auxiliary output line.

8A.4   TLM Analysis of Leading Edge Wave in PFXL

We may provide TLM analysis and arguments, parallel to those for the transformer, to obtain numerical and closed form expressions for the leading edge of the wave in a PFXL source.   We use the same PFXL design as before, when we obtained SPICE simulations, with the transformer section divided into ten cells as in Fig.8.15. Unlike the pure transformer, of course, the PFXL transformer is subject to a bias voltage and we add the output auxiliary line Z_o, and an input switch. The pulsing action is initiated by shorting out the $n=0$ node of the first cell. Following the shorting of the $n=0$ node we may trace the evolution of the wave as it proceeds down the transformer. Unlike the pure transformer, however, the wave energy is derived from preexisting standing waves in the PFXL, which are unleashed by the shorting of the $n=0$ node.

Immediately following the $n=0$ node activation , the reflection coefficient at this node becomes $B(0,1) = -1$ and thus an inverted wave of amplitude $-V_o/2$ is initiated in the forward direction of the $n=1$ cell, or

$$^{+}V^{1}(1) = B(0,2)(V_o/2) = -V_o/2 \tag{8A.16}$$

and we remind ourselves of the fact that the second argument in $B(0,2)$ indicates a backward wave incident on the $n=0$ node. The wave then encounters the $n=1$ node, which separates the $Z(1)$ and $Z(2)$ cells. The transmission coefficient at this node ,$T(1,1)$, is

$$T(1,1) = 2Z(2)/(Z(1)+Z(2)) = 2(11.75)/(7.25+11.75) = 1.237 \tag{8A.17}$$

where we inserted the specified impedance values. The transmitted wave, however, represents only a portion of the forward wave in the $n=2$ cell. To this we must add the backward wave, of amplitude Vo/2, and partially reflected at the $n=1$ inversion because of the negative mismatch. The reflection coefficient the $n=1$ node is

$$B(1,2)=[Z(1)-Z(2)]/[Z(1)+Z(2)]=[7,25-11.75]/[7.25+11.75]=-.2368 \tag{8A.18}$$

where the negative sign indicates the reflected wave is inverted, adding to the transmitted wave (which has already undergone an inversion because of the initial short at the $n=0$ node). The total forward wave in the $n=2$ cell is therefore

$$^{+}V^{2}(2) = T(1,1)(-V_0/2) + B(1,2)(V_0/2)= -1.474\ (V_o/2) \tag{8A.19}$$

The process is then continued on to the next cell in like manner. The above wave encounters the $n=2$ node, and part of this wave energy is transmitted into the $n=3$ cell. The transmitted wave is again augmented by the reflection of the (Vo/2) backward wave. Using this procedure we may obtain a closed form expression, but because of the fact that we are dealing with a number of discrete cells, the final expression for the wave output is somewhat lengthy. We then obtain the *leading edge* field in the final cell, V(10):

$$^+V^{10}(10)=(V_0/2)B(0,2)T(1,1)T(2,1)T(3,1)T(4,1)T(5,1)T(6,1)T(7,1)T(8,1)T(9,1)$$

$$+(V_o/2)B(1,2)T(2,1)T(3,1)T(4,1)T(5,1)T(6,1)T(7,1)T(8,1)T(9,1)$$

$$+(V_o/2)B(2,2)\,T(3,1)T(4,1)T(5,1)T(6,1)T(7,1)T(8,1)T(9,1)$$

$$+(V_o/2)B(3,2)\,T(4,1)T(5,1)T(6,1)T(7,1)T(8,1)T(9,1)$$

$$+(V_o/2)B(4,2)\,T(5,1)T(6,1)T(7,1)T(8,1)T(9,1)$$

$$+(V_o/2)B(5,2)\,T(6,1)T(7,1)T(8,1)T(9,1)$$

$$+(V_o/2)B(6,2)\,T(7,1)T(8,1)T(9,1)$$

$$+(V_o/2)B(7,2)\,T(8,1)T(9,1)$$

$$+(V_o/2)B(8,2)\,T(9,1)$$

$$+(V_o/2)B(9,2) \tag{8A.20a}$$

Eq.(8A.20a) may be written in a more compact form

$$^+V^{10}(10)=(V_o/2)\sum_{n=0}^{n=8}[\Pi B(n,2)T(n+1,1)T(n,1)...T(9,1)]+(V_o/2)B(9,2) \tag{8A.20b}$$

Once we know $V(10)$ we can calculate the leading edge output to R_L with the help of the circuit in App. 8A.3, with $Z_F = Z(10)$. There are two contributions to the pulse delivered to R_L the first is that transmitted from the $n=10$ cell, and the second is from the auxiliary line, with impedance Z_o. For this, we need the transmission coefficients emanating from $Z(10)$ and Z_o, designated by T_I and T_{II}. From the circuit in Fig.8A.3,

$$T_I = 2\,[(R_L+Z_O)/(R_L+Z_O+Z(10))]\,[R_L/(R_L+Z_O]$$

$$= 2R_L/(R_L+Z_o+Z(10)) \tag{8A.21}$$

$$T_{II} = 2\,[\,(R_L + Z(10))/(R_L + Z(10) + Z_O)]\,[R_L/(R_L + Z(10))]$$

$$= 2R_L/\,(R_L + Z_o + Z(10)) = T_I \qquad (8A.22)$$

The second bracket in Eq.(8A.21) accounts for the voltage division between R_L and Z_O. Similarly, the second bracket in Eq.(8A.22) accounts for the division between R_L and $Z(10)$. We note the interesting result that $T_I = T_{II}$. This is representative of the general result that the forward and backward transfer coefficients to a series load, separating two lines, are equal. In fact, if we view R_L as a series TLM line (with impedance R_L) then the equality of T_I and T_{II} may be regarded as a special case based on the symmetry relationships discussed in Section 3.3 and displayed in Table 3.4. Finally. the leading edge output delivered to R_L is

$$V_{OUT} = {}^{+}V^{10}(10)\,T_I + {}^{-}(V_o/2)\,T_{II} \qquad (8A.23)$$

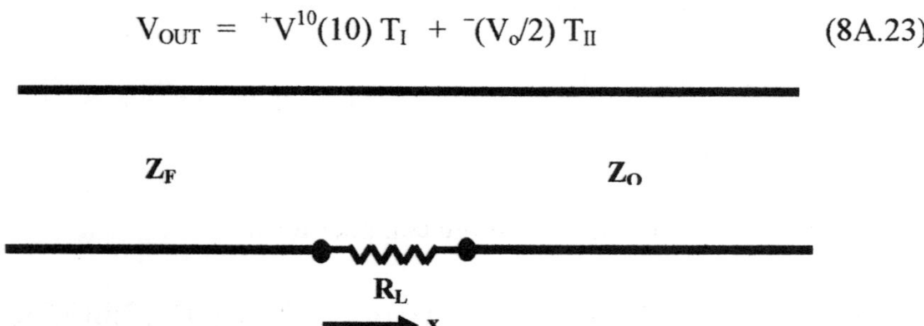

FIG. 8A.3 EQUIVALENT CIRCUIT SEEN BY LEADING EDGE OF PULSE AT FINAL SECTION OF PFXL TRANSFORMER. Z_F IS THE FINAL SECTION, Z_O THE AUXILIARY LINE, AND R_L THE LOAD.

Both terms in Eq.(8A.23) add constructively. A discussion of the signs may be worthwhile, again using the Figure in Fig.8A.3 with $Z_F = Z(10)$. ${}^{+}V^{10}(10)$ has a negative polarity as it impinges on R_L, due to the field inversion at the start of the chain. T_I will therefore deliver the pulse to R_L with its polarity in the negative x direction. $^-(V_o/2)$ on the other hand, has a positive polarity. Since $^-(V_o/2)$ is a backward wave T_{II} will deliver $^-(V_o/2)$ to R_L in the negative x direction as well. The two contributions therefore add.

Eq.(8A.23) should provide exactly the same result as the SPICE simulations in Figs.8.23-8.24 (i.e.,the leading edge in the SPICE should reproduce Eq.(8A.23)). Using the same transformer cells, and with Z_0 =50Ω, R_L=100 Ω, we obtain V_{OUT} =243V, which is identical to the simulation result, as expected.

As with the pure transformer, we should inquire whether a simple closed form result is possible if we allow the cell size of the PFXL to shrink to infinitesimal dimensions. This is indeed the case. As with the case of the transformer, we modify our notation, so that two neighboring cells are designated z and $z+\Delta z$. The leading edge voltage in $z+\Delta z$ is made up of two parts. One is the wave portion transmitted from the previous cell at z, and the other is the result of the backward wave reflected at the node separating the two cells. The transmission and reflection coefficients are

$$T(z) \ = 2(z+\Delta z)/(2z+\Delta z) \ \sim \ 1+ \Delta z/2 \tag{8A.24}$$

$$B(z) \ = \ \Delta z/ (2z+\Delta z) \ \sim \ \Delta z/2z \tag{8A.25}$$

where we have retained only first order terms. $V(z+\Delta z)$ is then

$$^+V(z+\Delta z) = \ ^+V(z) \, T(z) + \ ^-(V_o/2) \, B(z) \tag{8A.26a}$$

Strictly speaking , both contributions in the above are negative (which we have suppressed) due to the negative mismatch at either the first cell, which pertains to $^+V(z) \, T(z)$, or to the negative mismatch further down the chain, which pertains to $^-(V_o/2) \, B(z)$. Combining Eq.(8A.26a) with the expressions for T(z) and B(z), we have

$$^+V(z+\Delta z) = \ ^+V(z) + {}^+V(z)\ \Delta z/2z + {}^-(V_o/2)\ \Delta z/2z \qquad (8A.26b)$$

In differential form, Eq(8A.26) becomes

$$d^+V(z)/dz = {}^+V(z)/2z + (1/2)\ {}^-(V_o/2)/z \qquad (8A.27)$$

The solution of the above is

$$^+V(z) = \ -(V_o/2) + V_o(z/z_1)^{1/2} \qquad (8A.28)$$

where we have used the initial condition, $^+V(z) = V_o/2$ when $z=z_1$, i.e., at the beginning of the PFXL line the initial wave is that caused by the reflection from the short at the $n=0$ node. Note that the functional dependence of Eq.(8A.28) differs somewhat from that of a pure transformer. If we consider a pure transformer with an input of $V_o/2$, the output is given by $^+V(z) = (V_o/2)\ (z/z_1)^{1/2}$ Eq(8A.28) exceeds the transformer output by an amount $(V_o/2)[(z/z_1)^{1/2}-1]$ which is always greater than one. As mentioned previously, the enhanced output is due to the reflected wave in each cell, which add constructively with the forward transmitted wave.   Finally   , Eq.(8A.28) treats only the transformer section of the PFXL. At this point we need to take into account the augmented line Z_o at the output. This is done in exactly the same manner as was done when treating the transformer with numerable cells, using Eqs.(8A.21)-(8A.22). In the case of the infinitesimal cell transformer, we simply substitute the wave output for the final value of the transformer, z_f, given by

$$V(z_f) = \ -(V_o/2) + V_o(z_f/z_1)^{1/2} \qquad (8A.29)$$

The above is used in Eqs.(8A.21)-(8A.23) , replacing $z(10)$ with z_f and $V(10)$ with $V(z_f)$. Thus

$$T_I = 2\ R_L/(R_L+Z_O+z_f) \qquad (8A.30)$$

$$T_{II} = \ T_I \qquad (8A.31)$$

$$V_{OUT} = V(z_f)T_I + (V_O/2)T_{II} \qquad (8A.32)$$

where again both terms in Eq.(8A.32) add constructively to R_L, as outlined in the discussion following Eq.(8A.23).

It is of interest to compare the leading edge output from Eq.(8A.32) to that using the ten stage PFXL(either the SPICE simulation or equivalently, Eq.(8A.23)). Using $z_1 = 7.25 \ \Omega$ and $z_f = 47.75\Omega$, $R_L = 100\Omega$ and , $Z_o = 50\Omega$ we obtain the result $V_{OUT} \sim 256V$ which is about 5% higher than the SPICE result using the ten cell simulation done previously. As the number of cells is increased, of course, the difference between the exact result and the simulation grows smaller. The simulation is most valuable, however, when we wish to determine the entire pulse output, or when we add additional elements to the circuit.

REFERENCES

1.M. Weiner and A. Kim, "Optically Activated Traveling Wave Generators", in A. Rosen, F.Zutavern , *High Power Optically Activated Solid State Switches,* Artech House, Norwood, Mass., 1993.

2. G.Glasoe and L.Lebacqz, *Pulse Generators*, McGraw Hill, New York, 1948.

Index